Schools for Health and Sustainability

Venka Simovska • Patricia Mannix-McNamara
Editors

Schools for Health and Sustainability

Theory, Research and Practice

Springer

Editors
Venka Simovska
Department of Education
Aarhus University
Copenhagen, Denmark

Patricia Mannix-McNamara
Faculty of Education and Health Sciences
University of Limerick
Limerick, Ireland

ISBN 978-94-017-9170-0 ISBN 978-94-017-9171-7 (eBook)
DOI 10.1007/978-94-017-9171-7
Springer Dordrecht Heidelberg New York London

Library of Congress Control Number: 2014950063

Printed on acid-free paper

Springer is part of Springer Science+Business Media (www.springer.com)

Foreword

Watching parents take their children to school for the very first time is wonderful. The looks of both excitement and apprehension on their faces convey their feelings. There is much holding of hands, some laughter and usually a few tears. The expectations of parents and the experiences their child will receive that year, and in subsequent years, concern the attainment of knowledge and skills from the curriculum and opportunities for social interactions with other students and teachers. Their expectations rarely include health outcomes, although parents do value such attributes. Herein lays the dilemma for school health initiatives. School health is of low priority in a school education setting. It cannot compete with literacy, numeracy, scientific knowledge and reasoning and the social sciences for which schools are held to account, and it also competes with the creative arts. The same case can be made for sustainable development which the Brundtland Commission conceptualised as "development which meets the needs of the present without compromising the ability of future generations to meet their own needs." Yet health has always been part (albeit small) of the mandated curriculum in most countries for about 100 years. Ecological concepts and issues about sustainability are a more recent addition to school programs and have received some attention in the last 20–30 years.

The Health Promoting School (HPS) approach that grew out of the thinking behind the Ottawa Charter for Health Promotion has thrived internationally for nearly 30 years. It is an attempt to look at health in schools in a holistic way. Phrases such as 'whole school approach', and 'integrated school health' indicate an ecological approach to building the health and wellbeing of school students. There is a natural link with the sustainability initiatives which have been promoted, although somewhat spasmodically, in a number of countries and regions. Both focus on providing students with realistic experiences to gain knowledge; attain cognitive skills such as analysing, synthesising, evaluating, creating options; and accomplish 'action competencies' such as preparing meals, resolving conflicts, purchasing commercial products, developing realistic health and environmental policies within the school setting, and exploring what can be done in partnership with the local community.

There are many examples in the literature, albeit more from the health sector than from the sustainable development sector, about how schools and their communities have holistically addressed health and sustainable issues. Meta analyses show a whole school approach can produce gains in health given certain conditions. These have been summarised in publications that have interrogated the wealth of evidence. A solid body of evidence on sustainable development and schools is yet to be produced, and as such, one cannot make any conclusions about outcomes in this area. The health sector has been the main driver of school health initiatives and the HPS in particular. There have been many short and medium term programs using the HPS approach funded by Ministries of Health in countries across the globe. United Nation bodies such as the World Health Organization have auspiced the creation of evidence based guidelines and projects in both developed and developing countries which use a whole school approach to enhance the health and wellbeing of students. The International Union of Health Promotion and Education (IUHPE) has shown leadership and support, with financial assistance from the Center for Disease Control and Prevention in the USA, to develop a series of succinct evidence based documents to assist school personnel and education and health officials to be more strategic and effective in their attempts to address health. But there are still many barriers preventing wider acknowledgement of health and sustainability being embraced more wholeheartedly by schools. These barriers do not appear to have weakened in the last two decades. And the reasons for the existence of such barriers are quite understandable.

The prime focus of schools is to build educational outcomes. They are not places which should be expected to address morbidity, mortality and environmental indicators of the country or region. The factors that shape our health and wellbeing are influenced by our genetics; the families, society and cultures in which we live; government policies, practices and resource allocations; and the physical environment. Our behaviours are a result of all these factors and their interactions. Developments in Complexity Theory reinforce the need for a stronger focus on the dynamics of the school and those factors shaping teachers' thinking and practice, and their interactions with students both inside and outside the classroom. Schools are not a 'black box' into which agencies can put programs and resources and expect a reduction in x, y, or z in the health and sustainability fields. However, it is possible to achieve a set of action competencies in students, which will reduce, but not eliminate, some risk behaviours now and into the future. The evidence from the last two decades tells us that addressing social and emotional wellbeing is the most important health area which can achieve the best outcomes in not only that area, but in others such as experimentation in drugs and early sexual activity. School leaders and teachers are the gatekeepers of what can and cannot be done in schools. The evidence indicates that the quality of school leadership is fundamental to any successful HPS initiative and embedding it in the ongoing priorities and curriculum of the school. A critical mass of teachers open to change is also necessary to ensure successful implementation of any program/project using a HPS approach. However, there are often very few resources available to support the necessary capacity building of both school leaders and teachers.

Nevertheless, given all the barriers, there is still a strong body of evidence about the holistic approach to health related initiatives in schools. (As indicated earlier, solid evidence on gains in the sustainable development in schools is not yet present.) From this health evidence we can make a number of assertions and claims about what works and what does not. However, what is not clear is whether this evidence reaches schools in ways that teachers understand, whether it can be used to change the way health is addressed. To address these issues we need to focus on the school and the classroom in particular, to gain an appreciation of what happens in the day-to-day life of teachers and their students to gain an understanding about the realities facing them.

Teachers are under time constraints. The school day is finite and much time is directed at managing student logistics and behaviour. Arriving at and leaving class takes time. Classroom management, including coping with expected and unexpected behaviours, takes time. There is a curriculum to be taught with content from many fields. Teachers may also be under pressure to address political and/or parental priorities such as literacy and numeracy in greater scope, and there is a range of different learning styles and preferences in the student population. Some students rely more on auditory learning while others are more attuned to visual or experiential learning. Time is also absorbed by assessment and reporting and, in many cases, accounting to a wide constituency such as parents and education authorities about the attainment of curriculum standards. When it comes to the small health part of the curriculum competing priorities emerge again. Teachers and schools have to decide about the balance of time allocated to healthy eating and nutrition, physical activity, hygiene, sexuality, drugs (both legal and illicit), social and emotional health and safety (traffic, personal, environmental).

In some countries there are no or limited resources for teachers and students to use, whilst in others schools are bombarded with resources from health and environmental agencies. Recently the author was contracted to evaluate Traffic Safety resources to determine if they met the Traffic Safety Education Guidelines, which were based on HPS guidelines and associated evidence. In one region with a population of nearly three million people, over 30 resources from four different government departments, 9 NGOs and at least 15 commercial and community groups were promoted to schools. Only one met the Guidelines for evidence-based practice. It is not surprising then that teachers in many developed countries are confused about how to choose from the many programs and products at their disposal. In developing countries, the absence of resources is the main issue. Here, as in developed countries, teachers' knowledge of the scientific evidence of the factors that shape student health is limited. Resources can help to fill this gap. But sadly, tens of thousands of teachers across the world do not have access to them.

Financial resources to train teachers and support the implementation of HPS over a 5–7 year time period do make a difference to how schools can develop the building blocks of healthy and sustainable actions in their students. There is strong evidence about what makes an effective school, how program initiatives can be implemented and sustained, and how leadership within the school and local community facilitates program growth and development. We need to remind ourselves that teachers do become overwhelmed with new concepts and structures. For example, the recent focus by many in the health sector on health literacy may be distracting for schools. It is interesting to note that many of the components of health literacy being promoted by health sector personnel are the very same components of health education where there is at least 40 years of knowledge, understanding and acceptance by the education sector, particularly in schools. It has taken a generation for education professionals to become comfortable with the term health promotion. Health literacy has the potential to confuse the education sector given the importance the term 'literacy' has in their working lives.

We also need to be reminded about the focus of any school-based innovations in health and sustainable development. It should be the students who are at the top of the priority pyramid. How they learn, their interests and needs, their culture beliefs and values, and whether they are seen by their society as young people to be empowered or young people to be indoctrinated is where attention must be directed when we are writing guidelines, developing teaching and learning materials and designing and implementing programs in schools.

That first day at school is an exciting time for students – full of apprehension, making new friends, a taste of independence, and the beginning of years of discovery and creativity. The health of students and the sustainable environments in which they live are both fundamental to their learning outcomes. This is the main reason why we need to put considerable effort in ensuring health and sustainable development are linked and are seen as basic investments to support education of young people.

This book will challenge your thinking. The reader is provided with evidence based case studies and insights into how whole school approaches to health and sustainable development have developed and are in fact closely linked. The various chapters challenge our beliefs and practices and keep us asking questions about what actually happens in schools. The reality of the school setting is a welcome respite. A focus on the classroom and school is where we need to have our conversations and reflections about what is best for our students, and how we can work with them to design and implement actions which will benefit their health, their contributions to sustainable development, and above all, their education.

Burwood, VIC, Australia Lawrence St. Leger

Preface

Schools are unique and dynamic places. They pay a central role in the formation and development of young people. The importance of *how* young people are educated and *how* they are encouraged to live and learn cannot be underestimated. Often the emphasis in school policy development focuses on curriculum, knowledge production and the structuring of schooling. There is an emergent movement, however, that recognises that both the whole school environment and the ethos of schools need to be life affirming and health enhancing. Creating an environment where young people can learn holistically, through classroom teaching and also through the everyday life/culture of the school, is a core aim of the health promoting schools and of education for sustainable development. *Schools for Health and Sustainability: Theory, Research and Practice* presents recent research evidence specific to the development, implementation, evaluation and potentials of health promoting schools as well as their impact in terms of health, sustainability and education outcomes. The book draws together current international expertise and scholarship on the theory, research and practice of health promoting schools and elucidates the links between health education and education for sustainable development, capitalizing on their shared values, educational principles and common aims. It also advocates for the fostering of agency not only amongst school personnel but also amongst younger generations for health and sustainability and provides the reader with a new lens with which to discover health promoting schools and education for sustainable development. It invites the reader to look more deeply into both and to accompany the authors on a journey of discovery of the real potential for each to enhance the practice of schooling.

Schools for Health and Sustainability: Theory, Research and Practice has arisen from the Schools for Health in Europe Research Group (SHE Research Group) and its collaboration with the Health Education Research Network of the European Educational Research Association (EERA Network 8). The chapters contained here extend beyond European boundaries and include contributors from Australia and Reunion Island. The contributors, who all individually and collectively strive for health promoting schools, have held the values of inclusivity and empowerment at their core. The book attests to the value and potential of the association between health education and education for sustainable development. As you read, you will experience the belief in and commitment to schools that are promoting health and sustainability as distinctive and significant environments in which to live and learn. You may even recognise your potential to make a contribution to the field and in so doing, the development and potential of health promoting schools continues…

Copenhagen, Denmark
Limerick, Ireland
March 2014

Venka Simovska
Patricia Mannix-McNamara

Acknowledgments

We acknowledge the Editorial Advisory Board:

Goof Buijs, Senior Consultant, Dutch Institute for Healthcare Improvement (CBO) and the World Health Organisation Collaborative Centre for Health Promoting Schools, The Netherlands

Monica Carlsson, Associate Professor, Department of Education, Research Programme in Learning for Care, Sustainability and Health, Arrhus University, Denmark

Kevin Dadaczynski, Research Fellow, Centre for Applied Health Sciences, Leuphana University, Lüneburg, Germany

Saoirse Nic Gabhainn, Senior Lecturer, Health Promotion Research Centre and the World Health Organisation Collaborative Centre for Health Promotion Research, National University of Ireland Galway

Margaret Hodgins, Head of Department, Department of Health Promotion, National University of Ireland, Galway, Ireland

Bjarne Bruun Jensen, Director, Steno Health Promotion Centre, Denmark

Peter Paulus, Professor, Leuphana University, Lüneburg, Germany

Nina Grieg Viig, Pro-Dean of Research, Faculty of Education, Bergen University College, Norway

Oddrun Samdal, Vice-Rector for Education, University of Bergen, Norway

Zali Yeager, Senior Lecturer, College of Education, Victoria University, Melbourne Australia

Contents

Contributors

Maryvette Balcou-Debussche
maryvette.balcou-debussche@univ-reunion.fr
Maryvette Balcou-Debussche, Ph.D., Senior Lecturer in Education and Sociology at Reunion University/ESPé. She is a member of the LCF Research Centre (http://www.lcf-reunion.fr/) and of the scientific committee of UNIRéS network (http://unires-edusante.fr/). Focused on health education and therapeutic patient education, her research has resulted in numerous scientific publications.

IUFM de La Réunion, LCF-Icare
Université de La Réunion
Saint-Denis, France

Margaret M. Barry
margaret.barry@nuigalway.ie
Margaret M. Barry, Ph.D., is Professor of Health Promotion and Public Health and Head of the World Health Organization Collaborating Centre for Health Promotion Research at the National University of Ireland Galway. Professor Barry has published widely in the field of mental health promotion and works closely with policymakers and practitioners on the development, implementation and evaluation of mental health promotion interventions and policies at national and international level.

Health Promotion Research Centre
National University of Ireland Galway
Galway, Ireland

Matthias Barth
matthias.barth@rmit.edu.au
Matthias Barth is Professor of Learning and Teaching in Environmental Science at the University of Applied Science Ostwestfalen-Lippe in Germany, developing and researching programs to support the development of key competencies for a sustainable future. Previously he was a Senior Research Fellow at RMIT University,

researching implementation patterns of sustainability into higher education and related students' competencies.

Environmental Science
University of Applied Science Ostwestfalen-Lippe
Lemgo, Germany

Dimitri Batras
dimitri.batras@monash.edu; dbatras@attainedsuccess.com.au

Dimitri Batras is currently a Principal Consultant at Attained Success Consulting, Australia. His work involves assisting organisations to evaluate the success of their programs, and providing advice to government and non-government agencies regarding health promotion practice and healthy public policy. Dimitri is also a part-time Ph.D. candidate at Monash University's School of Public Health and Preventive Medicine where he previously worked as a Research Fellow.

Attained Success Consulting
Melbourne, VIC, Australia

School of Public Health and Preventive Medicine
Monash University
Melbourne, VIC, Australia

Nicole M.W.M. Boot
Nicole.Boot@ggdzl.nl

Nicole M.W.M. Boot, Ph.D., studied Health Education and Health Promotion at Maastricht University, the Netherlands. Since 2004 she has worked at the Regional Public Health Service South Limburg. She was involved with the developing and implementation of the School Beat approach. For her Ph.D. research Nicole evaluated the implementation of the School Beat method in South Limburg. At this moment Nicole is a Senior Policy Advisor at the Department of Knowledge and Innovation at the Regional Public Health Service South Limburg. School Health Promotion is still part of her daily work.

Academic Collaborative Centre for Public Health Limburg
Maastricht, The Netherlands

Monica Carlsson
monica@edu.au.dk

Monica Carlsson, Ph.D., is Associate Professor in Evaluation at Department of Education, Aarhus University, Denmark. She is a member of the Research Programme on *Learning for Care, Sustainability and Health*, is coordinating a module on Health Education, Health Promotion, Learning and Competence Development on the M.A. in Health Education and Health Promotion, and is a Head of the Doctoral School Programme in Curriculum Research and Education at Faculty of Arts, Aarhus University.

Department of Education
Aarhus University
Copenhagen, Denmark

Aleisha M. Clarke
aleisha.clarke@nuigalway.ie

Aleisha M. Clarke, Ph.D., is a Postdoctoral Researcher working at the Health Promotion Research Centre at the National University of Ireland, Galway. Aleisha is a qualified primary school teacher and she has worked on a number of research projects, including the evaluation of mental health promotion interventions and the development of online mental health resources to support young people.

Health Promotion Research Centre
National University of Ireland
Galway, Ireland

Nanne de Vries
n.devries@maastrichtuniversity.nl

Nanne de Vries is Professor in Health Promotion and Health Communication at Maastricht University, the Netherlands. Nanne de Vries is Associate Editor of the Dutch Journal of Medicine and a member of the editorial board of Psychology and Health. He serves on several research committees of NWO, ZonMW, and the Dutch Foundation for STI and AIDS. Furthermore, he is a member of the board of the Academic Centre for Public Health in Limburg, the Foundation for Scientific Research of the Dutch Association of Health Promotion, and the Dutch Foundation of Mental Health. Coming from the field of psychology, Prof. de Vries is highly interested in determinants and change of health behaviours, both in a fundamental and an applied sense.

Department of Health Promotion, CAPHRI
Universiteit Maastricht
Maastricht, The Netherlands

Kevin Dadaczynski
dadaczynski@leuphana.de

Kevin Dadaczynski, Ph.D., is a Research Fellow at the Centre for Applied Health Science at the Leuphana University of Lueneburg, Germany. He has a diploma degree in health promotion and a master's degree in health psychology/health education. His research interests are: mental health of pupils, teachers and principals, health education and health promotion in schools and quality development and assurance in (school) health promotion and prevention. He is involved in several national and international research projects.

Centre for Applied Health
Leuphana University of Lueneburg
Lueneburg, Germany

Daniel Fischer
daniel.fischer@uni.leuphana.de

Daniel Fischer is a Research Fellow at the Institute for Environmental and Sustainability Communication (INFU) at Leuphana University Lüneburg, Germany. He is a trained school teacher and holds a master's degree in school development. His Ph.D. research focuses on consumer education and socialization processes in schools.

Institute for Environmental & Sustainability Communication (INFU)
Leuphana University
Lüneburg, Germany

Edith Flaschberger
edith.flaschberger@lbihpr.lbg.ac.at
Edith Flaschberger, Ph.D., is a Senior Researcher at the Ludwig Boltzmann Institute Health Promotion Research in Vienna, Austria. Her research focus lies in school health promotion, especially in the areas of teacher education for school health promotion and mental health promotion for students.

Ludwig Boltzmann Institute Health Promotion Research
Ludwig Boltzmann Gesellschaft GmbH
Vienna, Austria

Marie-Renée Guével
Marie-Renee.Guevel@ehesp.fr
Marie-Renée Guével, Ph.D., is a Research Fellow at the EHESP and is a member of the 'health education in schools' research team (ACTé EA 4281), France. She completed her Ph.D. in 2012 on school health promotion in French primary schools. Her research was related to evaluation issues, on how to take into account the complexity of health promotion initiatives as well as the context in which they are implemented.

Department of Social and Human Sciences
EHESP (French School of Public Health)
Rennes, France

Lisa Gugglberger
lisa.gugglberger@lbihpr.lbg.ac.at
Lisa Gugglberger, Ph.D., is a Senior Researcher at the Ludwig Boltzmann Institute Health Promotion Research in Vienna, Austria. She mainly works in the area of health promoting schools, focusing on research on the implementation of health promotion.

Ludwig Boltzmann Institute Health Promotion Research
Ludwig Boltzmann Gesellschaft GmbH
Vienna, Austria

Jennifer Hennessy
Jennifer.Hennessy@ul.ie
Jennifer Hennessy, Ph.D., is a Lecturer in Education in the Department of Education and Professional Studies at the University of Limerick, Ireland. She has worked in initial and continuing teacher education since 2007. Her research interests include pedagogy and performativity, curriculum development and affective education.

Department of Education and Professional Studies
University of Limerick
Limerick, Ireland

Maria W.J. Jansen
Maria.jansen@ggdzl.nl
Dr. Maria W.J. Jansen studied Human Nutrition and started her career as a teacher in nutrition, patient education and epidemiology at the University of Applied Sciences, the Netherlands. Since April 1989 she has worked at the Regional Public Health Service South Limburg where she introduced multidisciplinary cooperation and inter-sectoral implementation as a regular working method for school health promotion and public health in general. In 2007 she was appointed as programme manager of the Academic Collaborative Centre for Public Health in which opportunities are created to collaborate between policy makers, practitioners and researchers to make school health promotion more effective.

Department of Health Promotion
Regional Public Health Service (GGD) South Limburg
Geleen, The Netherlands

Didier Jourdan
Didier.JOURDAN@univ-bpclermont.fr
Didier Jourdan, Ph.D., is Professor, Dean of the Faculty of Education of the University Blaise Pascal Clermont-Ferrand, France. He is also Adjunct Professor at the University of Limerick Ireland. During the last 10 years, he was the head of the 'health education in schools' research group (ACTé EA 4281). Professor Jourdan's research is focused on health education and health promotion in schools and concerns three main fields: teaching, learning and teacher training. He is one of France's leading health promoters and health educators. He is a special adviser to the Minister of Education and to the Minister of Health for Children and Adolescent Health Policy in France.

Faculty of Education
University Blaise Pascal Clermont-Ferrand
Ferrand, France

Andrew Joyce
andrew.joyce@mcri.edu.au
Dr. Andrew Joyce works in evaluation for the Centre for Community Child Health. He has a background in practice based research holding joint appointments between schools, health services, and academia.

Centre for Community Child Health
Murdoch Childrens Research Institute
Melbourne, VIC, Australia

Athina Karavoltsou
akaravol@yahoo.gr
Athina Karavoltsou is a health and drama educator from Greece. She holds a diploma in Mathematics and Theatre Studies, M. Ed. in Drama Education and M.Sc. in Health Education. She has participated in children's theatre groups as an actor and directed children's performances. She is now organizing Mental Health Education projects for children and teacher training programs, using a drama-based pedagogy, as a freelancer.

Athens, Greece

Rune Johan Krumsvik
Rune.Krumsvik@psych.uib.no
Rune Johan Krumsvik, Ph.D., is a Professor at the Department of Education at the University of Bergen, Norway. His main research interest is pedagogy, with a special interest in ICT and learning, formative e-assessment, class management and physical activity.

Department of Education
University of Bergen
Bergen, Norway

Sari Laine
sari.laine@uef.fi
Sari Laine, M.Sc., graduated from the Department of Nursing Science at the Faculty of Health Sciences of the University of Eastern Finland in 2012. She currently works at the Social Insurance Institution of Finland and is a doctoral student conducting scientific postgraduate studies at the Department of Nursing Science of the University of Eastern Finland with the aim of attaining the degree of Ph.D. in Health Sciences. Her areas of research interest include health promotion and leadership at workplaces.

Department of Nursing Science, Faculty of Health Sciences
University of Eastern Finland
Kuopio, Finland

Ulf Leo
ulf.leo@soclaw.lu.se
Ulf Leo, Ph.D. is a Senior Lecturer at the Sociology of Law Department at Lund University, Sweden. He is also working in the Swedish National School Leadership Training programme at Centre for Principal Development, Umeå University.

Sociology of Law Department
Lund University
Lund, Sweden

Mariken T. W. Leurs
mariken.leurs@rivm.nl
Mariken T. W. Leurs, Ph.D., is Deputy Director of Sports at the Ministry of Health, Welfare and Sports in the Netherlands. At the time of writing the article she was the Head of the Centre for Healthy Living at the Dutch National Institute for Public Health and the Environment.

Centre for Healthy Living Dutch Institute for Public Health
and the Environment (RIVM)
Bilthoven, The Netherlands

Katrine Dahl Madsen
kdma@edu.au.dk
Katrine Dahl Madsen, Ph.D., is Postdoctoral Fellow in the Research Programme on Learning for Care Sustainability and Health at the Department of Education, Aarhus University, Denmark. She is currently engaged within the Research Centre on Schools for Health and Sustainability. Her research revolves around education for sustainable development (ESD) with a special focus on how education for sustainable development takes form as curriculum thinking and teaching practice in primary and lower secondary schools.

Department of Education
Aarhus University
Copenhagen, Denmark

Patricia Mannix-McNamara
Patricia.M.McNamara@ul.ie
Patricia Mannix-McNamara, Ph.D., Senior Lecturer in Education in the Faculty of Education and Health Sciences at the University of Limerick, Ireland. She is Co-Director of the Research Centre for Education and Professional Practice (RCEPP). Patricia has expertise in both health and education and has authored and co-authored books, book chapters and journal papers spanning both fields. She has been a national evaluator for ENHPS since 1999 and is a core group member of the Schools for Health in Europe (SHE) Research Group.

Faculty of Education and Health Sciences
Research Centre for Education and Professional Practice
University of Limerick
Limerick, Ireland

Sharon Moynihan
Sharon.Moynihan@ul.ie
Sharon Moynihan (B. Ed. Physical Education) holds a degree in Physical Education, from the University of Limerick, Ireland. She is an advanced scholar at the University of Limerick. Her doctoral studies focus on health promoting schools and critically examine the importance of teacher education for health education.

Department of Education and Professional Studies
University of Limerick
Limerick, Ireland

Lone Lindegaard Nordin
lln@edu.au.dk
Lone Lindegaard Nordin, Ph.D. is Postdoctoral Fellow at the Research Programme on Learning for Care, Sustainability and Health at the Department of Education, Aarhus University, Denmark. She is involved in the Research Centre on Schools for Health and Sustainability. Her research focuses on implementation of health

promotion and education in schools, in particular implementation of policies and the ways it is affected by individual and institutional factors.

Department of Education
Aarhus University
Copenhagen, Denmark

Leena Paakkari
leena.paakkari@jyu.fi
Leena Paakkari, Ph.D., is a Senior Lecturer and a researcher in health education teacher training, within the Faculty of Sport and Health Sciences at the University of Jyväskylä, Finland. She is currently involved in conceptualising and assessing health literacy among the school-aged children.

Faculty of Sport and Health Sciences
University of Jyväskylä
Jyväskylä, Finland

Peter Paulus
paulus@uni.leuphana.de
Peter Paulus, Ph.D., works as a Professor at the Institute of Psychology at the Leuphana University of Lueneburg, Germany. His work focuses on the following areas: educational psychology, family psychology, health psychology, education, counselling, and health promotion. He heads the research of several nationwide projects relating to school health promotion: e.g. MindMatters – promotion of health in primary and secondary schools (since 2002); Competence Center: Mental Health in Education (since 2008); Whole day school and mental health (2010–2012).

Institute of Psychology
Leuphana University of Lueneburg
Lueneburg, Germany

Jeanine Pommier
Jeanine.Pommier@ehesp.fr
Jeanine Pommie, Ph.D., is a Public Health and Health Promotion Professor and Deputy Director of the Department of Social and Human Sciences of EHESP (French School of Public Health), France. She joined the Research Center on Policy Action in Europe (CRAPE), in January 2011. She coordinates for EHESP the Erasmus Mundus master degree EUROPUBHEALTH.

Department of Social and Human Sciences
EHESP (French School of Public Health)
Rennes, France

Crane Rogers
crane.rogers@univ-lyon1.fr
Crane Rogers is a geographer and research engineer at the Department of Education of University of Lyon, France. She is a member of the ACTE Research Centre

EA 4281, and coordinator of UNIRéS, a French network of researchers and educators in Health Promotion for Schools.

Université Lyon 1 IUFM de l'Académie de Lyon
Université de Lyon
Lyon, France

Louise Rowling
louise.rowling@sydney.edu.au
Louise Rowling is an Honorary Associate Professor in the Faculty of Education and Social Work at University of Sydney, Australia. She has established over a 25-year period a National and International reputation for research and publications on health promoting schools and school mental health promotion. She was the President of Australian Health Promoting Schools Association for five years, the Inaugural President of 'Intercamhs' the International Alliance of Child and Adolescent Mental Health and Schools and co-led the development and evaluation of MindMatters the Australian national secondary school mental health promotion initiative.

Faculty of Education and Social Work
University of Sydney
Sydney, NSW, Australia

Terhi Saaranen
terhi.saaranen@uef.fi
Terhi Saaranen, Docent, Ph.D., is a Senior Researcher and University Lecturer at the Department of Nursing Science of the Faculty of Health Sciences at the University of Eastern Finland. Her research interest is health education and health promotion, especially the development of school community staff's occupational well-being. For example she has experience of interventions and action researches with school community staffs and occupational health nurses in different school communities.

Department of Nursing Science
Faculty of Health Sciences
University of Eastern Finland
Kuopio, Finland

Oddrun Samdal
Oddrun.Samdal@uib.no
Oddrun Samdal, Ph.D., is Professor at the Department of Health Promotion and Development at the University of Bergen, Norway. Her main research interest is to study and evaluate interventions on how individual and environmental correlates interplay and relate to adolescent health behaviours and subjective well-being. She is the Data Manager and the Norwegian Principal Investigator for the WHO-collaborative study "Health Behaviour in School-aged Children". From 1993 to 2009 she was the Norwegian coordinator of the Health Promoting Schools.

Department of Health Promotion and Development
University of Bergen
Bergen, Norway

Elizabeth Senior
lsenior@EACH.com.au
Elizabeth Senior is a Senior Health Promotion Officer with EACH Social and Community Health Service, Victoria, Australia. She works with local schools using the Health Promoting Schools framework. She has 30 years of experience in the health and welfare field. She has a Diploma of Applied Science, a Social Work degree, a Grad Dip in Health Promotion and a Master of Public Health. She is currently the Secretary of the Australian Health Promoting Schools Association.

EACH Social and Community Health Service
Melbourne, VIC, Australia

Venka Simovska
vs@edu.au.dk
Venka Simovska, Ph.D., Professor and Director of the Research Programme on *Learning for Care, Sustainability and Health* at the Department of Education, Aarhus University, Denmark. Professor Simovska also leads the newly established Research Centre on *Schools for Health and Sustainability* at the Department. She is Chair of two international research networks: *Schools for Health in Europe (SHE) Research Group* and *Health Education Research* (network within the European Association of Educational research (EERA). Simovska has authored and co-authored a number of books, book chapters and journal articles within the field. Most recently, she has guest edited two special issues of *Health Education* focusing on processes and outcomes of health promoting schools.

Department of Education
Aarhus University
Copenhagen, Denmark

Marjorita Sormunen
marjorita.sormunen@uef.fi
Marjorita Sormunen, Ph.D., works as a Postdoctoral Researcher at the Department of Nursing Science, University of Eastern Finland, and is a Fulbright alumna from University of Florida, USA. Her research interests are school health promotion and health education, and particularly collaboration between home and school related to children's health.

Department of Nursing Science
Faculty of Health Sciences
University of Eastern Finland
Kuopio, Finland

Lawrence St. Leger
lstleger@bigpond.com
Professor Lawrence St. Leger began his professional life as a teacher. He taught in both state and independent systems in rural and urban secondary schools. Lawry was Head of the School of Nutrition and Public Health at Deakin University before undertaking the role of Dean of the Faculty of Health and Behavioural

Sciences for seven years. He is currently honorary professor at Deakin University Australia. School health has been Lawry's prime area of research for over 25 years. He has published widely and has been commissioned by the World Health Organisation (WHO) to write international evidence based guidelines for planning, implementing and evaluating school health initiatives. More recently Lawry, along with one of Europe's leading school health experts, Ian Young, has been commissioned to write more evidence based guidelines for health promoting schools and these are published in ten languages. Lawry continues to advise Ministries of Education and Health in many parts of the world including Canada, New Zealand, Hong Kong, Cambodia, Mongolia, Taiwan, Singapore and various Australian jurisdictions.

Deakin University
Burwood, VIC, Australia

Hege E. Tjomsland
Hege.Tjomsland@iuh.uib.no
Hege E. Tjomsland, Ph.D., is an Assistant Professor at the Department of Education at the University of Bergen, Norway. Her main research interest is school health promotion, with a particular interest in physical activity promotion.

Department of Education
University of Bergen
Bergen, Norway

Kerttu Tossavainen
kerttu.tossavainen@uef.fi
Kerttu Tossavainen, Ph.D., Professor, Head of the research group, has long-standing experience in intervention and action research in her special field, health promotion with children, families, and school nurses in different school communities, as follows: 1984–1988, the Eastern Finland Youth Project; 1989–1992, the National Healthy School Project; 1993–2007, the European Network of Health Promoting Schools (ENHPS); and since 2007, the School for Health in Europe (SHE) and a member of the SHE research core group. Professor Tossavainen is the Responsible Director of the AHIC (Addressing challenging health inequalities of children and youth between two Karelias) in years 2013–2014 funded by the European Union, the Russian Federation and the Republic of Finland.

Department of Nursing Science
Faculty of Health Sciences
University of Eastern Finland
Kuopio, Finland

Hannele Turunen
hannele.turunen@uef.fi
Hannele Turunen, Ph.D., Professor, Head of the Department of Nursing Science, University of Eastern Finland. She has expertise in both health promotion and health

education. She has been a member of the ENHPS partnership since 1997 in co-operation with Kerttu Tossavainen, the Finnish Centre for Health Promotion, and a nominated member of the SHE international research core group.

Department of Nursing Science
Faculty of Health Sciences
University of Eastern Finland
Kuopio, Finland

Per Wickenberg
per.wickenberg@soclaw.lu.se
Per Wickenberg, Ph.D., Professor in Sociology of Law, Lund University, Sweden. He is also Director of Studies and Head of Department in Sociology of Law. He has been national coordinator in the research network *Education and Sustainable Development*. Together with Ulf Leo he has been involved in the research project Implementation of education for sustainable development (2009–2012) with Uppsala, Örebro, Stockholm University.

Sociology of Law Department
Lund University
Lund, Sweden

Bente Wold
Bente.Wold@iuh.uib.no
Bente Wold, Ph.D., is a Professor at the Department of Health Promotion and Development at the University of Bergen, Norway. Her main research interests concern health promotion with young people, with a particular interest in positive youth development based on developmental, social and health psychology, as well as behavioural epidemiology.

Department of Health Promotion and Development
University of Bergen
Bergen, Norway

Ian Macrae Young
imyoung@blueyonder.co.uk
Ian Macrae Young is an international health promotion consultant, based in Scotland, with a background in high school teaching, biology, ecology and public health. He has been involved in the health promoting school movement from its inception in the 1980s, and he was formerly Head of International Development at NHS Health Scotland. In 2008 he was awarded an MBE in Queen Elizabeth's Honours list in recognition of his work in health promotion.

International Health Promotion Consultant
Edinburgh, Scotland

Part I
Health Promoting Schools and Education for Sustainable Development

Chapter 1
Schools for Health and Sustainability: Insights from the Past, Present and for the Future

Patricia Mannix-McNamara and Venka Simovska

Keywords Health promotion • Health education • Sustainability • Public health services • School-based

1.1 Introduction

The prioritisation of health education and of education for sustainable development in schools, from a common sense perspective, requires little in terms of justification as to its value or necessity. Schools are settings that can have a great impact on the health and emotional wellbeing of children and young people as they influence them at important stages in their lives (Deschesnes et al. 2014). Promoting, and learning about health, wellbeing and sustainability early in childhood clearly has long term gain both for the individual and for society through advancing health status, equity and quality of life for all. In building children's competences, i.e. their knowledge, attitudes and skills related to health and sustainability, they are encouraged to place significant value on their health and wellbeing, and to understand in a meaningful manner the link between sustainability and health issues, as well as the health of the planet. They are also encouraged to appreciate health and sustainability as conditions that they themselves can influence, whether it is to improve, maintain or enhance their health, or to improve the conditions for health and sustainable development in their proximate environments. Health education/promotion and education for

Patricia Mannix-McNamara, Ph.D. (✉)
Faculty of Education and Health Sciences, University of Limerick, Limerick, Ireland
e-mail: Patricia.M.McNamara@ul.ie

V. Simovska, Ph.D.
Department of Education, Aarhus University, Tuborgvej 164, 2400 Copenhagen, Denmark
e-mail: vs@edu.au.dk

V. Simovska and Patricia Mannix-McNamara (eds.), *Schools for Health and Sustainability: Theory, Research and Practice*, DOI 10.1007/978-94-017-9171-7_1,

sustainable development in schools also encourage students to become critical about their own attitudes and behaviours linked to health and sustainability and contribute significantly to reduction of unsustainable and risk behaviours.

Many countries globally have recognised the potential of the school as a setting that can significantly contribute to student health and wellbeing and have thus adopted some form of model of health promoting schools (Simovska and Jensen 2002; Whitman and Aldinger 2009). These models of practice include for example the 'healthy schools' approach that has been adopted in the UK and USA (Department of Health [UK] 2005; Environmental Protection Agency [USA] 2012), the 'good healthy school' approach in Germany (Paulus 2007; Paulus 2005) or the 'health promoting schools' approach, the more common European framework initiated by the WHO, CE and EC (WHO 1998; Buijs 2009). Regardless of variation in model or in implementation, common to all initiatives is recognition that real and meaningful engagement with health and wellbeing in schools cannot be ignored. The promotion of the health gain of nations is dependent on building solid foundations for health amongst populations. Schools, quite rightly, have been recognised as an important setting in which this can be actualised.

This book is situated within the conceptual landscape of the Health Promoting Schools initiative in Europe (Clift and Jensen 2005; Simovska 2012). As the title of the book indicates, however, the scope of the health promoting schools is widened to integrate education for sustainable development. The text takes the classic perspective on sustainable development, as defined by the World Commission on Environment and Development, also known as the Brundtland Commission. This definition states that sustainable development is "development that meets the needs of the present without compromising the ability of future generations to meet their own needs" (United Nations 1987: 11). Consequently, education for sustainable development is conceptualised through the following four features (UNESCO 2014):

- Education for Sustainable Development (ESD) allows every human being to acquire the knowledge, skills, attitudes and values necessary to shape a sustainable future.
- Education for Sustainable Development means including key sustainable development issues into teaching and learning; for example, climate change, disaster risk reduction, biodiversity, poverty reduction, and sustainable consumption.
- It also requires participatory teaching and learning methods that motivate and empower learners to change their behaviour and take action for sustainable development.
- ESD consequently promotes competencies like critical thinking, imagining future scenarios and making decisions in a collaborative way.

These features are clearly synergetic with the main principles, values and methods linked to health promoting schools and health education endorsed in this book.

The World Health Organization (WHO 1998: 1) defines a health promoting school as "one that constantly strengthens its capacity as a healthy setting for living, learning and working." In particular, according to the WHO, health promoting schools foster health and learning with a wide variety of measures. In addition, they engage across a range of stakeholders that include health and education officials, school principals, teachers, teachers' unions, students, parents, health providers and community leaders. Health promoting schools strive to provide a healthy environment across a broad range of activities that include curriculum, school physical and social environment, school health services, school/community projects and outreach, nutrition and food safety programmes, opportunities for physical activity and recreation, and programmes for counselling, social support and mental health promotion. Health promoting schools work through effective development and implementation of policies and practices that respect an individual's wellbeing and dignity, provide multiple opportunities for success, and acknowledge positive efforts and intentions as well as personal achievements. Finally, they strive to improve the health of all school personnel, families and community members as well as pupils; and they work in partnership with community leaders to help them understand how the community contributes to, or conversely undermines, health and education (WHO 2014). This is clearly a comprehensive agenda with consequences that reach much further than the traditionally understood school and/or curriculum boundaries. The strong focus on participation, partnership and collaboration in the WHO's conceptualisation of health promoting schools demonstrates the importance given to a comprehensive health promoting school agenda.

The emergence of Health Promoting Schools as a global strategy through which coherence in the implementation of health education and promotion can be achieved is significant. The concept of health promoting schools has been in use for over 30 years. Its emergence in Europe in the early 1980s became more coherently conceptualised at the WHO Health Promoting Schools Symposium in Scotland in 1986. The definition provided by Young and Williams has since become the foundation upon which the framework for health promoting schools was built (Young and Williams 1989). The definition provided was based on a holistic view of health and had three key components: (1) health education in the formal curriculum; (2) the "hidden" curriculum or school ethos; and (3) links with family and the local community (Inchly et al. 2000). Since this seminal work, much has been achieved in the drive to enhance health promoting schools in conjunction with the World Health Organisation. The guidance towards evaluation and performance indicators *Health-promoting schools: a resource for developing indicators* (Barnekow et al. 2006) has also added significantly to the conceptualisation and coherence of health promoting schools globally.

A significant development in furthering the success of the health promoting schools initiative was the establishment of the European Network of Health Promoting Schools (ENHPS) in 1991. The ENHPS, at its inception, was a tripartite activity, launched by the European Commission, the Council of Europe and the WHO Regional Office for Europe. It began with only seven countries, and subsequently has grown to become global in nature. Barnekow et al. (2006: 13) set out the aims of the health promoting school as follows:

- To establish a broad view of health;
- To give students tools that enable them to make healthy choices;

- To provide a healthier environment engaging students, teachers and parents, using interactive learning methods, building better communication and seeking partners and allies in the community;
- To be understood clearly by all members of the school community (students, their parents, teachers and all other people working in this environment), the "real value of health" (physical, psychosocial and environmental) in the present and in the future and how to promote it for the well-being of all;
- To be an effective (perhaps the most effective) long-term workshop for practising and learning humanity and democracy;
- To increase students' action competence within health, meaning to empower them to take action – individually and collectively – for a healthier life and healthier living conditions locally as well as globally;
- To make healthier choices easier choices for all members of the school community;
- To promote the health and well-being of students and school staff;
- To enable people to deal with themselves and the external environment in a positive way and to facilitate healthy behaviour through policies and to increase the quality of life.

These aims have significantly influenced the European understanding, research, and practices of health promoting schools.

Currently, school health promotion in Europe is organized through the Schools for Health in Europe (SHE) network, which has replaced the ENHPS's focus on schools with focus on policy and structural support. It now has at least 43 participating countries from the WHO European region, each represented by a national co-ordinator (Buijs 2009). The current status of the SHE network is summarised in two factsheets (http://www.schools-for-health.eu/she-network), based on background information obtained through a review of evidence (Young et al. 2013). Building on the previous work within the ENHPS and the International Union of Health Promotion and Education (IUHPE) (e.g. Clift and Jensen 2005; International Union of Health Promotion and Education 2009; St Leger et al. 2010), the SHE network endorses five core values (equity, sustainability, inclusion, empowerment and action competence, and democracy) and five pillars (whole school approach to health, participation, school quality, evidence, schools, and communities) as a common basis of the SHE approach to school health promotion (Buijs 2009). In addition to including sustainability among the key values, the Conference Statement following the 4th European Conference on Health Promoting Schools, which took place in Denmark in 2013, emphasises the support of education for sustainable development within the framework of the health promotion schools approach advocating that "the health of the people is closely linked with the health of societies and of the planet" (CBO 2013: 9).

Over the 20 years since inception, health-promoting schools have in practice been interpreted differently in global cultural, geographical and educational contexts, thus obtaining a wide range of meanings (Simovska and Jensen 2002). A variety of approaches, related underlying values and evidence of their effectiveness have been repeatedly debated in the literature (e.g. Green and Tones 2010; Simovska 2012). The gap between the analytical conceptualizations and the realities of practice throughout Europe and indeed more globally is worthy of further consideration. A systematic review of the effectiveness of the health-promoting schools (Stewart-Brown 2006) emphasizes that the programmes that are more likely to be effective are complex, multidimensional and embedded in more than one domain of school life. More recently Macnab and colleagues (2014), summarizing a consensus statement concerning the current challenges, strategies, and potential of health promoting schools, developed at the 2011 colloquium at the Stellenbosch Institute for Advanced Study in South Africa, suggest that educational initiatives, which are relatively simple, inexpensive and flexible, can be effective in positively influencing children's lifestyle and wellbeing (p. 170). Similar to Stewart-Brown (ibid) they conclude that educational programmes that are relevant resonate with students and engage school communities so that they come to own and sustain their programmes are most likely to be successful. Yet the practice remains dichotomous with most of the studies within the field focused on classroom-based or topic-based programmes neglecting the more wide-reaching features of the health-promoting schools approach, for instance school development, pupil participation, critical health literacy and empowerment (Griebler et al. 2014).

1.2 Health Promoting Schools and Education for Sustainable Development

Over the previous years of health promoting schools progress, questions pertaining to sustainability have often been mentioned, but seldom followed through or indeed more actively taken up for more serious integration within the health promoting school paradigm. However, closer examination of the seminal documentation of health promotion, the Ottawa Charter, as an example, evidences the sowing of the seeds of the issue of sustainability for health promotion. Even earlier, Alma Ata in 1978 was also ground-breaking in terms of bringing health and development together, while emphasising health as a social objective and a fundamental human right. In addition, by reconceptualising traditional understandings of health by moving away from top-down approaches focused solely on disease prevention and individual responsibility for lifestyle, to more positive, comprehensive, systems aware and empowering approaches, the Ottawa Charter (1986) placed significant focus on the pre-requisites for health. These pre-requisites are identified as peace, a stable eco system, social justice, and equity, as well as resources such as education, access to food and stable income – all concepts that are also inextricably linked to issues of sustainability. Sustainability is explicitly reflected in one of the background documents to the Ottawa Charter, written by Trevor Hancock (see Hancock 2009). Hancock has been a keen advocate for what he initially termed ecological sanity, and what is now more commonly known as sustainability (Hancock 1993). Hancock's emphasis on the importance of healthy built environments and his work on healthy cities have stressed the relationship between health promotion and sustainable development at city level (Hancock 2009) in the same way this book now advocates for the school and community. The underpinning rationale that health as an expression and component of human development needs to be perceived in an ecological way remains as relevant today, over 20 years since the publication of Hancock's call for an ecological and sustainable approach (Hancock 1993). Prioritising sustainability explicitly in one of the background documents for the Ottawa Charter, which focused on *creating environments for health*, meant in effect that a broader understanding of the concept of environment was initiated. This socio-ecological perspective on health envisions people as inextricably linked with their environments, with the understanding of *environment* to consist of the social, environmental and economic dimensions of life, which, in fact, correspond with the three pillars of sustainability – economic, environmental and social (United Nations 1987).

The shift in focus to incorporate attention to sustainability meant that sustainable development became implicitly embedded in the founding documents for health promotion (WHO 1986). This was reinforced by another influential policy document from the World Commission on Environment and Development, "*Our Common Future*" (United Nations 1987), which is the founding document for Education for Sustainable Development. The report also linked the incorporation of sustainable resources and ecosystems with the determinants of health. The importance of education as central thus became embedded in the discourses and understanding of sustainable development. The United Nations define education for sustainable development as: "a concept that goes far beyond environmental education. ESD is the educational process of achieving human development ("the three pillars of human development" proposed by UNDP: economic growth, social development, and environmental protection) in an inclusive, equitable and secure manner. It thus includes education for poverty alleviation, human rights, gender equality, cultural diversity, international understanding, peace and many more" (United Nations 2008: 1).

However, the impetus of sustainability evident in the founding policy documents of health promotion did not gather momentum for some time. This is reflected in the resolutions and declarations concerning health promotion that followed, where the emphasis on sustainability and education for sustainability is not evident. It was not until 2005 that the link between health promotion and sustainable development was again taken up, when the Sixth Global Conference on Health Promotion in Bangkok focused on four thematic tracks: (a) the new (global) context; (b) health-friendly globalisation; (c) partners; and (d) sustainability. The Bangkok Charter is explicit in outlining that the global context for health promotion has changed markedly since the development of the *Ottawa Charter.* It identified the critical factors that influence health as: the increasing inequalities within and between countries, new patterns of consumption and communication, commercialization, global environmental change and urbanization. There can be no doubt in reading the Bangkok Charter that real commitment to linking health promotion and sustainable global development was now taking centre stage. In the required actions called for, the charter specifically identifies the need to invest in sustainable policies, actions and infrastructure to address the determinants of health (World Health Organsiation 2005).

The Bangkok charter has been an important milestone in ensuring that health promotion practice is informed by and is cognizant of the inextricable link between sustainability and health. One would expect then, that health promoting schools would hold issues of sustainability at their core. However, in reality the developments relating to schools appear to largely have kept health and sustainability separated. Health promoting schools have served as an umbrella for a number of educational developments, policies and research that follow the principles and values of the Ottawa Charter. Despite the UN Decade of Education for Sustainable Development 2005–2014 (United Nations 2008) initiative, the developments within education for sustainable development have been more heterogeneous, with varying emphasis in different countries for example 'Green Schools'; in Ireland and Scandinavia or 'Sustainable Schools' in Australia. The actual links to health promoting schools have depended on the commitment and emphasis placed in individual countries rather than through a more effective model of focused intersectoral partnership. Yet such a model is needed. The impact of our rapidly evolving life circumstances on children's health and wellbeing and that of future generations cannot be currently predicted with any certainty (Davis and Cooke 2007). Indeed, the emergence in wealthy nations of increasing rates of childhood obesity, mental health problems, asthma and allergies is cause for deep concern. Even if our current lifestyles were ecologically and economically sustainable, they are hardly good for children's health (ibid). It is clear that a broader and more interlinked conceptualisation of health promoting schools that recognises the conceptual, ethical, and pragmatic synergies between health promoting schools and education for sustainable development - such as the whole-school approach, participation, social justice, equity, reciprocal links between health and sustainability, and the building of children's competences for the future complexities of the societies in which they will live. These issues are explored in more depth in the chapter by Madsen and colleagues later in this book.

The challenges of integration do not appear to lie with the rationale in interweaving education for sustainable development with education for health. The synergies and importance of this have already been established. Rather, it is the practical implications of implementation that appears to be currently at odds. One way that this can be better facilitated is to carefully examine and articulate the synergies between the aims, processes and content of health promoting schools and education for sustainable development in a manner that schools can meaningfully engage with. Schools are not a panacea or a place to 'cure all the ills' for society. Schools are often overloaded with political pressure to address the current hot political issues without due recognition that schools themselves may view their core mission differently and may be more focused on discipline specific (and consequently more measurable) educational outcomes within an already overcrowded curricula (Hennessy and Mannix McNamara 2012, 2013). Deschesnes and colleagues (2014: 209) citing Kremser (2010) and Simard et al. (2011) argue that because "of its multifaceted, integrated and concerted nature, the HS approach is inherently complex from a practical point of view and is a challenge with regard to its absorption within the core business of schools." In order for sustained and meaningful implementation of health promoting schools and education for sustainable development these concepts need to be framed within the core business of the school. Nor is it a case of asking schools to choose between them, but rather as Davis and Cooke (2007) advocate that researchers, practitioners and supporters of both initiatives work more closely together in recognition of shared goals in terms of desired educational and social outcomes. Greater convergence between health promoting schools and education for sustainable development can bring new energy and perspectives, and new partnerships, for example, with the environmental, climate change, and human rights initiatives (ibid). Such convergence could also yield greater sharing of resources, ideas, networks, and learning from each other's theories, empirical findings, perspectives and experiences in the move towards schools that are both *green and healthy* in a critical, empowering and socially responsible manner.

This book draws together a range of expertise in the field of health promoting schools and in education for sustainable development. The book is a result of the activities within the *Schools for Health Research Group* – a research network established in 2009 and constituted by more than 70 members from research institutions in 27 countries in Europe (information on SHE research group can be obtained at the following web address: http://www.schools-for-health.eu/she-network/research-group). Members of this network, many of whom are internationally renowned experts in the field of school health promotion, have contributed chapters to this book and have acted as peer-reviewers. The Schools for Health Research Group also operates as a Network titled *Health Education Research* within the European Educational Research Association (EERA). Within this research community, collaboration has been developed with the *Network on Environmental and Sustainability Education*, in the form of joint symposiums at the annual conferences within EERA (European Conference in Educational Research, 2012 and 2013). Members of this wider research community are also among the contributors of this book. The members of the international editorial advisory board reviewed all the chapters prior to the final reviews by the editors.

Schools for Health and Sustainability: Theory, Research and Practice contributes to the debate concerning wider-reaching dimensions of the health-promoting schools approach, including education for sustainable development. Its intention is also to add to the increasing evidence base in the field. Contributors address pertinent research questions and underlying theoretical perspectives of health-promoting schools research. They illuminate the existing evidence and explore contradictions and dilemmas in the development, implementation and evaluation of schools for health and sustainability. The chapters also examine the importance of this European as well as global school initiative from a comprehensive perspective that includes pupils, teachers, school management and staff, families and the local community members.

The book has been organised around four cohering themes. Part I focuses on ***Health Promoting Schools and Education for Sustainable Development.*** The chapters under this theme explicitly examine the potential interconnection between health promoting schools approaches and education for sustainable development. Ian Young makes this case in his chapter *Health Promotion and Sustainable Development in Schools*: *Historical Perspectives.* In it he examines in some detail the relationship between health promotion and sustainable development. He sets out the history of health promotion and sustainable development and explores ten themes which describe the common ground between health promotion, ecology and sustainable development. These themes are: history, ecological principles, whole school approaches, complexity, partnership-working, inequalities, school improvement, political issues, ethical issues and shared solutions. Young offers detailed practical examples to help elucidate the issues and makes the case for greater collaboration and even merging the two initiatives. In their chapter *Key Competencies: Reconciling Means and Ends in Education for Sustainable Consumption,* Daniel Fischer and Matthias Barth promote the case for sustainable consumption as a focal point of interest in the interplay of school-based promotion of health and sustainability. They point to the similarity in terms of goals of both approaches and advocate that in both discourses of health promotion and of sustainable development, education is commonly called upon as a powerful instrument to achieving both health and sustainability outcomes. Picking up on this theme Ulf Leo and Per Wickenberg, in *Under One Umbrella: Professional Norms Promoting Education for Sustainable Development at the School Level,* examine the role of school leaders and professional norms - between legal norms and professional action - and how the implementation of education for sustainable development is led, organised and realised. There is also interest in identifying the kinds of support, mechanisms as well as obstacles to implementation and norm setting exhibited by the school organisations in this regard. In the chapter *Linking Health Education and Sustainability Education in Schools - local transformations of international policy,* Katrine Dahl Madsen, Lone Lindegaard Nordin and Venka Simovska examine the relationships between international and national policies regarding sustainability and health promotion, which influence school-based health promotion and education for sustainable development in Denmark. Their analysis of international policy documents, as well as of research literature in both fields, indicates that school-based health education integrated in the concept of health promoting schools and education for sustainable development share a number of features, including cross-disciplinarity, participatory approaches, cultivating social imagination, and developing critical competences related to working with 'real life' health and sustainability issues. The discussion in this chapter explores the tendency of health and sustainability education in schools to be framed in national action plans. In so doing some critical educational aspects are lost by narrowing the concepts of health and sustainability to fit particular school subjects (e.g. physical education or science), losing the wider reaching dimensions of health promoting schools, such as the whole-school approach, critical competences, values and partnerships.

Part II is concerned with ***Partnerships, Standards and Change*** in health promoting schools. The chapters presented under this theme serve to give the reader insight into the complexity of health promoting schools from a situated and lived perspective.

Interestingly, a different perspective on education for sustainable development is adopted in this part of the book. Rather than explicitly addressed as a discrete entity, sustainability is now conceptualised within systemic integration of health promotion in the organizational structures of the school so that the outcomes are maintained in a sustained manner rather than the previously common single project based perspective. In their chapter entitled *Health Promotion in Dutch Secondary Schools: Promising Collaboration between School and Public Health Services,* Nicole Boot, Maria Jansen, Mariken Leurs and Nanne de Vries identify some challenges in prioritizing health promotion in schools. This chapter makes a call for investment in advocacy and school commitment, seeking connections between educational and health goals and shared responsibility between the educational and public health sector. Elizabeth Senior, Andrew Joyce and Dimitri Batras are also concerned with sustaining health promotion interventions in schools. In their chapter entitled *Becoming a Health Promoting School: Using a 'Change Agent' to Influence School Structure, Ethos and Ensure Sustainability,* they argue, although the health promotion model is useful in introducing and guiding health promotion activities, that without extra assistance, such as a dedicated health promotion officer, or 'change agent' who can motivate committed champions, changes to the ethos and the culture of the school will be difficult. Every school, they suggest is unique; there is no 'one size fits all' model and that the process and journey are just as important as the successes. Athina Karavoltsou explores the ways that drama in education methodology enhances adult learning, in this case through professional development activities aimed at improving the instructional practices of teachers in health promotion curricula. The chapter *Drama-based Learning for Teachers' Education in Health Promotion* explores adult learning theories, such as self-direction, transformation and emancipation as well as social cognitive theory in an attempt to explore synergistic ideas between drama and the philosophy of health promotion. Monica Carlsson provides a critical examination of competencies development in her chapter *Professional Competencies within School Health Promotion - Between Standards and Professional Judgment* and discusses them in relation to school-based health promotion. The chapter suggests caution as to the 'production logic' and economic values that are emphasized in the motivation of the project and in the knowledge base underpinning the competency-framework. Carlsson argues that by underemphasizing the potential of education, i.e. teaching and learning, and reducing changes at individual and group level to behavioral change, the formulations of competencies and standards are not in concert with essential values and approaches in school health promotion, and thus, their usefulness is somewhat problematic for professionals in this field. This cautionary tale of the 'production logic' and the performativity agenda is echoed by Sharon Moynihan, Jennifer Hennessy and Patricia Mannix-McNamara in their chapter *Health Education in the Context of Performance Driven Education: Challenges and Opportunities.* They advocate for the importance of health promoting schools but argue that the practice of schooling diverges significantly from the espoused ethos of holistic and personal development, with measurement, accountability and performativity taking centre stage. The significant influence of a consumerist agenda in education has meant that in the drive to educate for exam success, the development of affective education has been severely challenged. This has placed significant pressure on health promoting schools initiatives, including education for sustainable development. This chapter provides a critical analysis of the role of education and of the health promoting school within school systems where pedagogical practice is driven by a different value system, that of exam performativity and knowledge reproduction. The chapter makes recommendations in terms of the future development of a sustainable, affective curriculum and health promoting schools initiative.

Part III entitled ***Examining Implementation*** illuminates perspectives of implementation of health promoting schools in a variety of cultural contexts. Under this theme Oddrun Samdal and Louise Rowling suggest that to ensure efficient implementation of the health promoting school approach there is need to identify how the approach can best be implemented, which so far has not been sufficiently addressed within health promoting schools research. Their chapter entitled: *Implementation Strategies to Promote and Sustain Health and Learning in School,* carefully enunciates implementation components that will allow practitioners to understand the function of each component, and present theory based guidelines so each component can be utilised with fidelity. The chapter also discusses the common core of promoting health and learning in schools and demonstrates how similar implementation processes may be applied to achieve an efficient and sustainable change process for both health and learning. School leadership takes centre stage in the writing of Kevin Dadaczynski and Peter Paulus who identify the strategic importance of school leadership for the success of health promoting schools. In their chapter *Healthy Principals - Healthy Schools A Neglected Perspective to School Health Promotion*; they point to the need to understand how school principals impact upon teachers' health, and to understand the influence of principals on the implementation and overall success of health promoting activities in the school. Leena Paakkari in her chapter *Three Approaches to School Health Education as a Means to Higher Levels of Health Literacy* advocates for the central role that health literacy as an educational outcome plays in health promoting school initiatives. She points to the need to systematically address the various ways of approaching health education in schools in order to analyze how these approaches differ from each other. She discusses three approaches to school health education: the facts and skills approach, the individual thinking approach, and the personal growth and citizenship approach to be used in planning for learning experiences aimed at supporting the development of higher levels of health literacy. Maryvette Balcou-Debussche and Crane Rogers illuminate the contrasting social and cultural backgrounds that are a challenge for health promoting schools, given the importance of social and cultural influences on health-related behaviour, and the sensitive and personal nature of health topics addressed at school. In their chapter *Promoting Health Education in a Context of Strong Social and Cultural Heterogeneity: the Case of Reunion Island,* they discuss the importance of social context and cultural diversity in health promoting schools and advocate for the importance of an open, inclusive approach for sustainable health promotion in schools. More specifically focusing the attention to mental health and well-being, Aleisha Clarke and Margaret Barry, in their chapter

Implementing Mental Health Promoting Schools, examine the promotion of children's positive mental health within the health promoting schools framework. The chapter provides an overview of implementation components critical to health promoting school practice. Based on their research they call for the need to design evaluation studies that move beyond the conventional outcome-focused approach to a more systems-based approach that account for non-linear causality and seek to embrace and elucidate the inter-relationships, interactions and synergies within the whole settings approach.

The final section of the book part IV examines ***The Challenges of Evaluation and Evidence.*** Edith Flashberger and Lisa Gugglberger, examine what they term the core processes of schools, namely teaching and learning, and argue that research on school health promotion rarely includes attention to this central aspect of school life. In their chapter entitled *Health-Promoting Teaching Strategies in Schools: a Review of the Literature and Recommendations for Teacher Education,* they conceptualise health-promoting teaching and learning processes with due recognition of the interrelated nature of health and well-being on the one hand and teaching and learning on the other. They argue that to ensure sustainability of health promotion efforts and to make them more compatible to school life, it is advisable to integrate a health perspective in the core (teaching and learning) processes of schools. The concepts and discourses of evidence within schools for health and sustainability, and in particular the question of what counts as evidence is an important consideration for health promoting schools and education for sustainable development. Specific challenges in addressing evidence and research are also explored. Hege Tjomsland, Bente Wold, Rune Krumsvik and Oddrun Samdal pick up the theme of evaluation in their chapter *Evaluation Research in Health Promoting Schools and Related Challenges*. They argue that in order to ensure widespread dissemination of health promoting schools evidence, that the health promoting school approach can be a crucial vehicle for enhancing both students' health and educational outcomes is warranted. They call for study designs and methods that take into account the multifaceted, whole-school and context specific characteristics of health promoting schools. The chapter discusses specific challenges in health promoting school research, and they propose an evaluation design combining the advantages of different research methodologies to examine the health promoting school's effectiveness in creating "better schools through health". In their chapter *Mixed Methods' Contribution to the Evaluation of Health Promotion Initiatives in the School Setting*, Marie-Renée Guével, Jeanine Pommier, and Didier Jourdan, make the case for development of suitable approaches for evaluating health promotion in schools in order to produce useful evidence to enhance sustainability and transferability. They point to the trend among health promotion researchers to develop evaluation approaches that are able to measure the impact of initiatives. They also advocate the importance of deeper understanding of how this impact is obtained in order to inform the implementation of sustainable health promotion initiatives by practitioners and decision-makers from both health and education sectors. Thus they illustrate in their chapter the contribution of mixed methods to take into account the complexity of school health promotion initiatives to help address the challenges faced in the field especially those related to evaluation, sustainability and transferability. Terhi Saaranen, Kerttu Tossavainen, Marjorita Sormunen, Sari Laine, and Hannele Turunen point to the future by championing the need for strong theory for health promoting schools. In their chapter entitled *Developing and Testing a Health Promotion Theory – An Example of Creating a Model of School Staff's Occupational Well-Being* they call for the need for developing a theory that can be used to produce models applicable to workplace health promotion in school communities. They propose what they call 'a middle-level theory' and a content model for the promotion of school community staff's occupational well-being, which continues to be tested and developed. They demonstrate in the chapter how the theory contains the premises for planning comprehensively through four aspects, and it serves as a suitable model for implementing and evaluating the development of school staff's occupational well-being, which also affects the learning, health and well-being of children and adolescents.

1.3 Endnote

Health Promoting Schools are evolving well. They are now supported by strong policy frameworks, rich experience from practice and a growing body of research evidence. There remains significant progress to be made in terms of stronger interconnectivity with Education for Sustainable Development. Rigorous and wide-ranging evidence is required that embraces diverse research methodologies, stronger theoretical foundations and keen advocacy in the face of increasing neoliberal politics which effectively eclipses the place of education for health and sustainability in schools in favour of more performance driven indicators. This is despite the strong political emphasis placed on both health and sustainability as global social challenges which need to be urgently addressed if the costs of the global burden of chronic disease and climate change are to be avoided. The aim of this book is not to provide a ready-made 'how to do' Schools for Health and Sustainability; rather it is to contribute to the reflection, deliberations and debate and to examine models of inspiring practice from critical, rigorous and sound research perspectives.

References

Barnekow, V., Buijs, G., Clift, S., Jensen, B. B., Paulus, P., Rivett, D., et al. (2006). *Health promoting schools: A resource for developing indicators.* http://www.euro.who.int/ENHPS

Buijs, G. (2009). Better schools through Health: Networking for health promoting schools in Europe. *European Journal of Education, 44*, 507–520.

CBO. (2013). *Report of the 4th European conference on health promoting schools: Equity, education and health.* Retrieved March 22, 2014, from http://schools4health.dk/wp-content/uplAoads/2014/01/S4H-Conference-Report-2013-FINAL.pdf

Clift, S., & Jensen, B. B. (2005). *The health promoting school: International advances in theory, evaluation and practice*. Copenhagen: Danish University of Education Press.

Davis, J., & Cooke, S. (2007). Educating for a healthy, sustainable world: An argument for integrating health promoting schools and sustainable schools. *Health Promotion International, 22*, 346–353.

Department of Health UK. (2005). *National healthy schools: A guide for schools*. London: Department of Health.

Deschesnes, M., Drouin, N., Tessier, C., & Couturier, Y. (2014). Schools' capacity to absorb a healthy school approach into their operations, insights from a realist evaluation. *Health Education, 114*(3), 208–224.

Environmental Protection Agency USA. (2012). *Sensible steps to healthier school environments*. Washington, DC: National Service Center for Environmental Publications.

Green, J., & Tones, K. (2010). *Health promotion: Planning and strategies*. London: Sage.

Griebler, U., Rojatz, D., Simovska, V., & Foster, R. (2014). Effects of student participation in school health promotion: A systematic review. *Health Promotion International,* 4–12. doi:10.1093/heapro/dat090

Hancock, T. (1993). Health, human development and the community ecosystem: Three ecological models. *Health Promotion International, 8*, 41–47.

Hancock, T. (2009). *Equity, sustainability and healthy built environments, global research network on urban health equity.* Cuernavaca. http://www.ucl.ac.uk/gheg/GRNUHE/Events/Mexico/Hancock_-_for_GRNUHE.pdf

Hennessy, J., & Mannix McNamara, P. (2012). Educational experience, standardisation and instrumentalism: A century of limited change. *International Journal for Cross-Disciplinary Subjects in Education (IJCDSE), 2*(1), 951–955.

Hennessy, J., & Mannix McNamara, P. (2013). At the Altar of educational efficiency: Poetry, performativity and the role of the teacher. *English Teaching: Practice and Critique, 12*(1), 6–22.

Inchly, J., Currie, C., & Young, I. (2000). Evaluating the health promoting school: A case study approach. *Health Education, 100*, 200–206.

International Union of Health Promotion and Education. (2009). *Achieving health promoting schools: Guidelines for promoting health in schools*. Retrieved December 6, 2011, from http://iuhpe.org

Kremser, W. (2010). Phases of school health promotion implementation through the lens of complexity theory: Lessons learnt from an Austrian case study. *Health Promotion International, 26*(2), 137–147.

Macnab, A. J., Gagnon, F. A., & Stewart, D. (2014). Health promoting schools: Consensus, strategies, and potential. *Health Education, 114*(3), 170–185.

Paulus, P. (2005). From the health promoting school to the good and healthy school. In S. Clift & B. B. Jensen (Eds.), *The health promoting school: International advances in theory, evaluation and practice* (pp. 55–74). Copenhagen: Danish University of Education Press.

Paulus, P. (2007). 20 Years of health promotion research in and on settings in Europe -the case of school health promotion. *Italian Journal of Public Health, 5*(4), 248–254.

Simard, C., Deschenes, M., & Bujold, M. (2011). *Recension des e'ecrits sur les approaches globales en promotion de la santé à base scolaire durant la décennie 2000–2009* [Literature review on comprehensive school health approaches during the decade of the 2000s]. Québec: Institut national de santé publique du Québec.

Simovska, V. (2012). Editorial: What do health-promoting schools promote? Processes and outcomes of health-promoting schools. *Health Education, 122*(2), 84–87.

Simovska, V., & Jensen, B. (2002). *Models of health promoting schools in Europe*. Copenhagen: World Health Organization.

St Leger, L., Young, I., Blanchard, C., & Perry, M. (2010). *Promoting health in schools: From evidence to action*. Retrieved December 6, 2011, from http://iuhpe.org

Stewart-Brown, S. (2006). *What is the evidence on school health promotion in improving health or preventing disease and specifically: What is the effectiveness of the health promoting school approach?* Copenhagen: World Health Organisation.

United Nations. (1987). *Report of the World Commission on environment and development: Our common future*. Oxford: Oxford University Press.

United Nations. (2008). *United Nations decade of education for sustainable development 2005–2014*. Retrieved March 22, 2014, from http://www.desd.org/About%20ESD.htm

United Nations Educational, Scientific and Cultural Organization (UNESCO). (2014). *ESD Webpage*. Retrieved March 24, 2014, from http://www.unesco.org/new/en/education/themes/leading-the-international-agenda/education-for-sustainable-development/browse/2/

Whitman, C. V., & Aldinger, C. (Eds.). (2009). *Case studies in global school health promotion: From research to practice*. New York: Springer.

World Health Organization. (1986). *Ottawa charter for health promotion* (WHO/HPR/HEP/95.1). Geneva: WHO.

World Health Organization. (1998). *WHO's global school health initiative: Health-promoting schools*. Geneva: WHO.

World Health Organsiation. (2005). *The Bangkok charter for health promotion in a globalized world*. Retrieved March 22, 2014, from http://www.who.int/healthpromotion/conferences/6gchp/hpr_050829_%20BCHP.pdf?ua=1

World Health Organisation. (2014). *Global school health initiative*. Retrieved January 15, 2014, from World Health Organsiation http://www.who.int/school_youth_health/gshi/en/

Young, I., & Williams, T. (1989). *The healthy school*. Edinburgh: Scottish Health Education Group.

Young, I., St Leger L., & Buijs, G. (2013). *School health promotion: Evidence for effective action. Background paper SHE factsheet 2*. Utrecht: CBO. Retrieved March 22, 2014, from http://www.schools-forhealth.eu/uploads/files/SHE%20Factsheet%202%20background%20paper_final.pdf

Chapter 2
Health Promotion and Sustainable Development in Schools: Historical Perspective

Ian Macrae Young

Abstract This chapter examines in some detail the relationship between health promotion and sustainable development. It explores the existing literature and also draws on the experience of the author's professional practice as an ecologist, a science teacher, a health-promoter with the national health promotion agency in Scotland and also his work with international agencies. It sets out the history of the health promotion and sustainable development and explores ten themes which describe the common ground between health promotion, ecology and sustainable development. These themes relate to history, ecological principles, whole school approaches, complexity, partnership-working, inequalities, school improvement, political issues, ethical issues and shared solutions. The author offers detailed practical examples to help elucidate the issues. A case is made for greater collaboration and even merging the two movements within the health promoting schools approach. Practical suggestions on the nature of further research to inform this collaborative approach are made.

Keywords Collaboration • Health promoting schools • Sustainable development • School improvement

2.1 Introduction

Treat the earth well: it was not given to you by your parents; it was loaned to you by your children. We do not inherit the Earth from our Ancestors, we borrow it from our children.

North American First Nation's proverb

I.M. Young (✉)
International Health Promotion Consultant, Edinburgh, Scotland
e-mail: imyoung@blueyonder.co.uk

V. Simovska and Patricia Mannix-McNamara (eds.), *Schools for Health and Sustainability: Theory, Research and Practice*, DOI 10.1007/978-94-017-9171-7_2,

In this chapter I will draw on the existing literature on the relationship between health promotion and sustainable development. In addition, I will explore issues which arise from my original training in ecology and also issues from my professional practice with the national health promotion agency in Scotland and in work for international agencies such as The World Health Organization (WHO), and the International Union for Health Promotion and Education (IUHPE). I will set out ten themes which describe and explore the common ground between health promotion, ecology and sustainable development. I will also offer some detailed practical examples to help elucidate the issues, attempting to draw out what we know and suggest where further research may be helpful.

2.2 History of the Concepts and the Ecological Link

Creative activity could be described as a type of learning process where teacher and pupil are located in the same individual (Arthur Koestler).

It is essential to set out the historical development of the two concepts of health promotion and sustainable development as there are important common threads running through them, but also because some of the problems inherent in the two approaches working collaboratively are related to these historical issues. The concept that schools have a role in promoting the health of young people is not a new one (Young 1993). However, the modern view of health promotion in schools can only be fully understood in the context of wider developments in health promotion in general in the last three decades. A move towards a consensus on the meaning of the term health promotion was to a large degree the product of the work of the European office of the WHO. Their original discussion paper (WHO 1984) laid out the broad concepts and principles of health promotion and stimulated the debate that led to the development of the Ottawa Charter which defined health promotion as 'the process of enabling people to increase control over, and to improve, their health' (WHO 1986).

The ground-breaking Lalonde report in Canada, was the first time a government had set out a model of health that was moving from a disease-orientated one towards a social one (Lalonde 1974). Although it is true that over 150 years ago reformers such as Rudolf Virchow (1848) in Germany and Edwin Chadwick in Britain (1842) had been pioneers in Europe, setting out a social model of health and conceptualising the role of the state in promoting health. The Ottawa Charter for Health Promotion built on this European tradition and integrated current thinking on the nature of health promotion, drawing on progressive work in Canada, Australia and in parts of Europe. Early writing on health promotion appeared to an extent to be rather dismissive of an educational approach. For example the WHO's original discussion paper on health promotion (WHO 1984) referred to above, described health education as a core component of health promotion 'which aims at increasing knowledge and disseminating information related to health.' At the time this definition seemed narrow to educationalists as it ignored the critical, affective and action domains of learning and, in addition it implied that the subject exposed to this process of dissemination is merely a passive recipient of information. At the time many of us with an interest in pedagogy felt that there was a danger that the value of an educational approach was not only being misrepresented, but that health education could effectively disappear.

Before comparing the development of health promotion and sustainable development it is important to reflect on the origins of the science of ecology as this illustrates some interesting parallels and connections with health promotion. The term ecology ("Ökologie") was first coined by the German biologist Ernst Haeckel in 1866. (Haeckel 1866) (from Greek: οἶκος, "house"; -λογία, "study of") and it refers to the scientific study of the interactions of living organisms with each other and with their natural environment. Haeckel's broad concept is still relevant today. Modern ecologists study living systems, processes and adaptations at both the micro and macro level. They also study the movement of materials and energy through living systems and the environment, and they are interested in how ecosystems develop over time (successional development). As we know, ecologists also measure the variety (biodiversity) and abundance (biomass) of different species. Ecosystems create biophysical feedback mechanisms between living and non-living components of the planet. These feedback loops play a part in regulating and sustaining living communities of organisms as well as climate systems, and global geo-chemical systems.

Many health promotion professionals are aware of Virchow's first use of the term health promotion and many ecologists are aware of Ernst Haeckel's role in first proposing the term ecology. However, the significance of the links between these two scientists is not widely understood. Haeckel was actually a student of medicine under Virchow in Berlin in the 1850s and worked as his assistant for a short time. They later had a very public and vehement disagreement over interpretations of the nature of the scientific evidence for evolution (Zigman 2000). While the term ecology is not synonymous with sustainable development, the application of the principles of ecology has been a major influence in the development of the concept of sustainable development since the 1970s. Therefore, there was to some degree a personal historical link within the scientific powerhouse of nineteenth century Germany, between the two concepts of health promotion and sustainable development under discussion. Curiously the modern concepts of ecology and health promotion were reborn in the twentieth century after losing prominence. For example, Darwin didn't use Haeckel's new term, although he was describing living organisms in their natural environment. The term ecology became more prominent again in the 1920s although the English botanist, Arthur Tansley had kept the flame alive with a seminal paper in the first decade of the twentieth century (Tansley 1904). Health promotion reappeared later as an updated concept in the 1980s. Good ideas and even scientific discoveries do have a habit of disappearing and reappearing, which is another topic for another occasion!

The concept of sustainable development emerged from the ecology and environmental movements, which recognised the negative impacts of human growth and development on the environment and ecosystems. The term sustainable development had been in circulation in the 1970s after a report for the Club of Rome (Meadows et al. 1972), but it came to prominence through the Brundtland Commission (United Nations 1987). The commission's report, *Our Common Future*, defined sustainable development as "development which meets the needs of the present without compromising the ability of future generations to meet their own needs." Sustainable development as a concept received further attention at the United Nations Conference on Environment and Development in Rio de Janeiro in 1992, (UN 1992) which was the first international attempt to develop strategies for a more sustainable pattern of development. Representatives of 178 national governments, including more than 100 heads of state, and many organisations representing civil society attended the conference. At the summit, governments around the world made commitments to work for sustainable development. Subsequent international gatherings have taken place and many countries now have policies on sustainable development which take account of the Millennium Development Goals, the Doha Development Agenda of the World Trade Organisation, the Monterey Consensus on Financing for Development and the Plan of Implementation of the 2002 World Summit on Sustainable Development.

While these developments were unfolding in the sustainable development movement, those working in the public health sector were moving towards an ecological view of the health of *one* species, *Homo sapiens*. The Sundsvall Statement (Petterson and Tillgren 1992) which was the output from a significant health promotion conference in Sweden reflected this when it stated;

> Humankind forms an integral part of the Earth's ecosystem. People's health is fundamentally linked with the total environment. All available information indicates that it will not be possible to sustain the quality of life for human beings and all living species, unless drastic changes in attitudes and behaviours at all levels are adopted with regard to the management and preservation of the environment. (Sundsvall, Sweden, June 1991).

The phrase…. *and all living species*… is highly significant because it is the first time that a modern health promotion document had acknowledged that other living organisms (other than infective agents) were relevant to human health and health promotion! I emphasise *modern* health promotion document because it is clear that many First Nations and indigenous people who lived closer to the land had such concepts of human health and welfare being intimately linked to all living creatures and systems, which were developed over the last 10,000 years (Houde 2007; Gammage 2012; National Chief Shawn A-in-Chut Atleo 2012; Leonard et al. 2013). It could be argued that this view of the world had largely disappeared at the time of the industrial revolution and that in fact the anthropocentric view of some religions encouraged the relegation of other living organisms to the role of serving human needs only.

2.3 Schools and Ecological Models of Health

If we teach today as we taught yesterday, we rob our children of tomorrow (John Dewey).

It is worth recounting the above historical development, because it is important to understand that the health promoting school was born at a time when there was considerable scepticism that educational approaches in general or schools in particular could have much impact on the health of the population. While Scotland was not a founder member of the European Network of Health Promoting Schools (ENHPS) in 1991, it was in fact at the heart of the development in the previous decade. In 1986 The Scottish Health Education Group (SHEG), which was a collaborating centre for the WHO (European Office), was given the task of organising a European symposium. This took place at Peebles, Scotland over 6 days and was attended by 150 delegates from 28 member states. The symposium was entitled 'The Health Promoting School.' At this time Europe was effectively a divided continent with 32 states and there existed restrictions on travel from the former Eastern Bloc countries. However, with WHO assistance it was possible to get senior staff from the Soviet Union and almost all of the Eastern Bloc countries. At this event the concept of the health promoting school was developed and refined. In fact the name 'The Health Promoting School' was born in the planning for the event although the concept had been evolving for several years and the literature in Europe talked of 'the school as a health promoting institution' in the early 1980s (Williams 1982).

This symposium was significant because it offered WHO an opportunity to apply its developing theoretical model of health promotion to the setting of the school. From this event a report entitled 'The Healthy School' was produced on behalf of WHO in 1989 (Young and Williams 1989). The reason the name was amended was political at that time as WHO (Euro) was keen to make links with its new Healthy Cities project. The report, which was translated into more than ten European languages, described health promotion in schools as a "combination of health education and all the other actions which a school takes to protect and improve the health of those within it". However, the term health *promoting* school soon bounced back into the forefront and a pilot project supported by WHO in the Czech Republic, Hungary, Poland and Slovakia (Woynarowska and Sokolowska 2009) supported its spread in central Europe and in turn to many parts of the world. More modern ideas of the breadth of the health promoting school concept have since developed and further refined the thinking about the scope of the concept in Europe and beyond. In the USA and Canada the terminology used to describe the processes relating to health in schools was and is different. In the United States terms such as the 'healthful school environment' had been used since the early 1950s indicating that there was an awareness of the importance of wider influences on health in schools beyond the 'health instruction' of the classroom. In the early 1980s in the USA the term 'comprehensive school health programme' became the common term to encompass a broader approach. At that time this concept was considered to have the components of health instruction, the school health services and the school environment which was remarkably close to the European model in the same time period although using the different nomenclature of comprehensive school health.

In the 1990s this broader concept of 'comprehensive school health programmes' was further developed in the USA and Kolbe suggested the following components within this conceptual framework (Kolbe 1993).

- School health education;
- School health services;
- School health environment;
- School physical education programme,
- School food service programme;
- School psychology and counselling programmes;
- Programmes to protect and improve the health of staff;
- Integrated efforts of school and community agencies to improve the health of school and students.

These components closely parallel the European model set out in The Healthy School Report (Young and Williams 1989), which included a comparison between a traditional approach and a modern health promoting school approach.

The European model emphasised pupil participation more strongly than the USA approach but the European report also had some statements which now, with the advantage of research and hindsight, appear rather simplistic on the exemplar role of teachers and on the unquestioned role of self-esteem in determining health-related behaviours. The important role of schools in addressing inequalities in health was not emphasised in these early developments. The European model has further developed to emphasise equity and democracy at the core of its activities (WHO 1997) and this reflects the egalitarian approach that had its origins in earlier European history. However both models have much more in common than any differences, which is interesting because to some extent they evolved independently of each other with only limited cross-fertilisation compared to the international contacts that exist today through travel and the immediacy of electronic media.

2.4 The Development of Eco-schools and Sustainable Schools

The achievement of excellence can only occur if the organization promotes a culture of creative dissatisfaction (Lawrence Miller)

Although health promoting schools are based on an ecological view of health and health promotion, it could be argued that for many years traditional educational approaches were still prominent within the health promoting school model. To some extent school health remained focussed on narrow educational outcomes such as acquiring knowledge and developing skills at an individual student level. More social and environmental changes in schools such as the quality of the food and drinks offered or the provision of safe and secure storage of students bicycles, or the quality of the social environment were not usually monitored or measured in mainstream education in the 1980s and 1990s in many countries.

The growth of sustainable development education in schools in different countries has also taken place over the last 25 years. It has taken slightly different routes and this is reflected in the names given to it at various stages in different countries. For example Eco-schools in some European countries and parts of Africa and South America, (Pirrie et al. 2006); Green Schools in Scandinavia; Sustainable Schools in Australia, (AuSSI 2013); and Enviroschools in New Zealand (Jackson 2009). The United Nations Educational, Scientific and Cultural Organisation (UNESCO) is the lead agency for the United Nations Decade of Education for Sustainable Development (2005–2014), (United Nations 2002). UNESCO states that Education for Sustainable Development allows every human being to acquire the knowledge, skills, attitudes and values necessary to shape a sustainable future. UNESCO also sets out the scope and competencies of education for sustainable development to include key sustainable development issues within teaching and learning; for example, climate change, disaster risk reduction, biodiversity, poverty reduction, and sustainable consumption. It is suggested that it also requires participatory teaching and learning methods that motivate and empower learners to change their behaviour and take action for sustainable development. Education for sustainable development consequently promotes competencies like critical thinking, imagining future scenarios and making decisions in a collaborative way. Within this educational emphasis it has been an understanding that the whole life of the school should reflect an ecological and sustainable development approach and this parallels the whole school approach of health promoting schools. These different developments with different starting points, human health and the global environment, have a great deal in common and yet in many schools are perceived separately. There is an Irish story where a stranger stops and asks directions to a local person on how to get to a specific place. The local person replies "Well if I was going there I wouldn't start from here!" This raises a smile but it is also profound, as our starting points influence the nature of the journey but can also influence where we reach in our final destination. It could be argued that this is the case with health promoting schools and sustainable schools or eco-schools.

2.5 What Do These Developments Have in Common and What Are the Consequences of This?

I have identified ten links between health promotion and sustainable development and in parallel between health promoting schools and sustainable schools or eco-schools. The common issues can be listed as follows. Health promoting schools and sustainable schools:

- Share related but separate historical developments;
- Are underpinned by ecological principles;
- Work with complexity;

- Have similar views on the nature of learning and a whole-school approach;
- Require a partnership approach;
- Have an ethical approach which is broadly similar;
- Are concerned with inequalities at national and international level;
- Share a political dimension and their roles and relevance to mainstream education in schools may be contested;
- Share features with the school improvement movement's agenda;
- Have potential to offer shared solutions to education, health and ecological issues.

Let us examine each of these in turn in more detail. As I have outlined in Sect. 1.3, these two movements are linked in their *historical development*, although these historical links have not produced a great deal of practical collaboration in reality. This has not involved much conflict or many battles for territory, but has resulted in two separate developments, when in fact from the perspective of a setting such as schools, teachers and school managers may come to the view that it is inefficient to have two such developments which have significant common ground. Very few writers have commented on this separation therefore it is not routinely perceived as problematical, however there are exceptions to this. Davis and Cooke in Australia produced an eloquent argument for developing schools that are both green and healthy (Davis and Cooke 2007) and Jensen and colleagues in Denmark have brought together international colleagues to explore these issues (Jensen et al. 2000). However, formal integration of the two approaches under one title is still rare although a 'Green and Healthy schools' movement has now started in the USA in Wisconsin (http://eeinwisconsin.org) and in Kentucky (http://greenschools.ky.gov). In both these cases health appears as only one of many sustainable development themes. One of the few examples of continuous formal collaboration is from Wales where a partnership scheme (Pembrokeshire County Council 2011) as part of "Education for sustainable development and global citizenship" has representatives of Eco-schools Wales and Welsh Healthy Schools which meets once per term. The sustainable schools award in Wales has a healthy living component, although it has to be said that this progressive approach in Wales, while it has resulted in collaboration, has not resulted in the two programmes being integrated.

Both health and sustainable development have an underpinning *ecological conceptual basis*, although in the case of health this starts with the human species as its starting point and concern, whereas in the case of sustainable development the human species is not the sole concern, as all living organisms and the global environment are seen as interlinked. Exploring this ecological link is relevant to aspects of school health promotion as it provides an understanding of the changing priorities of school health promotion as local and global environments change (Young and Whitehead 1993). Some traditional models of public health, developed before modern ideas on health promotion were elucidated, were ecological in their nature, describing health and disease in terms of a dynamic interaction. This was a triangle between the host (human), the infective or injurious agent and the environment. It recognised that if one component of the system changed it would affect the others. If one considers the growth of allergies in school children in the wealthier western countries, one can see an example of what can happen to developing immune systems when this balance is disturbed by the creation of highly protective urban environments. If this is coupled with excessive use of antibiotics in childhood there is some evidence that it can produce immune systems which are under-developed and possibly result in attacks on the body's own systems, which, along with genetic factors, can produce conditions such as asthma and eczema (Cohet et al. 2004).

Similarly the mineral requirement for optimum human health is also an ecological and environmental issue. The ecologist Eugene Odum, in the first general textbook on ecology, (Odum and Odum 1953) reminded us that, unlike energy which is lost from systems, mineral elements just get recycled. Some of them such as lead or cadmium are not needed by the human body and are highly poisonous, but others are needed in either trace amounts such as molybdenum or copper in enzyme systems, or others such as calcium in larger amounts for our bones and teeth. Element such as calcium, iodine or fluorine cannot normally change into a different element. They can of course combine with other elements or radicals to form compounds such as calcium fluoride the basic mineral component in our teeth. However these minerals are essential to our health and yet in some environments e.g. fluorine in Scotland they are not readily available and, with the exception of two small areas of Scotland, (MacDonald and Ódochartaigh 2005) corrective measures need to be considered to redress this environmental deficiency. A similar problem exists with iodine deficiency in the environment in countries such as Azerbaijan (Markou et al. 2001). It is thought that some iodine reaches the land through complex gaseous processes with methyl iodide being produced by marine algae (Lovelock 1979). As with fluoride, the human body cannot create iodine or iodide, and has to acquire it from the environment, to then create the hormone of the thyroid gland. This deficiency is tackled by adding iodine supplements to salt and/or flour. These two examples, relating to the mineral requirements of our bodies, remind us of the extent to which the health of children and adults is linked to our environment. They are particularly important for the growing bodies of children, for example the damage done to growing teeth can have lifelong effects. The environment is the only place we can get these building blocks and they are either ingested in our food or in solution in our drinking water.

Similar examples of the role of the environment on human health could be viewed in terms of an ecological perspective on infectious diseases. The majority of human infections which challenge the health of children and adults are transmitted from animals in our environment. Zoonoses are any infectious diseases transmitted between species (in some instances, by a vector or intermediary host) from animals to humans or from humans to animals. In a study of 1,415 pathogens known to affect humans, 61 % were identified as zoonotic (Taylor et al. 2001). Many modern diseases started out as zoonotic diseases. It is hard to prove which diseases have jumped the species barrier from animals to humans, but the balance of evidence suggests that measles, smallpox, diphtheria, influenza, ebola and HIV are all examples of this phenomenon. School-age children are vulnerable to many of these diseases transmitted from animals because of the way children explore their environment and because their immune systems are still developing. While the origins of some of these diseases goes back to humans early domestication of farm animals (Diamond 1997), new emerging risks are appearing (Brown 2004). It is thought that these emergent diseases are due to a range of factors such as ecological disruption and the increasing density of human populations. The health of ecosystems and human health are intertwined both at a conceptual level and a practical level. I would suggest that the professional separation of human ecology and human health from the ecology of other species is not helpful when addressing the types of issues of mineral deficiencies and infections explored above.

Both health promotion and sustainable development have to work with *complexity*. This complexity exists at different levels, for example schools are complex social systems. Keshavarz et al. (2010) has applied concept of "complex adaptive systems" as a framework to better understand ways in which health promoting school interventions could be introduced and sustained. She concluded that schools exhibit most, but not all of the characteristics of social complex adaptive systems, and that this may help to explain some of the challenges of introducing and sustaining change in schools. An example of this complexity was found in an experimental study (Young 1993). The study concluded that the choices of the pupils in relation to where and what they ate at lunch time were influenced by a diverse range of issues such as social/friendship influences, school policy, internal school environmental factors, availability of school meals choices, the distance of the school from home and the distance of the school to external shops. It is clear that developing and implementing a health promotion policy in a school will only have a chance of being effective if there is some understanding of the many complex interactions in the life of a school.

Ecological systems are also complex and this is demonstrated in what happens when small changes are made in an ecosystem, for example, a behaviour change or an environmental change or the introduction of an alien species in relation to the health of living organisms including humans. In fact there is a journal entitled *Ecological Complexity* which is completely devoted to exploring this issue (Petrovskii 2004). Another level of complexity is that the socio-political agencies which can potentially support or inhibit these developments cut across many sectors, for example, the education, health, environment and commercial sectors. Finding policy solutions to health promoting schools requires more than the efforts of the education system.

Health promotion and sustainable development both favour an educational approach that is interested in developing not only knowledge and understanding but also in exploring attitudes and values, and in developing real-life competencies that help young people become engaged citizens who can make a difference through their actions (Simovska 2012). Both health promoting schools and sustainable schools require a *whole-school approach* which goes beyond the learning and teaching of the classroom to all aspects of school life including the social and physical environment of the school and its surrounding communities (Dahl et al. this volume).

A consequence of the above is that a *partnership approach* is essential because of the diversity of sectors with a genuine interest in health promotion and sustainable development in schools. This is easy to state but experience and evidence demonstrates that it is not easy to achieve the inter-sectoral collaboration required. This issue of inter-sectoral collaboration or partnership-working will be explored in more detail later in this chapter.

Both developments share common ground on the *ethics* which underpin their approach. For example respecting life, health, and lifelong learning and behaving in ways which reflect this, would be examples of common ground. However, there are also tensions between them; an example would be that the nutritional advice and recommendations common in health promotion may not always take account of the sustainability of the food sources being suggested (Young 2003). The priority in health promotion might be related to inexpensive and healthy food sources whereas in sustainable development other factors such as the sustainability of fish sources or the issue of 'food miles' or even the ethics of eating food from animal sources would be prominent. Tim Lang, (Lang 1992) has been a pioneer for over 20 years in stimulating policy debates on these issues. More recently he suggests that understanding the environmental impact of food systems challenges nutritionists to bring together thinking from the fragmented disciplines of life sciences, social sciences and environmental science if policy engagement and clarification is to occur (Lang and Barling 2013). I suggest that the ethical debate will only be meaningful if we move away from an individualistic model to an ecological one which will enable all to consider principles such 'first do no harm' being considered in relation to all species and the living and non-living environment.

Issues of *inequality* are central to both health promotion and sustainable development. This is true at a global level where wealthy western economies enjoy better health and within the sustainability domain, western countries are challenged by the fact that their economies are accused of reducing the sustainability of the planet's resources. The people who struggle to live in the most damaged parts of the world's environment, both urban and rural, are poor and suffer from the most health problems. This problem is exacerbated by the fact that they are less likely to get a basic education; therefore there is a global inequality aspect at the school level as well. Of particular importance to health and sustainable development is the inequality in educational opportunities between girls and boys, especially in developing countries. When women have greater control over resources in the family, they are more likely than men to allocate more resources to food, children's health care, and education—a finding from the World Bank which is evident in a diverse set of countries including Bangladesh, Brazil, Côte d'Ivoire, Ghana, Indonesia, and South Africa (Tembon and Fort 2008). This report also indicates that countries with higher levels of female secondary-school enrolment have lower infant mortality rates, lower fertility, lower rates of HIV and AIDS, and better child nutrition. Promoting gender equality in education can be viewed as a pre-requisite for achieving both sustainable development as well as a route to better health and well-being. Again it is clear these two issues of health promotion and sustainable development are inextricably linked. Michael Wilson suggested health cannot be possessed (Walker 2013), it can only be shared. He stated "There is no health for me without my brother. There is no health for me without Bangladesh". This concept is easily applied to sustainable schools and sustainability, as it is self-evident that for an action to be truly sustainable it should not be harming other living creatures, including humans, as well as the environment.

The ideas of health promotion and sustainable development share another feature and potential problem. They are both *contested and highly political* (see also Dahl et al. this volume). This is true at the fundamental conceptual level and also at the level of applying these concepts to be part of the business of schools. First at the conceptual level sustainable development sits uneasily with particular industry, commerce and economic measures. While there is some evidence of this changing, the traditional economic approaches are still dominant and this is illustrated by, for example, the economic measures of a country's output such as gross national production (GNP) taking no account of broader issues of environmental impact and the long term sustainability of particular forms of economic output. If one considers the role of multi-national companies in the food industry, many of those are being forced to start to consider the health and sustainability issues surrounding their products. This is highly political and the voices supporting health and sustainability, although committed, are under-resourced compared to the massive marketing and advertising budgets behind those selling unhealthy and unsustainable products to the public (Consumers Union 2005). A similar argument can be made on the impact of specific mineral extraction developments on human health and environmental sustainability. An example of working within such a highly political, contested area is explored later (Fig. 2.1).

At the level of schools, both health promotion and sustainable development are contested as they have to fight for their place in a pressurised and crowded curriculum and in the wider life of schools. This is also highly political because governments and policy makers may take a view of the curriculum as being solely about literacy, numeracy and preparation for the world of work. In this climate there would be a propensity to view health and sustainable development as peripheral to the main work of schools. In most countries which have had some success in developing health promoting schools and sustainable schools this has been a slow process and one which has been subjected to political forces which can stall or reverse developments (Young 2005). This is also an argument for health promoting schools and sustainable schools potentially strengthening their case by collaborating in their approaches to education departments, an issue I will return to in this chapter.

Both health promoting schools and sustainable schools have positive approaches which can improve the motivation and learning abilities of staff and students. This fits well with the progressive aspects of the *school improvement agenda*. The qualities of improving schools have been well documented in the educational research literature over the last 15 years. This suggests that highly effective schools in different cultures have features such as the following:

- A clear and focused vision;
- A safe and stimulating school environment;
- A climate of high expectations for student success;
- Frequent monitoring and review of student progress;

- A focus on achievement and celebrating student and school success;
- Head teachers or principals who provide excellent leadership;
- Strong home-school relations.

It could be argued that both health promoting schools and sustainable schools have also promoted these values relating to vision, leadership, celebrating success, the importance of the social and physical environment of schools, and the links with parents and the wider school community. Yet even in core functions of the school such as the physical structures that help to create a modern learning environment, we all have much to learn and informative research is surprisingly limited. Stephen Heppell (2005) has argued that:

> whereas traditionally, we have designed for productivity, processing large numbers of children through the effective use of buildings, designing a room for learning is very complex. No one knows how to prevent 'learning-loss' when you design a room "pedagogically", whereas we know lots about designing for minimum heat loss. (Higgins et al. 2005: 3)

However, we do have experience of the recent history of educational change and development and this suggests that even when there *is* good evidence of successful innovation and school improvement, these can be marginalized in the pursuit of quicker and politically popular approaches to raising educational attainment. Michael Fullan has explored the complexity of the change process, suggesting that schools are pulled in two directions, by stable and less stable forces, and that the dynamics of the successful organization are of 'irregular cycles and discontinuous trends' (Fullan 1993, 1999). This analysis could also apply to both health promoting schools and sustainable schools where national and regional commitment can and does vary from country to country and within countries at any given time. Progressive innovation may be buffeted by a political climate which at times can force a 'back to basics' agenda and the quality of an education system ends up being measured only on limited criteria relating to literacy and numeracy. It is interesting to note that the Programme for International Student Assessment (PISA) run by the Organisation for Economic Cooperation and Development (OECD) is an international assessment of student attainment focusing on testing the knowledge and skills required for participation in society and assessing the extent to which students can apply skills gained in school in everyday adult life at age 15. PISA has a tendency, in my view, to define educational success in terms of national economic benefits (OECD 2010) and this is exacerbated by the focus in media reporting of PISA comparisons on numeracy and literacy that can result in political pressures for a narrower view of the curriculum which in turn can undermine the broad focus on innovation which health promoting schools and sustainable schools wish to foster.

Health promoting schools and sustainable schools have shown willingness to build their own networks internationally and this has also been true of the school improvement movement to some extent. However, Fuller and Clark (1994), viewing from the perspective of the school improvement agenda, argue that those involved in international educational innovations overlook local culture at their peril. This has implications for the external support required by individual schools which needs to take account of the reality that leadership has to come from within the school.

It also has implications for international support which must avoid the assumption that innovative work in one cultural context can be transplanted to another. In my view the various international health promoting schools networks have been relatively good at recognising the importance of the uniqueness of individual school contexts and culture, but in the past have not always taken on board the lessons of the school improvement movement in the education sector. There have been important exceptions to this such as the innovative work of the EVA research in Europe which involved collaboration with the school improvement movement (Piette 2002). More recently there has been clearer recognition that health promoting schools can contribute to the education systems role in school improvement, (St Leger et al. 2010; Buijs 2009; Simovska and Carlsson 2012). Buijs stated:

> Health contributes to better learning. Increasingly, health and well-being are regarded as an entry-point for school improvement and school development.

Buijs also highlighted the need for the health sector to integrate their indicators of effectiveness and quality with the education sector. This has only happened in a limited number of cases to date in health promotion in schools, such as in Scotland where health promotion indicators of effectiveness have been incorporated into the general school improvement indicators of effectiveness (Scottish Health Promoting Schools Unit 2004).

Lastly, health promoting schools and sustainable schools have the potential to offer *shared solutions to education, health and ecological issues*. The promotion of healthy eating, with a higher vegetable component and a reduction of animal fats in the diet for health reasons, is broadly compatible with the ecological/sustainability issues of the energy inefficiency relating to the production of pork, meat and other animal food sources. One of the main issues of interest to those promoting sustainable development in the context of food production is the greater energy efficiency of a diet which is either vegetarian or predominantly vegetarian. A green plant such as grass (a primary producer) converts solar energy into carbohydrate through photosynthesis. When a cow eats the grass only approximately 4 % is converted into new animal flesh, approximately 60 % is lost in faeces, methane gas and urine, and the remainder is lost in respiration. This energy loss happens at every link in the food chain, therefore energy is used much more efficiently if there is a shorter food chain, for example humans eating vegetables instead of meat. People may choose to address this by eating less meat and considering the environmental impact of their food. Others may take the step of a completely vegetarian diet for environmental or ethical reasons.

Other important issues in the sustainability movement are the ethical issues of animal welfare in confined production areas, the use of pesticides and, related to that, the use of organic production methods. While the focus of health promotion and sustainable development is different, I would argue that they have more in common on this issue than is initially apparent. It is clear that health promotion and sustainable development share the core fundamental position that a varied diet which is high in vegetable sources would be better for both the environment and for human health. Most of the debate that separates the movements is in other domains, such as the benefits of organic food sources compared to non-organically produced food sources. Most studies have not found higher vitamin levels in organic food compared to non-organic produce but research (Smith-Spangler et al. 2012) found that food from organic cultivation is significantly less likely to be contaminated with pesticides than fruit and vegetables produced in non-organic ways, though the pesticide levels of all foods studied were within allowable limits.

Mainstream health promotion advice has tended not to address the organic cultivation issue, seeing it as largely irrelevant compared with what are perceived as the big issues on food intake relating to fat, sugar and dietary fibre. Perhaps health promoters need to acknowledge this and be more sensitive to the reality that people make food choices for a wide variety of reasons such as taste, texture, enjoyment, cost, social reasons, the environment, animal welfare practices and ethics. Health promotion has to fight for its place on that agenda but needs to acknowledge that the other interests are also important and have their own validity. We need to remind ourselves of the idea that the goods and services of both human beings and nature are inextricably coupled (Odum 1998).

With reference to a specific schools example of the interaction of health promotion and an environmental approach, a health promoting schools programme in Kosovo was developed to deal with a crisis of severe lead pollution. It showed it was possible for health promoters, environmentalists and education professionals to work together to improve the health and educational potential of children in Mitrovica using a health promoting schools model (Young and Tahirukaj 2009). Figure 2.1 summarises how the health promoting school model was utilised to integrate the education, health and environmental dimensions of this specific issue (see Fig. 2.1).

In addition to these specific examples there is new evidence emerging that the traditional separation of aspects of education in the school curriculum are artificial and unhelpful and this supports the diverse work of health promoting schools and sustainable schools. An example of this would be cognitive development in literacy and numeracy being viewed as separate and distinct from areas of the curriculum such as the promotion of physical activity. These views go back in time to the mind/body dualism of Descartes (1641) and are problematic when viewed in the context of modern understanding of the brain and learning. For example a research review concluded that a significant positive relationship exists between physical activity and cognitive function in children aged 4–18 years (Sibley and Etnier 2003). They noted that there is evidence that physical activity improves perceptual skills, intelligence quotient, and achievement in verbal and mathematic tests. Other studies (e.g. Hillman et al. 2008) suggest that new understanding of brain function in young people indicates that appropriate physical activity early in life can be of benefit to cognitive function during childhood and that this often extends to adult life. The authors' note that many physical activity requirements in schools have been reduced or eliminated to increase a student's academic performance and yet no evidence exists that the removal of physical activity from the curriculum has had benefits for academic achievement. As noted earlier both health promoting schools and sustainable schools have encouraged more physical activity in the course of the

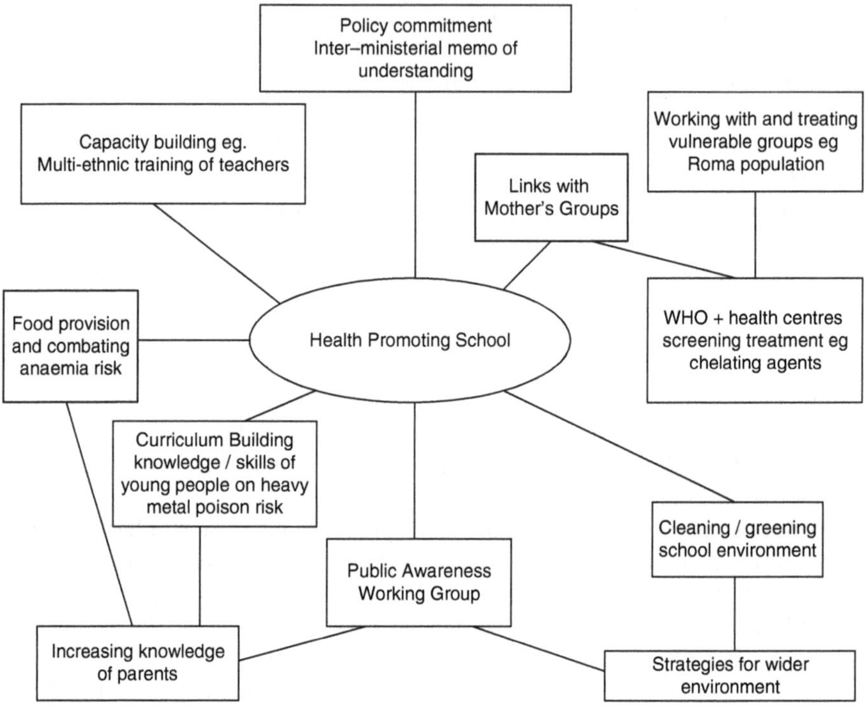

Fig. 2.1 Lead poisoning, children and mothers: Utilising the health promoting school model in Kosovo to develop an integrate approach incorporating education and environmental dimensions

school day and that, as well as health and environmental benefits (such as reducing car journeys), there are potential educational benefits.

Modern neurophysiology is revealing the importance of the limbic system in the brain and the importance of the emotions in learning. (OECD 2007). Young people are often highly motivated and have strong emotional commitment to the living world and the environment (Jensen et al. 2000). The sustainable school approach has the potential to harness this energy and I would postulate that increasing an individual's internal motivation will have positive benefits in other areas of the young person's curriculum. The OECD report also suggests that a stressful school environment will have significant negative effects on the learning ability of some young people because of the importance of the limbic system in processing emotions in the learning process. The benefits of learning together, rather than in isolation, are also referred to by OECD. The positive approach of both health promoting schools and sustainable schools in promoting social learning experiences, through project work in the context of a positive ethos in the school, also provides a good fit with modern understanding of the brain processes and learning.

In concluding, it is clear that health promoting schools and sustainable schools are intertwined in their approach and in their goals relating to the human species and the environment. Both movements share considerable potential to enhance the educational effectiveness of schools. It has been established for some time that at a population level good standards of education improve the health status of a population. According to a UN report, a cross-country comparison over time shows that increases in educational attainment precede improvements in health status (United Nations 2003). This consistent pattern over time makes it probable that there is a causal relation between education and health status. This could be because education leads to better economic status but if could also be related to education producing better health literacy, that is the ability to find, process and interpret information relevant to their health (Paakkari, this volume).

There is also evidence that better health in an individual and at a population level leads to better educational potential. I have referred to this above in describing the importance of physical activity and emotional health to learning. This is also true of the nutritional status of children in relation to learning. In addition, in developing countries the health of other family members affects educational enrolment, as healthy siblings and parents reduce the need for children to care for other family members at home (Bloom 2007). The evidence for causative mechanisms within health and education's interaction is not complete, but Bloom concludes that the interactions between them can create 'virtuous development spirals' and he believes that that national and international policies that take advantage of these interactions should be further developed and implemented. I would suggest that combining of the health promoting schools and sustainable schools movement has the potential to make a significant contribution to these virtuous development spirals.

2.6 What Are the Policy and Research Implications for Health Promoting Schools and Sustainable Schools?

There is nothing so useless as doing efficiently that which should not be done at all (Peter Drucker).

If it is accepted that these two developments have a great deal in common, then this has considerable implications for future policy at national and international level and also for future research which will be desirable to inform such policies. As resources for educational, health and sustainable development work are scarce, it is important to co-ordinate related work to avoid wasteful duplication of effort. The World Health Organisation has been highlighting the importance of inter-sectoral collaboration since the start of the modern health promotion movement in the 1980s (WHO 1986) and yet this type of partnership working is not the norm even between United Nations Agencies. The following issues create challenges for partnership-working which have to be overcome:

- The nature of professions and sectors which tend to defend their own territories;

- Misunderstandings of each other's professional roles; one profession tends to define other professions' roles more narrowly than reality;
- Professional concepts, language and jargon which can create misunderstanding and barriers;
- Professional enhancement linked to success within a profession, rather than due to recognition of inter-sectoral achievements IUHPE (2011);
- Political barriers;
- The nature of schools and their roles;
- Control of budget issues;
- Inequality issues.

Policy makers in health promoting schools and sustainable schools need to initiate discussions to review common goals and genuine differences and build trust around these issues. This will only work if they are open and honest about political agendas and budgetary control. If these initial steps are successful then the next stage would be to develop joint strategy for taking both movements forward either in an integrated way or as two complementary approaches. At present there is considerable overlap in the nature and process of each approach, and there is considerable content overlap. In most countries each development is aware of the other and there is no direct animosity, but in many cases partnerships are either non-existent or limited in their joint working. It is potentially wasteful of precious resources to ignore the common ground and interactions between education, health and the environment when policy is being formulated or put into action.

From the perspective of the busy teacher or school manager, it is very important that they perceive that related innovations, such as the health promoting schools movement and the sustainable schools movement, take account of each other's approach and they need to believe that innovators "have got their act together". An international review looking at the variety of the factors inhibiting fundamental change to traditional educational practices suggested that, in general, schools have weak networking and knowledge-sharing among teachers (OECD 2008). I suspect as a former high school teacher, that, this may be a reflection of the shear demanding immediacy of teaching. The school day for most teachers is busy and sometimes chaotic, and concomitantly there is little time for reflection and sharing issues with colleagues. Yet there is an acknowledged need for schools to have developed a sense of ownership in relation to innovations if they are to have any chance of being sustained and to become part of the life of the school. Perhaps the accusation of weak networking and knowledge-sharing is also applicable to those outside the schools who wish them to take on board a plethora of innovations that may overlap in a confusing way. Policy makers and those who wish to develop educational innovations have a duty to review who is working in a related field and to at least consider if they can work with them. If this is not done school authorities at national and district level and school principals will be more likely to respond in a protective or even negative manner to their approaches. The reality may be that it is difficult to create the optimum conditions for innovation and even when initiatives are backed by research, they may become marginalized by expedient and politically popular initiatives claiming to raise attainment levels.

This is the complex context in which health promoting schools and sustainable schools have to work. I suggest that they will be stronger together. It is desirable that successful new programmes will requires teachers to develop their own understanding of the meaning of the change though a process of reflection, but this reality is not an excuse for the outside agents of change to be confusing in their initial approach in relation to their role compared with other potential innovations.

Examining any of the following research issues would improve our understanding of the relationship between health promoting schools and sustainable schools and lead to better collaboration in the future.

- A review identifying any examples of systematic and sustained joint working between health promoting schools and sustainable schools in a country or region.
- A qualitative analysis of the factors that have promoted any such sustained partnerships and in the case of countries where this has not occurred, what are the reasons or barriers to such joint-working?
- A review of school managers and co-ordinators views on these two developments working together. This could involve schools which have been actively involved in one of the developments as well as schools involved in both developments.
- In conjunction with the above review it may be possible to design an experimental study comparing outcomes from schools involved in one of the programmes compared with schools involved in both programmes, although I would acknowledge that this type of study is very complex given the uniqueness of every school community and the potential problem of having many confounding variables.

Although answers have been suggested here as to what promotes and inhibits joint-working, much of the analysis comes from personal work experience. I suspect that many of the barriers to partnership working are almost universal because I have observed that the barriers exist in diverse cultures, but this view has not been tested. Indeed it would be surprising if there were not also different cultural issues at national and local level and therefore it is important to explore and test these ideas more formally in different contexts. One such area worthy of discussion in more detail is the dynamic between what a government education department may want and expect and the need for a degree of autonomy and responsiveness at the individual school community level. Given the different degrees of centralisation in different educational systems, this would imply that there will be a need for different approaches to integrating health promoting schools and sustainable schools which take account of the socio-political context in a national or state education system.

Being more effective health promoting and sustainable schools is not merely a matter of having better evidence on what is effective. Enough is known from the balance of evidence to get on with it. As Ron Labonte has pointed out when discussing globalization and health promotion (Labonte 2007) it is primarily about equity, ethics and survival. It would be more than careless if professional territorial disputes inhibit progress in this essential work in education. Let us create these 'virtuous development spirals'. Presumably all of us reading this book have benefited and will continue to benefit from good education, why would we deny this to the next generation?

References

Bloom, D. (2007). *Education, health and development* (American Academy of Arts and Sciences UBASE Occasional Paper, p. 5).

Brown, C. (2004). Emerging zoonoses and pathogens of public health significance, an overview. *Revue Scientifique et Technique, 23*(2), 435–442.

Buijs, G. M. (2009). Better schools through health: Networking for health promoting schools in Europe. *European Journal of Education, 44*(4), 507–520.

Chadwick, E. (1842). *Report on an enquiry into the sanitary conditions of the labouring population of Great Britain*. London: Poor Law Commission.

Cohet, C., Cheng, S., MacDonald, C., Baker, M., Foliaki, S., Huntington, N., Dowes, J., & Pierce, N. (2004). Evidence based public health policy and practice: Infections, medication use, and the prevalence of symptoms of asthma, rhinitis, and eczema in childhood. *Journal of Epidemiology and Community Health, 58*, 852–857.

Consumers Union. (2005). Out of balance: marketing of soda, candy, snacks and fast foods drowns out healthful messages. www.consmersunion.org. Accessed 24 Sept 2012.

Davis, J. M., & Cooke, S. M. (2007). Educating for a healthy, sustainable world: an argument for integrating Health Promoting Schools and Sustainable Schools. *Health Promotion International, 22*(4), 346–353.

Descartes, R. (1641/1996). Meditation VI. In *Meditations on the first philosophy* (J. Cottingham, Trans.). Cambridge: University of Cambridge Press.

Diamond, J. (1997). *Guns, germs and steel*. London: Chatto and Windus.

Fullan, M. G. (1993). The complexity of the change process. In *Change forces: Probing the depth of educational reform*. London: Falmer Press.

Fullan, M. G. (1999). *Change forces: The sequel*. Philadelphia: Falmer Press.

Fuller, B., & Clark, P. (1994). Raising school effects while ignoring culture? Local conditions and the influence of classroom tools, rules and pedagogy. *Review of Educational Research, 64*(1), 119–157.

Gammage, B. (2012). *The biggest estate on earth: How Aborigines made Australia*. Sydney: Allen & Unwin.

Haeckel, E. (1866). *Generelle Morphologie der Organismen: Allgemeine Grundzüge der Organischen Formen; Wissenschaft, Mechanisch Begründet durch die von Charles Darwin Reformierte Deszendenz-Theorie*. Berlin: G. Reimer.

Heppell, S. (2005) In Higgins, S., Hall, E., Wall, K., Woolner, P., & McCaughey, C. *The impact of school environments: A literature review*. London: The Design Council. http://www.design-council.org.uk/resources/. Accessed 20 Sept 2012.

Higgins, S., Hall, E., Wall, K., Woolner, P., & McCaughey, C. (2005). *The impact of school environments: A literature review*. London: The Design Council. http://www.design-council.org.uk/resources/. Accessed 20 Sept 2012.

Hillman, C., Erickson, K., & Kramer, A. (2008). Be smart, exercise your heart: Exercise effects on brain and cognition. *Nature Reviews Neuroscience, 9*(1), 58–65.

Houde, N. (2007). The six faces of traditional ecological knowledge: Challenges and opportunities for Canadian co-management arrangements'. *Ecology and Society, 12*(2), 34.

IUHPE (2011) 15èmes Journées Annuelles de Santé Publique du Quebec//15th Annual Public Health Days of Quebec (JASP) in Quebec – New understandings, better integration, sustainable and equitable actions: Schools that promote health, well-being and educational success in the next decade. http://jasp.inspq.qc.ca/english. Accessed 20 May 2013.

Jackson, S. (2009). *Enviroschools in New Zealand* (MAI Review, Intern Research Report 5). http://www.review.mai.ac.nz. Accessed 20 May 2013.

Jensen, B., Schnack, K., & Simovska, V. (Eds.). (2000). *Critical environmental and health education: Research issues and challenges*. Copenhagen: The Danish University of Education.

Keshavarz, N., Nutbeam, D., Rowling, L., & Khavarpour, F. (2010). Schools as social complex adaptive systems: A new way to understand the challenges of introducing the health promoting schools concept. *Social Science and Medicine, 70*(10), 1467–1474.

Kolbe, L. J. (1993). Developing a plan of action to institutionalize comprehensive School health programs in The United States. *Journal of School Health, 63*(1), 12–13.

Labonte, R. (2007). Globalization and health promotion. In D. V. McQueen & C. Jones (Eds.), *Global perspectives on health promotion effectiveness*. New York: IUHPE/Springer.

Lalonde, M. (1974). *A new perspective on the health of Canadians: A working document*. Ottawa: Government of Canada.

Lang, T. (1992). Food policy and public health, 7th Duncan Memorial Lecture. *Public Health, 106*, 91–125.

Lang, T., & Barling, D. (2013). Nutrition and sustainability: an emerging food policy discourse. In *Proceedings of the Nutrition Society* (pp. 1–12). Cambridge: Cambridge University Press.

Leonard, S., Parsons, M., Olawsky, K., & Kofod, F. (2013). The role of culture and traditional knowledge in climate change adaptation: Insights from East Kimberley, Australia. *Global Environmental Change, 23*, 623–632.

Lovelock, J. (1979). *Gaia: A new look at life on earth*. Oxford: Oxford University Press.

MacDonald, A. M., & Ódochartaigh, B. É. (2005). Baseline Scotland: An overview of available groundwater chemistry data for Scotland (British Geological Survey Commissioned Report). Swindon: The Natural Environmental Research Council.

Markou, K. B., Georgopoulos, N. A., Makri, M., Anastasiou, E., Vlasopoulou, B., Lazarou, N., Deville, L., Megreli, C., Vagenakis, G. A., Sakellaropoulos, G. C., Jabbarov, R., Kerimova, M. G., Mamedgasanov, R. M., Vagenakis, A. G. (2001) *Iodine deficiency in Azerbaijan after the discontinuation of an iodine prophylaxis program: reassessment of iodine intake and goitre prevalence in schoolchildren*. www.ncbi.nlm.nih.gov/pubmed/12186501. Accessed 22 Sept 2012.

Meadows, D. L., et al. (1972). *The limits of growth* (Report for The Club of Rome). New York: Universe Books.

National Chief Shawn A-in-Chut Atleo, Assembly of First Nations. (2012). Presentation to the House of Commons Standing Committee on the Environment and Sustainability. Ottawa.

Odum, E. P (1998). Ecology, a bridge between science and society. *Bioscience* 48.10. Oxford: University of Oxford Press.

Odum, E. P., & Odum, H. (1953). *Fundamentals of ecology* (1st ed.). Philadelphia: Saunders.

OECD. (2007). *Understanding the brain: The birth of a learning science*. Paris: OECD.

OECD. (2008). *21st century learning: Research, innovation and policy directions from recent OECD analyses*. Paris: OECD.

OECD. (2010). *The high cost of low educational performance and the long-run economic impact of improving PISA outcomes*. Paris: OECD.

Pembrokeshire County Council. (2011). *Pembrokeshire sustainable schools award scheme*. http://www.pembrokeshire.gov.uk/content.asp?nav=1626,559,1300&parent_directory_id=646. Accessed 24 Sept 2012.

Petrovskii, S. (Ed.). (2004). Challenges of ecological complexity. *Journal of Ecological Complexity, 1*, 1. Elsevier.

Petterson, B., & Tillgren, P. (Eds.). (1992). *The Sundsvall conference on supportive environments*. Stockholm: Karolinska Institute.

Piette, D. (2002). *Tracking down ENHPS successes for sustainable development and dissemination: The EVA2 project*. Copenhagen: WHO.

Pirrie, A., Elliot, D., McConnell, F., & Wilkinson, J. E. (2006). *The evaluation of eco schools Scotland* (SCRE Research Report No. 124). Glasgow: SCRE Centre.

Scottish Health Promoting Schools Unit. (2004). *Being well doing well: A framework for health promoting schools in Scotland*. Dundee: LTS Scotland.

Sibley, B. A., & Etnier, J. L. (2003). The relationship between physical activity and cognition in children: A meta-analysis. *Pediatric Exercise Science, 15*, 243–256.

Simovska, V. (2012). Case Study of a participatory health promotion intervention in school. *Democracy & Education, 20*(1):1–10.
Simovska, V., & Carlsson, M. S. (2012). Health-promoting changes with children as agents: Findings from a multiple case study research. *Health Education, 112*(3), 292–304.
Smith-Spangler, C., Brandeau, M. L., Hunter, G., Bavinger, J. C., Pearson, M., Eschbach, P. J., Sundaram, V., Liu, H., Schirmer, P., Stave, M., Olkin, I., & Bravata, M. (2012). Are organic foods safer or healthier than conventional alternatives? A systematic review. *Annals of Internal Medicine, 157*(5), 348–366.
St Leger, L., Young, I., & Blanchard, C. (2010). *Health promotion in schools: From evidence to action*. Paris: International Union for health promotion and Education.
Sustainable Schools in Australia. (2013). http://www.environment.gov.au/education/aussi/index.html. Accessed 20 May 2013.
Tansley, A. G. (1904). The problems of ecology: Research in British ecology. *New Phytologist, 3*, 191–200.
Taylor, L. H., Latham, S. M., & Woolhouse, M. E. (2001). Risk factors for human disease emergence. *Philosophical Transactions of the Royal Society B, 356*(1411), 983–989.
Tembon, M., & Fort, L. (Eds.). (2008). *Girls' education in the 21st century: Gender equality, empowerment, and economic growth*. Washington, DC: World Bank.
United Nations. (1987). *Report of the World Commission on environment and development: Our common future* (The Brundtland Report). New York: United Nations.
United Nations. (1992). *Conference on environment and development*. Rio de Janeiro: United Nations.
United Nations. (2003). *World development report 2003*. New York: United Nations.
United Nations General Assembly. (2002). *Proclamation of the years from 2005 to 2014 the Decade of Education for Sustainable Development* (DESD).
Virchow, R. (1848). Die Offentliche gesundheitsptlege (Mediciche Reform 5).
Walker, D. (2013). The Ministry of healing. *Resource Magazine*. www.resource-arm.net. Accessed 6 June 2013.
WHO. (1997). *The first conference of the European Network of Health Promoting Schools. The Health Promoting School, an investment in education, health and democracy: Conference report Thessaloniki, Greece*. Copenhagen: WHO Regional Office for Europe.
Williams, T. (1982). *Schools council project: Health education 13 to 18 teachers' guide*. London: Forbes Publications.
World Health Organization. (1984). *Health promotion, a discussion document on the concept and principles*. Copenhagen: WHO Regional Office for Europe.
World Health Organization. (1986). *The Ottawa charter*. Geneva: WHO.
Woynarowska, B., & Sokolowska, M. (2009). Poland: the Health Promoting School National Certificate. In C. Vince Whitman & C. Aldinger (Eds.), *Case studies in global school health promotion* (pp. 213–224). New York: Springer.
Young, I. (1993). Healthy eating policies in schools: An evaluation of effects on pupils' knowledge, attitudes and behaviour. *Health Education Journal, 52*(1), 3–9.
Young, I. (2003). Is healthy eating all about nutrition? *Nutrition Bulletin of British Nutrition Foundation, 27*, 7–12.
Young, I. (Ed.). (2005). *Promotion and education* (Vol. 12, pp. 3–4). Paris: IUHPE.
Young, I., & Tahirukaj, A. (2009). Kosovo: A health promoting schools approach to reduce the risks of lead poisoning and to establish cross-ethnic collaboration. In C. Whitman & C. Aldinger (Eds.), *Case studies in global school health promotion*. New York: Springer.
Young, I., & Whitehead, M. (1993). Back to the future: Our social history and its impact on health education. *Health Education Journal, 52*(3), 114–119, London.
Young, I., & Williams, T. (1989). *The healthy school*. Copenhagen: Scottish Health Education Group and WHO Regional Office for Europe.
Zigman, P. (2000). Ernst Haeckel and Rudolf Virchow: The dispute over the character of science in the controversy over Darwinism. *Medizinhistorisches Journal, 35*(3–4), 263–302.

Chapter 3
Key Competencies: Reconciling Means and Ends in Education for Sustainable Consumption

Daniel Fischer and Matthias Barth

Abstract Sustainable consumption is a focal point of interest in the interplay of school-based health promotion and sustainability. It calls for alternative ways to satisfy the objective needs of current society and future generations whilst respecting "planetary boundaries". Action is called for that protects and safeguards environmental conditions that allow all humans to live a good and healthy life. In the pursuit of sustainable consumption, education is widely ascribed a pivotal role as an instrument for disseminating more sustainable consumer behaviors. However, beneath this seemingly consensual surface the questions of which sustainability objectives are appropriate in an educational engagement with consumption issues and how they can be pursued is the subject of controversial debate. This controversy is addressed in this chapter. In a first step, it suggests the development of key competencies as a valid and legitimate objective that addresses both individual and societal needs. The authors systematically derive a framework of key competencies for sustainable consumption and discuss the framework's application to educational practice. Secondly, the chapter addresses the question of how formal and informal learning settings need to be designed in order to promote the acquisition of such competencies among students. A participatory whole-school approach to changing the "culture of consumption" in educational organizations is presented that was developed, implemented and evaluated in a transdisciplinary 3-year project. The chapter concludes with a discussion of synergies between the sustainability and health agendas for the emergence of innovative schools for the twenty-first century.

D. Fischer (✉)
Institute for Environmental & Sustainability Communication (INFU), Leuphana University, Lüneburg, Germany
e-mail: daniel.fischer@uni.leuphana.de

M. Barth
Institute for Competence Development, University of Applied Science Ostwestfalen-Lippe, Lemgo, Germany
e-mail: matthias.barth@rmit.edu.au

V. Simovska and Patricia Mannix-McNamara (eds.), *Schools for Health and Sustainability: Theory, Research and Practice*, DOI 10.1007/978-94-017-9171-7_3,

Keywords Sustainable consumption • Competence • Consumer competence • Consumer education • Education for sustainability

3.1 Consumption as a Mediator Between Health and Sustainability

The care for sustaining environmental, social and economic conditions that enable all human beings currently and in the future to satisfy (a set of yet to be specified) needs is the essential idea of sustainable development (Fischer et al. 2012). Consequently, the notion of care has been put at the heart of the UK's sustainable schools program, comprising care "for oneself, care for each other (across cultures, distances and time) and care for the environment (far and near)" (UK DfES 2006, p. 2). Based on the widely shared recognition that "the health of people, places and the planet are interdependent" (Orme and Dooris 2010, p. 425), the notion of caring for one's own well-being, for the well-being of other human beings, and for a healthy environment is also central to the health promotion movement. Another link between health and sustainability issues is the fact that "the causes and manifestation of unsustainable development and poor health are interrelated and frequently pose further interconnected challenges" (ibid.). This interconnectedness is also addressed in the context of the UN world decade on education for sustainable development (ESD), where health promotion is considered as one of the more important key themes (UNESCO 2003).

Despite these close linkages, synergies and joined-up approaches between the health and sustainability agenda in education are so far only poorly developed. On the contrary, it can be summarised that "work on sustainability and work on health have tended to happen in parallel rather than as integrated efforts" (Poland and Dooris 2010, p. 287).

This lack of integration is surprising in light of apparent parallels in the conceptual developments of health promotion and ESD that have both undergone a shift from rather moralistic to democratic paradigms (Schnack 2008). It is also surprising with respect to the programmatic consonance between practical approaches emerging from the two fields, particularly concerning the renewed interest in whole school approaches both in health promotion and in ESD that there have been a great variety of different implementation approaches. The emergence of health promoting schools approaches was characterized as "a dynamic process of contextual interpretation rather than a static result of the implementation of global principles" (Simovska 2012, p. 85). A similar diversity of approaches becomes apparent in the review of whole-school approaches in international ESD initiatives (Henderson and Tilbury 2004).

A cross-cutting theme of common interest at the interplay of school-based health promotion and sustainability is consumption (Fien 2000). In the health promotion field, there is a long tradition of viewing consumption patterns as potential risk factors for health outcomes. In such a pathogenetic perspective, consumption comprises issues as substance use (e.g. tobacco, alcohol), food and nutrition preferences or pathological (e.g. compulsive or addictive) shopping behaviours. From a more salutogenetic perspective, consumption patterns are closely associated with lifestyle choices that "affect our health through the 'power of resistance' [that] covers a number of personal skills and competencies, which can make a person healthy in spite of influences from a high-risk lifestyle" (Svedbom 2000, p. 170). In the ESD field, consumption is considered as a key driver of global change (The Royal Society 2012; WWF 2012; Randers 2012) causing severe sustainability problems such as exceeding "planetary boundaries" (Rockström et al. 2009) and a large proportion of the world's population suffering from critical deprivations (Raworth 2012).

Consequently, consumption issues feature prominently in whole-school approaches to sustainability. The UK's sustainable schools program, for example, describes eight doorways that schools can use as a starting point for an engagement with ESD. These comprise such issues as food and drink, school grounds and local well-being (UK DfES 2008). These themes feature prominently in other national and international approaches to sustainable schools, too (AGDEH 2005; Breiting et al. 2005; Transfer-21 2007).

Themes such as food policies, school ground design and physical, mental, and social well-being have been at the heart of a number of prominent health promoting schools initiatives, as different national examples within the European network of Health promoting schools (Jensen and Simovska 2002; Clift and Jensen 2005; Gray et al. 2006; Buijs 2009), the German speaking region (Nilshon and Schminder 2008; Paulus 2009; SNGS 2010), and international endeavours (IUHPE 2009; Simovska et al. 2010) as well as comparative studies (Carlsson and Simovska 2012; Simovska and Carlsson 2012) illustrate.

This conceptual paper conceives consumption as a key concern for an educational engagement with the sustainability and health agendas. While the notion of education for sustainable consumption has received considerable support particularly in the political sector, the question of what sustainability objectives are appropriate in consumer education and how they can be pursued is the subject of controversial debate. This controversy is addressed in the next section.

3.2 What Are We Striving For? Learning Objectives in Education for Sustainable Development

Education is among the most frequently used words in Agenda 21 and acclaimed as "indispensable to changing people's attitudes" in the transition to a sustainable society (United Nations Division for Sustainable Development 1992; Fien 2000). This stance has received strong political support in the past two decades. The United Nations declared a World Decade on "Education for Sustainable Development" (ESD) for the years 2005–2014 that seeks to integrate the principles and practices of sustainable development into all aspects of education and learning (Pigozzi 2010). The decade explicitly addresses sustainable consumption as one of its key themes and aims to develop "knowledgeable consumers who purchase goods with low life-cycle impacts and who use their purchasing power to support corporate social and environmental responsibility and sustainable business practices" (United Nations Educational 2005, p. 29). The final document of the Rio+20 conference reaffirms the political commitment to "improve the capacity of our education systems to prepare people to pursue sustainable development" (United Nations 2012, p. 44).

However, the inherent understanding of education as a means to achieve sustainable development and the tendency to aim at certain (consumer) behaviors has been the subject of critical discussion since the concept of ESD has been formally recognized at the United Nation's World Conference on Environment and Development in Rio de Janeiro in 1992 (Jickling 1992). Central to the debate is the question of how education should relate to the concept of sustainable development and what primary outcomes ESD should aspire to: the achievement of *education*, or the achievement of *sustainable development*.

The discourse is framed by two opposing positions that are deeply critical of each other (Wals et al. 2008; Sterling 2010). *Instrumental* approaches, on the one hand, start from the normative socio-political concept of sustainable development in ESD and consider sustainability as a legitimate objective that education should seek to contribute to. From this perspective, education and learning are interpreted as a means to achieving an end, for example, better sustainable consumption levels. *Emancipatory* approaches, on the other hand, focus on educational purposes. Proponents of this view see the function of sustainability as a learning context and hence as a means to support broader educational ends.

While the instrumental function of education (e.g. equipping the future workforce with skills required for a sustainable economy) is generally undisputed and regarded as indispensable (Bänninger et al. 2007), the utilization of education for achieving sustainable development objectives has been severely criticized. Prominent features of this criticism are the attempt to instrumentalize education for political purposes and a general tendency to 'educationalization'. According to Schnack (1996), approaches that seek to make educational institutions responsible for solving social or economic problems are both common and inappropriate. The argument is also made that short-term behavioral changes do not adequately prepare young people for mastering an unknown future of dynamic sustainability challenges (Scott 2009). On the other side, emancipatory approaches have been criticized as necessary but not sufficient responses to the challenges humankind is facing, because they carry the risk of denying or avoiding "a purposive or directive dimension" (Sterling 2010, pp. 514 f.).

To overcome such a polarized debate, steps have recently been made that are helpful in informing the further exploration of an educational response to the consumption challenge. Vare and Scott (2007) argue that it takes both types of learning, which they call ESD-1 and ESD-2. In a first step, learning of a more instrumental nature is needed as a tool to promote skills and information to change behaviors based on what is well known and agreed on at present (ESD-1). In addition it is learning for empowerment as the essential form of "building capacity to think critically about [and beyond] what experts say and to test sustainable development ideas" (ibid., p. 194) in the face of what is not adequately known and agreed on (ESD-2). Similarly, Sterling (2010) suggests conceiving instrumental and emancipatory approaches to ESD as "potentially and necessarily complementary, indeed co-dependent" (p. 523).

As a result of the search for learning objectives that consider such a complementary view, early on the concept of key competencies has gained ground in the debate on ESD. Competencies can be understood as "a roughly specialised system of abilities, proficiencies or skills that are necessary or sufficient to reach a specific goal" (Weinert 2001, p. 45). They deal with complex demands that necessitate the interplay of internal structures such as cognitive, emotional and motivational dispositions and represent a system of preconditions that enable individuals' self-organized learning processes (Klieme et al. 2007). The term '*key* competencies' represents a qualitative extension that highlights the significance of certain competencies: key competencies are relevant across different spheres of life and for all individuals and do not replace but rather comprise domain-specific competencies, which are necessary for successful action in certain situations and contexts (Rychen 2003; Barth and Michelsen 2013).

From 1997 to 2003 an interdisciplinary and international project, commissioned by the Organisation for Economic Cooperation and Development (OECD) took up the challenge to define and select key competencies that are necessary in today's society (Rychen and Salganik 2001, 2003). Based on a comprehensive literature review and the synthesis of a wide range of expert and stakeholder contributions, a framework of 'key competencies for a successful life and well-functioning society' was developed and discussed at two international symposia. The DeSeCo framework plays an authoritative role for contemporary educational research and was linked, for example, to the two international large-scale assessments of competencies PISA and ALL (see Rychen 2003). It identifies three categories of key competencies that are valid across different domains: (1) Interacting in socially heterogeneous groups, (2) acting autonomously, and (3) using tools interactively (Rychen and Salganik 2003).

In the ESD-literature, numerous articles and reports made significant progress in conceptualizing key competencies for sustainability. Although the approaches focus on different educational sectors such as secondary education (de Haan 2006) or higher education (Wiek et al. 2011) and different terms are used – such as skills (McKeown 2002), literacy (Parkin et al. 2004; Stibbe 2009), competencies (de Haan 2006; Wals 2010) or capabilities (Holdsworth et al. 2008), a broad consensus has emerged on the main aspects to be considered.

In a widely used approach of education for sustainable development in school education in the German speaking countries, de Haan (2006) describes the overarching educational objective of ESD as 'Gestaltungskompetenz' ('shaping competence'). It encompasses a set of key competencies that can be positioned in the three categories of key competencies from the DeSeCo framework and are expected to enable active, reflective and co-operative participation toward sustainable development. The term Gestaltungskompetenz is used to summarize those key competencies that form the "forward-looking ability to modify and to shape the futures of those societies we live in by active engagement for a sustainable development" (de Haan and Harenberg 1999, p. 62, translated by the authors).

This understanding of key competencies focuses on critical, self-determined and self-reflexive individuals while taking shared societal values into account at the same time. Thus, it has the potential to reconcile the tension between knowledgeable and autonomous individuals on the one hand and overall societal necessities on the other. It also serves as a valuable reference frame for more specific objectives in response to the consumption challenge that has the potential to overcome the dilemma of 'ego-centered consumership' versus 'eco-centered citizenship' (Benn 2004) or privately-oriented and self-serving consumer-citizens versus public-oriented and self-sacrificing citizen-consumers (Micheletti 2002).

3.3 Key Competencies for Sustainable Consumption

3.3.1 Key Competencies and the Context of Sustainable Consumption

Linking the concept of key competencies in ESD to sustainable consumption necessitates specification of these key competencies, which takes context-specific requirements into account. These requirements derive both from broader thematic debates (e.g. on dimensions, main themes and challenges in sustainable consumption) as well as from conceptions of consumer roles and behaviors with a greater focus on the individual.

From a broader perspective conceptually, the sustainability of individual acts of consumption can be assessed in two different ways (Fischer et al. 2012). In an impact-oriented approach, the focus is on the consequences of individual acts of consumption on the possibilities of other human beings to live a good life. From this perspective, individual acts of consumption like the purchase of a certain consumer good can be qualified as more sustainable than others according to their impact that can be measured against certain indicators. In an intent-oriented approach, it is not the impact of individual acts of consumption that is of interest, but the intentions with which individuals enact their consumer choices. Obviously, these choices may be made with the intention to contribute or not to contribute to the goal of sustainability, which is to create or maintain the possibilities of other humans to live a good life. The distinction between intent and impact has ramifications for education. On the one hand, education is challenged to promote the acquisition of competences to engage with the idea of sustainability in order to be able to intentionally base one's consumption choices on it. On the other, in order to be able to make informed choices, students need heuristics and knowledge about the actual impacts that different consumption choices have on environmental and social conditions.

In their discussion of the current state of research on sustainable consumption, Peattie and Collins (2009) identify a predominantly ecological interpretation of sustainable development as a major shortcoming. While it is widely agreed that sustainable development should seek to balance different dimensions, it is only recently that comprehensive educational approaches to sustainable consumption encompassing ecological, social, economic and cultural aspects have begun to be developed. This is not surprising given the fact that the integrative consideration of multiple dimensions with their respective objectives gives rise to the emergence of various dilemmas and conflicts of interests. It is therefore seen as a major challenge for promoters of sustainable consumption to enable individuals to consider carefully the pros and cons of consumption choices within and between different dimensions and "deal with the complexities, conflicts and compromises" (Peattie and Collins 2009, p. 112) involved in such comprehensive engagement with sustainable consumption. Hence, educational approaches seeking to promote sustainable consumption need to consider different dimensions and norms inherent in the idea of sustainability (i.e. in particular intra- and intergenerational justice, ecological limitations, satisfaction of needs) as well as their complex interaction that often result in conflicts and dilemmas of decision (Lundegård and Wickman 2007).

As for individual requirements, conceptions of the dual role of consumers acknowledge that individuals are not merely market players dealing with "the acquisition, consumption, and disposal of marketplace products, services, and experiences" (Macinnis and Folkes 2010, p. 905). Instead, the individual is given a wider recognition as a citizen and participant in everyday life who is of course influenced by the production and consumption of goods and services (Reisch 2004). In this context, the notion of 'consumer citizenship' extends the narrow concept of consumption as marketplace interactions into broader political contexts and employs more altruistic concerns (Sagoff 1996; Thoresen 2005). In particular, it conceives as a consumer duty the consideration of ethical aspects, responsible behavior towards others near and far, while acknowledging at the same time that consumers need to be effectively enabled to execute these duties (Schrader 2007).

Promoting sustainable consumer citizenship furthermore means to consider the individual's *diverse* consumer roles, including consumer behaviors both in- and outside the narrow limitations of marketplace interactions. Such a comprehensive approach to the conceptualization of consumer behavior and the sustainable consumer is offered by Stern (2000) in his typology of four different patterns of environmentally significant behavior (see Table 3.1).

Stern distinguishes between direct and indirect behaviors. While indirect consumer behaviors refer to individual actions of a more political nature (compared to the term consumer *citizen*), direct consumer behaviors relate to traditional individual marketplace actions in private households and organizations (*consumer* citizen). According to Jensen (2002), direct and indirect consumer behaviors can further be classified as either collective or individual. From this perspective, consumer activism and consumption-related behaviors in organizational settings would

Table 3.1 Types of sustainable consumer behaviour (According to Stern 2000)

Directness	Type	Examples
Indirect, mediating influence on the sustainability of consumption	Consumer activism	Active involvement in organisations and demonstrations committed to sustainable consumption
	Nonactivist consumer behaviors in the public sphere	Active kinds of citizenship, support or acceptance of public policies, social and viral marketing for sustainable consumption
Direct influence on the sustainability of consumption	Private-sphere consumption	Purchase of major household goods and services, use and maintenance of goods, waste disposal, purchasing patterns
	Other consumption-related behaviors	Individual influences on consumption-relevant actions of organizations

constitute rather collective forms of indirect and direct consumer behaviors, while non-activist consumer behaviors and private-sphere consumption instead focus primarily on individual consumer behaviors with an indirect or direct influence on the sustainability of consumption. Such analytical distinctions clearly help to disclose the variety of different forms of consumer behaviors and to accentuate different characteristics. The boundaries, however, between individual and collective actions are fluent and witness the emergence of new sites of political consumption. For example, empirical evidence shows that traditional forms of collective action through "interest articulating and interest aggregating" (Micheletti 2002, p. 227) are on the decline, giving rise to more individualized forms of collective action that build on everyday activism and responsibility-taking in geographically close settings. Hence, in Micheletti's perspective, private-sphere consumption can in fact be understood as a form of individualized collective action that can trigger political engagement. What follows from that is that a framework of key competencies for sustainable consumption should focus primarily on the questions to what degree and in which ways responsibility is taken, rather than on representations of existing classifications of specific forms and sites of consumer behaviors. Another distinction that needs to be considered with regard to direct consumer behaviors is that between *different phases* comprising consumer behaviors, reaching from problem recognition, information search, evaluation of alternatives, purchase decision to post-purchase behavior which includes the usage, maintenance and disposal of goods (Kotler et al. 2009). Research findings also indicate that these behaviors can be further divided according to the extent to which their underlying decision-making processes are guided either by cognitive control, emotional involvement or even reactive, spontaneous behavioral responses (Fischer and Hanley 2007). It follows from this account that a framework of key competencies for sustainable consumption needs to consider not only cognitive dispositions but must pay special attention to non-cognitive decision-processes and their underlying control mechanisms.

Table 3.2 Framework of key competencies for sustainable consumption

Act autonomously
1. Competency to reflect individual needs and cultural orientations
2. Competency to plan, implement, and evaluate consumption-related activities
Interact in heterogeneous groups
3. Competency to critically take on one's role as an active stakeholder in the market
4. Competency to communicate sustainable consumption
Use tools interactively
5. Competency to use, edit and share different forms of knowledge
6. Competency to use ICT interactively
7. Competency to think visionary and to consider interrelatedness

Furthermore, a framework of key competencies for sustainable consumption needs to account for both *direct* and *indirect* as well as *individual* and *collective* forms of consumer action.

3.3.2 A Synthesized Framework of Key Competencies for Sustainable Consumption

We propose a context-specific framework of key competencies for sustainable consumption that is based upon the threefold categorization of key competencies from the DeSeCo approach and is closely linked to the concept of Gestaltungskompetenz in ESD. This framework acknowledges the valuable insights into the requirements of the domain of consumption that existing approaches to consumer competency (Thoresen 2005; Grønhøj 2007; Lachance and Legault 2007; Bonnemaizon and Batat 2010) and consumer literacy (Young 2000; Wallendorf 2001) provide. Furthermore, it complements and re-orientates them in light of the specific demands of sustainable consumption. It is composed of seven key competencies, each of which consists of underlying cognitive and non-cognitive dispositions (see Table 3.2).

What makes this undertaking special is its approach to systemize existing competence lists, which often are critically labeled as "laundry lists" as they offer neither a classification nor an elaboration of what they consist of. Using the well accepted framework that was laid out in the DeSeCo-Project, necessary competencies can be depicted and described against the three dimensions that refer to the individual and its ability to act autonomously, to abilities for interacting with others and to use tools in its broadest sense. So while the key competencies that are elaborated in what follows are nothing completely new, it is the systematic framework that allows a further operationalization that makes this a noteworthy contribution.

A more detailed exploration of each competency illustrates how these dispositions relate to each other and shape between them the nature of each key competency.

Today's consumer societies are characterized by an increasing commoditization of needs and their satisfaction through market goods. The competency to *reflect individual needs and cultural orientations* refers to the ability of students to critically engage with these developments and to reflect their compatibility to their idea of a sustainable future (number 1 in Table 3.2). This includes the willingness and ability to explore and scrutinize one's own aspirations, wants and needs as well as established habits and practices of their satisfaction. In particular, it presupposes knowledge of how these preferences are culturally contextualized and shaped. In a social context, such reflective stance requires the capacity to criticize unsustainable trends and systems of provision and to accept criticism by others.

Planning, implementing and evaluating consumption-related activities in a sustainable way are processes crucial to sustainable consumption. There is a whole range of relevant behaviors, encompassing not only the individual's role in the market (e.g. the acquisition of goods or services, contact with companies' customer service, or consultation of peers in acquisition processes) but also aspects of citizenship, like participation in acquisition decisions in the public sector (number 2 in Table 3.2). In both types, knowledge about relevant methods and instruments has to be combined with the consumer's motivation and willingness to use this knowledge and to put it into practice. The ability to *plan* activities refers to the knowledge of how to *assess* in advance available resources, collaboration activities and consequences as well as possible side effects of different actions. It further covers the *skill in developing* and *applying* certain criteria to the selection and assessment of different options (e.g. in terms of quality or costs but also of the ecological and social impacts). Finally, it demands *critical reflection on one's own needs* and even the consideration of non-consumption as an option. The ability to *implement* encompasses the knowledge and practical skills to execute a specific activity, while the ability to *evaluate* enables one to assess whether the activity contributes towards the intended objective. Underpinning these cognitive abilities are important non-cognitive dispositions, such as a general motivation to become active both as a consumer and a citizen, a generally positive attitude towards sustainable consumption and the willingness to act responsibly in specific consumption-related activities.

The competency to *critically take on one's role as an active stakeholder in the market* enables students to contextualize their roles in the broader system of provision and consumption (number 3 in Table 3.2). It requires knowledge of system environments and of the roles, rights and duties of different actors within, and more particularly of opportunities to work towards changing these conditions into more sustainable ones. This includes skills to adopt a life-cycle perspective on the production, consumption and disposal of goods (and services), the ability to put oneself in the position of other actors as well as the willingness and capacity to forge strategic alliances with other actors to achieve common goals.

With this systems perspective, the role of a consumer citizen transcends the narrowly confined boundaries of individual actors who engage in market-based transaction of commodities. Moreover, it implies the development of change agency that requires the competency to *communicate the idea of sustainable consumption to others* (number 4 in Table 3.2). This presupposes knowledge of the ideas, values and concepts underpinning the notion of sustainable consumption and involves the cognitive, motivational and emphatic skills to process this knowledge for different communicative formats, situations and conversational partners and audiences.

In order to be able to make sustainable consumption choices and to act as a consumer citizen, students need to be able to *use, edit and share different forms of knowledge* (e.g. own practical experiences, word of mouth experiences, written facts, mental concepts) for different purposes (number 5 in Table 3.2). This requires the willingness to invest resources into the search for adequate information, the ability to evaluate the validity of the information, and to use it to inform one's own consumption-related decisions. As different knowledge may recommend different actions, the competency involves the readiness to endure tensions that arise from such seemingly contradictions. In a social perspective, it further reflects the motivation and ability to share knowledge with others.

Information and communication technology (ICT) plays an important role as a source of knowledge and of interaction about sustainable consumption issues. The competency to use this resource effectively does not only include the motivation and ability to search information and reflect on its validity, but also involves an *interactive* component that enables students to use ICTs to process information and pass it on to others. This interactive component also features the ability to use the potential of ICTs for social interaction and to critically assess the opportunities that ICTs are offering and the risks that are involved with their dissemination (number 6 in Table 3.2).

The competency to *think visionary and to consider interrelatedness* focuses on future-oriented thinking and comprises the students' awareness of their attitudes and values to intragenerational and intergenerational justice as well as their opportunities to contribute to a sustainable development of consumption and production patterns (number 7 in Table 3.2). This presupposes knowledge of the interlinkages between consumption and production systems, the ability to appraise the implications of one's own consumption choices for others today and in the future, as well as a general perception of the future as undetermined and shapeable.

With this framework we introduced seven distinctive but closely related key competencies. Their configuration represents an individual's capacity to enact sustainable consumption – directly and indirectly, individually as well as collectively. The framework facilitates elaboration on the dispositions constituting each of the competencies and how they interrelate and depend on each other. For example, in order to successfully communicate sustainable consumption (key competence 1), a person needs to some extent be able to reflect on his or her individual needs and cultural orientations (key competence 2). Thus, while the framework is a helpful tool to distinguish certain key competencies, in practice the manifold interrelations have to be taken into account.

3.4 Implications for the Design of Learning Settings

To consider these competencies and their development in learning and teaching in schools poses new challenges on the learning design. From a socio-constructivist perspective on learning, since the individual as an autonomous system cannot be 'changed' from the outside, it is the learning environment in which one is acting that needs to be designed in order to support intended learning outcomes (Kotzee 2010; Knuth and Cunningham 1993). It is the concept of *open learning environments* which acknowledges that learning takes place in manifold forms and depends of a variety of factors (Duffy 1993). Open learning environments are based on authentic learning situations, offer a rationale and a starting point and, even more important, offer a frame for the whole learning process. In this way, learning processes are created that enable exploration and offer learner-centred as well as collaborative tools so as to give control over the learning process to the learner.

Taking this into account poses new challenges to (formal) school education. Learning does not only happen in formal, "controlled" learning settings but also in informal learning settings be it intentional or non-intentional. The question remains how formal and informal learning can be systematically related to each other to establish a learning culture that enlarges the learning space and facilitates better learning opportunities for developing key competencies (Barth et al. 2007).

The design of learning environments that facilitate competence development in formal and informal learning is informed by three key principles of competence development:

The first principle is *self-directed learning*. This acknowledges the importance of autonomous and constructive learning processes in which knowledge is actively developed in a self-directed manner. The aim is to stimulate learning processes in which students construct their own knowledge base independently (Shin 1998; Zimmermann 2001). This principle is based on a view of learning which is not directly linked to teaching and which emphasizes the active *development* of knowledge rather than its mere *transfer*. Accordingly, facilitating learning has to focus on stimulating and offering starting points for learning rather than dictating certain learning outcomes. This becomes especially important if you consider the key competencies that we are addressing. For an individual to learn to reflect upon his or her individual needs, there must be adequate time and space to allow for introspection and self-directed learning.

As the acquisition of competencies is both an individual and a social activity, *collaborative learning* is the second important principle. In contrast to cooperation, which focuses on dividing and sharing tasks, collaboration involves joint learning processes with participation and empathy as critical factors. Successful collaboration increases both individual and collective knowledge, based on shared experiences and jointly accepted learning objectives. Learning design to facilitate competence developments takes this into account and emphasizes collaborative activities inside and outside the classroom and with different stakeholders. Learning in group processes in which different opinions and approaches are not only tolerated but appreciated, together with a critical reflection on shared experiences takes both cognitive and social-affective aspects into account and stimulates deep learning processes. As this directly addresses abilities to communicate in the context of sustainable consumption and to take one's role as an active stakeholder, the second learning principle again resonates with the intended key competencies.

The third principle of *problem-oriented learning*, finally, focuses on complex real-world situations and the development of creative solutions to facilitate the development of competencies. While transmissive learning processes often encounter problems because of their exclusive focus on factual knowledge, which cannot be used for action in specific situations, a transformative, problem-oriented approach is especially suited to support action-relevant procedural knowledge and skills (Steinemann 2003). Such learning is facilitated by authentic problems from the students' life world and different approaches and perspectives. Thus, the first two principles of self-directed learning and collaboration can be seen as preconditions for a problem-oriented approach. This third learning principle is first and foremost a general precondition for the development of key competencies as it supports a context-spanning consideration of different competencies.

A similar catalogue of quality criteria for the design of teaching-learning processes has been formulated by the ENSI and SEED networks. Resulting from a broad international debate, the catalogue of quality criteria can be considered as a widely consensual reference framework in the field of environmental education and ESD. In the field of teaching and learning processes the catalogue calls on teachers to provide settings that allow for cooperative, participatory, future-oriented, problem-based learning (Breiting et al. 2005).

The three key principles of competence development proposed before reflect these principles. In our perspective, the key principles with their emphasis on *learning* processes do not compete with, but rather constructively complement the rather teacher-centered criteria of the ENSI/SEED framework.

3.5 Facilitating Sustainable Consumer Learning in a Participatory School Development Approach

How can the principles discussed above be translated into an educational approach to transform schools into settings that promote the acquisition of consumer competencies and a more sustainable lifestyle among young adults?

This question was explored in the 3 year research and development project BINK (the German acronym stands for *educational institutions and sustainable consumption*). At the heart of the approach was the assumption that both formal and informal learning settings need to be systematically realigned towards the promotion of sustainable consumption. In order to bring about these changes, a participatory whole-school development approach was put in place in two secondary schools and two vocational schools. Change management teams were established that comprised representatives of all relevant groups of actors (e.g. administration, management, teachers, students, parents, training companies), each granted full voting rights. In workshops practitioners and researchers collaboratively planned and designed intervention measures that contribute to organizational changes by triggering formal and informal learning processes (Matthies and Krömker 2000; Nastasi et al. 2000).

The structure of the intervention process was designed according to the principles of competence development discussed before. It engaged students on two levels that both involved intensive cooperation (*principle of collaborative learning*): On the level of the change management team (CMT) which took charge of the entire intervention process in each organization and on the level of smaller intervention teams (IT) that were responsible for the design and delivery of single intervention measures. The selection of appropriate activities was informed by the analytical framework of educational organizations' *culture of consumption* that identifies relevant consumption-related formal and informal learning contexts in schools (Fischer 2011a). In school-based workshops, a steady-state-analysis of the organizational culture of consumption was performed and critical spots discussed. The object of inquiry was thus always the characteristics of the local organization in general and those organizational hot spots in particular that the local actors considered as crucial and problematic in terms of their effect on a sustainable consumer learning. In this way it could be ensured that the measures taken corresponded to local needs and could best be embedded by the local actors in their everyday routines and interactions (*principle of problem-oriented learning*). Finally, students had the opportunity to address those hot spots that they considered most relevant by developing and implementing their own intervention measures (*principle of self-directed learning*).

Both levels of participation (change management teams and intervention teams) provided opportunities for students to acquire and develop key competencies for sustainable consumption.

In the change management teams, different actors shared their perspectives on their organization's culture of consumption and their knowledge about and experiences with consumption (education) issues. Students participating in the CMTs were challenged to use, edit and share these different forms of knowledge that was further complemented by scientific knowledge provided by the academic project staff and scholarly resources (e.g. manuals, research summaries) (key competency 5). The CMT was also in charge of setting long-term visionary goals of a culture of sustainable consumption in their organization and monitoring progression towards this goal, which involved the reflection of the interrelatedness of single intervention measures and unforeseen effects (Fischer 2011b) (key competency 7). On the more strategic level, student members of the CMTs were involved in decision-making processes on organizational consumption issues (e.g. procurement, catering) which required taking on other consumer roles than that of a private consumer (key competency 3).

The intervention teams carried out smaller intervention projects on specific issues. At one school, a student initiative started their own student company that today provides other students with organic snacks and fair trade coffee during school breaks (key competency 2). At another institution students refurbished an abandoned room and founded the first swap-shop in town where used goods could be traded and exchanged free of charge. The objective of this project was to encourage a cultural shift towards sharing and using goods rather than owing them (key competency 1).

Finally, as a specific intervention measure implemented in each organization, a media intervention was carried out that challenged students to use media (in particular film) and to explore new and entertaining ways of communicating sustainable consumption messages (key competencies 4 and 6).

3.6 Conclusion

In this paper, the promotion of sustainable consumption was viewed as a key concern at the interplay of the health and sustainability agenda. In both discourses, education is commonly called for as a powerful instrument to achieving health and sustainability outcomes. We discussed the critical implications of utilizing education merely as an instrument for achieving societal goals and introduced the concept of key competencies as an alternative perspective. Based on a careful consideration of appropriate learning objectives we developed a set of key competencies for sustainable consumption.

The concept of key competencies for sustainable consumption reflects *educational* prepositions for the promotion of sustainable consumption in schools. In this endeavor, it resonates well with the discourse on health literacy. Here, too, it is argued that a merely instrumental focus on individual health outcomes should be given up in favor of broader educational objectives, namely the "critical competence of students to meet the demands of future society" (Simovska 2012, p. 86). The framework of key competencies for sustainable consumption has the potential to substantiate a more holistic and integrative view on educational outcomes in adjectival educations addressing consumption issues. It can serve educational practitioners and researchers as a lens to focus on such learning outcomes that are both crucially relevant and educationally legitimate in the quest for promoting sustainable consumption.

Moreover and beyond that, the framework can also be of use to inform school development processes. Corresponding with the concept of open learning environments, schools are not only places of formal tuition and training of skills, but everyday settings in which students live and learn (Fischer 2013). In this context, the framework of key competencies for sustainable consumption has the potential to offer guidance on how to design formal and informal learning environments in such ways that the potential of schools as settings that enhance consumer learning is more effectively tapped. An example for practical implications of such approach has been described in this paper: the framing of problems and fields of action as a starting point for school development processes should be designed as a collaborative endeavor that allows all actors to contribute with their own perspectives. In the implementation phase, participation structures need to be put in place that allow students to actively get involved in organizational change processes and develop their own projects. For this to happen, key stakeholders in charge of managing educational organizations need to commit to the promotion of their students' competencies as their core educational mandate and mission that permeates and informs all facets of school life.

References

Australian Government Department of the Environment and Heritage (AGDEH). (2005). *Educating for a sustainable future: A national environmental education statement for Australian schools*. Canberra: AGDEH.

Bänninger, C., Di Giulio, A., & Künzli, C. (2007). Schools and sustainable development. *GAIA – Ecological Perspectives for Science and Society, 16*(4), 267–271.

Barth, M., & Michelsen, G. (2013). Learning for change: An educational contribution to sustainability science. *Sustainability Science, 8*(1), 103–119.

Barth, M., Godemann, J., Rieckmann, M., & Stoltenberg, U. (2007). Developing key competencies for sustainable development in higher education. *International Journal of Sustainability in Higher Education, 8*(4), 416–430.

Benn, J. (2004). Consumer education between 'consumership' and citizenship. Experiences from studies of young people. *International Journal of Consumer Studies, 28*(2), 108–116.

Bonnemaizon, A., & Batat, W. (2010). How competent are consumers? The case of the energy sector in France. *International Journal of Consumer Studies, 35*(3), 348–358.

Breiting, S., Mayer, M., & Mogensen, F. (2005). *Quality criteria for ESD-schools: Guidelines to enhance the quality of education for sustainable development.* Vienna: Austrian Federal Ministry of Education, Science and Culture.

Buijs, G. (2009). Better schools through health: Networking for health promoting schools in Europe. *European Journal of Education, 44*(4), 507–520.

Carlsson, M., & Simovska, V. (2012). Exploring learning outcomes of school-based health promotion – A multiple case study. *Health Education Research, 27*(3), 437–447.

Clift, S., & Jensen, B. B. (Eds.). (2005). *The health promoting school: International advances in theory, evaluation and practice*. Copenhagen: Danish University of Education Press.

de Haan, G. (2006). The BLK '21' programme in Germany. A 'Gestaltungskompetenz'-based model for education for sustainable development. *Environmental Education Research, 12*(1), 19–32.

de Haan, G., & Harenberg, D. (1999). *Bildung für eine nachhaltige Entwicklung. Gutachten zum Programm von Gerhard de Haan und Dorothee Harenberg, Freie Universität Berlin*. Berlin: Bund-Länder-Komm. für Bildungsplanung und Forschungsförderung.

Duffy, T. M. (1993). *Designing environments for constructive learning*. Berlin: Springer.

Fien, J. (2000). Education for sustainable consumption: Towards a framework for curriculum and pedagogy. In B. B. Jensen (Ed.), *Critical environmental and health education. Research issues and challenges* (pp. 45–66). Copenhagen: Research Centre for Environmental and Health Education of the Danish University of Education

Fischer, D. (2011a). Educational organisations as 'Cultures of Consumption': Cultural contexts of consumer learning in schools. *European Educational Research Journal, 10*(4), 595–610.

Fischer, D. (2011b). Monitoring educational organizations' 'Culture of Sustainable Consumption': Towards a participatory initiation and evaluation of cultural change in schools and universities. *Journal of Social Science, 7*(1), 66–78.

Fischer, D. (2013). Promoting sustainable consumption in educational settings. In U. Schrader, V. Fricke, D. Doyle, & V. W. Thoresen (Eds.), *Enabling responsible living* (pp. 147–160). Berlin: Springer.

Fischer, A., & Hanley, N. (2007). Analysing decision behaviour in stated preference surveys: A consumer psychological approach. *Ecological Economics, 61*(2–3), 303–314.
Fischer, D., Michelsen, G., Blättel-Mink, B., & Di Giulio, A. (2012). Sustainable consumption: how to evaluate sustainability in consumption acts? In R. Defila, A. Di Giulio, & R. Kaufmann-Hayoz (Eds.), *The nature of sustainable consumption and how to achieve it. Results from the focal topic "From Knowledge to Action – New Paths towards Sustainable Consumption"* (pp. 67–80). München: Oekom.
Gray, G., Young, I., & Barnekow, V. (2006). *Developing a health-promoting school: A practical resource for developing effective partnerships in school health, based on the experience of the European Network of Health Promoting Schools*. Woerden: European Network of Health Promoting Schools.
Grønhøj, A. (2007). The consumer competence of young adults: A study of newly formed households. *Qualitative Market Research: An International Journal, 10*(3), 243–264.
Henderson, K., & Tilbury, D. (2004). *Whole-school approaches to sustainability: An international review of sustainable school programs*. Canberra: Australian Government Department of the Environment and Heritage & Australian Research Institute in Education for Sustainability (ARIES).
Holdsworth, S., Wyborn, C., Bekessy, S., & Thomas, I. (2008). Professional development for education for sustainability. How advanced are Australian universities? *International Journal of Sustainability in Higher Education, 9*(2), 131–146.
International Union for Health Promotion and Education (IUHPE) (2009). *Achieving health promoting schools: Guidelines for promoting health in schools*. St. Denis: IUHPE.
Jensen, B. B. (2002). Knowledge, action and pro-environmental behaviour. *Environmental Education Research, 8*(3), 325–334.
Jensen, B. B., & Simovska, V. (Eds.). (2002). *Models of health promoting schools in Europe*. Copenhagen: WHO Regional Office for Europe.
Jickling, B. (1992). Why I don't want my children to be educated for sustainable development. *The Journal of Environmental Education, 23*(4), 5–8.
Klieme, E., Hartig, J., & Rauch, D. (2007). The concept of competence in educational contexts. In J. Hartig, E. Klieme, & D. Leutner (Eds.), *Assessment of competencies in educational contexts* (pp. 3–22). Cambridge, MA: Hogrefe & Huber Publishers.
Knuth, R. A., & Cunningham, D. J. (1993). Tools for constructivism. In T. M. Duffy, J. Lowyck, & D. H. Jonassen (Eds.), *Designing environments for constructive learning* (pp. 163–188). Berlin: Springer.
Kotler, P., Keller, K. L., Brady, M., Goodman, M. R. V., & Hansen, T. (2009). *Marketing management*. Harlow: Pearson/Prentice Hall.
Kotzee, B. (2010). Seven posers in the constructivist classroom. *London Review of Education, 8*(2), 177–187.
Lachance, M., & Legault, F. (2007). College students' consumer competence. Identifying the socialization sources. *Journal of Research for Consumers, 13*, 1–21.
Lundegård, I., & Wickman, P. (2007). Conflicts of interest: An indispensable element of education for sustainable development. *Environmental Education Research, 13*(1), 1–15.
Macinnis, D. J., & Folkes, V. S. (2010). The disciplinary status of consumer behavior: A sociology of science perspective on key controversies. *Journal of Consumer Research, 36*(6), 899–914.
Matthies, E., & Krömker, D. (2000). Participatory planning – A heuristic for adjusting interventions to the context. *Journal of Environmental Psychology, 20*(1), 65–74.
McKeown, R. (2002). *Education for sustainable development toolkit* (Version 2). Knoxville: Energy, Environment and Resources Center, University of Tennessee.
Micheletti, M. (2002). Consumer choice as political participation. *Statsvetenskaplig Tidskrift, 105*(3), 218–234.
Nastasi, B. K., Varjas, K., Schensul, S. L., Silva, K. T., Schensul, J. J., & Ratnayake, P. (2000). The participatory intervention model: A framework for conceptualizing and promoting intervention acceptability. *School Psychology Quarterly, 15*(2), 207–232.
Nilshon, I., & Schminder, C. (2008). *Die gute gesunde Schule gestalten: Stationen auf dem Weg der Schulprogrammentwicklung* (3rd ed.). Gütersloh: Verl. Bertelsmann-Stiftung.

Orme, J., & Dooris, M. (2010). Integrating health and sustainability: The higher education sector as a timely catalyst. *Health Education Research, 25*(3), 425–437.

Parkin, S., Johnston, A., Buckland, H., Brookes, F., & White, E. (2004). *Learning and skills for sustainable development*. London: Guidance for Higher Education Institutions.

Paulus, P. (2009). *Anschub.de – ein Programm zur Förderung der guten gesunden Schule*. Münster/New York/München/Berlin: Waxmann.

Peattie, K., & Collins, A. (2009). Guest editorial: Perspectives on sustainable consumption. *International Journal of Consumer Studies, 33*(2), 107–112.

Pigozzi, M. J. (2010). Implementing the UN Decade of Education for Sustainable Development (DESD). Achievements, open questions and strategies for the way forward. *International Review Education, 56*(2–3), 255–269.

Poland, B., & Dooris, M. (2010). A green and healthy future: The settings approach to building health, equity and sustainability. *Critical Public Health, 20*(3), 281–298.

Randers, J. (2012). *2052: A global forecast for the next forty years*. White River Junction: Chelsea Green Pub.

Raworth, K. (2012). *A safe and just space for humanity. Can we live within the Doughnut?* (Oxfam Discussion Paper). Oxford: Oxfam.

Reisch, L. A. (2004). Principles and visions of a New Consumer Policy. Discussion paper by the Scientific Advisory Board for Consumer, Food, and Nutrition Policies to the German Federal Ministry of Consumer Protection, Food, and Agriculture. *Journal of Consumer Policy, 27*, 1–42.

Rockström, J., Steffen, W., Noone, K., Persson, Å., Chapin, F. S., Lambin, E. F., Lenton, T. M., Scheffer, M., Folke, C., Schellnhuber, H. J., Nykvist, B., de Wit, C. A., Hughes, T., van der Leeuw, S., Rodhe, H., Sörlin, S., Snyder, P. K., Costanza, R., Svedin, U., Falkenmark, M., Karlberg, L., Corell, R. W., Fabry, V. J., Hansen, J., Walker, B., Liverman, D., Richardson, K., Crutzen, P., & Foley, J. A. (2009). A safe operating space for humanity. *Nature, 461*(7263), 472–475.

Rychen, D. S. (2003). Key competencies. In D. S. Rychen & L. H. Salganik (Eds.), *Key competencies for a successful life and well-functioning society* (pp. 63–108). Cambridge: Hogrefe & Huber Publishers.

Rychen, D. S., & Salganik, L. H. (Eds.). (2001). *Defining and selecting key competencies*. Seattle: Hogrefe & Huber Publishers.

Rychen, D. S., & Salganik, L. H. (Eds.). (2003). *Key competencies for a successful life and well-functioning society*. Cambridge: Hogrefe & Huber Publishers.

Sagoff, M. (1996). *The economy of the earth. Philosophy, law, and the environment*. Cambridge: Cambridge University Press.

Schnack, K. (1996). Internationalisation, democracy and environmental education. In S. Breiting & K. Nielsen (Eds.), *Environmental education research in the Nordic countries* (pp. 7–19). Copenhagen: The Royal Danish School of Educational Studies.

Schnack, K. (2008). Participation, education, and democracy: Implications for environmental education, health education, an education for sustainable development. In A. Reid, B. B. Jensen, J. Nikel, & V. Simovska (Eds.), *Participation and learning. Perspectives on education and the environment, health and sustainability* (pp. 181–196). Dordrecht: Springer.

Schrader, U. (2007). The moral responsibility of consumers as citizens. *International Journal of Innovation and Sustainable Development, 2*(1), 79–96.

Schweizerisches Netzwerk Gesundheitsfördernder Schulen (SNGS) (2010). *Qualitäts-Kriterien Gesundheitsfördernde Schule*. Luzern: SNGS.

Scott, W. (2009). Judging the effectiveness of a sustainable school. A brief exploration of issues. *Journal of Education for Sustainable Development, 3*(1), 33–39.

Shin, M. (1998). Promoting students' self-regulating ability. Guidelines for instructional design. *Educational Technology, 1998*, 38–44.

Simovska, V. (2012). What do health-promoting schools promote? Processes and outcomes in school health promotion. *Health Education, 112*(2), 84–88.

Simovska, V., & Carlsson, M. (2012). Health-promoting changes with children as agents: Findings from a multiple case study research. *Health Education, 112*(3), 292–304.

Simovska, V., Dadaczynski, K., Grieg Viig, N., Tjomsland, H. E., Bowker, S., Woynarowska, B., Ruiter, S. d., & Buijs, G. (2010). *HEPS tool for schools: A guide for school policy development on healthy eating and physical activity*. Woerden: NIGZ.

Steinemann, A. (2003). Implementing sustainable development through problem-based learning: Pedagogy and practice. *Journal of Professional Issues in Engineering Education and Practice, 129*, 216–224.

Sterling, S. (2010). Learning for resilience, or the resilient learner? Towards a necessary reconciliation in a paradigm of sustainable education. *Environmental Education Research, 16*(5–6), 511–528.

Stern, P. C. (2000). New environmental theories. Toward a coherent theory of environmentally significant behavior. *Journal of Social Issues, 56*(3), 407–424.

Stibbe, A. (Ed.). (2009). *The handbook of sustainability literacy. Skills for a changing world.* Totnes: Green Books Ltd.

Svedbom, J. (2000). A comparison of different approaches to health education – Some reflections on the role of health education in relation to inequality in health. In B. B. Jensen (Ed.), *Critical environmental and health education. Research issues and challenges* (pp. 167–183). Copenhagen: Research Centre for Environmental and Health Education of the Danish University of Education

The Royal Society (2012). *People and the planet*. London: The Royal Society.

Thoresen, V. (2005). *The consumer: A fellow human being*. Hamar: Trykkeri AS.

Transfer-21 'Quality and Competencies' Working Group (2007). Developing quality at "ESD Schools" quality areas, principles & criteria. Berlin: Freie Universität Berlin.

United Kingdom Department for Education and Skills (UK DfES) (2006). *Sustainable schools for pupils, communities and the environment: Government response to the consultation on the sustainable schools strategy*. Nottingham: UK DfES.

United Kingdom Department for Education and Skills (UK DfES) (2008). *s3: Sustainable school self-evaluation: For primary, middle and secondary schools*. London: UK DfES.

United Nations Educational, Scientific and Cultural Organization (UNESCO) (2003). *United Nations decade of education for sustainable development (2005–2014) framework for the international implementation scheme*. Paris: UNESCO.

United Nations (2012). The future we want. Resolution adopted by the General Assembly on 27 July 2012. New York: United Nations.

United Nations Division for Sustainable Development (1992). Agenda 21. Rio Declaration on Environment and Development. http://www.un.org/esa/sustdev/documents/agenda21/english/agenda21toc.htm. Accessed 25 May 2012.

United Nations Educational, Scientific and Cultural Organization (UNESCO) (2005). *International implementation scheme: United Nations decade of education for sustainable development (2005–2014)*. Paris: UNESCO Publishing.

Vare, P., & Scott, W. (2007). Learning for a change: Exploring the relationship between education and sustainable development. *Journal of Education for Sustainable Development, 1*(2), 191–198.

Wallendorf, M. (2001). Literally literacy. *Journal of Consumer Research, 27*(4), 505–511.

Wals, A. E. J. (2010). Mirroring, Gestaltswitching and transformative social learning: Stepping stones for developing sustainability competence. *International Journal of Sustainability in Higher Education, 11*(4), 380–390.

Wals, A. E. J., Geerling-Eijff, F., Hubeek, F., van der Kroon, S., & Vader, J. (2008). All mixed up? Instrumental and emancipatory learning toward a more sustainable world: Considerations for EE policymakers. *Applied Environmental Education and Communication, 7*(3), 55–65.

Weinert, F. E. (2001). Concept of competence. In D. S. Rychen & L. H. Salganik (Eds.), *Defining and selecting key competencies* (pp. 45–66). Seattle: Hogrefe & Huber Publishers.

Wiek, A., Withycombe, L., & Redman, C. L. (2011). Key competencies in sustainability: A reference framework for academic program development. *Sustainability Science, 6*(2), 203–218.

World Wide Fund for Nature (WWF) (2012). *Living planet report 2012: Biodiversity, biocapacity and better choices*. Gland: WWF.

Young, B. (2000). The child's understanding of promotional communication. *Young Consumers: Insight and Ideas for Responsible Marketers, 2*(3), 191–203.

Zimmermann, B. J. (2001). Theories of self-regulated learning and academic achievement. An overview and analysis. In B. J. Zimmerman & D. H. Schunk (Eds.), *Self-regulated learning and academic achievement. Theoretical perspectives* (2nd ed., reprint, pp. 1–35). Mahwah: Erlbaum

Chapter 4
Under One Umbrella: Professional Norms Promoting Education for Sustainable Development at the School Level

Ulf Leo and Per Wickenberg

Abstract The aim of this chapter is to discuss the link – *the professional norms* – between legal norms and actions regarding education for sustainable development based on an empirical study. The study has a special interest in the role of school leaders and how the implementation of education for sustainable development (ESD) is led, organised and realised. There is also an interest in identifying the kinds of support mechanisms and obstacles to implementation and norm setting exhibited by the school organisations. The theoretical framework follows theories in sociology of law where the relation between law and society is the focus of the research. Education for sustainable development is regulated in the Swedish national curriculum and syllabi, thus constituting *legal norms* and national goals, which the professionals have to interpret and put into action.

The empirical study was conducted in three phases. It started with a web-based questionnaire for all teachers and principals at three upper secondary schools. In the second phase, each principal was interviewed individually. The third phase used focus groups to interview the principals in the leadership groups. The results indicate that professional norms are set when principals and teachers experience expectations from each other, from students and from policy documents. The school principal has a crucial role in these norm setting processes. If principals become more aware of existing norms in the organisations, and how norms can be changed, that knowledge can support them in change efforts. An expression used by one principal, that they were lacking a "shared umbrella", is a metaphor for the lack of shared norms to guide ESD.

Keywords Professional norms • School level • Principals • Education for sustainable development (ESD) • Implementation

U. Leo, Ph.D. (✉) • P. Wickenberg, Ph.D.
Sociology of Law Department, Lund University, Lund, Sweden
e-mail: ulf.leo@soclaw.lu.se; per.wickenberg@soclaw.lu.se

V. Simovska and Patricia Mannix-McNamara (eds.), *Schools for Health and Sustainability: Theory, Research and Practice*, DOI 10.1007/978-94-017-9171-7_4,

4.1 Introduction

Education, sustainable development and *health* are closely linked in the outcome document, "The Future We Want" (2012), of the United Nations Conference on Sustainable Development, 2012, "Rio+20". The document states that education and training in sustainability at all levels are keys to strengthening and supporting sustainable development (p. 44).

The aim of the study presented is to identify and analyse professional norms as a means of illuminating the implementation of *education for sustainable development* (ESD) at the school level. We have a special interest in the role of school leaders and how they lead, organise and realise school development. The central research question is: What *professional norms* do school leaders highlight in relation to ESD? The study uses a norm perspective to explore the links between education for sustainable development *and* professional norms in order to discuss the implementation of ESD. The study also identifies and analyses common professional norms in schools that promote ESD and how these norms are initiated and distributed in the schools investigated.

4.2 Education for Sustainable Development (ESD)

The first use of the term "sustainable" in the modern sense was by the Club of Rome in 1972 in its epoch-making report, "The Limits to Growth" (Meadows et al. 1972). Describing the desirable "state of global equilibrium", the authors used the word "sustainable" and that was a real eye-opener to many people in the environmental debate at that time. "Sustainable development" was then politically and globally launched by the UN appointed "Brundtland Commission" as development that meets the needs of the present without compromising the ability of future generations to meet their own needs (World Commission on Environment and Development 1987, p. 40). The concept is based on a holistic view of the needs, situations and problems of people and their societies. The principle is that the three dimensions of *economic*, *social* and *environmental* conditions and processes are to be integrated, interdependent and mutually reinforcing (ibid.). This holistic view is something teachers and principals in the study presented often refer to, along with the problems they have to combine the *economic*, *social* and *environmental* conditions. Although this concept of sustainable development was contested from the start and critically debated by many as being vague and self-contradictory (e.g. Lafferty and Langhelle 1995; Jacobs 1999; Luke 2005), it is still widely used in many contexts all over the world, such as in education, production, business and planning.

At the 1992 United Nations Conference on Environment and Development (UNCED) in Rio de Janeiro, *education for sustainable development (ESD)* was identified as one of the most important challenges faced by the schools. One outcome was the adoption of the Agenda 21 action plan on sustainable development by the 178 countries present. Education was regarded as a substantial part of Agenda 21. An interesting fact is that the word "education", alone or in combination, occurs 486 times in this UN document (Wickenberg 1999). This has led to a progressive development of environmental education as an ongoing process, which in the Scandinavian countries has the potential to be ahead of mainstream developments in education and the schools (Breiting and Wickenberg 2010).

In 2004, UNESCO's report in advance of the United Nations Decade of Education for Sustainable Development (DESD) 2005–2014 specified six characteristics of ESD: (1) Education is to be interdisciplinary and holistic, incorporating teaching for sustainable development into the entire curriculum rather than making it a separate subject. (2) Education is to be based on underlying values that can be examined, debated, tested and applied. (3) Critical thinking and problem solving are to create a sense of confidence when dilemmas and challenges to sustainable development are addressed. (4) The processes are to be shaped by diverse teaching methods. (5) The schools are to have a participatory decision-making culture in which students help determine how they learn. (6) Instruction is to be locally relevant, examining both local and global problems (UNESCO 2012).

Environmental education, health education, and education for sustainable development can each be seen as "efforts that are driven by a shared ideal of improving the world we live in now, and for the future" (Schnack 2008, p. 181). This is a part of the discussion as to whether education should be understood as an end in itself or as bringing about social change. Environmental education, health education, and education for sustainable development have the potential to teach pupils in a broader, humanistic sense involving democracy and participation. However, ever since the concept of ESD was launched, there has been a critical and intense policy debate about the ideological basis and the forms and content of this kind of education (e.g. Jickling and Spork 1998; Wals and Jickling 2000).

4.3 Theoretical Framework: Professional Norms Guiding ESD

Norms play an important role in human interaction. They reduce uncertainty about how to act in different situations, they set standards and specify appropriate behaviour, and they place different expectations on different people (e.g. students, teachers, and principals) in an organisation (Giddens 1989). During the last few decades, theories and methods for studying norms and norm supporting structures at the individual and organisational level have been developed in the discipline of sociology of law at Lund University (Hydén 2002; Wickenberg 1999; Svensson 2008; Leo 2010; Svensson and Larsson 2012). Sociology of law focuses on legal and other norms, and offers a different perspective in studies that seek to explore, interpret and understand changes in the education system.

Far from all actions are governed by norms. Elster describes Durkheim's social aspect, "homo sociologicus", where people are pushed forward by actions guided by social norms, in contrast to Adam Smith's "homo economicus", where people are guided by instrumental rationality with the hope of future reward. Elster believes that it is difficult to draw a line between the two and know when norms and rationality govern actions. He puts forward a theory that Western rationality is not found in all cultures and that rationality in certain contexts may be a result of socialisation. Elster argues that instrumental rationality can be a social norm among other norms (Elster 1992, p. 97). One of the functions of a norm is that it reduces complexity, but at the same time, we sometimes end up in situations where competing norms guide our conduct. What is it that makes us follow one norm or the other? According to Elster, self-interest and instrumental rationality become important, and what happens when norms are activated is a very complex issue.

The social psychologists Aarts and Dijksterhuis (2003) studied how situational norms arise and are activated in situations related to specific environments, such as how we lower our voices when we visit a library, or even at the mere thought of a future visit to the library. Situational norms represent the general opinion on how to act in certain situations. The most common way to learn these norms is to study how other people act and then do the same. This theory assumes that people want to be appreciated and avoid disapproval from others. It is the influence of the surroundings and expectations from the environment that create a norm and a norm is learnt in a social context; it can also be learnt through verbal communication (Aarts and Dijksterhuis 2003).

Another factor that contributes to norms being difficult to study is that people are not aware they are following norms because they are internalised: "The operations of norms are at a large extent blind, compulsive, mechanical and unconscious" (Elster 1992, p. 100). Rommetveit describes the internalisation of a norm as a subtle change that occurs when a persistent social pressure is gradually perceived as an obligation. He sees it as an aspect of socialisation (Rommetveit 1955, p. 56). For example, are you aware that you should turn off your mobile phone before the meeting or do you do it unconsciously? Are you thinking that you should lower your voice when you enter the reading room at the library? If you do either of these unconsciously, it is probably an internalised norm that governs your behaviour.

The concept "norm" is used in many different ways, both scientifically and in everyday parlance. In this study, a sociology of law based definition of the concept of norms is used that meets the needs of both the social and legal sciences: Norms are *action instructions* that are *socially reproduced* and represent the *individual's perception of the expectations surrounding their own behaviour* (Hydén and Svensson 2008).

4.3.1 Norms Are Action Instructions

Norms are imperatives and thus direct actions, which is the essence of norms. This study's starting point is the *legal norms* from the national policy documents on education for sustainable development. The Swedish *Curriculum for the Upper*

Secondary School 2011, states that all students have the right to an environmental *and* sustainable development perspective in the teaching and learning processes.

> An environmental perspective in education provides students with insights so that they can not only contribute to preventing harmful environmental effects, but also develop a personal position to major global environmental issues ... Education *shall* illuminate how the functions of society and our ways of living and working can best be adapted to create the conditions for sustainable development (Curriculum for the Upper Secondary School 2011, pp. 4–5) (authors' italics).

In our norm perspective, legal norms are imperatives that are supposed to initiate various actions; a pattern of similar actions is an indication of the existence of a social or professional norm.

4.3.2 Norms Are Reproduced

The norms of interest from a sociology of law perspective are the ones that occur in a social context. They have social connections, social impacts and are communicated in a social community (Wickenberg 1999, p. 262). The action instructions must be communicated and disseminated in a social community to live up to this essential attribute. In this study, the focus is on the *professional* norms that guide education for sustainable development in the schools. The professional norms of teachers are reproduced in teacher training programmes, in texts for teachers, in meetings with other teachers and so on. Norms are generated in a similar way for school principals through their training, mentoring and in discussion with other principals (Leo and Wickenberg 2013).

4.3.3 Norms and Expectations

Social pressure plays a major role in setting norms. For example, collective expectations influence individuals to engage in correct or culturally desirable behaviour (Durkheim 1982[1895]; Rommetveit 1955). According to the *theory of planned behaviour*, the strength of norms can be measured by studying perceived social pressure that an individual experiences (Ajzen and Fishbein 1980; Ajzen 1991).

4.4 Methods

The three Swedish upper secondary schools selected for the study are located in big cities. They were chosen as examples of schools with the ambition to offer education for sustainable development as a special profile for the entire school. All the principals considered the choice to be not only a matter of adapting to the national goals in the national curriculum, but to also have a special school profile that promotes ESD. The principals and teachers described the schools as "popular" with a large number of applicants and had students with high admissions scores.

The study was designed to collect data in relation to the norm perspective with most of the questions focusing on: *actions*, to be able to identify patterns of actions that indicate norms; c*ommunication*, to be able to examine reproduction of norms; and *expectations* to be able to identify from whence the pressure derives and how strong the external expectations are.

The study was conducted in three different phases by use of a multi-methods approach to collect quantitative and qualitative data, starting with a web-based questionnaire. The questionnaire was distributed to all ten principals and eight responded (two were on leave). It was also distributed to 225 teachers and 145 responded. The questionnaire consisted of seven items. Five gathered information about the respondents' teaching experience and background. One was open ended: "Please provide three motives as to why you as a teacher/principal work with sustainable development". The motives play an important role in helping us to understand what is "behind" the actions and norms.

The last item in the questionnaire was a Likert scale question: "Where do the expectations come from that you as a teacher/principal should work with sustainable development?" The respondents were asked to rate how strongly they felt the expectations came from students, colleagues, principals, the school board, local policy documents, national policy documents, the media, friends and family on a six-point scale where 0 represented "no expectations" and 5, "strongest expectations".

In the second phase, the ten principals were interviewed in person individually. The purpose was to expand upon and validate the results of the questionnaire, to explore why they wanted sustainable development to be a special profile, how sustainable development affected the way they ran their schools, and what approach they took to these questions. We went on to explore the support mechanisms and obstacles posed by their organisations by asking how the principals communicated, and how they viewed their roles as leaders in the implementation of education for sustainable development.

The third and final phase used focus groups to carry out an in-depth dialogue in the form of a group interview on one occasion about the preliminary results of the questionnaires and individual interviews. The groups were made up of the principals in the leadership groups at two of the schools. The third, smaller school had one principal and one part-time vice principal and no leadership group, which is why no focus group interview was conducted there.

4.5 Findings

From the sociology of law perspective, it is interesting to see how people use policy documents as a support. In this case, it is a matter of how the principals use the six characteristics of education for sustainable development from the UNESCO report to create legitimacy in their school development efforts (UNESCO 2012) and to set goals for this development. In the individual interviews and in the focus groups, the principals talked about the importance of

- An interdisciplinary and holistic education;
- Incorporating teaching for sustainable development into the entire curriculum rather than making it a separate subject;
- Education based on underlying democratic values;
- Critical thinking and problem solving to create a sense of confidence when addressing dilemmas and challenges to sustainable development;
- Diverse teaching methods;
- A participatory decision-making culture in which students help determine how they learn; and
- Instruction that is locally relevant, examining both local and global problems.

4.5.1 Why Teach ESD?

One question from the survey was open ended: "Please provide three motives as to why you as a teacher/principal work with sustainable development". The outcome shows a pattern of motives derived from outside the organisation, inside the organisation and from the individual's personal driving forces.

Both principals and teachers refer to the Education Act and the national curriculum. Principals in the focus groups refer to the responsibility of Swedish principals regulated in the curriculum. They also link their motives to what is happening in the surrounding world and the future for young people.

> One reason can be found in our governing documents. There is much that is there but we have chosen to prioritise ... We believe in sustainable development. This has become more important to me over the past few years when you see what's happening in the outside world. I think it is a matter of concern for young people who wants a world that works in the future (Upper secondary principal, focus group interview).

The teachers also refer to the syllabi for their respective subjects. They recognise the top-down legal norms but this answer is always combined with other concerns. Many of the answers are multifaceted, often linking social concerns with environmental concerns.

> Working with sustainable development involves working in a large-scale context with different perspectives that provide a greater understanding of the occurrence of problems and possible solutions in society and in nature (Upper secondary school teacher, questionnaire).

Both principals and teachers refer to many different social concerns, stating that sustainable development involves local and global problems related to equity, solidarity or action competence issues for the future. There are environmental concerns with an emphasis on ecological or environmental dilemmas, such as climate change and the use of natural resources. Educational considerations are also involved, where principals and teachers use ESD as a means of encouraging an interdisciplinary approach and holistic thinking in classroom instruction.

> To work on sustainable development represents a perspective of thinking that promotes co-operation between different subjects and a holistic approach to problems in society (Upper secondary school teacher, questionnaire).

Personal interest and involvement is reflected in answers like: "ESD is important for the future of my children and grandchildren".

The principals expressly stated that they wanted to implement education for sustainable development at their schools. As mentioned before, a common starting point for the principals was that it is not only about adapting to national goals in the national curriculum, but also about having a special profile that promotes education for sustainable development.

The principals also differ in another sense from the teachers by adding another perspective as to why they want to implement ESD. They see themselves as actors in a competitive school market and a number of them mentioned the importance of having an appealing profile to attract students. They stated that the young people they wanted to attract to their schools were interested in issues about the future, the environment and global concerns.

> The challenge is to capture the interests of the individual, to have a profile that is possible to develop and of interest to students. It is also about marketing (Upper secondary principal, individual interview).

It is a reality for principals and schools in many countries like in Sweden to compete on the school market. "Marketing" is a relatively new concept for the principals and today about 40 % of the upper secondary schools are independent and about 25 % of the students chose independent schools (Swedish National Agency for Education 2013).

One result of the study is that ESD in schools often consists of temporary projects, elective courses, theme days and other occasional events. Initiatives usually come from a teacher or student groups and ordinarily focus on environmental issues. The principals have introduced the practices of classroom observation and mentorship to disseminate knowledge among teachers in order to reach a more holistic approach, integrating environmental, social and economic perspectives for all students. The efforts to implement ESD at the school level were described as lacking "a shared umbrella". Perhaps the feeling of being "under one umbrella" is established when there are shared professional norms guiding education for sustainable development in the school (Leo and Wickenberg 2013).

4.5.2 Action Instructions, Meta-norms and Underlying Norms Guiding ESD

Norms are imperatives and thus direct actions, which is the essence of norms. Norms appear in hierarchical systems, from the highest overall principles to those that are applied in specific situations (Therborn 2002). We have identified a number of professional norms that are presented in the discussion and conclusions, but to give an example we start with an overarching norm, a meta-norm from the study: "We should teach sustainable development at our school". A meta-norm is an effective way of setting a norm and subsequently establishing other underlying norms (Axelrod 1986; Persson 2010; Leo 2010). The overarching norm in this case does not provide much guidance to the actions of the teachers, thus underlying norms have to be established in line with the local context of the school. According to the principals in the study, it varies considerably whether teachers actually include ESD in their teaching and follow the norm. It varies from almost all teachers in one school, to 15 out of 70 in another school, according to the principals in the interviews. This means that the teaching of the majority of the teachers in the other schools is *not* guided by the legal norms on sustainable development in the national curriculum, *or* professional norms guiding teaching for ESD.

4.5.3 Communication and Arenas to Establish, Reproduce and Disseminate Norms

Common discussions and reflections in practice among the professionals in the schools is one way to create and establish new action instructions and new norms. Interpersonal communication is vital to setting and disseminating norms. One principal stated that communication at the school was the most difficult challenge she faced, and the principals interviewed did not offer a clear perspective on how they used communication to attain particular objectives. Research shows that there is a lack of awareness among principals about what good communication involves (Törnsén 2009). According to Ärlestig (2008), "organisational communication blindness" interferes with the conversations in which principals engage. This study confirmed Ärlestig's conclusion that the communication of principals is not uniformly distributed and is often in response to teacher initiatives.

One problem identified is the lack of defined concepts from which to proceed when talking about sustainable development, or as one teacher expressed it: "Sustainable development becomes an empty vessel that can be filled with whatever content you choose". There is no single definition of what ESD is in two of the schools and the following quote is a good example:

> We haven't discussed what it would mean for us personally. We haven't got that far. What does it stand for? Do we have different perceptions? I don't know whether we are in agreement, but I haven't noticed that we disagree either. We haven't put words to what we are in agreement about (Upper secondary principal, focus group interview).

A common definition of ESD is a crucial starting point for its implementation at the school level. This statement is supported by the example from one school in the study, where the principals and teachers had put considerable effort into defining what they called "key concepts" related to ESD. The principal stressed the need for a common understanding on topics like "a holistic approach in education", "social dimension of ESD" and so on. The indications are that the opportunity for in-depth dialogue has improved accordingly, thereby leading to more and stronger underlying professional norms for teaching sustainable development. Perhaps it is not the definition in itself that is most important, but the discussions that aim to create some kind of common understanding of why, what and how to do it. In this school, the entire organisation was structured to work with environmental education and sustainable development. The teachers and principals discussed a holistic approach to learning and reported that dialogue proceeded from mutually defined key concepts. As a result, the teachers worked in interdisciplinary teams with decentralised responsibility for content and scheduling. Delegated responsibility served as a means of providing students with meaningful influence on classroom instruction. As a result, interdisciplinary projects were a widespread, established working method at the school, enforced by the following professional norm: We should have an interdisciplinary approach to teaching and learning.

A concrete method to stimulate broad discussions for defining concepts is that schools put together their own local policy documents.

> It's like H. says, these are the key issues, future issues affecting everything we work with. All we are working with needs to have that perspective. It's in the curriculum; this is what we *should* work with. It ties things together. What is important is that young people get an overall view … (Upper secondary principal, focus group interview).

Such an effort has been made at all three schools and attracted different degrees of participation. Based on the results from the interviews, we find that principals need to be more aware of the ways that communication can improve, as well as the need for ongoing dialogue aimed at defining key concepts so that the staff can engage in in-depth discussions about sustainable development as a way to create new professional norms.

4.5.4 Strong Expectations Create Strong Norms

The following question was in the web-based questionnaire: "Where do the expectations come from that you as a teacher/principal should work with sustainable development?" The respondents were asked to rate how strongly they felt the expectations came from students, colleagues, principals, the school board, local policy documents, national policy documents, the media, friends and family on a six-point scale where 0 represented "no expectations" and 5, "strongest expectations".

An analysis of the results indicates that the strongest expectations came from national and local policy documents. From a national perspective, this is desirable since the requirements from the national curriculum, *constituting the legal norms*, are supposed to create expectations for the professionals in the schools; professional norms are supposed to be derived and become a part of the institution. The teachers also experienced strong expectations from the principals and the results indicate that the implementation of education for sustainable development is a top-down process to many of the teachers, proceeding from policy documents and the principals.

The teachers at two of the schools experienced low expectations from their fellow teachers, students, the student council, and parents. This is argued to be an indication that professional norms about ESD in everyday practice may be weak within the schools. The data from the teacher questionnaire suggest that two of the schools appear to be at the beginning of the implementation phase of ESD due to the low expectations from colleagues and students. Interviews with teachers and principals indicate that groups of teachers collaborate on courses and projects and develop professional norms that shape the sustainable development effort.

The results from the third school differ. It is a new institution and the staff were and still are hired because they want to work with ESD and with an environmental, participatory approach to interdisciplinary work. The school's point of departure is a common vision – that ESD is a special profile guiding all school work – and the principals and teachers at this school experience strong expectations from various directions, both top-down from policy documents and principals but also bottom-up from students, parents and other teachers. This indicates the existence of several widespread professional norms guiding ESD in the school.

All of the principals experienced strong expectations from their students. They described it as a need to be a role model for the school's profile. An interesting result is the relatively low expectations the majority of principals experienced from the teachers and the teachers' expectations for each other.

> I understood that teachers feel that we are pushing a lot and we do it with the support of the governing documents and what comes from above. I thought that they would expect a little more of each other (Upper secondary principal, focus group interview).

The principals in our study experienced strong expectations from the other principals in the school's leadership group to be engaged in sustainable development issues. They stressed the importance of strong support from the leadership group. As a result, the principals may have been producing and reproducing special professional norms within their particular domain (Leo 2010). An area for improvement that the principals stressed was that communication directed at students and teachers would be clearer when the leadership group was united.

4.6 Discussion

Many studies have been published about educational development and improvement but this one is unique with its focus on professional norms to illuminate the implementation of education for sustainable development. Implementation processes are complex, and a highly simplified approach to discussing improvement efforts at schools is to proceed from the concepts of initiation, implementation, institutionalisation and dissemination (Miles et al. 1987; Blossing 2008).

4.6.1 From an Initiation Phase to Institutionalisation and Dissemination

One of the schools studied has focused on education for sustainable development ever since opening 7 years ago. The teachers there were thus hired for this purpose and the initiation phase began immediately for all of them. During the initiation phase, the members of an organisation create a common understanding of what is new and they experiment with different kinds of behaviour in accordance with our norm perspective – they communicate about what is and is not working, and expectations emerge. Since opening, this school has gone through an implementation phase during which the teachers and principals maintained a clear focus on the ESD profile and continually shared their experiences by

> … formulating concepts, ensuring that the entire staff were involved in what ESD would mean and what teaching about this is (Upper secondary principal, individual interview).

Based on the results of this study, it is argued that *new professional norms* for teaching sustainable development are set during the implementation phase. The behaviours of teachers and principals become patterns and *action instructions* – "I should act like this". Their behaviour is affected by the various expectations that come from students, other teachers and principals, and the individual action instructions become professional norms that are communicated and disseminated during the implementation phase. Among the norms that emerged at the school were: "We should take an interdisciplinary, project-oriented approach"; "We should give students a lot of influence over classroom instruction".

Due to the perseverance of the teachers and principals in one of the schools, the implementation phase transitioned to the institutionalisation phase after 3–5 years, and the effort has become routine. Thus, professional norms have been established and have grown strong. The final dissemination phase enables other schools to learn about the improvement efforts because this school has been the object of countless field trips from around the country over the years (Leo and Wickenberg 2013).

4.6.2 Principals and Teachers as Change Agents Promoting ESD

One may follow a norm out of *identification* with the norm source and its values (Therborn 2002, p. 869). Ellickson uses the term "change agents" for the people who first transmit new norms (Ellickson 2001, pp. 35–75). To bring about change at schools so that new norms are set requires enthusiasts, movers or change agents who challenge old norms, who want to stake out new paths and establish new norms. Change agents can be very dedicated, enthusiastic, and deeply involved in organisations, "burning" for a certain cause they strongly believe in. They are sometimes called "souls of fire" (Wickenberg 1999, 2013). According to Ellickson, change agents are self-motivated leaders who favour changes and do not need special esteem rewards in order to challenge the existing norms. They can be seen as charismatic change agents with their own internal driving forces working for social change, and with special leadership abilities supplying the establishment of new norms. From the start, self-motivated leaders act at the forefront in supplying a new norm (Ellickson 2001). In our study, the enthusiasts or change agents can also be seen as vital links between the legal norms and the new local professional norms in schools. These dedicated change agents in addition act as mediators bringing their personal commitment from their lifeworld into their professional role in the school system, using the concepts of Habermas (1987; Wickenberg 1999).

One question involves the role that principals play in building professional norm supporting structures and whether they may also serve as change agents with whom others can identify such that new professional norms are set. Legal norms in the Education Act, national curriculum and syllabi govern education for sustainable development, and change agents are required to interpret and establish them as new professional norms (Wickenberg 1999, p. 451; Leo 2010, p. 68). The results of the study show that there are a number of enthusiasts and change agents at the schools and that they are motivated by both internal and external factors.

One of the most important tasks of a principal, as expressed in the interviews, is to provide teachers with the tools they need. The principals talk about supporting teachers through the management of financial resources, scheduling and other structural matters.

> Above all, we try to support and assist teachers in their duties, thereby creating space for it to be sustainable, in order to work with sustainable development. It's difficult, it's tricky, and it's complicated. The schedules are quite compact, student groups are different; it's a difficult task (Upper secondary principal in focus group interview).

Principals also speak of the importance of having a "vision that can be communicated" and that can always be referred to during the development efforts.

> Ultimately, we hope there will be a situation where management and staff have a common interest; we do this together, we have the same goal, and we help each other (Upper secondary principal, focus group interview).

4.6.3 School Leadership in ESD

A number of researchers have stressed the role of principals in educational development and classroom learning (Pont et al. 2008; Hattie 2008; Leithwood and Day 2008). Our study focuses on the role of principals as "change agents". A common ingredient in a number of studies has been the role of the fundamental, guiding values that principals articulate, as well as the significance of the principal's purpose for the change effort (Fullan 2001; Leithwood et al. 2002; Starratt 2004; Day 2007). According to UNESCO's report to the United Nations, education is to be based on underlying values that can be examined, debated, tested and applied

(UNESCO 2012). The principals in the study strive for consensus and there appears to be a lack of open discussion in the schools as to why ESD should bring change and improve the teaching and learning processes.

Hargreaves and Fink define sustainable educational leadership based on the same principles as those that apply to sustainable development:

> Sustainable educational leadership and improvement preserves and develops deep learning for all that spreads and lasts, in ways that do no harm to and indeed create positive benefit for others around us, now and in the future (Hargreaves and Fink 2006, p. 17).

Such leadership, which has a clear purpose and objective, proceeds from the primary task of the schools in accordance with the legal norms from the Education Act and the national curriculum. A long-term approach is required by which leadership and responsibility percolate throughout the organisation. This is also defined as *distributed leadership*, which involves participation and influence on the part of teachers and students (Hargreaves and Fink 2006; Harris 2004, 2011; Day 2007). One component of sustainable leadership is to forge strong professional school cultures, which in this study is defined as professional norms. Fullan defines *educational sustainability* as "the capacity of a system to engage in the complexities of continuous improvement consistent with deep values of human purpose" (Fullan 2005, p. 114). Thus, a key ingredient of sustainable leadership is the ability to handle complexity and constant change as moulded by differing values.

The principals who were interviewed for the study said that they wanted to facilitate the improvement efforts of the organisation, including assumption of responsibility by the teachers for scheduling, local finances and other new areas. Their wishes are not easy to fulfil given that the teachers feel as though they are being saddled with additional tasks in addition to the heavy burden they already carry.

The principals in the interviews talk about how they delegate and want to share responsibility with the teachers, and that also involves delegating power since distributed leadership is a process that is not directly linked to a particular leader. According to Harris (2008), formal leaders (principals in our study) are gatekeepers for distributed leadership. In other words, principals provide the opportunity for distributed leadership, which should facilitate changes in leadership, given that it is not linked to particular people.

In accordance with our norm perspective, a distributed leadership can lead to a new distribution of power and responsibility that creates new social pressure and stronger and mutual external expectations from students, teachers and principals. As a result, a distributed leadership can lead to strong widespread professional norms deriving from a local bottom-up perspective.

4.7 Conclusions – Under One Umbrella

The aim for the schools in this study is to have education for sustainable development as a special profile, one that concerns all students, teachers, and principals. The expression used by one principal, that they were lacking a "shared umbrella" is a metaphor for the lack of shared professional norms guiding ESD. It is also seen as an expression of an ongoing process to initiate, implement and institutionalise professional norms that guide ESD at the school level. Based on the overarching meta-norm, "We should teach sustainable development at our school," new underlying professional norms can be established. One example is the norm "We should teach an interdisciplinary, cross curriculum". This professional norm is promoted by the principals and it is strongly related to the principals' motives for developing the teaching and learning processes in the school.

If teachers are guided by the professional norm of interdisciplinary teaching, they work in teams and projects, with a holistic view of education. This study argues from the empirical material that principals are using the concept of education for sustainable development as a means to make changes in the organisational structure and culture to improve the education as a whole, to improve teaching methods in the classrooms in favour of students' participation with a democratic perspective of education. Some examples of the identified *professional norms* of the principals that guide their leadership to implement and institutionalise ESD are that principals should work with the *culture* of the school to:

- Create and constantly point out the common vision of ESD in the school, that ESD is a special profile guiding all school work.
- Create and support social arenas for dialogue on a common understanding of the different concepts linked to ESD.
- Distribute leadership creating a school culture with professional norms that support initiatives and influence from teachers *and* students.
- Promote a view of education based on the Education Act (1985 and 2010) and the national curriculum. Currently many teachers focus primarily on the syllabi of their respective subjects.
- Ensure that there are local policy documents that interpret the tasks in the curriculum to support the change work and to set new professional norms.

Principals should work with the *structure* of the school to:

- Manage the resources to promote education for sustainable development in terms of arranging schedules to promote cross-curricular activities.
- Organise the teachers in different teams promoting cross-curricular activities.
- Allocate budget for in-service training for teachers on topics related to ESD, study visits, and co-operation with external organisations.

For the principals, the leadership group is mentioned as the strongest support, and is considered to be a forum that enables the creation of special professional norms for principals that guide their leadership. In the same way, it is also clear that teachers working in teams support the creation of professional norms that guide teaching.

This study does not examine the hierarchy of norms in greater depth. From the empirical data the study concludes that the schools have differing *action instructions* and that differing norms are established in the various school cultures. It also concludes that the potential for *reproduction* of professional norms varies among the schools and that reproduction is a function of the arenas for communication encompassed by an organisation. In a recent study on leading sustainable primary schools, Kadji-Beltran et al. (2013) stress the importance of professional learning for principals, such as regarding ways of generally empowering staff, encouraging critique of current approaches and the exploration of alternative possibilities for curriculum, pedagogy and policy. Our conclusion, from a norm perspective, is that there is a need for special training for principals and opportunities for principals to meet and discuss at a deeper level what they do as leaders. Principals have to meet other school leaders to be able to reproduce special professional norms that guide them in their role as principals.

Principals play a major role in this respect when it comes to building structures for encounters and dialogue and all principals in our study see this as a key problem. The schools that are in the implementation phase do not have clear internal communication about why they want sustainable development to be a special profile, what is meant by sustainable development or how sustainable development can be taught. The most important thing may not be finding a universal definition of sustainable development. A more useful objective would be for principals, teachers and students at each school, on a local basis and depending on the context in which they find themselves, to come up with their own definitions of the concepts on which they want classroom instruction to be based. According to the study results, a starting point in the process of implementing ESD in schools is to lead and support dialogue within the school on defining vital concepts in order to adapt them to the local school context. Furthermore, in this study it is argued that the *expectations* on behaviour from the surroundings affect professional norm setting and that the process also varies from school to school. The results indicate that sustainable professional norms are established when principals and teachers experience expectations from each other, from policy documents and from students.

Finally, some principals and teachers in the study refer to policy documents to establish legitimacy in the implementation of education for sustainable development, but only in general terms and not directly linked to an objective for the school or for a particular subject. As sociologists of law, it is interesting to find that principals do not use *the legal norms* to a greater extent to gain legitimacy in their implementation of ESD at the school level. One finds resistance in the study from some teachers who claim that it is up to the principles to set the agenda – not them. This in turn motivates teachers to work with ESD. One can ask if these teachers are aware, or influenced by the task of the school from the Curriculum for Upper Secondary School, 2011. It also reveals a top down approach to the implementation of ESD.

Research on environmental education and education for sustainable development has focused on the interested and engaged teachers, the ones already following professional norms regarding ESD. There is a need for further research to examine the implementation and dissemination of professional norms that guide ESD to find out how ESD can reach *all* teachers and students. The principals in the study use ESD as a goal in itself and as a way to improve their schools and bring all individuals in the institution "under one umbrella".

References

Aarts, H., & Dijksterhuis, A. (2003). The silence of the library: Environment, situational norm and social behaviour. *Journal of Personality and Social Psychology, 84*(1), 18–28.

Ajzen, I. (1991). The theory of planned behavior. *Organizational Behavior and Human Decision Processes, 50*, 179–211.

Ajzen, I., & Fishbein, M. (1980). *Understanding attitudes and predicting social behavior.* Englewood Cliffs: Prentice-Hall.

Ärlestig, H. (2008). *Communication between principals and teachers in successful schools.* Doctoral dissertation, Umeå University, Umeå, Sweden.

Axelrod, R. (1986). An evolutionary approach to norms. *The American Political Science Review, 4*, 1095–1111.

Blossing, U. (2008). *Kompetens för samspelande skolor, om skolorganisationer och skolförbättring* [Skills for interacting schools, on school organisations and school improvement]. Lund: Studentlitteratur.

Breiting, S., & Wickenberg, P. (2010). The progressive development of environmental education in Sweden and Denmark. *Environmental Education Research, 16*(1), 9–37.

Day, C. (2007). What being a successful principal really means. *Educational Leadership and Administration, 19*, 13–24.

Durkheim, E. (1982). *The rules of sociological method and selected texts on sociology and its method.* New York: Free Press [original, 1895. Les règles de la méthode sociologique].

Education Act (Swedish Code of Statutes 1985:1100).

Education Act (Swedish Code of Statutes 2010:800).

Ellickson, R. (2001). The evolution of social norms: A perspective from the legal academy. In M. Hetcher & K. D. Opp (Eds.), *Social norms* (pp. 35–75). New York: Russell Sage Foundation.

Elster, J. (1992). *The cement of society: A study of social order.* Cambridge: Cambridge University Press.

Fullan, M. (2001). *Leading in a culture of change.* San Francisco: Jossey-Bass.

Fullan, M. (2005). *Leadership and sustainability.* Thousand Oaks: Corwin Press.

Giddens, A. (1989). *Sociology.* Cambridge: Polity Press.

Habermas, J. (1987). *The communicative action: Vol. II. Lifeworld and system: A critique of functionalist reason.* Oxford: Polity Press.

Hargreaves, A., & Fink, D. (2006). *Sustainable leadership.* San Francisco: Jossey-Bass.

Harris, A. (2004). Distributed leadership and school improvement: Leading or misleading? *Educational Management Administration Leadership.* doi:10.1177/1741143204039297.

Harris, A. (2008). Distributed leadership: According to the evidence. *Journal of Educational Administration, 46*(2), 172–188.

Harris, A. (2011). Distributed leadership: Implications for the role of the principal. *Journal of Management Development, 31*(1), 7–17.

Hattie, J. (2008). *Visible learning: A synthesis of over 800 meta-analyses relating to achievement.* London: Routledge.

Hydén, H. (2002). *Normvetenskap* [Science of norms]. Lund: Lund University.

Hydén, H., & Svensson M. (2008). The concept of norms in sociology of law. In P. Wahlgren (Ed.), *Law and society, scandinavian studies in law* (Vol. 53, pp. 15–32). Stockholm: Stockholm Institute for Scandinavian Law.

Jacobs, M. (1999). Sustainable development as a contested concept. In A. Dobson (Ed.), *Fairness and futurity: Essays on environmental sustainability and social justice.* Oxford: Oxford University Press.

Jickling, B., & Spork, H. (1998). Education for the environment: A critique. *Environmental Education Research, 4*(3), 309–327.

Kadji-Beltran, C., Zachariou, A., & Stevenson, R. B. (2013). Leading sustainable schools: Exploring the role of primary school principals. *Environmental Education Research, 19*(3), 303–332.

Lafferty, W. M., & Langhelle, O. (Eds.). (1995). *Bærekraftig utvikling. Om utviklingens mål og bærekraftens betingelser* [Sustainable development. On development goals and sustainable conditions]. Oslo: Ad Notam Gyldendal.

Leithwood, K., & Day, C. (2008). The impact of school leadership on pupil outcomes, school leadership and management. *Formerly School Organisation, 28*(1), 1–4.

Leithwood, K., Jantzi, D., & Steinbach, R. (2002). *Changing leadership for changing times*. Buckingham: Open University Press.

Leo, U. (2010). *Rektorer bör och rektorer gör* [Principals should and principals do]. Doctoral dissertation, Lund University, Lund, Sweden.

Leo, U., & Wickenberg, P. (2013). Professional norms in school leadership: Change efforts in implementation of education for sustainable development. *Journal of Educational Change*. doi:10.1007/s10833-013-9207-8. Published online: 30 March 2013.

Luke, T. W. (2005). Neither sustainable nor development. Reconsidering sustainability in development. *Sustainable Development, 13*, 228–238. published online 16 August 2005 in Wiley InterScience (www.interscience.wiley.com). doi:10.1002/sd.28.

Meadows, D. H., Meadows, D. L., Randers, J., & Behrens, W. W. (1972). *The limits to growth*. New York: Universe Books.

Miles, M. B., Ekholm, M., & Vanderberghe, R. (1987). *Lasting school improvement: Exploring the process of institutionalization*. Leuven/Amersfoort: Acco.

Persson, L. (2010). *Pedagogerna och demokratin* [Teachers and democracy]. Doctoral dissertation, Lund University, Lund, Sweden.

Pont, B., Nusche, D., & Moorman, H. (2008). *Improving school leadership, Vol. 1 policy and practice*. http://www.oecd.org/education/preschoolandschool/41200009.pdf. Accessed 15 Aug 2012.

Rommetveit, R. (1955). *Social norms and roles: Explorations in the psychology of ending social pressures*. Oslo: Universitetsforlaget.

Schnack, K. (2008). Participation, education, and democracy: Implications for environmental education, health education, and education for sustainable development. In *Participation and Learning* (pp. 181–196). New York: Springer. doi: 10.1007/978-1-4020-6416-6_11.

Starratt, R. (2004). *Ethical leadership*. San Francisco: Jossey-Bass.

Svensson, M. (2008). *Sociala normer och regelefterlevnad: trafiksäkerhetsfrågor ur ett rättssociologiskt perspektiv* [Traffic safety in the perspective of sociology of law]. Doctoral dissertation, Lund University, Lund, Sweden.

Svensson, M., & Larsson, S. (2012). Intellectual property law compliance in Europe: Illegal file sharing and the role of social norms. *New Media Society, 14*(7): 1147–1163. doi:10.1177/1461444812439553.

Swedish National Agency for Education. (2011). *Curriculum for the upper secondary school*. Stockholm: Swedish National Agency for Education.

Swedish National Agency for Education. (2013). *Skolor och elever i gymnasieskolan, läsåret 2012/13* [Schools and students in upper secondary school, 2012/13]. Registration no. 71-2013:28. http://www.skolverket.se/publikationer?id=2982/. Accessed 19 May 2013.

The future we want. Outcome document of "Rio+20" – United Nations Conference on Sustainable Development. (2012). United Nations General Assembly. A/RES/66/288. Resolution adopted by the General Assembly. Sixty-sixth session. Agenda item 19. 123rd plenary meeting, 27 July 2012. http://sustainabledevelopment.un.org/futurewewant.html. Accessed 18 May 2013.

Therborn, G. (2002). Back to norms! On the scope and dynamics of norms and normative action. *Current Sociology, 50*, 863–880.

Törnsén, M. (2009). *Successful principal leadership: Prerequisites, processes and outcomes*. Doctoral dissertation, Umeå University, Umeå, Sweden.

UNESCO. (2012). *Shaping the education of tomorrow: 2012 full-length report on the UN decade of education for sustainable development*. DESD Monitoring & Evaluation – 2012. UNESCO EducationSector.Author:ArjenE.J.Wals.http://unesdoc.unesco.org/images/0021/002166/216606e.pdf. Accessed 19 May 2013.

Wals, A. E. J., & Jickling, B. (2000). Process-based environmental education seeking standards without standardizing. In B. B. Jensen, K. Schnack, & V. Simonovska (Eds.), *Critical environmental and health education: Research issues and challenges* (pp. 127–149). Copenhagen: The Danish University of Education, Research Centre for Environmental and Health Education.

Wickenberg, P. (1999). *Normstödjande strukturer: miljötematiken börjar slå rot i skolan* [Norm supporting structures: the environmental theme is taking root in school]. Doctoral dissertation, Lund University, Lund, Sweden.

Wickenberg, P. (2013). Souls of fire, change agents, and social norms. In M. Baier (Ed.), *Social and legal norms*. London: Ashgate Publishing (in print, November 2013).

World Commission on Environment and Development. (1987). *Our common future*. Nairobi: United Nations Environment Programme. http://www.un.org/documents/ga/res/42/ares42-187.htm/. Accessed 15 Aug 2012.

Chapter 5
Linking Health Education and Sustainability Education in Schools: Local Transformations of International Policy

Katrine Dahl Madsen, Lone Lindegaard Nordin, and Venka Simovska

Abstract This chapter addresses the relationship between international and national policies regarding sustainability and health promotion which have the potential to affect school-based health education/promotion and education for sustainable development in Denmark. Based on policy mapping and analysis, the focus is on the transformation processes that occur during the transition from international policy frameworks to the national context. The chapter considers the consequences of these transformation processes for educational practices within schools in light of the current major reform of basic general education in Denmark with its aims of ensuring overall school improvement, increasing pupil wellbeing and improving academic outcomes. Analysis of international policy documents, as well as of research literature shows that school-based health education (HE) and education for sustainable development (ESD) share a number of features. These include a whole-school approach, cross-disciplinarity, participatory approaches, cultivating social imagination, and developing critical competences related to working with 'real life' health and sustainability issues. The discussion in this chapter focuses on the common tendency that when health and sustainability education in schools are framed in national action plans, certain critical educational aspects are lost by narrowing the concepts of health and sustainability to fit particular school subjects (e.g. physical education or science), and defining outcomes solely in terms of individual lifestyle factors. This neglects the importance of working with broader social values and the complexity of the interplay between individual and society in relation to both health and sustainability.

Keywords Health education • Education for sustainable development • Policy • Health-promoting schools

---

K.D. Madsen, Ph.D. • L.L. Nordin, Ph.D. • V. Simovska, Ph.D. (✉)
Department of Education, Aarhus University, Tuborgvej 164, 2400 Copenhagen, Denmark
e-mail: kdma@edu.au.dk; lln@edu.au.dk; vs@edu.au.dk

V. Simovska and Patricia Mannix-McNamara (eds.), *Schools for Health and Sustainability: Theory, Research and Practice*, DOI 10.1007/978-94-017-9171-7_5,

5.1 Introduction

Health and sustainability have been identified among the major societal challenges of the twenty-first century, and the school is considered as a key arena for health education/promotion and education for sustainable development. Both fields are underpinned by high-level policy documents, charters, declarations and agreements between and within governments. For example, within sustainability discourses, climate change is an issue addressed with political urgency on a global level. Similarly, policy discourses on health, characterized by slogans such as "global burden of disease", and "health in all policies", are intense, pointing to escalating health problems such as obesity, mental health issues and a range of chronic conditions. Based on the assumptions that there can be no sustainable development without learning (Scott and Gough 2003) and no health promotion without health education (Green and Tones 2010), schools can be considered one of the key sites for working educationally with the promotion of health and sustainability. It is essential that schools incorporate both approaches if they are to respond to global and complex societal challenges and foster children's competences to deal with these challenges in creative, socially responsible and productive ways.

Although research within the individual fields of health education/promotion and education for sustainable development in schools is well developed (e.g. Clift and Jensen 2005; Simovska 2012a, b; Firth and Smith 2013), and even sometimes combined in joint publications, pointing to the shared educational principles and values (Jensen et al. 2000; Reid et al. 2008), few attempts have been made to combine the two fields and thereby capitalize on the synergies. Davis and Cooke (2007) and Davis and colleagues (2010), for example, have argued in favour of integrating the Australian Health Promoting Schools and Sustainable Schools initiatives. Similarly, Dooris has written about combining the concepts of health promoting and sustainable universities (e.g. Dooris 2012), while Patrick and colleagues have discussed the interconnectedness between humans and the natural environment, and the core competences required if health promotion is to address the health challenges linked to climate change (Patrick et al. 2012). In this chapter we aim to contribute new layers to this debate by exploring the policy background framing both fields and discussing some of the transformations and their consequences related to health education/promotion and education for sustainable development practices in Danish primary and lower secondary schools.

5.2 Rationale for the Study

There are at least three reasons why it is important to place a spotlight on the policy background for educational practice related to the themes of health and sustainability: First, the common denominator for health and sustainability, as well as related school practices, is that they are value-laden and shaped by a number of policies at global, national, regional and local levels. Second, research points to a persistent gap between, on the one hand, political aims and targets concerning health promotion and education for sustainable development and, on the other hand, the treatment of these topics in school pedagogical practice. This is true for the prescribed teaching and learning processes as well as for the everyday life or culture of the school (Stevenson 2007a; Jourdan 2011; Samdal and Rowling 2013; Nordin 2013). Third, both concepts can be considered to be "essentially contested" - that is socially constructed and open to diverse, often conflicting, interpretations. According to the criteria that Green and Tones (2010) propose, the concepts of health and sustainability can be seen as contested because:

- They are complex, ambiguous and value-laden;
- Their definitions are vague, their meaning depends on the sociocultural, historical and political contexts;
- Their different interpretations are mutually competitive, involving emotional reactions; and
- They hold a degree of authority and credibility.

These characteristics, naturally, pose certain common challenges for schools when it comes to determining the key pedagogical questions such as the aims, content, teaching strategies and desired outcomes of education for health and sustainability.

Against this background, we discuss the findings from the mapping and analysis of selected international and national policy documents, as well as other documents influencing the work with health education/promotion and education for sustainable development in primary and lower secondary education in Denmark. The focus is on identifying similarities and commonalities in the interpretations of the concepts of (promoting) health and sustainability within the policy documents, particularly in terms of the consequences these interpretations have for the educational aims, content, pedagogical strategies and expected outcomes at a national level.

5.3 Methodology

The overall method belongs to the genre of mapping and critical conceptual analysis of documents. We searched the websites of selected key international organisations within both fields for documents of relevance for school practices concerning health and sustainability. At international level the search was focused on international organizations, primarily UN, WHO and EU. Some documents published by the International Union for Health Promotion and Education (IUHPE) are also included as this network collaborates closely with WHO in the area of school-based health promotion and education. Also, the conference resolutions from the four European Conferences on Health Promoting Schools (Thessaloniki, Greece; Egmond, the Netherlands; Vilnius, Lithuania and Odense, Denmark) are included, as well as a Nordic document of relevance for education for sustainable development. The mapping is focused on the period from the end of the 1980s, as this period marks the emergence of both fields within broader international discourse, following the publication of the Ottawa Charter (WHO 1986) for Health Promotion and the Brundtland Report (UN 1987), until 2013.

In Denmark, health education and education for sustainable development constitute mandatory but transversal themes to be integrated across the boundaries of subject and year group within primary and lower secondary education. National learning objectives have been drawn up for both themes and implemented either in subject-specific or more general curriculum guidelines. Therefore, the search at national level focused on policy and strategy documents, school curricula, national curriculum guidelines and inspiration teaching material published by the Danish Ministry of Education and the Ministry for Children and Young People, during the same period. Relevant documents published by the Danish Ministry of Health were also taken into consideration.

The search strategy included a general screening of publications on the websites of the above-mentioned international and national bodies, and a keyword-based search of the above-mentioned websites. The keywords used (in English and Danish) were: *sustainable development, education for sustainable development, sustainability, climate change, the environment and environmental education, health, health education, health promotion.*[1] To be included, documents had to meet one of the following criteria:

- The document is central in framing the fields of health promotion and sustainable development in general and is therefore relevant for school practices;
- The document explicitly mentions health promotion and/or sustainable development in relation to schools, either in the title or in a specific section.

5.3.1 Analytical Framework

The analysis is based on the premise that there are active processes of reinterpretation and "translation" or transformation of policy taking place, not only on the path from international to national level, but also along the trajectories to regional (i.e. municipality) and further to local (i.e. school) levels. The analytical approach is inspired by studies of educational exchange focusing on how supranational discourses are re-contextualized in local settings (Beech 2006; Moos 2009). This suggests a vertical perspective in the analysis, exploring how health education/promotion and education for sustainable development in schools are constituted and (re)interpreted at different levels. The basic premise is that societal problems and issues are not given but constructed in social processes embedded in a specific

[1] We also consulted two experts in the field, Jeppe Læssøe for the field of education for sustainable development and Monica Carlsson for health education and health promoting schools. Their contribution to the mapping is gratefully acknowledged.

socio-historical context. The identification of problems creates the framework for selecting what is relevant and what is not (Schön 1983). At the same time, as Rein and Schön point out, the very definition of a problem points towards certain solution strategies:

> Problem setting is a judgment about the problematic situation – that is, a diagnosis that also contains the prescription of directions for actions (Rein and Schön 1981, p. 238).

This suggests that processes of negotiation in defining the issues at stake, as well as their solutions, are always present. The metaphor of "translation" indicates that something will be lost and something added in a dialectic process of negotiation, re-conceptualization and re-contextualization. In the words of Gherardi:

> The metaphor of translation is a way to describe movements between different forms of knowledge and cultural practices, but also of technology and artefacts. It has both a geometric and semiotic meaning: Translation is both the movement of an entity in space and time and its translation from one context to another – as in translating from one language to another, with the necessary transformation of meaning that this always implies (Gherardi 2006, p. 62).

The analytical categories are based on the key educational questions concerning schools' pedagogical practices: *what* is the content of the teaching (the concepts of health and sustainability); *how* should teaching be done (pedagogical strategies); and *why* are content and pedagogical strategies relevant and appropriate (values and purpose) (Schnack 2003; Biesta 2010). In addition, contextual factors are also taken into account; the context is interpreted as something *created* in the process rather that something that *is* (Gherardi 2006; Schön 1983). Thus, the analytical categories and related questions used in the document analysis include:

- *Aims*: How are the learning objectives related to health and sustainability named and framed?
- *Context*: Which local frames are created through the work with health promotion/education and education for sustainable development in schools, and what are the wider societal conditions in which these frames are established?
- *Pedagogical strategies*: Which pedagogical strategies and intervention initiatives are suggested with a view to promoting health and sustainable development?
- *Competences*: What are the expected outcomes, both in terms of general educational outcomes and health and sustainability related knowledge and skills?

The analysis builds on previous discussions within critical educational approaches to health promoting schools and education for sustainable development research, both in Denmark (e.g. Jensen and Simovska 2005; Carlsson et al. 2009; Læssøe et al. 2009; Breiting and Wickenberg 2010) and internationally (Jensen et al. 2000; Green and Tones 2010; Nutbeam 2008; Clift and Jensen 2005; Porter 2006; Reid et al. 2008; Scott and Gough 2003; Lotz-Sisitka 2007; Wals 2010; Simovska 2012c; Carlsson and Simovska 2012; Bonnett 2013). Within these perspectives, health and sustainability are interpreted as concepts that relate to both individual behaviour and factors affecting the individual's lifestyle, including physical, social and cultural environment. Both health and sustainability are perceived as being characterized by an uncertain and unstable body of knowledge which, from an educational perspective, demands specific approaches to teaching and learning, and point to the importance of working pedagogically with cross-disciplinarity, sociological imagination, critical reflection, and change of perspectives as epistemological principles (Stevenson 2007b). Further, the interaction between the individual on the one hand and the physical and social environments at local and global levels on the other is central within both fields (Kickbusch 1997). Schools are viewed as active players, employing the whole-school environment and local systems of meaning, as well as more general values of democracy, equity and social justice, in teaching and learning aimed at the development of pupils' competence related to both health and sustainability. In this way, schools can extend teaching beyond the classroom and also work with local communities.

5.4 Findings and Discussion

A number of policy documents have been identified in the mapping, and these are presented in chronological order in Table 5.1. The documents are categorized as relevant for: school-based education for sustainable development; health promotion/education at school; or, both areas. The international documents are presented first followed by the national documents.

Table 5.2 summarizes the main findings in the analytical categories illuminated with keywords. These findings serve as a basis for discussing the transformations from the international to the national frameworks. The remaining sections will present the analysis following the main analytical categories. For each analytical category, we first discuss the international and then the national discourses.

5.4.1 Aims and Context

5.4.1.1 International Framework: Initiation of a Shared Value Basis and a Need for Social Change

As shown in Table 5.1, two documents published in the mid-eighties are highly influential in shaping the fields of education for sustainable development and health promotion: The Brundtland report (UN 1987) and the Ottawa Charter (WHO 1986), respectively. Along with the UN Convention on the Rights of the Child (UN 1989), these documents signify a rupture in relation to the existing international frameworks and a basis (initiation) for subsequent developments within their respective fields. We intentionally use the term "rupture", as suggested by Foucault (1969) to signal discontinuity: a fracturing of the linear, evolutionary process of history. It could be argued, in line with Porter (2006), Scott and Gough (2003), and Wickenberg (1999), that these two documents helped establish professional orientations and

Table 5.1 Chronological overview of international and national documents of relevance for school-based health education/promotion and education for sustainable development in Denmark

International publications	
1980–1990	**Education for sustainable development**
	Our Common Future 1987. The World Commission on Environment and Development, UN
	Health promotion
	The Ottawa Charter for Health Promotion, WHO 1986
	Adelaide Recommendations on Healthy Public Policy, WHO 1988
	Common
	Convention on the Rights of the Child, UN 1989
1990–2000	**Education for sustainable development**
	Agenda 21, chapters 25 & 36 1992, Rio. United Nations Environmental Programme (UNEP)
	Health promotion
	Sundsvall Statement on Supportive Environments for Health, WHO 1991
	The Salamanca Statement and Framework for Action. UNESCO 1994
	Jakarta Declaration on Leading Health Promotion into the twenty-first century, WHO 1997
	Conference resolution. The Health Promoting School – an investment in education, health and democracy, 1st Conference of the European Network of Health Promoting Schools, Thessaloniki-Halkidiki, Greece, 1–5 May WHO 1997
	Health 21 – Health for All in the twenty-first century, WHO/European, 1998
2000–2010	**Education for sustainable development**
	Haga Declaration, Baltic 21, 2000. Baltic Sea States' Declaration on Environment and Sustainable Development
	The UNECE Strategy for Education for Sustainable Development 2005
	UN Decade 2005–2014 for Education for Sustainable Development, UNESCO
	The EU Sustainable Development Strategy, 2006
	UNESCO Strategy for Action on Climate Change, 2008
	Bonn Declaration, UNESCO, 2009
	Learning from each other: the UNECE Strategy for ESD, UNECE, Geneva, 2009
	Health promotion
	Health Promotion: Bridging the Equity Gap, WHO 2000
	The Egmond Agenda. A tool to help establish and develop health promotion in schools and related sectors across Europe. The 2nd European Conference on health Promoting Schools, Egmond aan Zee, the Netherlands, 25–27 September 2002/WHO 2002
	The Bangkok Charter for Health Promotion in a Globalized World, WHO 2005
	The Nairobi Call to Action, WHO 2009
	Achieving health promoting schools: guidelines for promoting health in schools, IUHPE 2009
	Better Schools through Health, the 3rd European Conference on health Promoting Schools The Vilnius Resolution 2009
	Common
	United Nations Millennium Declaration, UN 2000
	Recommendation of the European Parliament and of the Council on key competences for lifelong learning, EU 2006
	Improving competences for the twenty-first century: An Agenda for European Cooperation on Schools. EU, 2008
	Council conclusions of 12 May 2009 on a strategic framework for European cooperation in education and training. ("ET 2020"), 2009

(continued)

Table 5.1 (continued)

International publications	
2010–	**Education for sustainable development**
	Climate Change Education for Sustainable Development, UNESCO, 2010
	Education for Sustainable Development, Conclusions of the Council, EU, 2010
	Learning for the future – competences for education for sustainable development, UNECE, 2012
	Health promotion
	The new European Policy for Health – Health 2020 Vision: Vision, values, main directions and approaches, WHO/Europe 2011
	Early childhood education and care: providing all our children with the best start for the world of tomorrow. Conclusions of the Council, EU, 2011
	Facilitating Dialogue between the Health and Education Sectors to Advance School Health Promotion and Education, IUHPE, 2012
	Odense statement, the 4th European Conference on Health Promoting Schools, Equity, Education and Health, CBO 2013
	Common
	Council conclusions on the social dimension of education and training, EU, 2010
	A Resolution to Promote Health, Equity and Sustainable Development in Schools, IUHPE 2012
National publications	
1990–2000	**Education for sustainable development**
	'A touch of green' ['Det grønne islæt'], foreword to the Danish Act on primary and lower secondary education, Danish Ministry of Education 1993
	Objectives and central areas of knowledge and proficiency [Formål og centrale kundskabs- & færdighedsområder]. Danish Ministry of Education 1994. Biology, science and technology, social studies etc.
	Health promotion
	Health and sex education and family studies (Sundheds- og seksualundervisning og familiekundskab). Danish Ministry of Education 1994; 1999
	Common
	Students' all-round development [Elevernes alsidige udvikling)]. Danish Ministry of Education, 1994
2000–2010	**Education for sustainable development**
	Local Agenda 21, Danish Ministry of the Environment, 2000
	Common Objectives [Fælles Mål] 2009, Danish Ministry of Children and Education; history, social studies, home economics, biology. etc.
	Education for Sustainable Development – national strategy for the United Nations Decade 2005–2014. Danish Ministry of Education 2009
	Health promotion
	Healthy for Life: National health targets and public health strategies 2002–2010. [Sund hele livet, de nationale mål og strategier for folkesundheden 2002–10]. Danish Government 2002

(continued)

Table 5.1 (continued)

National publications	
	Healthy food and physical activity in schools [Sund mad og fysisk aktivitet i skolen], Danish Ministry of Education 2004
	The government's programme for children's health [Regeringens indsats for børns sundhed]. Danish Ministry of Health 2007
	Inspiration for health education in primary and lower secondary education [Inspiration til folkeskolens sundhedsundervisning]. Danish Ministry of Education 2008
	Common
	Revision of the Danish Act on primary and lower secondary education 2003
	Revision of the Danish Act on primary and lower secondary education 2006
	Teacher training [Læreruddannelsen] 2006
	Students' all-round development. Common Objectives, subject booklet 47 [Udvikling af elevernes alsidige udvikling. Fælles Mål, Faghæfte 47]. Danish Ministry of Education 2009
2010–	**Education for sustainable development**
	The ESD portal, EMU. Danish Ministry of Children and Education 2012
	Health promotion
	Health and sex education and family studies. Common Objectives, subject booklet 21 [Sundheds-seksualundervisning og familiekundskab. Fælles Mål, Faghæfte 21] Danish Ministry of Education 2009
	Physical activity and exercise in primary and lower secondary education [Fysisk aktivitet og motion i folkeskolen], Danish Ministry of Education 2010
	Common
	New Nordic School [Ny Nordisk Skole], Danish Ministry of Children and Education 2012
	Reform of Act for Primary and Lower Secondary School [Lov om ændring af lov om folkeskolen og forskellige andre love. Lov Nr. 1640.] Danish Ministry of Education 2013

norms within education for sustainable development and health promotion. Discontinuity is evident in the introduction of new concepts and redefinition of other key concepts. For example, in the Ottawa Charter, the biomedical health concept is supplemented by an eco-holistic health concept, integrating the dimensions of positive wellbeing, emotional, social, sexual and spiritual health, as well as societal determinants of health alongside individual lifestyle. Furthermore, the concept of *settings* is introduced and recommended as a new approach to health promotion. Health is seen as created and lived by people within the settings of their everyday life; where they learn, work, play and love (WHO 1986). Similarly, in the Brundtland Report, the definition of sustainable development emphasizes the interrelation of socio-economic, cultural and environmental issues, rather than focusing solely on economic or technical perspectives. The assumption in both documents is that the development of the societies they address is unsustainable and does not promote health. Therefore, both documents emphasize the need for social change, underlining the important role of education, learning and competence development.

Table 5.2 Local transformation of international policy on health and sustainability

	Aims	Context	Strategies	Competences
International documents				
Initiation; shared values	Social justice	Rising global interdependency complexity, inequities	Collaboration	Co-production of knowledge
	Equity	Limited global resources	Facilitation	Multidimensional knowledge
	Solidarity	Health problems	Participation	Action-oriented competences
	Democracy	Environmental problems, climate change	Empowerment	
			Re-orientation of services and education	
			Setting approach	
Dissemination	Implementation	Global targets	Networks	Cross-disciplinarity
	Scaling up	Indicators	Partnerships	Knowledge based on the synergy between policy-research-practice
	Streamlining	International networking	Decentralization	Change (problem solving) orientation
	Cost effectiveness	Common standards	Innovative examples	
		Evidence imperative	Cross-sectorial work (but also)	
			Strong national governance	
			Measurement	
			Evaluation	
			Best practices	

National documents				
Transformation/Translation	Physical activity	Nordic welfare	Voluntarism	Behavior change, Attitudes and skills
	Healthy diet	Economic crisis	Cross-disciplinarity	Action competence
	Climate change education	School reform	Sharing of experience on local level	Limited participation
	Evidence-based practice	Focus on subjects	Project work	
	Focus on individual behaviour	Measurement/assessment	External partners working with schools	
		Lack of teacher professional development		
		"What works" culture		

These views of health promotion and sustainable development are framed within the context of the great challenges of the twenty-first century, as summarized in Table 5.2: growing interdependence brought about by globalization, climate change, unsustainable consumption, social inequality, poverty, chronic disease and inequity in global health. The response to these challenges is based on a foundation of shared values, endorsing equity, democracy, solidarity and social justice which are also supported by the UN Convention on the Rights of the Child, as a document influencing both fields. Health and sustainability are treated as interrelated and overlapping issues; unsustainable consumption, for example, is directly linked to (a lack of) access to clean drinking water, and cross-references are often made between health promotion and sustainable development.

5.4.1.2 Danish Context: A 'might' Instead of a 'must'

The policy frameworks in Denmark for school-based education for sustainable development and health education/promotion differ in the sense that health education has its own national curriculum guidelines (Danish Ministry of Education 2009a), with clear content, aims and teaching strategies, whereas sustainability is integrated in a number of subjects. The aims expressed in the current curriculum guidelines for health education are more or less in line with the principles found in the Ottawa Charter. Although in the time of working on this chapter they are being revised within the school reform process, the assumption is that the key aim will remain the same. It is formulated as follows:

> Teaching should contribute in every way possible to the development of pupils' ability to take a critical stance and act, both individually and in cooperation with others, to promote their own and others' health. (Danish Ministry of Education FH21 2009a, p. 4).

This passage stresses the importance for health of pupils' ability to act, whether individually or as part of a community. Clearly, this approach is linked to enablement as outlined in the Ottawa Charter. The focus is less on lifestyle and more on competence development and joint action in support of better health.

Furthermore, the curriculum explicitly proposes that the whole-school environment comprise the framework for health education and health promotion. This means that in addition to teaching and learning processes, the school leadership and the physical and psychosocial environment should be considered important if schools are to be learning communities conducive to promoting health. Again, this is consistent with the Ottawa Charter and the introduction of the concept of settings as important for health promotion.

However, not all seems to be in harmony with the international policy framework. Although health education in Denmark is among the compulsory topics integrated within the curriculum, it is a topic with no centrally allocated hours. Consequently, it is up to the local authorities (municipalities), with responsibility for schools, as well as to the individual school leader and teachers to decide when and how to integrate health education and health promotion in classroom teaching or in the everyday whole-school practices. As such, there is a risk that health education gets lost in the mix by attempting to do 'too much and too little' at the same time. This risk is confirmed by recent research on health education and promotion in Danish schools which indicates that, although health education is a compulsory topic, many teachers are not aware of the existence of the national curriculum guidelines for *Health, sexual and family education,* and, even when they are, they are not familiar with its content (Nordin 2013; Smidt 2012). This is not helped by the fact that health education and promotion are not included in the compulsory subjects for pre-service teacher education (Høj et al. 2011), nor in the systematic in-service professional development, despite the expectation that all teachers in Denmark, regardless of subject specialisation, are supposed to be able to teach health education. The practice field has mainly developed through externally financed educational interventions aimed at health promotion initiated by municipalities (Nordin 2013) and other organizations in the local community following national recommendations by the Ministry of Health (Justiniano et al. 2010), or through international initiatives (Simovska et al. 2012; Simovska 2013). Often, the aims and desired outcomes in such interventions clash with the broader values advocated for in policy documents and with the clear educational agenda. Most interventions have clear objectives for health and health behaviour, which are to be achieved over the course of a limited period of time, limiting the possibilities for working with participatory and action-oriented educational strategies as suggested in the national curriculum guidelines.

With education for sustainable development, the situation is somewhat different. Sustainability is part of the curriculum within a number of subjects; e.g. science and technology, social studies, history, geography, home economics, health education and design. In each case, sustainability learning objectives are outlined in accordance with the overall aims for the subject. The aims presented in the curriculum guidelines for social studies, for instance, include:

> The teaching should lead to the development of the requisite knowledge and skills among pupils to discuss sustainable development in light of economic growth and the environment (Danish Ministry of Education, FH5 2009b, p. 5).

The concept of *sustainable development* has secured a foothold in existing national curricula; however the ideas and principles of *education* for sustainable development are not unfolded. Furthermore, as is the case with health education, there is no systematic teacher training within education for sustainable development in Denmark, neither pre- nor in-service. Another similarity to health education is that the practice field within education for sustainable development has mainly developed through sporadic educational development projects and 'theme weeks' in schools (Læssøe et al. 2009). The difference is that, in the case of education for sustainable development, the developments are not primarily initiated by municipal departments, but mainly by environmental NGOs. Consequently, recent research has shown that Danish schoolteachers still express uncertainty about the meaning of the concept of sustainable development and appropriate teaching strategies (Madsen 2013; Breiting and Schnack 2009).

When considering curriculum guidelines it is important to also take into account the issues of assessment and evaluation. As neither health education nor education for sustainable development are part of compulsory national testing and examinations, the responsibility for working with these issues, as mentioned above, is left to municipalities, school leaders and individual teachers. In principle, municipalities, schools and teachers in Denmark have considerable autonomy in terms of planning and developing pedagogical approaches and local teaching plans under the umbrella of the common learning objectives within national guidelines (Danish Ministry of Education 2009a, b). However, with the growing influence of national and international comparative assessments, quality assurance mechanisms, and the culture of "what works", educational areas which, while mandatory, are not subject to evaluation or examination are prone to neglect. There is a risk that working with health education and education for sustainable development in schools is perceived as something which '*might* be included', rather than something which '*must* be included' in the core pedagogical practice.

At the same time, even when it does exist, evaluation within these two fields of practice can be challenging given the dominant evaluation and evidence discourses (Simovska and Carlsson 2012). From a critical educational perspective, which underpins both health education and education for sustainable development, it is problematic to focus evaluation solely on the sort of narrowly defined learning outcomes which lend themselves to measurement and performance comparison (e.g. knowledge and/or skills) while neglecting related values, comprehensive competences and critical awareness. In other words, in order to remain consistent with the values endorsed in both international and national policy documents in both fields, the approach to evaluation within health education and education for sustainable development needs to broaden the question of "what works" by asking "what works for whom", as well as "how does it work and in which circumstances" (see Carlsson and Simovska 2009). This is clearly in contrast with an increasingly rigid evaluation culture within education and the imperative for standardized "evidence-based practice", as emphasized in the national guidelines and priorities related to the on-going school reform in Denmark (Danish Ministry of Education 2013).

Thus, the analysis shows that the Danish national curriculum for health education can only partially be regarded as a norm-supporting structure (Wickenberg 1999), particularly in terms of its conceptual and ideological foundations. The curriculum's aims and content are indeed based on an eco-holistic health concept, which includes the physical, psychological and emotional dimensions, as well as the socio-cultural determinants and living conditions, as introduced in the founding documents for health promotion. The fact that sustainable development is integrated within a number of subjects could be seen as conducive to an interdisciplinary approach to this topic in schools, although, unlike health education with its separate curriculum guidelines, without a medium for outlining such an approach in practice. On the other hand, the lack of support, whether in terms of resources, inclusion in the quality indicators for schools, or through initial or in-service professional development of teachers, could be seen as a norm-hindering structure (Wickenberg 1999), particularly in terms of resources that would allow implementation of the broader objectives and values outlined in the key international documents in both areas. The level and effectiveness of the implementation of the recommendations stated in the Ottawa Charter and Brundtland report therefore depend on the 'translation' process into local practice. Consequently, those who are actually involved in health education and education for sustainable development in schools, for example municipal education and health consultants, headmasters and individual teachers, can also be seen as "policy makers" on a micro, practice level (Lipsky 2010). While this could be seen as a positive opportunity, it still requires systematic support for schools and teachers; support which appears absent from the reforms proposed by the Danish Ministry of Education.

5.4.2 Pedagogical Strategies

5.4.2.1 International Action Plans: Dissemination and Re-orientation of Education

While the Brundtland Report and the Ottawa Charter formed a bedrock of values upon which to base subsequent health promotion and education for sustainable development efforts, the international action plans which followed can be seen as attempts to flesh out and popularise these values by outlining strategies and approaches, and suggesting methods - both on an organizational level and as teaching and learning approaches. These plans call for a re-orientation of existing education systems in line with the overall aim of social change. These documents include, for example, Agenda 21 (UN 1992); UN 2008; UNECE 2012; Sundsvall statement on Supportive Environments for Health (WHO 1991) and Jakarta Declaration on Leading Health Promotion into the twenty-first century (WHO 1997a) (see Table 5.1). The call for reorientation of education is best illustrated with the following excerpt from the UNECE Strategy for Education for Sustainable Development:

> ESD demands a reorientation [of education] away from focusing entirely on providing knowledge towards dealing with problems and identifying possible solutions (UNECE 2009, p. 18).

While referring to education for sustainable development, this outline of necessary changes is equally valid for health education. The action component is visible in the suggested shift from the transmission of knowledge towards problem solving and identifying solutions. Suggesting that existing education systems are not conducive to what Biesta (2010) would call subjectification (as additional educational dimension to qualification and socialization), at least in relation to sustainable development, another document from the same organization states:

> ... at present, education often contributes to unsustainable living. This can happen through a lack of opportunity for learners to question their own lifestyles and the system and structures that promote these lifestyles. It also happens through reproducing unsustainable

models and practice. The recasting of development, therefore, calls for the reorientation of education towards sustainable development. (UNECE 2012, p. 6).

Both passages above point towards the need for more comprehensive educational efforts that would support the development of critical competences related to sustainable development understood in a broader sense than solely economic growth.

In a similar vein, the recommendations and action plans following the Brundtland Report and the Ottawa Charter suggest that *participation and empowerment* are among the key strategies conducive to both sustainability and health promotion (Table 5.2). Empowerment is described at individual and/or group levels; participation is defined in a number of different ways, ranging from participation of the target groups in the formulation of aims (for sustainability or health) to participation as taking part in carrying out predetermined aims. As such, participation of youth at various levels of decision-making is stressed in Agenda 21, pointing to the aim of *'...encouraging the involvement of youth in project identification, design, implementation and follow-up'* (UN, Agenda 21, Chapter 25.9 g). In line with this, the Jakarta Declaration on health promotion points to participation as essential – both in decision-making and in education (WHO 1997a).

From a critical education perspective, one can argue that the perspectives on participation reflected in these international documents could signify both 'symbolic' and 'real' participation, the latter involving a certain redistribution of power in decisions and frames regarding the lives of the target groups, including pupils in schools (Simovska 2012a, b, 2013). This latter view of participation, suggesting a considerable degree of redistribution of power and expertise, also challenges universal models and approaches, as well as detailed planning and a traditional understanding of experts as *'... masters over a body of knowledge and its relevant techniques'* (Fischer 2000, p. 29).

Furthermore, the international documents within both fields emphasize the setting approach as a way forward. The setting strategy advocates including many aspects of school life in health promotion and sustainability work – from school management to the school environment to teaching practices to school building and gardens. The concepts of the Green Flag award for Eco-Schools in Denmark, coordinated by the Danish Outdoor Council, and The Schools for Health in Europe (SHE) internationally (Buijs 2009) could be seen as collaborative examples endorsing the setting approach. This is reiterated in all the conference resolutions and statement of the network, from its formation in 1997 in Thessaloniki, Greece to 2013 in Odense, Denmark. The latest conference statement within the SHE network, the Odense Statement, was published while this chapter was being written (CBO 2013). The statement is based on the proceedings of the 4th European Conference on Health Promoting Schools, which took place in Odense, Denmark in October 2013. The Odense Statement reaffirms the key values, aims and strategies from the previous work within the health promoting schools in Europe (e.g. empowerment, the whole school approach and participatory teaching strategies), with a renewed focus on research. Additionally, it is noteworthy that the Odense Statement explicitly links the health promoting schools initiative with education for sustainable development. Section B, point 4 states:

> … [Health Promoting Schools offer] support to education's contribution to sustainability – the health of the people is inextricably linked with the health of societies and of the planet (CBO 2013).

The UN Decade of Education for Sustainable Development (2005–2014), launched in 2005, with UNESCO as the lead UN agency, unfolds principles, approaches and concrete methods of *education* for sustainable development aimed at both informal and formal education. An empowerment-oriented approach is stressed, focusing on ownership, critical thinking, collaboration, action and social imagination, reflecting critical and emancipatory ideas, or ideals, of education. This is highlighted in the following extract from the UNESCO website presenting education for sustainable development:

> It [ESD] also requires participatory teaching and learning methods that motivate and empower learners to change their behaviour and take action for sustainable development. Education for sustainable development consequently promotes competencies like critical thinking, imagining future scenarios and making decisions in a collaborative way (UNESCO 2012).

Similarly, the action plans suggested in the international documents following the Ottawa Charter (see Table 5.1) propose approaches and strategies specifically in relation to health education and health promotion. As part of the Health Promoting Schools initiative, the IUHPE suggests the following nine prerequisite conditions for development: supporting policies, support from school management, cross-sector coordination between groups at the schools, analysis of existing health promoting work, clear aims and strategies at the schools, development of a charter for the schools' work with health promotion, celebration when milestones are reached, teachers' professional development, and, finally, recognition that changes take time – it might take 3–4 years to establish a health promoting school (IUHPE 2009, pp. 1–2).

Additionally, the analysis shows that the international documents within the fields of both education for sustainable development and health education/promotion which follow in the wake of the Ottawa Charter and Brundtland Report reflect a mixture of two, sometimes contradictory strategic approaches – on the one hand, local collaborative strategies and participation; on the other hand, global indicators, measurements and best practice (Table 5.2). Two tendencies can be identified in this respect:

(a) Appeals to the responsibility of national governments to formulate clear aims and strategies, as well as measurable outcomes for health promotion and sustainable development. At the same time, governments are expected to draw on softer governance approaches which give weight to collaborative processes within states, NGOs, companies, educational institutions and research institutions, and emphasise community-based approaches and knowledge exchange through network learning. This is reflected, for instance, in the UNECE strategy for ESD which stresses the importance of Ministries of Education in developing aims and strategies within the field while also underlining the key role played by local actors and local schools and the need to ensure the participation of all relevant stakeholders (UNECE 2012). Along the same lines, only this time in relation to health promotion, WHO states that:

> …governments have special responsibilities to guarantee basic and universally accepted human rights, support democratic and participatory processes, and create infrastructures and conditions which support action to address the determinants of health (WHO 2000, p. 18).

(b) An emphasis on efforts to identify common approaches and assessment tools across national borders and local communities by focusing on the development of indicators, evaluation schemes, collections of good examples and the identification of universal 'best practices'.

A shift in terminology and argumentation can be seen, especially within health promotion, from underlying social change based on arguments of equity and justice, to a growing focus on 'effective implementation' based on 'evidence based knowledge' and 'best practice', as exemplified in the Jakarta Declaration (1997) and the Bangkok Charter for Health Promotion in a Globalized World (2005). We agree with Porter's (2006) characterisation of the development from the Ottawa Charter to the later Bangkok Charter as moving from *…a 'new social movements' discourse of eco-social justice in Ottawa to a 'new capitalism' discourse of law and economics in Bangkok* (Porter 2006, p. 75).

Nevertheless, despite the growing focus on common standards and measurement, there is also an emphasis evident within both health promotion/education and education for sustainable development on a wide array of diverse approaches and methods, stressing local relevance, the importance of local actors and the socio-historical context. The diversity of the suggested approaches and methods is highlighted in a number of collections and catalogues of examples from all over the world within both health promotion and ESD, as seen, for instance, in the catalogue (UNU 2007) of examples of education for sustainable development at the global RCE webpage (www.ias.unu.edu), or in the book of examples from practice which was published following the 3rd European Conference of Health Promoting Schools in Vilnius, Lithuania (Buijs et al. 2009). These examples do not necessarily emphasize "best" practices, but feature innovative, challenging and inspiring examples to learn from.

The importance of diverse methods of practice embedded in the context, but also of sound research and evaluation, is explicitly addressed by the WHO in the report Health Promotion: Bridging the Equity Gap (WHO 2000), based on the Fifth Global WHO Health Promoting Conference in Mexico. This is best illustrated by the following excerpt:

> …it is difficult to determine a simple and universally agreed set of rules of evidence for health promotion. 'Evidence' is inevitably bound to social, political and cultural context, and will be related to the method of action, process of change and measure of outcome which are valued by the population affected by actions to promote health (WHO 2000, p. 18).

Thus, ambiguous messages are once again reflected in the international documents, giving the responsible government bodies the leeway to interpret and transform their content in different ways within national policy, which, in turn, is re-interpreted at local levels prior to realisation within classroom and/or school practices.

5.4.2.2 Danish Context: Risk of Losing the Key Educational Aspects

In 2009, the Danish (centre-right) government presented their national ESD strategy, 'Education for Sustainable Development – a strategy for the United Nations Decade 2005–2014' (Danish Ministry of Education 2009c). In line with the international policies discussed above, in the introduction, the strategy focuses on integrating sustainability education within all relevant school practices. However, in later passages, education for sustainable development is framed primarily in terms of a more general focus on the natural sciences, with reference to 'solid scientific knowledge' and 'science Bildung'.[2]

The interdisciplinary approach and the notions of empowerment, participation and collaborative approaches which are introduced in the international documents within ESD are only vaguely reflected in the Danish ESD strategy. Nature, technology and health are key terms in the strategy, and three primary objectives are stressed: (1) personal responsibility for sustainable development, (2) development should be based on a solid natural science foundation, and (3) economic growth, *ideally*, should not affect future generations or people living on the other side of the world (Danish Ministry of Education, National Strategy for the UN Decade of ESD 2009c).

Nevertheless, it would be fair to say that some efforts have been made on a national level to encourage more comprehensive and critical approaches to education for sustainable development. For example, the Danish Ministry of Education developed a website in 2012 introducing the broad concept of ESD, the international frameworks, links to key stakeholders, descriptions of field trips and collections of 'good examples' (www.ubuportalen.dk). In this way, interested schools or teachers can find resources and inspiration if they want to focus on education for sustainable development. However, as mentioned previously, with no time specifically allocated to ESD as part of the curriculum and with no formal examinations, one can argue, in line with Breiting and Wickenberg (2010), that education for sustainable development in Danish schools has been more or less restricted to a relatively small number of enthusiastic teachers, champions, or 'fiery souls' as they are called in Denmark.

In contrast, health promotion/education has had a high profile in Denmark, politically speaking, during the last decade: national action plans published by both the Ministry of Health and the Ministry of Education address health education in schools (see Table 5.1). In these documents health promotion is embedded within a framework where both parents often work full-time, and it is normal that children and young people spend a lot of time on computers etc. Thus, health promotion is predominantly related to the risks of 'lifestyle diseases', such as obesity and type 2 diabetes, emphasizing the so-called KRAM-factors (food, smoking, alcohol and physical activity), as particularly reflected in the following publications (Table 5.1):

---

[2]The term 'Bildung' refers to the German educational tradition of "Didaktik" emphasizing the broader educational aims of raising critically aware citizens, as developed by the German philosopher Wolfgang Klafki, among others.

- *Healthy for life – aims and strategies for public health 2002–2010* (Danish Government 2002),
- *Healthy food and physical activity in school* (Danish Ministry of Education 2004) and
- *Physical Activity and exercise in schools* (Danish Ministry of Education 2010).

In this sense, the approaches and methods found in national action plans and strategies related to health promotion and health education in Denmark, although consistent with the general values suggested in the Ottawa Charter, can be seen as endorsing disease prevention and individual lifestyle change rather than positive eco-holistic health promotion and critical health education. The approaches suggested in the national policy documents could be seen as using the school as an arena to reach large numbers of children and young people, and to work with predefined interventions aimed at health behaviour change. The settings approach, which treats schools and local communities as active partners in health promotion and education, that are involved in the process of formulating aims, strategies and priorities within the context of everyday school life and the main educational priorities of the school (Green and Tones 2010; Dooris 2012; Jensen 2012; Mathar 2013), seems to be "lost in translation".

In summary, the analysis of the national documents providing the framework for school practice within the fields of health education/promotion and education for sustainable development shows that, while basically consistent with the values endorsed in international documents, the national policy framework is not conducive to broader, comprehensive concepts of health and sustainability and a focus on their socio-historical contexts. By emphasizing narrowly defined concepts of sustainability and health, the focus remains on an individual rather than social change, which, consequently, seems to restrict the possibilities for fostering the educational outcomes such as critical action competences of children and young people.

5.4.3 Competences

5.4.3.1 International Framework: Co-production of Knowledge

Analysis of the international documents in relation to the various competence categories (Table 5.2) shows that an opening of the classroom towards the local community and co-production of knowledge in collaboration with external actors and organizations, e.g. local sports clubs, cultural institutions, 'green guides' and local farmers, is characteristic of both fields. As formulated in Agenda 21:

> Schools should involve schoolchildren in local and regional studies on environmental health, including safe drinking water, sanitation and food and ecosystems and in relevant activities, linking these studies with services and research in national parks, wildlife reserves, ecological heritage sites etc. (UN, Agenda 21, Chapter 36.5 e).

This facilitates cross-disciplinary and problem-based learning, working with 'real life' issues, multiple perspectives and experimental teaching approaches. Knowledge is seen as closely related to action, and competence development is related to experience, participation and action taking, as expressed in the following passage from the UNECE strategy for education for sustainable development:

> ...[the aim is to] *equip people with knowledge of and skills in sustainable development, making them more competent and confident and increasing their opportunities for acting for a healthy and productive life in harmony with nature and with concern for social values, gender equity and cultural diversity.*' (UNECE 2009, p. 16).

The analysis shows that the international action plans and guidelines within both health education/promotion and sustainability reflect a dual view of knowledge. On the one hand, the documents emphasize local knowledge and diverse ways of knowing and learning. On the other hand, they promote expert knowledge and expert-defined goals related to health promotion and sustainability. In other words, the policies point to the importance of individual knowledge about "healthy lifestyles" and "sustainable behaviour". The view on learning and competence development which is reflected in these documents could be seen as a combination of accumulative learning (adding new knowledge to already established cognitive schemes), and accommodative learning (challenge the existing pre-conceptions and understandings) (Piaget, 1946 in Illeris 2009), emphasizing cognitive insight, engagement, self-reflexivity, visions and critical sense. The development of competences and knowledge is related to both pupils and teachers in schools. The teacher also becomes learner, and teachers are described as 'agents of change', whereby the development of teachers' competences amounts to an 'empowerment' of teachers. This is seen, for instance, in the UN publication 'Learning for the Future – Competences in Education for Sustainable Development' (UN 2012). Key competences for teachers are here divided into four main categories: 'learning to know, learning to do, learning to live together and learning to be' (UNECE 2012, p. 13). An overall aim is professional teacher development through a critical, self-reflective practice.

5.4.3.2 Danish Context: Competence – For What?

The current Danish government proposed a reform of basic general education in 2012, to be implemented from August 2014. The reform was initially branded as *New Nordic School*, and with the slogan "Academic improvement of the school". The Ministry of Education published a *Manifesto for the New Nordic School* (Danish Ministry of Education 2012) outlining the main focal points of the reform. One of the 10 points in the manifesto explicitly stressed sustainability:

> By its teaching, pedagogical practice and exemplary conduct in the daily work and activities in the institutions, make children and young people co-creators of a democratic and sustainable society – socially, culturally, environmentally and economically (Danish Ministry of Education 2012, p. 10).

Interestingly, following the broad political agreement among most of the parties in the parliament, the notion of sustainability disappeared from the school reform (Danish Ministry of Education 2013). The terms sustainability, sustainable development, and education for sustainable development are all missing from the final agreement. Health, on the other hand, is mentioned a number of times, primarily in relation to physical activity and healthy diet. Although the broader concept of wellbeing is also mentioned a number of times, including the imperative that schools promote the wellbeing of all pupils, there is a clear indication that wellbeing will primarily be the subject of monitoring and documentation rather than of initiatives, whether at the whole-school or classroom level, to address determinants of psychological wellbeing and the links between wellbeing and learning outcomes.

Thus, in the Danish context, the co-production of knowledge through democratic processes of teaching and learning does seem to be reflected in national policy. However, the content of this knowledge, and the competences deemed necessary, seemingly remains rooted in the narrow categories of health behaviour (that is, knowledge about risks and behavioural change to avoid these risks). The educational concept of action competence, which implies multi-dimensional knowledge, including visions related to social changes and actual experience with initiating change, seems to be absent. The concept of sustainability and related competences and knowledge are not visible at all. The next step in the school reform process will be to translate the overall objectives into specific learning objectives for each subject. It remains to be seen whether sustainability, as well as a broader understanding of health education, will be included here.

5.5 Concluding Reflections

Based on the assumption that societies in their present form do not contribute sufficiently to sustainable and health-promoting development, the need for structural change of current school systems so as to support the development of critical competences among children and young people has been emphasized in the international policies and action plans following the Ottawa Charter and the Brundtland Report. However, such changes are more easily expressed in policy documents, in the form of intentions, calls for action, suggestions and recommendations, than as specific practices, as stressed by Stevenson (2007b), among others. A 'gap' is apparent between political intentions and aspirations on the one hand and local everyday practices on the other (Nordin 2013). Both health education/promotion and education for sustainable development compete with a number of other issues to be addressed in a busy school day, where demands by both students, parents and school management need to be met; for example, the introduction of more detailed requirements for lesson planning and quality assessment combined with a greater demands on efficiency and a reduction in the available hours for preparation (Jourdan 2011; Stevenson 2007b).

One criticism of the concepts of sustainable development and health promotion is that their broadness limits their usefulness through a process of dilution. There is a danger that sustainability and health promotion are applied as umbrella terms for a wide array of visions and initiatives about 'a better world' and 'the good life', which are not necessarily directly linked to either health or sustainability. Thus, working with either issue presents a challenge to schoolteachers in terms of establishing boundaries for relevant content.

Health promotion has been defined as *health policies X health education* (Green and Tones 2010), indicating that the one cannot fully function without the other. Similarly, sustainable development could be defined as *sustainability policies X sustainability education*. Within such definition, it can be seen as problematic if the stated aims, visions, ambitions and strategies in the policies are not followed by clear and specific efforts to strengthen the educational component in the form of explicit curriculum guidelines and priorities, in-service professional development for teachers, space for experimentation and sharing knowledge and experience. In a Danish context, strong national discourses on climate change, green growth, science education and lifestyle diseases risk reducing the fields of health education and education for sustainable development to a question of natural science education, green skills, physical activity and healthy eating to stay healthy and match labour market demands. Although mental wellbeing is mentioned in the new school reform document, it appears that the focus remains on narrowly defined lifestyle indicators, monitoring and measurement.

There seems to be a need for developing new supporting structures (Wickenberg 1999) able to help explore and generate local experiences with health education/promotion and education for sustainable development at municipal and school levels. Especially in relation to health and sustainability, which are considered among the major global societal challenges, international and supranational organizations have an interest in formulating specific recommendations and guidelines within education, even though education is normally a matter of national jurisdiction. As stressed by Kickbusch (1997), the tendency is that these international policies are expanding their domain, thereby often interfering with national policies. Referring to the WHO action plans, Kickbusch states:

> We will clearly see these types of agreements increasing, sometimes undermining national standards, sometimes going far beyond them (Kickbusch 1997, p. 279).

In the terminology of Moos (2009), international guidelines, recommendations, reports and statistics can be seen as 'soft laws'. Opposite to 'hard laws', which are legally binding, soft laws are characterized by the use of persuasion with a view to influencing the norms in a specific setting or country. However, the real influence of these international policy documents and guidelines on national education policies has been questioned in the fields of both school-based health promotion and education for sustainable development. Sterling (2000), for instance, claims that the education community has not truly responded to the calls within sustainability/environmental education and health education/health promotion:

> …we cannot expect environmental, health and other forms of 'education for change' to be effective, if they are working from a marginalized status, that is, if the dominant conception of the purpose and goals of education as a whole are largely unsympathetic to the changes called for in the international mandate (Sterling 2000, p. 254).

As pointed out by Kickbusch (1997), the international policies, guidelines and agreements often lack decision-making structures at the global level and work in various ways. Linkages between the international, national and local levels take form as various initiatives - stretching from structural initiatives, for example the European Union's education exchange programmes; to associations, for example the International Union for Health Promotion and Education (IUHPE); to networks, such as the Schools for Health in Europe Network (SHE); to collections of inspiration materials presenting 'good examples' on e-learning platforms; and international conferences aimed at exchanging knowledge and experiences. As such, the international guidelines and initiatives within education for sustainable development and health education/promotion could be said to form "*norm supporting structures*", as emphasized by Wickenberg (1999, 2004). The support is provided on two levels – on a symbolic level by highlighting, pooling resources, and prioritizing health and sustainability, and on a specific level by providing inspiration, knowledge sharing opportunities and examples of practice. Based on the framework of critical education, this support could provide space and time for joint experimentation, self-reflective practices, co-learning and collaborative processes, problem solving and opening of the school towards the local community.

In Denmark, the existing policy framework seems to narrow down the fields, as a number of action plans and strategies within health and sustainability do not fully reflect the *educational* ideas and approaches called for in some of the international policies. The analysis shows that the links between international and national policies are most clearly reflected in the curriculum guidelines for health education whereas the calls for action on social change and re-structuring of the education system do not seem to be reflected in the existing school framework or in the reform documents.

However, it is fair to acknowledge that some policy frameworks do exist which could form the foundation for future work on health education and education for sustainable development, such as the aim in the current school reform to open the school to the local community and the surrounding world, and the potential for integration of sustainability and health in relation to the specific subjects in the current reformulation of curricular aims for all the subjects.

If health education/promotion and education for sustainable development in schools are to go beyond the work of a few 'fiery souls', and be anchored systematically within the everyday life of the whole school, supporting structures, in the form of clear prioritization of the fields in national policies, followed by adequate pedagogical initiatives and approaches, including formative and realistic evaluation, seem necessary. In this sense, future research would do well to look not only at what is 'lost' when international policies are translated into a national context, but also what is 'gained' in the interplay with local practices and meaning construction at school level.

Acknowledgements The authors gratefully acknowledge the research funding within the AU IDEAS PILOT CENTRES scheme, granted by Aarhus University Research Foundation for the period 2012–2016. The AU IDEAS funding scheme aims to support visionary and innovative interdisciplinary research ideas over a period of 5 years.

References

Beech, J. (2006). Redefining education transfer: International agencies and the (Re)production of educational ideas. In J. Sprogøe & T. Winter-Jensen (Eds.), *Identity, education and citizenship – Multiple interrelations*. Frankfurt am Main: Peter Lang GmbH.

Biesta, G. J. J. (2010). *Good education in an age of measurement: Ethics, politics, democracy*. Boulder: Paradigm Publishers.

Bonnett, M. (2013). Normalizing catastrophe: Sustainability and scientism. *Environmental Education Research, 19*(2), 187–197.

Breiting, S., & Schnack, K. (2009). *Uddannelse for bæredygtig udvikling i danske skoler. Erfaringer fra de første TUBU-skoler i Tiåret for UBU* [Education for sustainable development in Danish Schools. Experiences from the first DESD schools]. Copenhagen: Research Program on Health and Environmental Education, The Danish School of Education.

Breiting, S., & Wickenberg, P. (2010). The progressive development of environmental education in Sweden and Denmark. *Environmental Education Research, 16*, 19–37.

Buijs, G. (2009). Better schools through health: Networking for health promoting schools in Europe. *European Journal of Education, 44*(4), 507–520.

Buijs, G., Jociuté, A., Paulus, P., & Simovska, V. (Eds.). (2009). *Better schools through health: Learning from practice*. Case studies of practice presented during the Third European Conference on Health Promoting Schools, held in Vilnius, 15–17 June 2009. Netherlands: Netherlands Institute for Health Promotion.

Carlsson M. S., & Simovska, V. (2009). *Fokus på evidens inden for sundhedsundervisning og sundhedsfremme* [Evidence in health education and health promotion].

Carlsson, M. S., & Simovska, V. (2012). Exploring learning outcomes of school-based health promotion: A multiple case study. *Health Education Research, 27*(3), 437–447.

Carlsson, M., Simovska, V., & Jensen, B. B. (Eds.). (2009). *Sundhedspædagogik og sundhedsfremme. Teori, forskning og praksis* [Health pedagogy and health promotion: Theory, research and practice]. Aarhus: Aarhus University Press.

CBO. (2013). *The conference report and the Odense statement*. The 4TH European Conference on Health Promoting Schools, Equity, Education and Health. Retrieved March 24, 2014, from http://schools4health.dk/

Clift, S., & Jensen, B. B. (Eds.). (2005). *The health promoting school: International advances in theory, evaluation and practice*. Copenhagen: Danish University of Education Press. on behalf of the European Network of Health Promoting Schools.

Danish Government. (2002). *Sund hele livet, de nationale mål og strategier for folkesundheden 2002–10* [Healthy for life – Aims and strategies for public health 2002–2010]. Retrieved March 24, 2014, from http://www.sum.dk/Aktuelt/Publikationer/~/media/F7BABB17699E42B4A11623E137D73D0C.ashx

Danish Ministry of Children and Education. (2012). *Ny Nordisk Skole* [New Nordic School]. Retrieved March 24, 2014, from http://nynordiskskole.dk/

Danish Ministry of Education. (2004). *Sund mad og fysisk aktivitet i skolen* [Healthy food and physical activity in schools]. Retrieved March 24, 2014, from http://pub.uvm.dk/2004/sundmad/

Danish Ministry of Education. (2008). *Inspiration til folkeskolens sundhedsundervisning* [Inspiration for health education in primary and lower secondary education]. Retrieved March 24, 2014, from http://pub.uvm.dk/2008/sundhedsuv/

Danish Ministry of Education. (2009a). *Sundheds-seksualundervisning og familiekundskab. Fælles Mål, Faghæfte 21* [Health and sex education and family studies. Common objectives, subject booklet 21]. Retrieved March 24, 2014, from http://www.uvm.dk/

Danish Ministry of Education. (2009b). *Samfundsfag, Fælles Mål, Faghæfte 5.* [Social studies. Common objectives, subject booklet 5]. Retrieved March 25, 2014, from http://uvm.dk/Service/Publikationer/Publikationer/Folkeskolen/2009/Faelles-Maal-2009-Samfundsfag

Danish Ministry of Education. (2009c). *National Strategy for the UN Decade of ESD*. Retrieved March 24, 2014, from http://pub.uvm.dk/2008/ubustrategi/

Danish Ministry of Education. (2010). *Fysisk aktivitet og motion i folkeskolen* [Physical activity and exercise in primary and lower secondary education]. Retrieved March 24, 2014, from http://www.uvm.dk/Service/Publikationer/Publikationer/Folkeskolen/2010/Fysisk-aktivitet-og-motion-i-folkeskolen

Danish Ministry of Education. (2012). *UBU portalen* [ESD webpage]. Retrieved March 14, 2014, from www.ubuportalen.dk

Danish Ministry of Education. (2013). *Lov om ændring af lov om folkeskolen og forskellige andre love.* Lov Nr. 1640 [Reform of Act for Primary and Lower Secondary School]. Law No. 1640. Retrieved March 24, 2012, from http://www.uvm.dk/

Danish Ministry of Education/The Danish UNESCO Commission. (2014). *UBU portalen* [ESD Webpage for teachers]. Retrieved March 24, 2014, from http://www.ubuportalen.dk/

Danish Ministry of Health. (2007). *Regeringens indsats for børns sundhed* [The government's programme for children's health]. Retrieved March 14, 2014, from http://www.sum.dk/~/media/Filer%20-%20Publikationer_i_pdf/2007/boern_og_unge_sundhed.ashx

Davis, J., & Cooke, S. (2007). Educating for a healthy, sustainable world: An argument for integrating health promoting schools and sustainable schools. *Health Promotion International, 22*, 346–353.

Davis, J. M., Cooke, S. M., Blashki, G., & Best, A. (2010). Healthy children, healthy planet: The case for transformative education in schools and early childhood from an Australian perspective. *International Public Health Journal, 2*(4), 561–570.

Dooris, M. (2012). Settings for promoting health. In L. Jones & J. Douglas (Eds.), *Public health: Building innovative practice*. London/Milton Keynes: Sage/Open University Press. ISBN: 9781446207741.

Firth, R., & Smith, M. (2013). Editorial: As the UN Decade of Education for Sustainable Development comes to an end: what has it achieved and what are the ways forward? *The Curriculum Journal, 24*(2), 169–180.

Fischer, F. (2000). *Citizens, experts, and the environment. The politics of local knowledge*. Durham/London: Duke University Press.

Foucault, M. (1969). *The archaeology of knowledge*. London: Routledge.

Gherardi, S. (2006). *Organizational knowledge. The texture of workplace learning*. Oxford: Blackwell Publishing.

Green, J., & Tones, T. (2010). *Health promotion: Planning and strategies* (2nd ed.). London: Sage.

Høj, B. B., Esmer, S., Bertelsen, K., &Paustian, P. (2011). ;*Sundhedsundervisning i læreruddannelsen – hvordan står det til?* [Health education in the initial teacher education: State of the art], *Det nationale videncenter KOSMOS* [National Knowledge Center KOSMOS]. Retrieved March 24, 2014, from http://www.vicekosmos.dk/

Illeris, K. (Ed.). (2009). *Contemporary theories of learning. Learning theorists in their own words*. London/New York: Routledge.

IUHPE. (2009). *Achieving health promoting schools: Guidelines for promoting health in schools.* Retrieved March 24, 2014, from http://www.dhhs.tas.gov.au/__data/assets/pdf_file/0011/115895/guidelines_for_health_promoting_schools1.pdf

Jensen, B. B., Schnack, K., & Simovska, V. (red.) (2000). *Critical Environmental and Health Education: Research Issues and Challenges.* Copenhagen: Royal Danish School of Educational Studies.

Jensen, J. M. (2012). Sundhedsfremmende settings [Health promoting settings]. In: V. Simovska & J. Magne Jensen (Eds.), *Sundhedspædagogik i sundhedsfremme* [Health pedagogy in health promotion]. Copenhagen: Gads Publisher.

Jensen, B. B., & Simovska, V. (2005). Involving students in learning and health promotion processes: Clarifying why? what? and how? *Global Health Promotion, 12*(3–4), 150–156.

Jourdan, D. (2011). *Health education in schools. The challenge of teacher training*. Saint-Denis: Inpes, Coll. Santé en action.

Justiniano, B., Albeck, C., & Simovska, V. (2010). Rum og fysisk aktivitet i børnehøjde. [Space and physical activity in the eyes of the children] *Cursiv, 5*, 101–128.

Kickbusch, I. (1997). Think health. What makes the difference? *Health Promotion International, 12*(4), 265–272.

Læssøe, J., Schnack, K., Breiting, S., & Rolls, S. (2009). *Climate change and sustainable development: The response from education*. A cross-national report from international alliance of leading education institutes. Copenhagen: The Danish School of Education, Aarhus University Press.

Lipsky, M. (2010). *Street-level Bureaucracy. Dilemmas of the individual in public services*. New York: Russell Sage Foundation.

Lotz-Sisitka, H. (2007). An opening dialogue with think pieces and feature articles on learning in a changing world. *Southern African Journal of Environmental Education, 24*, 8–18.

Madsen, K. D. (2013). Unfolding education for sustainable development as didactic thinking and practice. *Sustainability, 5*, 3771–3782.

Mathar, R. (2013). *The concept of whole school approach. A platform for school development with focus on sustainable development* (Concept Paper). Retrieved March 24, 2014, from http://esd-expert.net/assets/130314-Concept-paper-ESD-Whole-school-approach-general-introduction.pdf

Moos, L. (2009). A general context for new social technologies. *Nordic Studies in Education, 29*(1), 79–92.

Nordin, L. L. (2013). *Fra politik til praksis: Implementering af kommunale sundhedsfremmeprojekter med fokus på kost og fysisk aktivitet fra et lærerperspektiv* [From policy to practice: The implementation of municipality health promoting projects focusing on healthy food and physical activity.]. Ph.D. dissertation, Aarhus University, Copenhagen.

Nutbeam, D. (2008). The evolving concept of health literacy. *Social Science & Medicine, 67*(12), 2072–2078.

Patrick, R., Capetola, T., Townsend, M., & Nuttman, S. (2012). Health promotion and climate change: Exploring the core competencies required for action. *Health Promotion International, 27*(4), 475–485.

Porter, C. (2006). Ottawa to Bangkok: Changing health promotion discourse. *Health Promotion International, 22*, 72–79.

Reid, A., Jensen, B. B., Jutta, N., & Simovska, V. (2008). Participation and learning: Developing perspectives on education and the environment, health and sustainability. In A. Reid et al. (Eds.), *Participation and learning: Perspectives on education and the environment, health and sustainability*. Dordrecht: Springer.

Rein, M., & Schön, D. A. (1981). Problem setting in policy research. In C. H. Weiss (Ed.), *Using social research in public policy making*. New York: Lexington Books.

Samdal, O., & Rowling, L. (Eds.). (2013). *The implementation of health promoting schools: Exploring the theories of what, why and how*. London/New York: Routledge Falmer.

Schnack, K. (2003). Action competence as educational ideal. In T. Bonna, E. D. William, W. Hongyu & F. P. William (Eds.), *The internationalization of curriculum studies*. New York: Peter Lang Publishing Group.

Schön, D. (1983). *The reflective practitioner – How professionals think in action*. New York: Basic Books, Inc.

Scott, W., & Gough, S. (2003). *Sustainable development and learning. Framing the issues*. London: Routledge.

Simovska, V. (2012a). Editorial. Processes and outcomes in school health promotion: Engaging with the evidence discourse. *Health Education, 112*(3), 196–199.

Simovska, V. (2012b). Editorial. What do health-promoting schools promote? Processes and outcomes of health-promoting schools. *Health Education, 122*(2), 84–87.

Simovska, V. (2012c). Case study of a participatory health promotion intervention in school. *Democracy & Education, 20*(1), article 4.

Simovska, V. (2013). The role of health education in addressing the health divide: Evidence from two European health-promotion projects employing a participatory and action-oriented education approach. In: V. Barnekow, B. B. Jensen, C. Currie, A. Dyson, N. Eisenstadt, & E. Melhuish (Eds.), *Improving the lives of children and young people* (Vol. 2, Childhood, s. 2–13). Chapter 2.1. Copenhagen: WHO, Regional Office for Europe.

Simovska, V., & Carlsson, M. S. (2012). Health-promoting changes with children as agents: Findings from a multiple case study research. *Health Education, 112*(3), 292–304.

Simovska, V., Dadaczynski, K., & Woynarowska, B. (2012). Healthy eating and physical activity in schools in Europe: A toolkit for policy development and its implementation. *Health Education, 112*(6), 513–524. doi:10.1108/09654281211275863.

Smidt, L. (2012). *Undersøgelse af seksualundervisningen i grundskolen* [Exploring sex education in primary and lower secondary education]. Sex og Samfund. Retrieved March 24, 2014, from http://www.sexogsamfund.dk/Default.aspx?ID=31130

Sterling, S. (2000). The significance of system thinking to environmental education, health education and beyond. In B. B. Jensen, K. Schnack, & V. Simovska (Eds.), *Critical environmental and health education. Research issues and challenges*. Copenhagen: Research Centre for Environmental and Health Education, The Danish University of Education.

Stevenson, R. B. (2006). Tensions and transitions in policy discourse: Recontextualizing a decontextualized EE/ESD debate. *Environmental Education Research, 12*(3–4), 277–290.

Stevenson, R. B. (2007a). Editorial. *Environmental Education Research, 13*(2), 129–138.

Stevenson, R. B. (2007b). Schooling and environmental/sustainability education: From discourses of policy and practice to discourses of professional learning. *Environmental Education Research, 13*(2), 265–285.

United Nations. (1987). *Report of the World Commission on environment and development: Our common future*. Oxford: Oxford University Press.

United Nations. (1989). *Convention on the Right of the Child*. Retrieved March 24, 2014, from http://www.ohchr.org/en/professionalinterest/pages/crc.aspx

United Nations. (1992). *Agenda 21*, chapters 25 & 36. United Nations Environmental Programme (UNEP). Retrieved March 24, 2014, from http://sustainabledevelopment.un.org/content/documents/Agenda21.pdf

United Nations. (2008). *United Nations Decade of Education for Sustainable Development 2005–2014*. Retrieved March 22, 2014, from http://www.desd.org/About%20ESD.htm

United Nations Economic Commission for Europe. (2009). *Learning from each other: The UNECE strategy for ESD*. Geneva: UNECE. Retrieved March 24, 2014, from http://sustainabledevelopment.un.org/content/documents/798ece5.pdf

United Nations Economic Commission for Europe. (2012). *Learning for the future – Competences for education for sustainable development*. Retrieved March 24, 2014, from http://www.unece.org/environmental-policy/areas-of-work/education-for-sustainable-development-esd/education-for-sustainable-development-esdpublicationshtml/education-for-sustainable-development/2012/competences-for-esd/docs.html

United Nations Educational, Scientific and Cultural Organization (UNESCO). (2012). *ESD Webpage*. Retrieved March 24, 2014, from http://www.unesco.org/new/en/education/themes/leading-the-international-agenda/education-for-sustainable-development/browse/2/

United Nations University. (2007). *Regional Centres of Expertise (RCE)*. Retrieved March 24, 2014, from http://www.ias.unu.edu/sub_page.aspx?catID=108&ddlID=183

Vilnius Resolution. (2009). *Better schools through health*. The third European conference on health promoting schools, 15–17 June 2009, Vilnius. Retrieved February, 2014, from http://ec.europa.eu/health/ph_determinants/life_style/mental/docs/vilnius_resolution.pdf

Wals, A. (2010). DESD we can? Some lessons learnt from two mid-DESD reviews. *Global Environmental Research, 14*, 109–118.

Wickenberg, P. (1999). *Normstödjande strukturer. Miljötematiken börjar slå rot i skolen* [Norm supporting structures. Environmental issues anchored in schools]. Lund. Studies in Sociology of Law, 5.

Wickenberg, P. (2004). Norm supporting structures – In environmental education and education for sustainable development. In P. Wickenberg et al. (Eds.), *Learning to change our world? Swedish Research on Education and Sustainable Development*. Lund: Studentlitteratur.

World Health Organization. (1986). *Ottawa charter for health promotion* (WHO/HPR/HEP/95.1). Geneva: WHO.

World Health Organization. (1988). Adelaide recommendations on healthy public policy. Retrieved March 24, 2014, from http://www.who.int/healthpromotion/conferences/previous/adelaide/en/

World Health Organization. (1991). *Sundsvall statement on supportive environments for health*. Sundsvall. Retrieved March 24, 2014, from http://www.who.int/healthpromotion/conferences/previous/sundsvall/en/

World Health Organization. (1997a). *Jakarta declaration on leading health promotion into the 21st century*. Retrieved March 24, 2014, from http://www.who.int/healthpromotion/conferences/previous/jakarta/declaration/en/

World Health Organisation. (1997b). *Conference resolution. 1st conference of the European Network of Health Promoting Schools*. Copenhagen: WHO Regional office for Europe.

World Health Organisation. (2000). *Health promotion: Bridging the equity gap*. Mexico. Retrieved March 24, 2014, from http://www.who.int/healthpromotion/conferences/previous/mexico/en/

World Health Organisation. (2002). *The Egmond Agenda. A tool to help establish and develop health promotion in schools and related sectors across Europe*. Egmond aan Zee, the Netherlands, 25–27 September 2002//ENHPS Technical Secretariat. Copenhagen: WHO Regional Office for Europe.

World Health Organisation. (2005). *The Bangkok charter for health promotion in a globalized world*. Retrieved March 22, 2014, from http://www.who.int/healthpromotion/conferences/6gchp/hpr_050829_%20BCHP.pdf?ua=1

World Health Organisation. (2009). *The Nairobi call to action*. Retrieved March 22, 2014, from http://gchp7.info/nc2ace.pdf

Part II
Partnerships, Standards and Change

Chapter 6
Health Promotion in Dutch Secondary Schools: Promising Collaboration Between School and Public Health Services?

Nicole M.W.M. Boot, Maria W.J. Jansen, Mariken T.W. Leurs, and Nanne de Vries

Abstract The attention and need for evidence of school health promotion in the Netherlands are growing. This chapter examines the extent to which the School Beat approach being implemented in secondary schools in Southern Limburg, and what are the success factors, shortcomings and bottlenecks of sustained school health promotion? Eighteen health promoting schools were followed during a period of 4 years. A multi-method research approach was used including questionnaires and interviews. The theoretical framework of this research builds on the Diagnose Sustainable Collaboration (DISC) model which was developed by Leurs (Remediaal 4(1):3–8, 2003).

Secondary schools do consider health promotion as important. But, because of the high workload in schools and the lack of legal commitment health promotion has not been given high priority in schools, so far. Despite this position schools try to address health with support of their regional Public Health Service. For successful

N.M.W.M. Boot, Ph.D. (✉)
Academic Collaborative Centre for Public Health Limburg,
Maastricht, The Netherlands
e-mail: Nicole.Boot@ggdzl.nl

M.W.J. Jansen
Department of Health Promotion, Regional Public Health Service (GGD) South Limburg,
Geleen, The Netherlands
e-mail: Maria.jansen@ggdzl.nl

M.T.W. Leurs, Ph.D.
Centre for Healthy Living, Dutch Institute for Public Health and the Environment (RIVM),
Bilthoven, The Netherlands
e-mail: mariken.leurs@rivm.nl

N. de Vries
Department of Health Promotion, CAPHRI, Universiteit Maastricht,
Maastricht, The Netherlands
e-mail: n.devries@maastrichtuniversity.nl

V. Simovska and Patricia Mannix-McNamara (eds.), *Schools for Health and Sustainability: Theory, Research and Practice*, DOI 10.1007/978-94-017-9171-7_6,

implementation of school health promotion the appointment of a school health promotion advisor is crucial. Competence requirements for the implementation of school health promotion are related to organizational and educational knowledge. There is no natural connection between health and educational goals in schools. Investment in advocacy, school commitment, seeking connections between the educational and health goals and shared responsibility between the educational and public health sector can make a difference. Currently the attention for school health promotion policy is increasing. Nevertheless only little evidence-based school health policy is available. This research provides recommendations for the implementation of school health policies in the Netherlands and internationally.

Keywords School health policy • Health promotion • Secondary schools • School Beat approach • Collaboration

6.1 Introduction

Children's health has long been a frequently discussed topic in the world of education (Aggleton et al. 2000). This started with the development of rules pertaining to the hygiene of students and classrooms in the early 1800s, followed by international attention for school health services around the middle of the twentieth century and a change from health education to health promotion in the 1980s (Leurs 2008), culminating in the framework of the Health Promoting School. The Health Promoting School (HPS) concept combines health and educational goals through collaborative partnerships but also targets the physical and social environments in and around school, supported by community activities and health services (St Leger 1999; WHO 1998a, b). Broader frameworks for school health promotion are supposed to generate even greater health gain for children (Lister-Sharp et al. 1999; Stewart-Brown 2001, 2006; St Leger and Nutbeam 1999, 2000), and higher academic performances (Suhrcke and de Paz Nieves 2011). This HPS approach has been adopted in many countries (Marx and Wooley 1998; Clift et al. 2005; St Leger 2004), including the Netherlands (Leurs et al. 2002, 2005a; Bos et al. 2010).

To strengthen the position of health promotion in and around schools, a new method for school health promotion has been developed since 2002 in the Netherlands, commonly referred to as the School Beat approach ("Schoolslag" approach). This method places greater emphasis on the establishment and monitoring of sustained intersectoral collaborative support for comprehensive school health promotion (Leurs et al. 2005b).

This chapter will discuss the development of the Dutch interactive School Beat approach for Health Promoting Schools, detailing its success factors, shortcomings and impediments for intersectoral collaboration, based on a case study in the southern part of the Netherlands. The chapter will also detail opportunities for successful and sustained implementation of comprehensive school health promotion.

6.2 The School Beat Method: What to Do

The School Beat approach is a six-step method that thoroughly and systematically elaborates on the *content* of the Health Promoting School framework. The main question during the six steps procedures that characterize School Beat is about *what to do*, underpinned by why, when and by whom. The six steps are delineated as follows:

Step 1: Determining the health needs at the school
Step 2: Setting health promoting priorities
Step 3: Identifying important and modifiable determinants
Step 4: Designing the health promoting school plan
Step 5: Implementation
Step 6: Evaluation

The first step involves determining the school's health needs based on available epidemiological data about the students, as well as data on their academic performance and other relevant information obtained from the school. Step two involves setting health promotion priorities based on the information gathered in the first step. The opinions of the school team, parents and students all play an important role. In step three, important and modifiable determinants of priority problems are identified by the school, in order to select possible activities focusing on students, staff, school, school policy and the community. In step four, the school writes a school health plan on the basis of all the previous steps. In order to structure the program and the choice of activities, the (WHO) Health Promoting Schools approach and the corresponding American Coordinated School Health Program (Marx and Wooley 1998) were adapted to the Dutch situation (see Fig. 6.1).

Fig. 6.1 The "Schoolslag" interpretation of the Healthy School Model

Step five involves the implementation of the school health plan and activities. Activities can be implemented through (1) the school curriculum, teaching, and learning; (2) the school environment, and organization; and (3) school-community partnerships and services. The final step consists of an evaluation to measure the interventions's output and outcomes. A comprehensive description of the School Beat method has been published by Leurs (2008).

6.3 The DISC Monitor: How to Do

Special attention to the *process* is the rationale for the monitor to 'DIagnose Sustainable Collaboration' (DISC) which was developed by Leurs (2003). The main questions when using the DISC monitor are about *how to do, i.e. how to collaborate,* underpinned by why, when and by whom. Collaborations and partnerships are essential to Health Promoting Schools as schools deal with multi-party problems on a continuous basis (Andis et al. 2002; De Leeuw 1989; Gray 1989). Intersectoral collaboration in which people from different disciplines and cultures are expected to work together. This is clearly not without a distinct set of challenges (Allensworth 1987; Padgett et al. 2004; Van Eyk and Baum 2002). Interest in the process and prerequisites of collaboration in organizing Health Promoting Schools, is rising (Deschesnes et al. 2003, 2010; El Ansari and Weiss 2006). Underlying theories and principles of organizational change are widely available from other sectors (Cummings and Worley 2001; De Caluwé and Vermaak 2003; Van Raak 1998) and can serve to effectively inform health promotion implementation in schools.

Based on existing knowledge from other sectors and disciplines, the School Beat-partners developed what has become known as the DISC monitor (Leurs et al. 2005a). The DISC monitor specifies collaborative support at three levels: perceptions (what is the opinion of the collaborative organizations about collaboration?); intentions (do the parties intend to collaborate?); and actions (are the parties actually going to collaborate?). The DISC monitor functions to describe the state of an intersectoral collaboration at a given moment in time, aiming to reveal opportunities and impediments for change. A thorough analysis of the current status of the collaboration supports the selection of suitable change strategies to enhance the development of the collaboration (De Caluwé and Vermaak 2003). The monitor was first studied in an exploratory study in 2004. This study clearly supported the value of using a systematic approach to monitoring the process of intersectoral collaboration between organizations from public health, welfare, mental health and addiction and governments (Leurs et al. 2008). Some adjustments were made to improve the DISC monitor and this version was used in a follow-up study of Boot (Boot 2011). The DISC monitor enables practitioners to diagnose the progress of an innovation, *i.e.* the School Beat method, from adoption to implementation, routinization and sustainability (Fig. 6.2).

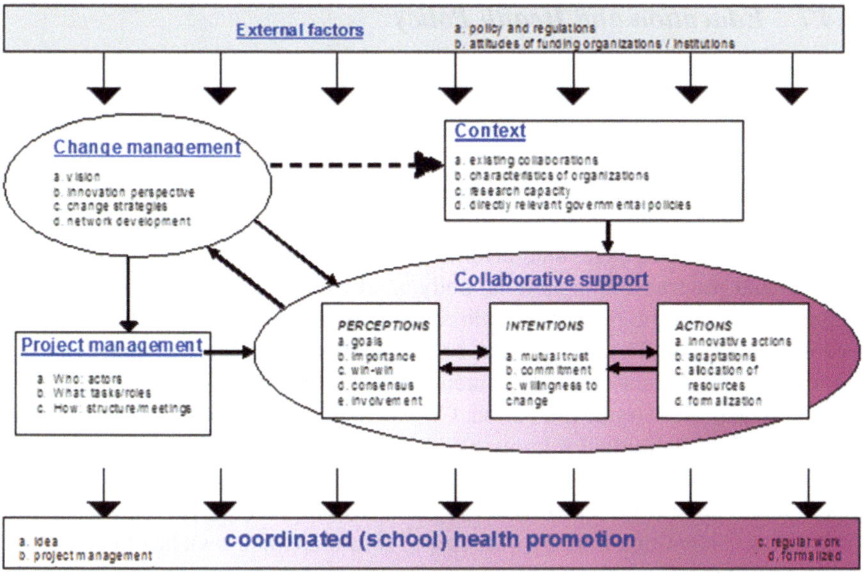

Fig. 6.2 The DIagnosis of Sustainable Collaboration (DISC) model (Leurs et al. 2005b)

6.4 The Dutch Example

The case study of the Southern Limburg region of the Netherlands, conducted in 2008–2011 serves to illustrate well the current situation of Health Promoting Schools in the Netherlands. The Southern Limburg region covers an area of 66,056 km^2, has an approximate population of 600,000, and has 25 secondary schools with approximately 40,000 students. The School Beat approach was introduced in Southern Limburg in 2002. The main aim of the program was to reduce adolescents' health risk behavior by guiding secondary schools in establishing planned and systematic school health promotion policies (Boot 2011). Since its inception in 2002, the implementation process has been combined with evaluative studies, to monitor progress and identify the need for adjustments. In 2008, a follow-up study began to give insights into characteristics of systematic school health promotion in the Netherlands.

The following research questions were addressed:

1. To what extent are the six steps of the School Beat approach being implemented at secondary schools in Southern Limburg?
2. Is there a relationship between the level of school satisfaction with the assistance offered by the HPS advisor, and progress in the implementation process of the School Beat approach?
3. What factors influence successful intersectoral collaboration and implementation of the School Beat approach in secondary schools?

6.4.1 Education and Health Policy

The Dutch Ministry of Education, Culture and Science is responsible for secondary education in the Netherlands. To this end, the Ministry has formulated core objectives related to language, arithmetic, science, society, art, culture, exercise, sports, recently supplemented with healthy living (for vocational schools only), sexuality and sexual diversity (SLO 2007; MBO Raad 2010). Prime focus however is on language and arithmetic. The objectives are translated into educational programs by a national body and then delegated to the local level of school boards. In the Netherlands there are 356 of such boards, managing 1,290 secondary schools. The main task of school boards is to create the conditions that enable individual schools to implement the national core objectives (SLO 2007; VO-Raad and Ministry of Education, Culture and Science 2008; VO-Raad 2009). At the level of individual schools, the national objectives have to be translated into practical activities and included in each school's own policy, according the operational conditions established by the school board (Turkenburg 2008). Within these conditions, schools are free to formulate their own health promoting policies and aims.

Although health promotion is not yet a standard part of the educational objectives, schools are nevertheless obliged by their municipalities to establish a care structure for individual students with health and social problems (Bosdriesz and Berkenbosch 2003; Wijsmuller 2002). To this end, each school has to appoint a pupil care coordinator (referred to as PC coordinator below) to guide students who encounter individual problems that affect their school achievements and behavior, supported by a care team of representatives of various health and youth care organizations (in the domains of welfare, crime, alcohol and drugs, mental health). This is an extra-curricular position, funded by the local government. This approach is mostly health-care driven, with little attention given to health promotion (Van Der Steenhoven and Van Veen 2009). The school care structure is the responsibility of the Ministry of Health, Welfare and Sport. This Ministry also mandates local governments to monitor the health status of young people, assess care needs and provide health education, advice and counseling, which may legitimize the role of the school care organizations in introducing health promoting policies in secondary education (Ministry of Health, Welfare and Sport 2011).

In summary, at national level two separate ministries are responsible for respectively education policy and health policy. At local level school boards and local governments are responsible for respectively education policy and health policy. In this interplay of responsibilities, collaborative implementation of school health promoting policies depends on the informal goodwill of the parties involved. The DISC monitor then can be a useful tool to facilitate practitioners to unravel the collaboration process between the different partners participating in the School Beat innovation.

6.4.2 Joined Forces

Health promotion does not have a structural position within the Dutch educational system. The position of school health promoting policies is somewhere between the legislative spheres of the Ministry of Health, Welfare and Sport and the Ministry of Education, Culture and Science. Since Dutch legislation does not link health and education, such a link is also lacking in the way these two fields are organized. Recently, the two ministries have shown more interest in intersectoral collaboration, for example in the national prevention programme of the Ministry of Health, Welfare and Sport to be established during 2013.

Although schools feel responsible for their students' health, they have limited resources available to promote it. In addition many prevention organizations approach schools, offering programs to solve social and health problems, so the educational system is swamped by requests. Due to this workload, schools often at best give some *ad hoc* attention to health promotion, but mostly focus on individual pupil care as far as health is concerned (Bos et al. 2010). In response to this situation, all regional organizations in Southern Limburg (The Netherlands) specialized in health, welfare, safety, individual pupil care and youth care joined forces in 2002. This group, coordinated by the regional Public Health Service (in Dutch: GGD) aimed to improve the role of Health Promoting Schools (Leurs et al. 2002, 2005a). All organizations were able to offer services to the schools based on their government financing; schools only had to pay for materials that might be used in lessons and for their own staff time (e.g. for meetings with a HPS advisor). All 25 secondary schools in the South Limburg region participate in the School Beat method since 2002.

Coordinated by the local GGD a so-called Health Promoting School advisor (HPS advisor) is appointed for each school from among its own staff. The assistance offered by the HPS advisor generally has two main components, the first relating to the actual introduction and elaboration of the steps of the School Beat method (referred to below as *content* assistance), and the second relating to identifying and advising on factors in the organizational context, such as external factors, change management, context, project management and collaboration (referred to below as *process* assistance). In total 12 HPS advisors were appointed to work with the 25 schools.

While shaping the school's health promoting activities, the HPS advisor cooperates with the school but also provides them with links to other agencies and organizations, matching needs to programs and interventions that are available (e.g. within the judicial, mental health or welfare systems). As such, collaboration is an important aspect of the School Beat approach.

6.5 Method

To answer the research questions, health promoting schools in Southern Limburg were followed over a period of 4 years using a multiple method research approach involving questionnaires and interviews. The PC coordinators (see Sect. 4.1 above),

Table 6.1 Response

Function	Questionnaires	Response	%
PC coordinators	50	39	78
School managers	165	59	36
Teachers	1,592	399	25
HPS advisors	18	12	67

school managers, teachers and HPS advisors of the participating schools filled out a questionnaire to measure the progress of the School Beat steps (response see Table 6.1). The questionnaire consisted of 20 statements addressing the implementation of the six steps of the School Beat approach. A step was considered to have been fully implemented if all statements regarding that step were answered affirmatively by both the representatives of the school and the HPS advisor. Furthermore, nine (five-point Likert-type) questions were asked about satisfaction with the *content* assistance offered by the HPS advisor (Cronbach's alpha = 0.896). Ten (three-point) questions referred to the level of satisfaction with the *process* assistance of the HPS advisor (Cronbach's alpha = 0.883).

Subsequently, group interviews were held in each of the 18 schools, represented by their PC coordinator, the school manager, and their school's HPS advisor. The main purpose of the group interviews was to get more detailed information about the process of implementation and collaboration, using the DISC monitor. Interviews were transcribed verbatim and analyzed using NVivo 8.

To illustrate the real discrepancies, we present the interview results of the two extreme cases in terms of the degree of implementation, *i.e.* the best and worst cases. This selection was based on the questionnaire using the 20 statements on the School Beat steps. As the worst case we selected the school that had implemented the fewest School Beat steps, while the best case was the school that had implemented the most steps.

6.6 Results of the School Beat Study

6.6.1 Implementation of the School Beat Steps

Table 6.2 presents major results. All the statements relating to step 1 (determining health needs), step 2 (setting priorities) and step 5 (implementation) were answered affirmatively by more than half of the schools, with one exception: implementation of the school health promoting plan in step 5 was reported by four schools only.

Steps 1 and 2 had been completed by more schools than the subsequent steps. The number of schools that had implemented steps 1–6 in the intended order decreased going from step 4 to steps 5 and 6. None of the schools reported having fully implemented the School Beat approach as originally intended.

6.6.2 *Satisfaction with the HPS Advisor and Progress in Implementation*

To answer the second research question, we focused on the role of the HPS advisor. Table 6.2 shows that schools that took an active interest in health and safety were more satisfied, in terms of both the assistance to content and process offered by the HPS advisor. The schools that had implemented the activities of step 4 (designing the school health promoting plan) tended to be more satisfied with the content assistance by their advisor, except regarding the presence of a school health promotion plan and the description of research results and activities in the school

Table 6.2 Scores on satisfaction with the *content* assistance (CA) and *process* assistance (PA) offered by the HPS advisors of schools that completed a particular step of the School Beat method and schools that did not complete that step

Statements for each step of the "Schoolslag" approach (At our school...)	Groups	Implementation School level N=18 N (yes)	Satisfaction with CA Range −2 to 2 N	Mean	Satisfaction with PA Range 0–2 N	Mean
Step 1 Determining the health needs at the school						
Attention is being devoted to health and safety	Completed	15	12	0.98**	12	1.91**
	Not completed	3	3	−1.30	2	1.40
There is a prevention team in place	Completed	12	11	0.85	11	1.82
	Not completed	6	4	0.08	3	1.93
There is a framework for prevention activities	Completed	11	11	0.78	11	1.82
	Not completed	7	3	1.01	3	1.90
Step 1 Determining the health needs at the school	Completed	7	7	1.00	7	1.91
	Not completed	11	8	0.33	7	1.77
Step 2 Setting health promotion priorities						
Priorities for health promotion have been set	Completed	16	13	0.84	13	1.83
	Not completed	2	1	0.67	1	1.90
Priorities are based on research	Completed	13	12	0.88@	12	1.84@
	Not completed	5	0	@	0	@
Priorities have been discussed at team meetings	Completed	12	10	0.75	10	1.80
	Not completed	6	3	1.04	3	1.97
Step 2 Setting health promotion priorities	Completed	9	8	0.80	8	1.78
	Not completed	17	3	1.04	3	1.97
Steps 1+2 completed	**Completed**	**4**	**4**	**1.06**	**4**	**1.87**
	Not completed	**14**	**10**	**0.50**	**9**	**1.82**
Step 3 Identifying important and modifiable determinants						
There is a working group to examine the priorities	Completed	8	7	1.09	7	1.91
	Not completed	10	8	0.25	7	1.77
Activities are coordinated within the team	Completed	8	7	1.15	7	1.91
	Not completed	10	3	0.26	3	1.63

(continued)

Table 6.2 (continued)

Statements for each step of the "Schoolslag" approach (At our school…)	Groups	Implementation School level N=18 N (yes)	Satisfaction with CA Range −2 to 2 N	Mean	Satisfaction with PA Range 0–2 N	Mean
Step 3 Identifying important and modifiable determinants	Completed	4	4	1.37	4	1.95
	Not completed	14	8	0.25	7	1.77
Step 1+2+3 completed	**Completed**	**2**	**2**	**1.11**	**2**	**1.90**
	Not completed	**16**	**12**	**0.59**	**11**	**1.82**
Step 4 Designing the school health plan						
There is a school health plan	Completed	5	5	1.13	5	1.94
	Not completed	13	8	0.63	8	1.76
Health needs have been described	Completed	7	7	1.05*	7	1.94
	Not completed	11	5	0.31	5	1.70
Research results have been described	Completed	5	5	1.11	5	1.98
	Not completed	13	6	0.43	6	1.73
Priorities have been described	Completed	7	7	1.05*	7	1.94
	Not completed	11	5	0.31	5	1.70
Strategies have been described	Completed	7	7	1.05*	7	1.94
	Not completed	11	5	0.31	5	1.70
Activities have been described	Completed	7	6	1.04	6	1.93
	Not completed	11	4	0.45	4	1.62
Step 4 Designing the school health promoting plan	Completed	3	3	1.37	3	2.00
	Not completed	15	10	0.66	10	1.78
Step 5 Implementation						
A school health plan is being implemented	Completed	4	4	1.25*	5	2.00
	Not completed	14	6	0.43	6	1.75
Activities are being implemented	Completed	17	14	0.83@	14	1.84@
	Not completed	1	0	@	0	@
Priorities are being incorporated in routine practice	Completed	10	9	0.86	9	1.82
	Not completed	8	6	0.31	5	1.88
Planned activities are being discussed at team meetings	Completed	14	11	0.76	11	1.80
	Not completed	4	2	0.77	2	2.00
Step 5 Implementation	Completed	1	1	1.66	1	2.00
	Not completed	17	11	0.32	10	1.81
Step 6 School-based evaluation						
The implementation of the school health plan is being evaluated	Completed	5	5	1.20*	5	1.98
	Not completed	13	5	0.31	5	1.72
The quality of the implementation of the school health plan is being evaluated	Completed	3	3	0.22	3	1.60
	Not completed	15	7	0.98	7	1.96
Step 6 School-based evaluation	Completed	1	1	0.89	1	2.00
	Not completed	17	9	0.74	9	1.83

**Difference significant at the p<.01 level (2-tailed)

*Difference significant at the p<.05 level (2-tailed)

@ t-test was not possible due to the small number of respondents in one of the groups health promotion plan. Schools that had implemented and evaluated the school health promotion plan reported more satisfaction with the content assistance than schools that had not done so.

It appears that satisfaction with content assistance is more predictive of the degree of implementation than satisfaction with process assistance.

6.6.3 Factors Influencing Successful Intersectoral Collaboration and Implementation

To answer the third research question, we focused on the organizational aspects, using the main factors of the DISC model as a conceptual framework. As mentioned in the methods section, we only report the results of the two extreme cases, the schools that implemented most respectively least of the School Beat steps.

6.6.3.1 External Factors

Schools have no legal obligation to implement health promotion. Overall, the schools in our study stated that their legal obligation was to offer high quality education to children. In addition, they were already faced with many extra tasks, so they were overloaded with work, and teachers and school managers already felt stretched to the limit. Since they were already facing a high workload and wanted to safeguard the quality of the education, they were quite apprehensive about taking on extra tasks. Although the schools might feel responsible for the health of their students, they could no longer count on the willingness of individual teachers to help improve student health, and without the cooperation of teachers, it was difficult to implement a Health Promoting School. In this respect, there was no difference between the best and worst cases.

6.6.3.2 Context

If collaborating parties had had positive experiences in previous collaborative processes and felt supported, they had a more open attitude towards the sustained collaborative process supporting inter-sectoral health promotion. Linking the tasks of the PC coordinator to health promotion activities seemed to have a positive influence on the implementation of a Health Promoting School. There is a strong relation between these two task fields. When health care is at stake individual help is needed to prevent deterioration of behavioral or social problems, while for health promotion extra assistance at school level is needed for instance to plan and coordinate school and community projects; health promotion programs for students and staff; nutrition and food safety programs; opportunities for physical education and recreation; and programs for counseling, social support, and mental health promotion.

The latter can better be addressed at group level. This is not only more efficient, but is also justifiable, as it gives schools the opportunity to intervene at an earlier stage, before problems arise and thus, to really promote health. Our findings suggest that a positive collaborative context supports the adoption and implementation of the School Beat innovation.

Example

In one school, the task of coordinating the health promoting school activities was combined with the function of the PC coordinator.

The PC coordinator already cooperated with various health care organizations participating in the School Beat approach. The PC coordinator had a strong sense of responsibility for the health of the students. Whereas health care focuses on individuals, health promotion is primarily aimed at the collective. Individual students who are at risk or have problems were discussed in the care team, and the input of the care team was essential for setting health promotion priorities.

6.6.3.3 Change Management

Change management and leadership skills are necessary to stimulate school managers and teachers, and to guide the implementation and collaboration process. The respondents reported that the leadership offered by the HPS advisor had encouraged them in their task of implementing a health promoting school. Both the best-case and the worst-case schools reported being satisfied with the assistance offered by the HPS advisor. With regard to leadership skills in the schools, the best-case school had explicitly assigned the task of coordinating health promotion activities to one of its team members, the PC coordinator. This person was in frequent contact with the school manager to report on the most important activities. The PC coordinator was responsible for overseeing the process as a whole, as well as for the coordination and implementation of the School Beat steps, and the HPS advisor supported the PC coordinator in fulfilling this task. The worst-case school had not explicitly assigned the task of coordinating health promotion activities to anyone. There was no strong leader in the school who had time to engage in a health promoting school program. Our analysis of this aspect of the DISC monitor suggests that the presence of a leader has a favorable influence and the absence of a leader an unfavorable effect on the implementation and collaboration process.

6.6.3.4 Project Management

Innovations often start as a project in a project management structure. The best-case school had organized the project management structure by appointing a special HPS team under the supervision of the PC coordinator, who also had the task of coordinating the health promoting policies. This HPS team assisted the PC coordinator in implementing activities in and around the school. The school staff was organized in year teams, one team for each school year. The year teams were responsible for the coordination of teaching and other activities in their year. Each year team had its own representative in the HPS team, who was responsible for the coordination between the prevention priorities and activities of their own year team and those of the other year teams. The HPS team, which met every 3 months, included not only teachers but also representatives from the student council and the parents' council. In addition to the meetings of the HPS team, there were monthly meetings between the PC coordinator and the coordinator of each year team, at which various aspects of health promoting policies were discussed and coordinated.

The worst-case school had started its own HPS team in 2002, under the supervision of the school manager. The team consisted of a biology teacher, a physical education teacher, parents and students. After a period of reorganization, health promotion activities were given lower priority, and the HPS team ceased to exist. In 2008, a new team was started, again under supervision of the school manager. This team consisted of the school manager and two other people: the PC coordinator and the HPS advisor. There was no interaction with most of the school's teaching staff, students or parents, and there was no organizational structure for health promotion at the school.

Embedding health promotion in the school management structure appears to have a positive impact on the implementation of school health promotion.

6.6.3.5 Collaborative Support

In the best-case school, the intentions were positive and actions were undertaken by the school management to implement health promotion (see the example below). In the worst-case school, the school manager was also positive and was taking care of the coordination of health promotion activities. However, he had neither extra time available for this task, nor a working budget for health promotion activities. The result was that health promotion was not given a high priority. According to the school manager, enthusiasm for health promotion could not be expected as long as no extra time was allocated to such activities.

It appears that schools need opportunities to act upon their intentions; positive perceptions and intentions are not enough.

Example
The school manager was positive about the collaboration and had taken action to organize the collaboration between the HPS advisor and the school. The PC coordinator had been allocated extra time for the coordination of health promotion activities. Although the government does not provide permanent funding for health promotion activities in schools, the school manager had allocated a small working budget of 2000 Euros to each year team for health promotion activities. In addition to the standard teaching hours, each teacher had extra-curricular hours allocated for tasks like organizing the school theater, student counseling or prevention activities. The team coordinator was responsible for assigning these extra tasks, one of which was to represent the year team in the HPS team and assist the PC coordinator in the coordination and implementation of health promotion activities.

6.7 Reflecting on the School Beat Case Study

The above results show that only one school had implemented the School Beat approach as intended, although even they had not fully implemented it: they had completed steps 1–5, including all related activities. The other schools seemed to have selected those activities that fitted their operational procedures at specific moments. The results show that more practical steps, like setting priorities for health promotion and implementing activities, were more likely to be completed than the more complicated structuring steps, such as writing a school health promoting plan, describing health needs, describing research results and describing priorities. The schools reported relatively more activities from step 5, that is, the practical work of implementing health promoting plans.

The original idea behind the School Beat approach was to create a rational, more or less linear process which schools can follow, step by step, to implement school health policies (Leurs et al. 2005a), but this linearity was not reflected in the actual implementation of the approach. An explanation could be that schools are not obliged to develop preventive health policies: health promotion is not a mandatory part of the school curriculum. As a result of this autonomy, schools tend to implement only those elements of the School Beat approach that fit their opportunities and operational procedures. The linear model of the School Beat approach as a whole does not seem to fit in with their preferences. This case study allows the conclusion that the process should be put into practice in an iterative, cyclic process. Flexibility is necessary in assisting schools. This finding has important implications for the HPS advisors. It requires flexibility in professional attitude and skills to tailor health promotion activities to the context of the school. Furthermore, both the assistance to content regarding knowledge and professional skills about health promoting schools, and the assistance to the collaboration process regarding constructive personal relationships had a positive impact on the degree of satisfaction expressed by the schools with the work of their HPS advisor. Assistance on demand appeared to be the key to success, within the limits of health promoting schools; this is precisely the main task of the HPS advisor.

The results of the DISC-based evaluation study reaffirm the importance of leadership, the need for project management and communication with school and external partners. Remarkably, the external factors were not decisive, which probably is due to the extremely high workload in schools, making it impossible to add tasks with respect to a health promoting school (which are not compulsory). We may conclude that intersectoral collaboration takes time and requires leadership to advocate the importance of health promoting schools, adequate project management and intense communication between all partners involved, to accomplish a shared outcome *i.e.* health and educational benefits for children, benefits for school staff, parents, community organizations and community members.

Our study showed the added value of appointing a HPS advisor to schools. We have strong indications that without such support schools would not proceed with establishing a school health policy. From a primary school study in the same region (Leurs et al. 2007) we learned that regional public health organizations are well appreciated by schools for their school health support. Schools respect and appreciate the expertise of public health organisations in this regard. Additionally, from the interaction with secondary schools we learned that schools are keen on having one health partner, *i.e.* HPS advisor, who is able to act as a kind of 'filter' for the information overload experienced by schools.

Creating a Health Promoting School means applying a new way of thinking. It means finding opportunities to develop policies, practices and structures that include health promotion in everything done by the school and community. It means working together under clear leadership, with everyone having a chance to express ideas or opinions and then agreeing to collaborate toward common goals. It also means professional empowerment of support organizations at regional/local level, to enable true partnerships between support organizations and schools. And at the regional and national level it means simplification, integration and translation of health content and interventions to meet the needs and requirements of schools. There is no blue print: a health promoting school excels in using existing evidence and tailoring to the needs and strengths of each individual school!

Acknowledgements This research has been supported by the Academic Collaborative Centre for Public Health Limburg, a collaborative between 18 municipalities in South Limburg, the regional Public Health Service (GGD Zuid Limburg) and Maastricht University Medical Centre (MUMC+). The research was conducted at the School for Public Health and Primary Care: CAPHRI, Department Health Promotion, of Maastricht University. The study [projectnumber 7125.0001] was funded by the Netherlands Organisation for Health Research and Development (ZonMw).

References

Aggleton, P., Rivers, K., Mulvihill, C., Chase, E., Downie, A., Sinkler, P., et al. (2000). Lessons learned: Working toward the National Healthy School Standard. *Health Education, 100*(3), 102–110.

Allensworth, D. D. (1987). Building community support for quality school health programs. *Health Education, 18*, 32–38.

Andis, P., Cashman, J., Oglesby, D., Praschil, R., Adelman, H., Taylor, L., & Weist, M. (2002). A strategic and shared agenda to advance mental health in schools through family and system partnerships. *International Journal of Mental Health Promotion, 4*(4), 28–35.

Boot, N. (2011). *Gezondheidsbevordering en voortgezet onderwijs, verstandshuwelijk of echte liefde? Een onderzoek naar de implementatie van schoolgezondheidsbeleid binnen scholen voor voortgezet onderwijs* [Health promotion and secondary education: Marriage of reason or true love? A study of the implementation of school health policies at secondary schools]. Maastricht: Universiteit Maastricht.

Bos, V., De Jongh, D., & Paulussen, T. (2010). *Gezondheidsbevordering en preventie in het onderwijs. Stand van zaken, effectiviteit en ervaringen van GGD'en en scholen* [Health promotion and prevention at schools: state of affairs, effectiveness and experiences of public health services and schools]. Bilthoven: RIVM.

Bosdriesz, M. & Berkenbosch, W. (2003). *Jeugdzorgadviesteams voor het basisonderwijs. Methodiek voor de aansluiting van basisonderwijs en jeugdzorg* [Youth care advisory teams for primary education. A method to link primary education with youth care]. Utrecht: NIZW.

Clift, S., & Jensen, B. B. (Eds.). (2005). *The health promoting school: International advances in theory, evaluation and practice*. Copenhagen: Danish University of Education Press.

Cummings, T. G., & Worley, C. G. (2001). *Organization development and change* (7th ed.). Mason: South-Western College Publishing.

De Caluwé, L., & Vermaak, H. (2003). *Learning to change*. Thousand Oaks: Sage Publications.

De Leeuw, E. (1989). *The sane revolution. Health promotion: Background, scope, prospects*. Assen/Maastricht: Van Gorkum.

Deschesnes, M., Martin, C., & Hill, A. J. (2003). Comprehensive approaches to school health promotion: How to achieve broader implementation? *Health Promotion International, 18*(4), 387–396.

Deschesnes, M., Trudeau, F., & Kébé, M. (2010). Factors influencing the adoption of a health promoting school approach in the province of Quebec, Canada. *Health Education Research, 25*, 438–450.

El Ansari, W., & Weiss, E. S. (2006). Quality of research on community partnerships: Developing the evidence base. *Health Education Research, 21*(2), 175–180.

Gray, B. (1989). *Collaborating: Finding common ground for multi-party problems*. San Francisco: Jossey Bass.

Leurs, M. T. W. (2003). Gezondheidsbevordering in het onderwijs: de schoolSlag-uitdaging [Health promotion in education: The schoolBeat-challenge]. *Remediaal, 4*(1), 3–8.

Leurs, M. (2008). *A collaborative approach to tailored whole-school health promotion. The Schoolslag study*. Maastricht/Amersfoort: Maastricht University.

Leurs, M., Jansen, M., & Ruiter, M. (2002). Coordinated School Health Program: het Gezonde School Model van de toekomst? [CSHP: The healthy school model for the future?]. *Tijdschrift voor Gezondheidswetenschappen, 80*(7), 471–473.

Leurs, M., Jansen, M., Schaalma, H., Mur-Veeman, I., & De Vries, N. (2005a). The Tailored Schoolslag approach: New concepts for health promotion in schools in The Netherlands. In S. Clift & B. B. Jensen (Eds.), *The health promoting school: International advances in theory, evaluation and practice* (1st ed., pp. 89–107). Copenhagen: Danish University of Education Press.

Leurs, M. T. W., Schaalma, H. P., Jansen, M. W. J., Mur-Veeman, I. M., St Leger, L. H., & De Vries, N. K. (2005b). Development of a collaborative model to improve school health promotion in the Netherlands. *Health Promotion International, 20*(3), 296–305.

Leurs, M., Bessems, K., Schaalma, H., & De Vries, H. (2007). Focus points for school health promotion improvements in Dutch primary schools. *Health Education Research, 22*(1), 58–69.

Leurs, M. T. W., Mur-Veeman, I. M., Van der Sar, R., Schaalma, H. P., & De Vries, N. K. (2008). Diagnosis of sustainable collaboration in health promotion – a case study. *BMC Public Health, 8*, 1–15.

Lister-Sharp, D., Chapman, S., et al. (1999). Health promoting schools and health promotion in schools: two systematic reviews. *Health Technology Asssessment, 3*, 22.

Marx, E., & Wooley, S. F. (1998). *Health is academic – A guide to coordinated school health programs*. New York: Teachers College Press.

MBO Raad. (2010). *Loopbaan en burgerschap in het mbo* [Careers and citizenship at MBO vocational schools]. Utrecht: Libertas.

Ministry of Health, Welfare and Sport. (2011). *Landelijke nota gezondheidsbeleid Gezondheid dichtbij* [National health policy memorandum – "Closer to health"]. Den Haag: Ministery of welfare and sport.

Padgett, S., Bekemeier, B., & Berkowitz, B. (2004). Collaborative partnerships at the state level: Promoting systems changes in public health infrastructure. *Journal of Public Health Management and Practice, 10*(3), 251–257.

SLO. (2007). *Concretisering van de kerndoelen. Kerndoelen voor onderbouw VO* [Specifying the national core objectives. Core objectives for the lower years of secondary education]. Enschede: Ministerie van OC&W.

St Leger, L. H. (1999). The opportunities and effectiveness of the health promoting primary school in improving child health – A review of the claims and evidence. *Health Education Research, 14*(1), 51–69.

St Leger, L. (2004). What's the place of schools in promoting health? Are we too optimistic? *Health Promotion International, 19*(4), 405–408.

St Leger, L., & Nutbeam, D. (1999). Evidence of effective health promotion in schools. In D. Boddy (Ed.), *The evidence of health promotion effectiveness. Shaping public health in a new Europe* (pp. 110–122). Brussels: European Union.

St Leger, L., & Nutbeam, D. (2000). A model for mapping linkages between health and education agencies to improve school health. *Journal of School Health, 70*(2), 45–50.

Stewart-Brown, S. (2001). Evaluating health promotion in schools: Reflections. In I. Rootman, M. Goodstadt, B. Hyndman, D. V. McQueen, L. Potvin, J. Springett, & E. Ziglio (Eds.), *Evaluation in health promotion – Principles and perspectives* (pp. 272–284). Copenhagen: World Health Organization.

Stewart-Brown, S. (2006). *What is the evidence on school health promotion in improving health or preventing disease and, specifically, what is the effectiveness of the health promoting schools approach?* Copenhagen: WHO Regional Office for Europe.

Suhrcke, M., & de Paz Nieves, C. (2011). *The impact of health and health behaviours on educational outcomes in high income countries: A review of the evidence*. Copenhagen: WHO Regional Office for Europe.

Turkenburg, M. (2008). *De school bestuurd. Schoolbesturen over goed bestuur en de maatschappelijke opdracht van de school* [Managing schools – School managers on good governance and school's societal tasks]. Den Haag: Sociaal en Cultureel Planbureau.

Van Der Steenhoven, P., & Van Veen, D. (2009). *NJI-monitor. Leerlingenzorg en zorg- en adviesteams in het voortgezet onderwijs 2008* [Student care and care teams in secondary schools]. Utrecht: Nederlands Jeugd Instituut.

Van Eyk, H., & Baum, F. (2002). Learning about interagency collaboration: Trialling collaborative projects between hospitals and community health services. *Health & Social Care in the Community, 10*(4), 262–269.

Van Raak, A. J. A. (1998). *Zorgvernieuwing: een kwestie van routine. Een studie naar de vorming van interorganisationele netwerken en naar systeemveranderingen in de thuiszorg vanuit interactionistisch perspectief* [Innovations in care: A matter of routine. A study into the development of inter-organizational networks and system changes in home care from an interactionistic perspective]. Dissertation, Maastricht University, Maastricht.

VO-Raad. (2009). *Ambitie aan zet! Jaarverslag voortgezet onderwijs 2008/2009* [Time for ambition! Annual report on secondary education, 2008/9]. Utrecht: VO-Raad.

VO-Raad, & Ministry of Education, Culture & Science. (2008). *Kwaliteitsagenda Voortgezet Onderwijs. Teken voor kwaliteit. Afspraken voor een beter voortgezet onderwijs 2008–2011* [A quality assurance agenda for secondary education. Go for quality. Agreements to improve the quality of secondary education]. Utrecht: VO-Raad, Ministry of Education, Culture & Science.

WHO. (1998a). *Local action creating health promoting schools. Model charter for a health-promoting school.* Atlanta: Division of Adolescent and School Health, National Center for Chronic Disease Prevention and Health Promotion, Centers for Disease Control and Prevention.

WHO. (1998b). *WHO's global school health initiative*. Division of Adolescent and School Health, National Center for Chronic Disease Prevention and Health Promotion, Centers for Disease Control and Prevention, Atlanta, Georgia, USA, UNAIDS; and the Johann Jacobs Foundation, Zürich.

Wijsmuller, M. (2002). Zorgstructuur rond het onderwijs. Problemen eerder signaleren en adequater oppakken [Health care structures to assist schools. Identifying and addressing problems at an earlier stage]. *Spectrum Actueel, 4*, 10.

Chapter 7
Becoming a Health Promoting School: Using a 'Change Agent' to Influence School Structure, Ethos and Ensure Sustainability

Elizabeth Senior, Andrew Joyce, and Dimitri Batras

Abstract Schools have long been sites for health promotion. Commencing with classroom lessons, schools have witnessed many projects and programs including the health promoting school model. Many authors indicate that this model is the most successful in achieving long term changes within a school, other authors report that implementation and sustainability are difficult to achieve. This chapter will examine the journey of a primary school as it works to implement the health promoting school model. The authors will demonstrate that, although the health promotion model is useful in introducing and guiding health promotion activities, without extra assistance, such as a dedicated health promotion officer, or 'change agent' who can motivate committed champions, changes to the ethos and the culture of the school will be difficult. Every school is unique; there is no 'one size fits all' model. Therefore, professionals working with schools need to meet the school at its point of need, rather than following a standard format. The process and journey are just as important as the successes. Influencing the organisation of the school is essential if the changes are to be sustained. Organisational change theories are used to support the practical examples.

E. Senior (✉)
EACH Social and Community Health Service, Melbourne, VIC, Australia
e-mail: lsenior@EACH.com.au

A. Joyce
Centre for Community Child Health, Murdoch Childrens Research Institute, Melbourne, VIC, Australia
e-mail: andrew.joyce@mcri.edu.au

D. Batras
Attained Success Consulting, Melbourne, VIC, Australia

School of Public Health and Preventive Medicine, Monash University, Melbourne, VIC, Australia
e-mail: dimitri.batras@monash.edu; dbatras@attainedsuccess.com.au

V. Simovska and Patricia Mannix-McNamara (eds.), *Schools for Health and Sustainability: Theory, Research and Practice*, DOI 10.1007/978-94-017-9171-7_7,

Keywords Health promoting schools • Organisational change • Critical friend • Diffusion of innovations

7.1 Introduction

Schools have long been sites for health promotion (Veselak 2001; Clift and Jensen 2005). During the nineteenth century the term 'school hygiene' was used to describe problems of school sanitation and construction (Veselak 2001). There was early recognition that children could not learn if they were in unsanitary crowded buildings. This led to the establishment in 1927, of the American Association of School Physicians. Interest in the Association grew so rapidly that in 1936, the school opened its membership to all professionals interested in promoting school health (American School Health Association 2012). It would appear that many health professionals saw the answer to health problems by 'getting into the schools'.

The next phase was health education which consisted of health, safety, exercise and narcotics (Veselak 2001). In the United Kingdom in the 1960s and 1970s, concerns about the physical health of students led to schools including health education in their curricula. The next move was to broaden the health education into the environment (Denman 1999). In 1986, the Ottawa Charter was adopted at the First International Conference on Health Promotion. This was a response to the growing expectations for a global new public health movement (World Health Organisation 1986). The Ottawa Charter recognized that health requires 'up stream' foundations such as shelter, food, social justice and equity as prerequisites. There was recognition that the things that affect health lie outside the conventional concerns of health professionals (Baum 2002). In 1986, a symposium, hosted by WHO and entitled 'The Health Promoting School' was held in Scotland. This symposium offered WHO the opportunity to apply its theoretical model of health to the school setting (Young 2005). Reflecting a move to addressing the social determinants of health, 'The Health Promoting School' (HPS) was described as 'a combination of health education and all the other actions which a school takes to protect and improve the health of those within it' (Young 2005). Naming six thematic fields as areas for change, the main aim is to combine traditional classroom education with improvements in the social and physical environment of the school (St Leger and Young 2009).

Improvements in the school environment suggest that organizational change processes are required and there has been a growing understanding that in order to change schools a new theory to underpin the work is essential. While HPS provides a broad framework for action there is mixed evidence that schools are able to implement this approach (Stewart-Brown 2006). Many authors indicate that the multimodel, whole school approach, espoused by the HPS framework, is most effective in producing long term changes to students' attitudes (International Union for Health Promotion and Education 2009; St Leger et al. 2007; Allensworth and Kolbe 1987; Clift and Jensen 2005). However, other authors have reported that successful implementation and sustainability have been difficult to achieve (Lynagh et al. 1997; Nutbeam et al. 1993). Rather than seeing schools as static entities, they are complex, ever changing systems. Keshavarz et al. (2010) describes them as 'social complex adaptive systems'. This resulted in a move away from seeing the school as a supportive environment for health promotion to viewing the school as a structure or 'ecosystem' which will respond to change as programs are implemented. Understanding system change as well as measuring individuals change becomes part of the of goal health promotion (Bond et al. 2001).

This chapter will explain how change management theories and processes were used to guide implementation of the HPS model within a primary school setting. The case study is from a whole of school health promotion program conducted at Bayswater North Primary School (Senior 2012). Bayswater Nth Primary School (BNPS) is situated in a State funded Community Renewal area in Melbourne, Australia. Community Renewal programs target neighbourhoods in areas experiencing growth or decline or facing difficulties such as falling employment, poor access to services or run down community facilities (Maroondah City Council 2009).

Due to the school's geographic situation within the Community Renewal area, the local community health service EACH Social and Community Health (EACH), which is a partner organization in the Community Renewal project, approached the principal of the school to discuss the idea of a partnership between the two organizations. EACH is a large multi-site community health service, with a site situated close to BNPS. EACH employs around 700 people assisted by 300 volunteers. The initial partnership agreement between EACH and the school focused on implementing the Health Promoting Schools model as described in Guidelines for Promoting Health in Schools; Version 2 (International Union for Health Promotion and Education 2009). Initially the Health Promotion Officer (HPO) at BNPS intended to use the Health Promoting Schools model to initiate multi-model health promotion projects run across the school alongside the capacity building framework developed by NSW Health (2001). The capacity building model identified action areas for building capacity to promote health such as: organisational change and development, workforce development and partnerships.

These capacity building strategies have a strong emphasis on the orientations and skills of managers, with an assumption that organisational change will occur as a result. Both the HPS and capacity building models do not articulate in detail the process by which this change occurs. As the work got underway, there was a growing realization that understanding the processes of change was more important than running programs (Butler et al. 2001). While still utilizing the HPS Model as the overarching framework to guide implementation of the work going on in the school, the HPO began to draw on specific organizational change and school health promotion theorist's view of schools as complex adaptive systems (Butler et al. 2001; Gibbs and Panayiotis 2008).

The chapter will firstly outline what is meant by a 'change agent' and how this particular role has been conceptualised and used within school based health promotion practice. In addition two organisational theories of change will be summarised: Lewin's (1997c) theory on the process of organisational change and

Diffusion of Innovations (Rogers 2003). From this basis, the chapter uses Lewin's (1997c) 'three-step' model for achieving organisational change, 'unfreezing', 'moving' and 'refreezing,' to structure the organisational change process that took place within the case study. Ideas from the literature on the role of a 'change agent' and Diffusion of Innovations will be used to explain critical points of change within these sections of the chapter. The final sections of the chapter discuss to what extent the 'change agent' model described in this case study is transferable and sustainable within other schools and the level of support required for someone in the 'change agent' role.

7.2 Critical Friend

The role of the 'critical friend' has been described as that of a 'change agent'. The 'critical friend' might be a project officer, a health professional who may or may not have an education background. The 'critical friend' is a "trusted person who will ask provocative questions and offer helpful critiques" (Costa and Kallick 1993, p. 51). The Gatehouse Project was a primary prevention programme, run in selected Victorian schools between 1996 and 2002. It included both institutional and individual focused components to promote the emotional and behavioural wellbeing of young people in secondary schools. Within the Gatehouse Project the 'critical friend' was a vital part of the programme and assisted in building the capacity of the school and facilitating the process of organizational change (Bond et al. 2001).

Butler et al. (2001) identify four key components that a 'critical friend' is involved in: conceptualizing the intervention as an ongoing process of change (not a product to be 'done'), facilitating the change process (not just training and technical assistance), bringing an in-depth understanding of the educational context and health and wellbeing, and assisting schools to integrate the work within their core business. The 'critical friend' does not offer a packaged solution for schools, but focuses on engagement with all members of the school community, relationships and structural change. Over the past years, collaboration, development of networks and learning teams have become more common. In the school literature there is evidence that this sort of collaboration can improve schools teaching, learning and the overall environment (Ainscroft and West 2006).

The benefit of the 'critical friend' has been acknowledged in the widely adopted 'KidsMatter' programs. KidsMatter is an Australian Primary School Mental Health Initiative focusing on implementing a systems approach to planning and implementation as opposed to adoption of a particular program (Australian Psychological Society 2014). The first step in implementing the KidsMatter program is to establish an Action Team which contains a 'critical friend'. The 'critical friend' brings an external perspective to the team. This external person is often a regional education sector staff member or community agency staff (Australian Psychological Society 2014). The KidsMatter pilot project was able to demonstrate improved emotional and social health outcomes for children (Slee et al. 2009).

The role and even title of the 'critical friend', is not without its critics. Swaffield and MacBeath (2005) express unease with the *double entendre* of 'critical friend'. They feel that the ambiguity of the word 'critical' can raise unease that it does not translate to other cultures or languages well. Different models cast the 'critical friend' into different roles. Eddy (2006) and O'Connor and Ertmer (2006) describe a 'critical friend' as a colleague and mentor, another teacher who provides graduates and new teachers an opportunity to learn from more experienced colleagues. Ainscroft and West (2006) argue that there has been much confusion over the role of the 'critical friend'. They postulate that if we are interested in seeing changes in schools, the 'critical friend' should have a close relationship with the school staff. They see the 'critical friend' as a person who is not part of the school staff, however who is closely involved with the school, as envisaged in the "Kids Matters" and Gatehouse programs. In these terms, the 'critical friend' is a friend to the school as a whole. The entry point may be the head teacher, however, as the relationship with the school develops, the 'critical friend' begins to work with a wider range of teachers and is seen as a supportive, yet challenging facilitator (Swaffield and MacBeath 2005; MacBeath 1999).

Fullan (2006) identifies that a good 'critical friend' is one that provides a different perspective, or new eyes. He argues that school leaders need to widen their sphere of engagement by interacting with other schools and people outside the education sphere, for example, a 'critical friend'. From this perspective, it is therefore important that the 'critical friend' is someone from outside the immediate school system. The 'critical friend' has been described as a 'detached outsider' who can provide an alternative viewpoint (Swaffield 2007). In Butler et al. (2001) a note from a 'critical friend's diary, discussed the difficulties of negotiating the politics of power relationships within the school. As a detached outsider, it is possible to transcend the interpersonal issues and tensions that come with working in a school community. The HPO in this case example undertook the 'critical friend' role from the external viewpoint of being employed through a community health service. The perceived benefits and drawbacks of this role will be discussed during the chapter.

The title 'change agent' might be more apt in describing the work of a person attempting to change the structure, ethos and culture of a school or any organization. Between 1973 and 1978, The Rand Change Agent Study was undertaken to determine the ways people thought about planned change in education. Fifteen years later the major findings of the study were reviewed to find out how change in schools actually happens (Laughlin 1990). Rand found that Federal policies played a major role in project adoption by schools; however adoption did not ensure successful implementation or the sustainability of the project, nor did access to seed funding, and extensive resources. Failure also occurred when the on-going and sometimes unpredictable support that teachers needed was unavailable and when schools were required to use packaged approaches. What mattered most was local capacity within the school and will. Rand found that outside assistants 'change agents' who were sensitive to the local issues facing the school, understood and could work with the fluid unpredictability within the school environment could be extraordinarily successful in changing people and practice (Laughlin 1990). Baker and associates (1991) also found that schools that have local support are more likely to improve as against those who had no external support.

However, just being a 'critical friend' in a school is not enough to elicit change. Numerous authors recommend that it is essential to have the support of head teachers (Inchley et al. 2006; St Leger 2005) and a group of champions or as St Leger (2005) describes them, activists. Having the 'support of head teachers' does not just mean a cursory nod to the HPS program. The school leadership needs to be committed to changing the school. It is helpful if the principal or assistant principal act as champions on the HPS committee. This will be the group who have more direct contact with the 'critical friend' and will be the champions who drive the changes forward. The HPS committee, made up of parents, teachers and students will be the representative champions for their peer groups. Butler and colleagues (2001) point out that many conversations conducted with the 'critical friend' will be repeated with other participants. It is these champions who will do this 'talking up' of the HPS model within the school community, and be the early adopters of change. Hawe and associates (1997) discuss the importance of reach of an intervention into a population in regards to its success. As will be discussed, at BNPS it was the assistant principal who acted as a major champion, taking ideas generated at the HPS meetings, back to the staff and ensuring the ideas were discussed widely within the school. Effectively utilising the roles of a 'critical friend' and 'health promotion champions' requires an understanding of organizational processes. Being able to work as a change agent within a school setting makes it paramount to understand some of the forces that can promote change within this setting.

In summary, a 'change agent' can perform a highly important role in assisting an organisation such as a school through a change process. Being sensitive to the local context in which they are operating while at the same time cognisant of the change that is required provides a unique perspective that has been well utilised in school based research. Together with a 'health champion' that is normally someone internal to the organisation, they provide the impetus to ensure that the school community is supportive and engaged in the change process and can steer this process towards productive outcomes. The next two sections outline some of the foundation theories of organisational change that have guided research and practice in this field. Firstly, Lewin's (1997c) theory of organisational change will be introduced and then the theory of Diffusion of Innovation (Rogers 2003) will be briefly summarised. These two theories both guided the practice outlined subsequently in this chapter and also provide an explanation for some of the change that occurred in the school once initial momentum was achieved.

7.3 Group Work and Forces for Change

From the late 30s until the late 40s Lewin's groundbreaking work shed light on group dynamics and forces within organizations and their impact on the outcomes of change initiatives. Similar to Keshavarz and colleagues (2010) who describe schools as complex ever-changing systems, Lewin (1997a, b, c) viewed change as a constant in any form of group work. It is the forces within and around the group (or school) which influence practice and outcomes. According to Lewin's (1997c) view of group dynamics, the school setting would be viewed as a broad group setting and the formal and informal sub groups within the school setting, for example; the mathematics teachers, the staff who leave the building at 12:35 pm to smoke cigarettes during lunch, and the senior staff team, are groups with their own set of dynamics and culture. In order for change to occur and succeed in the broad school setting it needs to succeed at a group level. Lewin (1997c) developed the 'three-step' model for achieving organisational change.

Lewin's 'Three-Step' Model

Stage 1. Unfreezing	The current state in which the school is in compared to where it would like to be to achieve its change agenda
Stage 2. Moving	Moving towards the new direction by piloting and implementing the initiative to achieve the school's objective
Stage 3. Refreezing	The school has reoriented the systems and structures to embed the new healthier way of working

7.4 Diffusion of Innovations

Diffusion is the communication of 'new ideas.' Fundamental to diffusion is a level of uncertainty for recipients of the communication because of the newness of the idea (Rogers 2003). According to Rogers (2003) organisational innovativeness relates to characteristics that are about individuals, organisational structure, and external organisational factors. A range of variables within these three broad categories positively or negatively influence an organisation's innovativeness.

The diffusion process within organisations follows a set of sequential stages. Rogers (2003) describes these stages, (1) agenda-setting; (2) matching the concept to the identified problem; (3) redefining/restructuring the innovation to make it fit within the organisation and the organisation fit the innovation; (4) clarifying the innovation by finding meaning and putting it into more widespread use; (5) routinizing phase is when the innovation is embedded into common practice within the organisation. The pace of innovation adoption is determined by how synergistic the innovation is with the priorities and ideologies of the organisation. Knowledge of the school organisation, its characteristics, values and potential congruence with a whole of school health promotion approach, is critical in pursuing this process of change. The following figure outlines the diffusion process using the BNPS case study and the detail of these changes will be outlined in the following sections (Fig. 7.1).

To illustrate in more detail the application of these theories, Lewin's three stages of change are used to describe the process by which BNPS became a health promoting school. Within these three broad stages the application of the other theories are described.

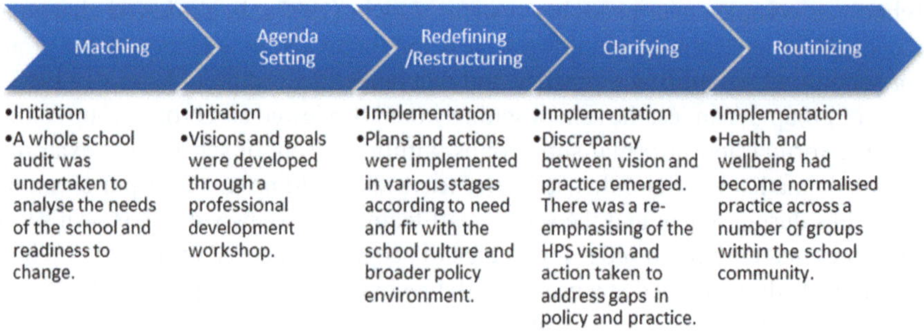

Fig. 7.1 The diffusion process using the BNPS case study (Adapted from Rogers 2003)

7.4.1 Unfreezing

School staff were initially cautious about the introduction of a HPO. Like many teachers around the world, teachers in Australia report that their workload has increased dramatically in the last 10 years. Not only do they report an increase in workload, but also an increase in complexity and the roles that they are required to fill (Easthope and Easthope 2000; Kyriacou 1987). Apart from increased workload, at least one author identifies that teachers can be suspicious of 'outsiders' coming into the school (Butler et al. 2001). This presents a difficulty for the HPO, as, to change the culture and ethos of a school requires the HPO to become 'embedded' within the school community (Butler et al. 2001). One of the early messages communicated to staff and administration was that a support person could 'lighten the load' and provide expertise to support schools in their process of becoming a health promoting setting (Armstrong 2011). When it became clear to teachers that the HPO was a resource who could be drawn upon to assist their work, rather than make more work, attitudes toward the HPO warmed significantly.

In early 2009, EACH Social and Community Health Service signed a 3 year partnership agreement with BNPS. This immediately sent a signal to the school that the HPO and the EACH were committed to the school for a number of years, and saw the intervention as an on-going process of change, not a project with a start and finish date. The HPO also worked hard at establishing relationships with the teachers. This involved making an effort to attend morning tea to talk with the teachers.

Critical to future success of the health promotion approach was understanding the culture and ethos of the organization, particularly the beliefs and values of leaders within the school community (Schein 2004). Time was spent understanding the values and beliefs of some of the more experienced teachers and those with capacity to shift opinion within the school environment. Many of the teachers have been at the school for an extensive period of time. It was important to develop relationships of a personal nature with as many of the teaching staff as possible. This made the morning tea-time, a vitally important part of the HPO's time at the school. MacBeath and Jardine (1998) confirm the importance of 'symbolic acts' within the school.

Who you sit with, and informal conversations can send out strong signals to the school staff. To some extent there was an element of centralization of power within the school environment whereby a few individuals within the school had the ability to influence the capacity of the school to adopt new ideas (Rogers 2003). Time was also spent getting to know some of the less experienced teachers and groups within the school community such as parents and students. This included forming relationships with teachers known to be less enthusiastic about the idea of the HPO so as to understand their perspectives.

The Health Promoting Schools Committee was then developed and it included 'members' of the various groups within the school. This meant including a range of experiences, ages, and teachers from different year levels, administration and parents. This aided in facilitating network connections and increased the degree to which new ideas could be implemented (Rogers 2003). The committee was also critical in making sure the HPS model was structured in a way that suited the organization, which is a critical element in ensuring sustainability (Rogers 2003).

One of the first tasks undertaken was analyzing some of the needs of the school, which is characteristic of good health promotion practice and essential in the unfreezing stage (Lewin 1997c). Using an audit adapted from the HPS Toolbox (Brisbane North Public Health Unit 2001), the school community of parents, teachers and students were asked what they liked best about the school and what they would like to see changed. Student focus groups and a professional development day were also undertaken, the process and results of which are described in Senior (2012).

Lewin (1997c) suggests that motivation is important in bringing about planned change within groups of people, when group decisions are made they can be quite powerful with respect to adherence to the decision made by the group. The school audit provided the motivation for change to commence. The school community voiced dissatisfaction with a number of areas of the school and once the audit was completed, there was anticipation that things would change. For example, when first arriving at the school, many teachers did not think that the canteen needed modification. However, after talking with teachers, working with the HPS committee and bringing in food related programs, teachers began to see that the canteen had an educative role within the school. They also began to accept that 'we can't be a health promoting school with an unhealthy canteen'.

Rogers (2003), in the Diffusion of Innovations model, identified that the 'agenda setting' stage is crucial in the identification of a need for innovation. The innovation needs to be tailored to fit the organizations need. This was achieved at the HPS Visioning Day. School teachers, support staff, and parents on the HPS Committee were invited to a daylong professional development workshop where the results of the audit were discussed, the philosophy of the health promoting school model was examined and staff had a chance to dream about the type of school they would like to work in. This day involved working to challenge the perceptions of the current state of the school, before bringing in new ideas to improve it. It was a time to examine the school and its community and ask how things could be done differently. At the end of the day 60 % of the staff identified that they had a good understanding of the HPS model, 30 % partially and 5 % not at all. In response to this, a HPS activities report has been placed as a permanent item on the staff agenda to keep the staff up to date on what is happening around the school and give them a chance to contribute. The success of the implementation of HPS depends largely on the teachers and their capacity to implement it (St Leger 1998).

7.4.1.1 Summary of Unfreezing

A key aspect of the unfreezing stage was developing relationships at an executive and staff level. A formal partnership agreement was developed with the health service and school, and the HPO worked very hard at developing relationships with staff. One key strategy was communicating to staff that she was a resource who could assist in reducing the workload for staff. Another strategy was spending time getting to know staff at morning tea and lunch-time and over time developing a sense of the values and beliefs of staff in the school. A Health Promoting Schools Committee was developed ensuring that it included 'members' of the various groups within the school. Conducting an audit of the school's health promotion activities and policies and organizing an 'HPS Visioning Day' were some of the early initiatives. All these developmental initiatives help to foster a shared understanding of the need to implement some changes to move the school in a health promoting direction.

7.4.2 Moving

When the audit was completed the school principal and assistant principal who are part of the HPS committee were committed to the plans that were drawn up which focused initially on canteen and healthy eating, physical activity, and staff health. In 2009, the Department of Education and Early Childhood Development (DEECD) annual health and well-being survey results showed that the school was below the state average in a number of areas in relation to health and welfare. To add to this, the school numbers were dropping. The school leadership saw becoming a HPS as part of a strategy to lift the survey results of the school, improve the culture and ethos and make the school 'the school of choice' for parents in the area.

As well as creating discordance between current canteen practice and the idea of being a health promoting school, there were legislative pressures acting as external forces that assisted the change management process (Lewin 1997b). Recent changes in legislation in regards to bans on selling confectionary in primary school canteens also worked in favour of developing a healthy canteen. The canteen manager's lease on the canteen building was up for renewal. The principal insisted on an overhaul of the canteen menu, removing the 'red' foods and having predominantly 'green' foods. This was conditional on the lease being renewed.

Programs are implemented within a political context (Rowling and Samadal 2011). At the time of the HPS program being introduced into BNPS, nationally there was much talk in the media around policies that concern diet and exercise. National and state support through public policy is essential for health promotion (Leeder 1997). State governments have been examining issues such as confectionary sold in schools, student's sedentary lifestyles and barriers to healthy eating and exercise. The policy changes introduced at the school have mirrored work that is currently being done nationally on these topics.

Plans for a school running track were implemented to support the increasing emphasis on physical activity. On behalf of the school, the HPO received a grant which enabled the school to construct a running track. EACH provided funds to purchase more lunch-time sports equipment for the school. The sports teacher commenced running lunch-time games for students. This focus on physical activity resulted in the administration reviewing the school's physical activity policy and it was revealed that the grade 3–6 students were not allocated the required amount of time for physical activity as advised by DEECD. This resulted in the policy being re-written to include three extra hours per week of physical activity for the grade 3–6 students. This allowed for an extra 120 h of physical activity per school year. Being seen as compliant with legislative requirements was a driver for change (Department Education and Early Childhood Development 2009).

Rogers (2003) recognized that during the re-defining stage, the organizations structure is modified to fit with the innovation, and at BNPS this process seemed to lead to the creation of an environment open to innovation. For example, in addition to the HPS framework, the school began to implement the Tribes Learning Community Program (Tribes) and Restorative Practices. The environment had been established whereby there was less resistance to implement new health and well-being programs. Tribes is a process which seeks to create a positive school learning environment (Gibbs 2001). Specific agreements in regards to behaviour are promoted throughout the school. Students learn a set of collaborative skills which are also to be practiced by the teachers and administrative staff. With the school adopting the HPS framework, Tribes has fitted well into the direction that the school was moving.

In tandem with the introduction of Tribes, the school began to provide teachers with training in the Restorative Practices behaviour management philosophy. Restorative Practices focuses on problem solving and repair of damaged relationships following an incident (Shaw 2007). Training in the Restorative Practices method was also available for parents to attend. Students began to request that teachers use this method to solve problems. Student leaders were up-skilled to be 'Peace Makers' who solved problems within the school yard using this system.

Restorative Practices and Tribes programs may have had a greater impact on the structure of the school than the Nutrition and Physical Activity strategies. The former programs were more challenging to introduce and possibly the most significant. Acceptance of the change to the structure by the staff was vital as it led up to what eventually became embedded into common practice. For example, the principal now recruits new staff members who are attracted to the ideologies of Restorative Practices and will embrace the approach it brings. There was much resistance to this method of dealing with classroom issues amongst the existing staff initially. In October 2010 a survey was conducted to determine the attitudes of the teachers to the adoption of the Restorative Practices philosophy. At this time the Restorative Practices program had been in place at the school for approximately 12 months. Fifty-six percent of the teaching staff responded to the survey. Out of this 56 %, only 46 % agreed that Restorative Practices had improved the school environment. However, after 1 year of 'clarifying' (Rogers 2003) during which, Restorative Practices was adopted by more teachers and gradually became imbedded, the survey was repeated. The repeat survey had a response rate of 95 %. Fifty-eight percent of teachers now agreed that Restorative Practices had benefited the school environment. There are still many teachers who are unsure if the environment has improved and 5 % of the staff disagree, however the figures show that slowly the teachers are beginning to accept the Restorative Practices as policy. The resistance to this policy has decreased as the principal and assistant principal have promoted it tirelessly. The assistant principal has continually used case histories and stories to emphasise the success of the method. Evaluating against outcomes for all students would need to be undertaken to validate these case study reports.

Both Tribes and the Restorative Practices process lend themselves to the HPS model. Both employ a whole of school approach and work to build social capital in the school, striving for a positive culture. Both Tribes and Restorative Practices cater to the school community's mental health and well-being. St Leger (2005) notes that school organisation has an impact on student's health and well-being. These programs provide a framework for structural change within the school by modifying how the teachers interact with the students. The programs help by creating a supportive school environment that is conducive to learning.

The NSW Health (2001) capacity building framework identifies the importance of leadership at different levels, literature on the role of 'champions' in the change process indicates that they do not necessarily need to be content experts, but rather they need to have credibility in the organisation and the ability to 'market' the initiative (Martinsons 1993). This may explain why the efforts of the principal and assistant principal were beneficial for the movement towards HPS, their position of leadership enabled them to communicate the benefits of the HPS model as well as publicly recognise those who were exhibiting the desired behaviours. Furthermore, this highlights the benefit of the HPS 'content expert' playing the role of the 'critical friend' alongside the staff who can drive the change internally. Similarly, Rogers (2003) identified that the 'change agent' i.e. the EACH HPO, is different to the recipients of an innovation with regards to technical competence and this provides them with credibility. Perhaps the fact that the HPO was visible within the school and staff were aware of her expertise together with the support of the principal and assistant principal who were credible in 'Education' and insiders too was key to the success of the change process.

7.4.2.1 Summary of Moving

There were a number of key areas the school identified that needed changing such as annual student survey results and the types of foods being sold at the canteen. There was also some media attention and recent legislative change at a State level related to nutrition and physical activity that helped create a more positive climate for pursuing a change agenda within the school. Programs were implemented for physical activity and the focus on this issue highlighted the need for school policy change, the school was not currently meeting Department guidelines on amount of time devoted to physical activity within the school curriculum. These activities and policy alterations created a broader momentum for change within the school that made possible the introduction of new programs aimed at improving the emotional and social health of the student population. At this stage of the HPS process the school administration were instrumental in pursuing these new initiatives.

7.4.3 Refreezing

A number of changes that took place are now part of the organizational norm within the school (Lewin 1997c). When appointing new teachers to the school the principal now makes a point of appointing teachers who have experience and training in Restorative Practices. The school achieved 'Kids- Go for your life' accreditation. Kids- Go for your life was a state-wide initiative that supports early childhood and primary school services, as well as local communities, health professionals and families in the promotion of healthy eating and physical activity for children. The program was based on six key messages around health targets such as drinking water and engaging in active play. When schools met the various requirements, they received the Kids-Go for your life award. In 2012, the program was superseded by the Victorian Prevention and Health Promotion Achievement Program (State Government Victoria 2012).

Initially there was apathy in regards to participation in the Kids-Go For Your Life award. MacBeath and Jardine (1998) point out that willingness to participate in change will generally not be found across the entire school. The majority of schools will experience resistance from some members of staff. Lewin (1997c) identifies that the signs of refreezing are when the changes are incorporated into everyday life and the changes are internalized. The policy changes were discussed at staff meetings and championed by the principal and assistant principal. A number of staff, who had traditionally used sweets as a reward for good behaviour in the class-room, were challenged by the 'no sweets' policy. The assistant principal emphasised why the change needed to occur and created a compelling message. Ultimately the policy had to be accepted due to the 'School Confectionary Guidelines' (Department education and early childhood development 2006). Although not universally embraced, the policy change went ahead, driven by the school leadership and the

HPS committee. Numerous authors (Lister-Sharp et al. 1999; International Union for Health Promotion and Education 2009; Williams et al. 1996) emphasise the importance of have school leadership involved in policy change.

The school staff and the majority of parents have accepted the new school manifesto, this is characterised by policy changes in regards to the canteen only selling healthy food and, staff not using sweets as a reward, however this has taken time. With the changeover of canteen managers, the Parents Association was quick to emphasise that whoever took the role on would need to abide by the healthy canteen policy.

Hawe (1994) sees schools as 'ecosystems' that respond to change with an intervention or a program. This system level change is difficult to document and when working in a school, almost imperceptible. It is only when group behaviour begins to change separate from being driven by policy or leadership that it is obvious that group norms are changing. Parents' new resistance to selling fund-raising chocolate is such an example. Parents began to raise concerns in regards to the annual fund-raising chocolate drive. A group of parents refused to participate in selling the chocolates as they felt that it was a conflict of interest with a health promoting school. Subsequently the parents association asked the HPO for ideas in regards to healthy fund-raisers, so that money could be raised without selling chocolates. The parents association also requested information about healthy food to be offered at the end of year function for the grade 6 students. The association was keen to uphold the idea of being a health promoting school and therefore was cautious about offering only 'red' foods to students. This is a clear example of the change in values and behaviours that is reflective of cultural change having taken place within the school community (Schein 2004).

Having established a 'Health Promoting' culture within the school, members of the school community started initiating more activities. A school vegetable garden was initiated by two enthusiastic teachers at the school. Members of the HPS committee began to explore the idea of setting up a Fruit and Veggie Co-op. BNPS is situated in a food desert. The ratio of fast food to fresh food outlets is 17:1. Public transport in the area is poor (Johnson et al. 2009). To build on the cooking demonstrations being offered to parents, using fresh produce, the committee indicated that they would like to make it easy for parents to purchase fresh fruit and vegetables and a Fruit and Veggie co-op commenced with eleven families signed up. This rose to 22 families by the end of the year. The art teacher agreed to work with the students to produce posters to promote the newly introduced wraps into the canteen. The posters were subsequently featured in the local paper. The HPO and a parent from the HPS committee attended a 'Greening up Your Canteen' workshop. The school prep teachers with a team of health professionals from EACH ran a Preps Dads breakfast. Along with a healthy breakfast, the fathers received health information and had their blood pressure taken. All these examples illustrate the change in culture and how enthusiasm can spread once new ideas are tried and seen to be successful (MacBeath and Jardine 1998; Schein 2004).

In the unfreezing stage it was identified that staff were dissatisfied with their own health and would like this to be an emphasis within the HPS approach. A staff

Pilates class was organized and initially subsidized by EACH. When the subsidy was finished, the teachers were happy to pay for this to continue. Forty percent of the teaching staff joined the classes which still continue to this day. Chrusciel (2008) identifies that in order for people to be accepting of the change, and indeed become early adopters, they need to see that they will benefit in some manner. The staff health program helped by affirming that the move to a HPS would also benefit the staff. The changes were not just about the students, but that the good health and happiness of the staff was just as important. By including the staff in the audit, the health promoting schools visioning day and regular chance for input at staff meetings, the staff were connected to the change process. Subsequent to the Pilates program, the Staff Health Term was introduced. Exercise equipment was placed in the staff room and friendly competition ensued. Ninety percent of staff were involved in a Staff Health Term team program, and 85 % of staff participated in staff health checks (Victorian Workcover Authority 2012). The idea that the school staff need to look after their own health and make it a priority has become embedded within the school psyche.

According to Rogers (2003), the easier it is for a group to see the innovation in practice, the more likely it is to be adopted. The Staff Health Term was a very visible outworking of the drive to become a HPS. The equipment was present in the staff room for a term and staff actively participated. This engendered much discussion on health, fitness, diet and exercise. Rogers (2003) also stressed the importance of peer-to-peer conversations in spreading ideas. Slowly the staff began to adapt and change to thinking of themselves as a health promoting school. Morning tea in the school staff room now consists of homemade yogurt, and stewed fruit. Staff decided to get rid of the biscuits as BNPS is a health promoting school. As schools are social complex adaptive systems 'freezing' is never a permanent situation (Keshavarz et al. 2010). However changes that occurred within the school have been accepted and become the new norm.

This has taken around 3 years to achieve and is still a work in progress. The success behind it has been due to a multi-model program which has addressed every area of the school. The commitment of the principal and assistant principal has been essential in achieving this culture change. The Principal continually talks about the school being a health promoting school. It is publicized in the school newsletter and frequently referred to within staff meetings.

7.4.3.1 Summary of Refreezing

At this stage there were a number of steps taken to embed the HPS within school policies and structures. In addition changes to the health and well-being environment were being driven by parents and teachers, whereas previously it had relied on the 'change agent' and school administration to initiate change. As examples parents began to raise concerns in regards to the annual fund-raising chocolate drive, requested information about healthy food to be offered at the end of year function for the grade 6 students, and a school vegetable garden was initiated by two teachers. These are all examples of a cultural change process where health and well-being were now seen as core business of the school and it also resulted in staff focusing on and making environmental improvements relevant to their own health and well-being.

7.5 Achievements and Limitations

In the 'Moving' section of this article, it was identified that the 2009 DEECD annual health and well-being survey results showed that the school was below the state average in a number of areas in relation to health and welfare. Since the introduction of the HPS model and its attendant programs, the school has seen an improvement in all areas of student engagement and wellbeing as identified in the annual DEECD Student Attitudes to School Survey.

In 2009 on the DEECD survey, School Connectedness was listed in the bottom 25 % of the State. In 2010 it moved into the second quartile and in 2011, it moved into the third quartile and above the State average for first time in 3 years. Student motivation was listed in the bottom half of the State in 2009, and 2010, however moved into the top half in 2011. In 2009 Connectedness to Peers was listed in the 2nd quartile, in 2010 just above the State mean and in 2011 moved into the 4th quartile, placing the school in the top 25 % of the State. Classroom Behaviour has moved from the first quartile in 2009, to the top 10 % of the State in 2011. The Staff Opinion Survey mirrors the data of the Attitudes to School Survey.

Other data collected such as behaviour records and group discussions with school staff validates this improvement in school behaviour. One of the significant limitations from an evaluation perspective is that while changes to policies relevant to physical health behaviours have been documented, there has been no ongoing monitoring of children's nutrition and physical activity levels which have also been priorities of the school. This is something that is planned for future. And of course as a single case study threats to validity such as differences in student cohort cannot be ruled out.

The school now faces the challenge of maintaining this good work. The Principal and senior staff are very aware, that without continued input to the programs that have been put in place that have yielded the improved data, the gains can be easily lost. Recommendations to DEECD from the school have included training of new teachers in Tribes, Restorative Practices, and an emphasis on including these as regular topics for discussion at staff meetings, so that skills can be frequently updated. Initiatives such as health promotion in particular the importance of mental health promotion needs to remain at the fore front of school operations. Rowe et al. (2007), reinforces the recognition of the significance of partnerships in the school community, and in particular highlights the influence of the relationships between students, school staff, partnering organizations and parents. One of the challenges is that, with changes in staff, these hard won relationships can be lost overnight unless reinforced by the entire school community.

7.6 Replicating the Model in Other Schools, Where to from Here?

The advantage of an internal/external 'change agent' model is that it can facilitate a whole of school health promotion approach. The case study detailed in this chapter was based on the HPO having one a day a week to devote to the role. This enabled the HPO to occupy a unique position in the school. Having an office at the school one day a week, attending morning tea, meetings etc. gave the HPO entrance to the teachers world, however, as the HPO was not a member of staff, she was also able to maintain a distance between the workings of the school and the HPO role. This has been part of the success of the change agent role.

The importance of an extra pair of hands around the school should also not be underestimated. Teachers feel overworked, under pressure (Kyriacou 1987; Easthope and Easthope 2000), and pulled in many directions. As well as teaching, many now find themselves taking on welfare roles and instigating health promotion projects. Having a worker who is prepared to assist the staff with these duties, is seen as a benefit to the teaching staff and the school. If the HPO is attached to a community health service, as is the HPO in this article, they have the added advantage of being able to draw on a wide variety of health professionals who can assist the school community. Examples of this include: nutritionists from EACH assisted with canteen reform and healthy cooking classes, nurses from EACH assisted with health checks at the Fathers Day breakfast, and an EACH disability group provide maintenance and gardening assistance in the school grounds.

A small study conducted in the south eastern suburbs of Melbourne surveyed either principals, assistant principals or leading teachers about how they currently structured their health promotion work and whether they saw merit in a paid HPO position. There were 15 respondents to the survey (37.5 %) and there was strong support for this model:

> the role of health promotion within the school is expanding, requiring greater resources - especially staff time. At present this is 'added' onto other roles often results in less than adequate provision of support. A paid coordinator would be an extremely valuable asset to the school.
>
> Schools do not have the money to fund any other bodies, we have enough trouble stretching the resources without finding and other things ??? that is thrown at us...Maybe we could share a body between 2 or 3 schools?

Another benefit of the HPO, not having a teaching degree and not being on the school staff, is that he/she cannot be co-opted into taking classes. One principal warned that if a staff teacher was given a percentage of their time to spend on health promotion activities, they would run the risk of been seen as an 'emergency fill in' when other teachers needed to be absent from their classes. If schools were given discretion over how the health promotion resources were allocated it might not be used to support a whole of school approach.

> We need more staff in schools. If the health promotion coordinator planned and implemented the health and PE program as well as an extra staff member yes. If not we need more staff for smaller class sizes. A welfare officer would be of more help - that's health promotion as well

While the majority of schools could see the benefit in a HPO coordinating and linking together whole school health and well-being programs, they had competing priorities that meant any additional resources would not be directed towards funding a health coordinator role. The results did indicate that few schools would be willing to provide the necessary resources to assist school health promotion, which leaves these schools vulnerable to inconsistent delivery of programs. Further, if they were provided with the extra resources to facilitate whole of school health promotion they might use the resources to fund welfare programs only without consideration of a broader social and environmental approach. The external/internal model whereby the HPO is responsible both to a health agency requiring settings based determinants approaches and to a school to facilitate this change can help ensure that a whole of school health promotion model is the goal being pursued. The challenge would be resourcing that level of support. In the KidsMatter pilot project there were eight project officers who were the 'critical friends' to the 100 participating schools (Slee et al. 2009). In the States with larger populations the project officer could have 20 schools and much fewer in smaller States and Territories. While schools can potentially see the benefit of such a position how this could be funded remains uncertain. Further research is required on sustainable models of support for school health promotion.

7.7 Training and Support for HPOs Based in Schools or Other Settings

It has been recommended by Butler et al. (2001) that a 'critical friend' in a school needs to have a teaching background. It can be argued, however, that it is not the professional background that is important; it is the personal interaction, along with the skills that the 'critical friend' brings to the position. The role has been described as being one that is dynamic, requiring a high level of flexibility (Butler et al. 2001). Ideally the person has skills in health promotion activities such as data collection, conducting surveys, needs analysis and the skills to implement and evaluate programs. However they also need skills in regards to navigating the relationships in the school, opening up dialogue, raising questions, encouraging and keeping the momentum of the program going. Boot et al. (2010) emphasises the importance of both constructive personal relationships combined with professional skills that are seen to be valued by the school.

There are a number of challenges within a change agent role described by Faubert (2009) which are similar within the school experience. The balance between focusing on process and building long-term capacity versus producing short term outcomes that can generate good will and momentum is a challenge. The requirement to meet the funding bodies' requirements on health promotion versus the 'bottom-up' approach of working with the school to identify and respond to their articulated needs is another tension. Faubert (2009) discussed that while there are these ongoing tensions one of the strengths in the dual role is being able to provide a distanced perspective while at the same time becoming immersed within the community. Training and support for change agents around some of these issues is one of her recommendations which are equally applicable for a health promotion practitioner based in the school system (Faubert 2009).

In summary, one of the lessons from this work is that project planning models are of secondary importance relative to an understanding of organizational change theories and practices. While core competencies for health promotion practitioners have a strong emphasis on partnerships and capacity building (James et al. 2007), how these skills are taught and developed in undergraduate and postgraduate degrees requires further research. They may be taught in the context of program planning rather than the broader context described in this chapter of being a 'change agent.'

7.8 Conclusion

A number of authors (Boot et al. 2010; Laughlin 1990) indicate that in regards to becoming a HPS the process and journey that the school experiences are just as important as the successes. Each school is unique in its own right, with its own, needs and strengths. Due to the lack of long term evidence there is a movement away from giving settings such as schools packages and expecting them to implement it as designed (Hawe et al. 2009a). Rather, viewing interventions as events that can alter the function and structure of a setting/system may offer more opportunity for long term sustainable gains in health (Hawe et al. 2009b). While this hypothesis needs to be tested (Hawe et al. 2009b), the experience of this case study supports this literature that understanding and influencing organisational change is fundamental to improving the health promotion capacity of an organisation. Schools will show the most interest in elements that are a pressing need for the school. If schools do not see a need or something as a priority, obtaining the goodwill and agreement from the school staff will be difficult. A school's core business is teaching. The HPO needs to demonstrate that health promotion interventions will improve the learning environment in the school. The HPO should be able to work constructively with the school, using the framework to meet the school at its point of need and therefore gain acceptance of the program.

Schools also need assistance to implement the HPS model. Ideally they need a professional from a health or education background who can become the 'critical friend' at the school. Boot et al. (2010), in the Dutch 'Schoolbeat' program and Bond et al. (2001) in the Australian Gatehouse project both agree that having assistance on demand was a key part of the success of the programs. In regards to health promotion, the 'critical friend' needs to have the professional skills to assist the school in implementing structural health promotion programs, policy development and the ability to work with the staff to change the ethos and culture of the school. However, assistance only is not enough. The relationship of the 'critical friend' to the school is paramount to the acceptance of the HPS framework. Both Boot et al. (2010) and Butler et al. (2001) stress that a close relationship with the school with resulting positive feelings is also important for mutual trust. The 'critical friend' has no standard role description. The role requires a high level of skill and flexibility and the ability to draw on a repertoire of actions, depending on the context of the school (Butler et al. 2001). Further research is required on how staff could be trained to work in such roles and potential funding mechanisms that could ensure equitable access to this resource.

References

Ainscroft, M., & West, M. (2006). *Improving urban schools, leadership and collaboration*. Buckingham: Open University Press.

Allensworth, D., & Kolbe, L. (1987). The comprehensive school health program: Exploring an expanded concept. *Journal of School Health, 57*(10), 409–412.

American School Health Association. (2012). *American School Health Association*. http://www.ashaweb.org/i4a/pages/index.cfm?pageid=3279. Accessed 20 May 2012.

Armstrong, E. (2011). *Principals and health sector perspectives on establishing health promoting schools in Victoria*. Melbourne: La Trobe University.

Australian Psychological Society. (2014). KidsMatter Australian Primary Schools Mental Health Intitative. http://www.psychology.org.au/publications/inpsych/kidsmatte

Baker, B. J., Curtis, D. B., & Benson, W. (1991). *Collaborative opportunities to build better schools: A study on the impact of educational reform on the schools*. Normal: Illinois Association for Supervision and Curriculum Development

Baum, F. (2002). *The new public health*. South Melbourne: Oxford University Press.

Bond, L., Glover, S., Godfrey, C., Butler, H., & Patton, G. (2001). Building capacity for system-level change in schools: Lessons from the Gatehouse Project. *Health Education and Behavior, 28*, 368.

Boot, N., van Assema, P., Hesdahl, B., & de Vries, N. (2010). Professional assistance in implementing school health policies. *Health Education, 110*(4), 294–308.

Brisbane North Public Health Unit. (2001). *Healthy promoting school toolbox: A toolbox for creating healthy places to learn, work and play*. Brisbane: Central Public Health Unit Network Queensland Health.

Butler, H., Krelle, A., Seal, I., Trafford, L., Drew, S., Hargraves, J., et al. (2001). *The critical friend. Facilitating change and well-being in school communities*. Camberwell: ACER Press.

Chrusciel, D. (2008). What motivates the significant/strategic change champion(s)? *Journal of Organizational Change Management, 21*(2), 148–160.

Clift, S., & Jensen, B. B. (2005). *The health promoting school: International advances in theory, evaluation and practice*. Copenhagen: Danish University of Education Press.

Costa, A., & Kallick, B. (1993). Through the lens of a critical friend. *Educational Leadership, 51*(2), 49–51.

Denman, S. (1999). Health promoting schools in England – A way forward in development. *Journal of Public Health, 21*(2), 215–220.

Department Education and Early Childhood Development. (2006). *School confectionary guidelines*. http://www.education.vic.gov.au/management/schooloperations/healthycanteen/confect.htm#2. Accessed 5 Apr 2012.

Department Education and Early Childhood Development. (2009). *Improving school sport and physical education in your school*. Melbourne: Department education and early childhood development.

Easthope, C., & Easthope, G. (2000). Intensification, extension and complexity of teachers workload. *British Journal of Sociology of Education, 21*(1), 43–58.

Eddy, C. (2006). You've got a friend. *Teacher, 174*, 12–15.
Faubert, C. (2009). Tensions and dilemmas experienced by a change agent in a community-university physical activity initiative. *Critical Public Health, 19*(1), 71–86.
Fullan, M. (2006). The future of educational change: System thinkers in action. *Journal of Educational Change, 7*(3), 113–122.
Gibbs, J. (2001). *Tribes learning communities- KidsMatter- Australian Primary Schools Mental Health Initiative*. http://www.kidsmatter.edu.au/primary/programs-guide/tribes-tlc/. Accessed 31 Mar 2012.
Gibbs, P., & Panayiotis, A. (2008). Understanding friendship between critical friends. *Improving Schools, 11*, 213–225.
Hawe, P. (1994). Capturing the meaning of 'community' in intervention evaluation: Some contributions from community psychology. *Health Promotion International, 9*(3), 199–210.
Hawe, P., Noortb, M., King, L., & Jordensa, C. (1997). Multiplying health gains: The critical role of capacity-building within health promotion programs. *Health Policy, 39*(1), 29–42.
Hawe, P., Bond, L., & Butler, H. (2009a). Knowledge-for-Action theories can inform evaluation practice. What can complexity theory add? *New Directions in Program Evaluation, 124*, 89–100.
Hawe, P., Shiell, A., & Riley, T. (2009b). Theorising interventions as events in systems. *American Journal of Community Psychology, 43*(3–4), 267–276. doi:10.1007/s10464-009-9229-9.
Inchley, J., Muldon, J., & Currie, C. (2006). Becoming a health promoting school: Evaluating the process of effective implementation in Scotland. *Health Promotion International, 22*(1), 65–71.
International Union for Health Promotion and Education. (2009). *Achieving health promoting schools: Guidelines for promoting health in schools*. Version 2 of the document formerly know as 'Protocols and Guidelines for Health Promoting Schools'. www.iuhpe.org/images/PUBLICATIONS/THEMATIC/HPS/Hpsguidelines_ENG.pdf
James, R., Howat, P., Shilton, T., Hutchins, C., Burke, L., & Woodman, R. (2007). *Core health promotion competencies for Australia 2007*. Western Australian Centre for Health Promotion Research, the centre for behavioural research in cancer control, Curtin University and the national heart foundation.
Johnson, C., Lewis, E., McDonald, G., Sampimon, E., Liu, F. Y., & McDonald, K. (2009). *Outer east community food access research project*. www.oehcsa.org.au/library-resources/occess-nutritions-food/outer-east-community-occess-research-project
Keshavarz, N., Nutbeam, D., Rowling, L., & Kavarpour, F. (2010). Schools as social complex adaptive systems: A new way to understand the challenges of introducing the health promoting schools concept. *Social Science and Medicine, 70*, 1467–1474.
Kyriacou, C. (1987). Teacher stress and burnout: An international review. *Educational Research, 29*(2), 146–152.
Laughlin, M. (1990). The Rand change agent study revisited: Macro perspectives and micro results. *Educational Researcher, 19*, 11.
Leeder, S. (1997). Health-promoting environments: The role of public policy. *Australian and New Zealand Journal of Public Health, 21*(4), 413–414.
Lewin, K. (1997a). Field theory and learning. (Originally published in 1942) In K. Lewin (Ed.), *Resolving social conflicts and field theory in social science* (pp. 212–230). Washington, DC: American Psychological Association.
Lewin, K. (1997b). Defining the "Field at a Given Time". (Originally published in 1943) In K. Lewin (Ed.), *Resolving social conflicts and field theory in social science* (pp. 200–211). Washington, DC: American Psychological Association.
Lewin, K. (1997c). Frontiers in group dynamics. (Originally published in 1947.) In K. Lewin (Ed.), *Resolving social conflicts and field theory in social science* (pp. 301–336). Washington, DC: American Psychological Association.
Lister-Sharp, D., Chapman, S., Stewart-Brown, S., & Sowden, A. (1999). Health promoting schools and health promotion in schools: Two systematic reviews. *Health Technology Assessments., 3*, 1–207.

Lynagh, M., Schofield, M., & Sanson-Fisher, R. (1997). School health promotion programs over the past decade: A review of the smoking, alcohol and solar protection literature. *Health Promotion International, 12*(1), 43–60.
MacBeath, J. (1999). *Schools must speak for themselves: The case for school self-evaluation.* London: Routledge.
MacBeath, J., & Jardine, S. (1998). I didn't know he was ill- the role and value of the critical friend. *Improving Schools, 1*(1), 41–47.
Maroondah City Council. (2009). *Bayswater Nth community renewal.* http://svc041.wic012v.server-web.com/BayswaterNorthCommunityRenewal.aspx. Accessed 11 Feb 2012.
Martinsons, M. G. (1993). Cultivating the champions for strategic information systems. *Journal of Systems Management, 44*(8), 31.
NSW Health. (2001). *A framework for building capacity to improve health.* Sydney: Health Department.
Nutbeam, D., Macaskill, P., Smith, C., Simpson, J., & Catford, J. (1993). Evaluation of two school smoking education programmes under normal classroom conditions. *British Medical Journal, 306*, 102–107.
O'Connor, D. L., & Ertmer, P. A. (2006). Todays coaches prepare tomorrow's mentors: sustaining results of professional development. *Academy of Educational Leadership, 10*(2), 97–112.
Rogers, E. M. (2003). *Diffusion of innovations.* New York: Free Press.
Rowe, F., Stewart, D., & Patterson, C. (2007). Promoting school connectedness through whole school approaches. *Health Education, 107*(6), 524–542.
Rowling, L., & Samadal, O. (2011). Filling the black box of implementation for health-promoting schools. *Health Education, 111*(5), 347–362.
Schein, E. H. (2004). *Organizational culture and leadership.* San Francisco: Jossey-Bass.
Senior, E. (2012). Becoming a health promoting school: Key components of planning. *Global Health Promotion, 19*, 23–31.
Shaw, G. (2007). Restorative practices in Australian schools: Changing relationships, changing culture. *Conflict Resolution Quarterly, 25*(1), 127–135.
Slee, P. T., Lawson, M. J., Russell, A., Askell-Williams, H., Dix, K. L., Owens, L., et al. (2009). *KidsMatter primary evaluation final report.* Bedford Park: Centre for Analysis of Educational Futures. Flinders University of South Australia.
St Leger, L. (1998). Australian teachers' understanding of the health promoting schools concept and the implications for the development of school health. *Health Promotion Journal of Australia, 20*(2), 223–235.
St Leger, L. (2005). Protocols and guidelines for health promoting schools. *Promotion and Education, 12*(3–4), 145–147.
St Leger, L., & Young, I. M. (2009). Creating the document 'Promoting health in schools; from evidence to action'. *Global Health Promotion, 16*, 69–71.
St Leger, L., Kolbe, L., Lee, A., McCall, D., & Young, I. (2007). School health promotion – Achievements, challenges and priorities. In D. J. C. McQueen (Ed.), *Global perspectives on health promotion effectiveness.* New York: Springer.
State Government Victoria. (2012). *Kids-Go for your life. Information for professionals.* http://www.goforyourlife.vic.gov.au/hav/articles.nsf/pracpages/Kids_Go_for_your_life?open. Accessed 31 Mar 2012.
Stewart-Brown, S. (2006). *What is the evidence on school health in improving health or reventing disease and specifically, what is the effectiveness of the health promoting schools approach?* Copenhagen: WHO Regional Office for Europe.
Swaffield, S. (2007). Light touch critical friendship. *Improving Schools, 10*(205). doi:10.1177/1365480207077845.
Swaffield, S., & MacBeath, J. (2005). School self-evaluation and the role of a critical friend. *Cambridge Journal of Education, 35*(2), 239–252.
Veselak, K. (2001). Historical steps in the development of the modern school health program. *Journal of School Health, 71*(8), 369–372.

Williams, P., Weston, R., McWhirter, J., James, E., Moore, D., Coulter, B., et al. (1996). Health promoting schools: Lessons from working intersectorally with primary schools in Australia. *Health Education Journal, 55*(3), 300–310.
Victorian WorkCover Authority. (2012). *WorkHealth check research*. http://www.vwa.vic.gov.au/safety-and-prevention/health-and-wellbeing/research/charts. Accessed Apr 2012.
World Health Organisation. (1986). *Ottawa charter for health promotion*. Paper presented at the first international conference on health promotion, Ottowa, Canada, 31 January 2012.
Young, I. (2005). Health promotion in schools: A historical perspective. *IUHPE-Promotion and Education, 12*(3–4), 112–117.

Chapter 8
Drama-Based Learning for Teachers' Education in Health Promotion

Athina Karavoltsou

Abstract This chapter is ultimately about the ways that drama in education methodology enhances adult learning, in this case through professional development activities aimed at improving the instructional practices of teachers in health promotion curricula. It will explore the theoretical connections, between a drama in education based pedagogy, adult education and health promotion. Adult learning theories, such as self-direction, transformation and emancipation as well as social cognitive theory will be examined, in an attempt to explore synergistic ideas and goals they share with drama and the philosophy of health promotion. It will be claimed that teacher education through drama is extended beyond knowledge and skill development and addresses the personal identities and moral purposes of teachers, the cultures and contexts in which they work. Moreover, drama will be discussed as an experiential and education-in-negotiation format of professional development, in which teachers' critical input for providing direction and meaning of planned initiatives and their outcomes, is strongly encouraged and cultivated. The Vygotskian paradigm will be proposed as a comprehensive epistemology for teacher education and a Foucauldian perspective will be offered as a promising theoretical basis for the elaboration of such a paradigm.

Keywords Drama in education • Adult education • Health promotion • Teacher training

A. Karavoltsou (✉)
Athens, Greece
e-mail: akaravol@yahoo.gr

V. Simovska and Patricia Mannix-McNamara (eds.), *Schools for Health and Sustainability: Theory, Research and Practice*, DOI 10.1007/978-94-017-9171-7_8,

8.1 The Need for Effective Teacher Education in Health Promotion

Traditionally, school-based curricula are provided to students by teachers, making teachers key to effective implementation. Consequently, teacher training is regarded as essential for the effective implementation in schools of any innovative, teacher-provided curriculum (Tortu and Botvin 1989; Cameron 1991; Glynn 1989; Smith et al. 1993). Even though prevention programs are a well-researched area, the importance of training prevention professionals tends to be overlooked (Dusenbury and Falco 1995; Finn and Willert 2006; Mclaughlin and Vacha 1993; Patterson and Czajkowski 1979). Training provided for potential implementers of prevention programs is a crucial element in the success of program implementation and intervention influence. Furthermore, successful training can minimize the chances of committing a statistical error – i.e. correctly concluding that the prevention program is ineffective for the wrong reason (Basch et al. 1985; Helitzer et al. 2000; Schwartz and Carpenter 1999; Wade 2001). Some research has shown that in-service training is positively related to more complete implementation and, in some cases, enhanced student outcomes (Connell et al. 1985; Ross et al. 1991). Other studies (Basch et al. 1985) have shown no relationship between implementation fidelity and cognitive outcomes. Multiple studies have demonstrated that when teacher training is overlooked or ineffective, the result is implementation failure (Tortu and Botvin 1989; Sobol et al. 1989; Flay et al. 1995; Botvin et al. 1989). In fact, implementation failure and degradation are problems encountered even in otherwise successful school-based behavior change programs that included teacher training (Rohrbach et al. 1993; McCormick et al. 1995; Olson et al. 1993).

Despite the obvious and important prerequisite of effecting (implementation) behavior of teachers, which must occur before we can succeed in positively influencing children's health risk behaviors, and despite the public health interest in school-based programs, teacher training has received little attention as a public health issue. Teaching as a profession is not unlike other practitioner enterprises, in that it requires constant maintenance and upgrading of knowledge and skills in order to meet the changing demands of the marketplace, in this case, the classroom. Overall, "The purpose of staff development is to bring about change in the beliefs, attitudes, and classroom practices of teachers, and ultimately to bring about changes in student learning outcomes" (McBride et al. 1994, p. 36). However, an in-depth look at what teachers' lived experience of professional development is like and the supports and constraints of their environmental conditions on implementation of what they learn in professional development, has received little attention in the published research.

In contrast to the many advantages associated with using schools as a vehicle for prevention programs, schools present natural barriers to effective implementation. Barriers can originate with the teacher or the system (e.g. lack of administrative support). Both types of barriers can inhibit a positive teacher reaction to a new curriculum and undermine its subsequent implementation and long-term acceptance

(Levenson-Gingiss and Hamilton 1989). Behavior change curricula typically address the psychosocial factors that promote the target behavior (e.g., smoking). Therefore, they involve innovative learning activities that may require the use of new or nontraditional teaching styles with which teachers are unfamiliar or uncomfortable (Botvin et al. 1990). Also, because health promotion is not a required subject at all grade levels or in all schools, there is often no natural "niche" for health promotion curricula and no teacher prepared to teach them (Tappe et al. 1995; Smith et al. 1995). Furthermore, for many schools, there is no incentive to allocate limited time and resources to health promotion curricula (D'Onofrio 1989; Taggart et al. 1990; Smith et al. 1995). Teachers may resist or even actively oppose assuming new work demands unrelated to their existing teaching objectives.

There is a plethora of evaluative studies that illustrate what constitutes quality or effective professional development, based on immediate participant feedback and how teachers implement learning into their classrooms (Abrami et al. 2004; Boyle et al. 2005; Bryant et al. 2001; Donnelly et al. 2005; Firth 1977; Florio-Ruane 2002; Guskey 2002; Johnson 2006; Supovitz and Turner 2000; Willinsky 2005). It is well recognized that the use of established theory is essential to the successful design and development of behavior change curricula (Parcel 1984; Evans 1988; McCaul and Glasgow 1985; Sussman 1991; Sussman et al. 1995). Indeed, recent studies have recognized the importance of incorporating theory into all aspects of project design, including teacher training (Stone et al. 1996; McGraw et al. 1994; Perry et al. 1996). There is a wide spectrum of theoretical models (Glickman et al. 2007; Gordon 2004) that describes effective professional development. Nonetheless, specifics of how teacher training might incorporate theory in general, and motivational strategies in particular, have been largely absent from the literature.

This paper is ultimately about the ways that drama in education (DIE) methodology enhances adult learning, in this case through professional development activities aimed at improving the instructional practices of teachers in health promotion curricula. In accord with Guskey and Huberman's (1995) synthesis on the literature of effective professional development, it will be claimed that teacher education through drama is extended beyond knowledge and skill development and addresses "...the personal identities and moral purposes of teachers, the cultures and contexts in which they work" (p. 14). It will also be argued that drama allows for "...personal involvement in reflexive learning" and "...involves the development of insights into experience, new mindsets and attitudes", all themes that arose from Carroll's analysis (2005, p. 206), of what elementary teachers consider important in an effective professional development model. More than a thousand mathematics and science teachers in U.S.A reported that they wanted to be part of professional development that considers their needs (Gordon 2004) and has opportunities for active learning (Garet et al. 2001). Hence, DIE's approach will be discussed as an experiential and education-in-negotiation format of professional development, in which teachers' critical input for providing direction and meaning of planned initiatives and their outcomes, is strongly encouraged and cultivated.

In the following sections, I examine the arts, in general, as a learning medium and the basic characteristics of a drama-based-pedagogy, as a theoretical framework for teachers' professional development, in health promotion. Adult learning theories, such as self-direction, transformation and emancipation as well as social cognitive theory are also discussed, in an attempt to examine synergistic ideas and goals they share with drama and the philosophy of health promotion.

8.2 Arts as a Way of Learning in Adult Education

Art is seen as a way of knowing that 'releases the imagination' (Greene 1995, p. 27), allowing creative learning to emerge. Once we allow the creativity to flow through us, 'we become agents in our own learning process' and '…learning becomes an active process rather than a passive one' (Greene 1995, p. 31). According to the position statement, on the philosophy of health education, of the American Association of Health Education (2008), knowledge should be taught in a manner that facilitates an understanding of current realities and fosters a willingness among students to accept today's information as usable, but anticipate that later discoveries, perceptions or political realities may significantly change the usefulness of that knowledge. Learners will gain their security, finally, not in given and fixed bodies of knowledge, but in the skill of knowledge acquisition and the ability to analyze and apply it. Individuals must be given an opportunity to choose which information to believe, and which behaviors to perform. If not given such opportunities they may revolt, become apathetic, or deliberately live in opposition to what is known and what is taught. In other words, the ability to learn how to learn is considered equally important in both disciplines.

Barone and Eisner (1997, p. 23), pioneers in the area of arts-based research, agree that 'rationalist modes of inquiry have served to suppress artistic modes of expression'. Eisner (1995, p. 2) has also stated that 'artistically crafted work also has the capacity to put us in the shoes of those we do not know'. This level of understanding can create a sense of empathy, which opens us up to even greater learning. Moreover, art not only engages us at an intellectual level; it evokes feelings, intuitions, and even bodily sensations. Paying attention to these modes of experiencing can result in deeper knowledge. According to Dirkx (2001), it is important to pay attention to the emotional aspects of our learning process. Awareness of our emotional state comes to us in the form of images. These images help us connect our inner self with the outer world, which is what Dirkx (2001, p. 69) refers to as 'soul work'. The concept of "emotional intelligence", as Goleman (1995) uses it, includes the abilities to rein in emotional impulses, to read another's innermost feelings, to handle relationships smoothly and to motivate oneself. "These are the capabilities that are going to determine our success in family life, in careers, with friends, as citizens," he says. "These are the abilities that make us people" (Goleman 1995, p. 156). The foundations of health behavior most certainly require these positive attributes.

During the last decade, the climate in adult education has shifted significantly in relation to the role and recognition of the arts as key learning processes which attend to multiple ways of knowing: notably affective, cognitive, spiritual, imaginative and somatic (see Stuckey 2009). Despite their apparent neglect in mainstream educational circles (see Kazemek and Rigg 1997; Rademaker 2003; Knowles and Cole 2008), it seems that the arts are finding a growing niche in adult education, where art is seen as a way of knowing that 'releases the imagination' (Greene 1995, p. 27), allowing creative learning to emerge, 'providing rich multi-sensory experiences that engage the whole mind-body-emotional system' (Dickinson 2002).

The full measurement of value in health education is its production of good for "me," not for "my eyes" or "my weight" or "my body." In modern health education the concept of "body" disappears. So does the concept of health education as good for any fragmented part. In its place is the evaluation of the effect of health education upon the individual as a person. Read and Stoll (2010), in their discussions of the implications of a holistic paradigm in health promotion, assert that health education is essentially an operation, a means of producing something, and thus it may neither employ, rely upon, nor use any "tangent of interest" which assumes the individual to be a "plurality", that is, made up of several separate and unrelated systems or parts. Operationism [in this instance health education] cannot rightly be used to justify the excursion onto tangents of interest that leave the organism fragmented into a plurality. Taking the discussion a bit further, the writers advocate for health educators to explore a closer cooperation between the empirical findings of psychoneuroimmunology (the evidence that thought and emotion affect the immune system at cellular and subcellular levels), and the logical, analytical concepts and reasoning expressed in current holistic thinking as it applies to health behavior. They also stress the need for this new science, as is of crucial importance to health educators, especially in professional preparation.

The arts are attributed with expanding the boundaries for knowing, learning, and comprehending culture 'holistically, naturally, and creatively, thus deepening understanding of self, others and the world' (Lawrence 2005, p. 3; see also English et al. 2003; Tisdell 2003; Song 2009). Further studies testify to the role of the arts in developing multi-skilling, enhancing the grasp of difficult concepts, aiding memory retention, extending attention spans, increasing concentration, and enjoyment of learning (Dickinson 2002; Yorks and Kasl 2002; Merriam et al. 2007; Manning et al. 2010). Art is thus used as a pedagogical means to access learning in curricular subjects as diverse as English language acquisition, cultural diversity (Clover 2006; Wesley 2007), community building and activism (Branagan 2005; Archer-Cunningham 2007; Kinloch 2007; Scher 2007; Clover 2010) and social justice (Clover and Stalker 2007; Trounstine 2007).

Increasingly, adult educators and community organizations are turning to art-based educational approaches to respond to social issues and 'to speak out and be heard', connecting 'vision to action' (Aprill and Townsell 2007, p. 62; see also Thompson 2002; Clover 2006; Lawrence 2005; Grace and Wells 2005, 2007). The arts, though not necessarily represented in mainstream literature, have frequently been a part of adult education in the context of social change. One of the most noted examples of emancipatory learning in the arts is the use of popular theatre, and especially the use of Augusto Boal's (1979) 'Theatre of the Oppressed'. Boal's work was developed in the tradition of Paulo Freire (1970) in order to empower participants to actively engage with and rehearse solutions to problems in their life. Within the critical tradition in adult education, Marcuse (1969) also turned to the liberating power of art as an avenue for social change. Marcuse believed that for an altered consciousness to develop, it is necessary for the adult to experience a fundamental estrangement from commonly accepted ways of thinking and feeling. Immersion in artistic experience is one way to induce a revolutionary estrangement from everyday life, a distancing from normality that nurtures the tendency to political critique. Art then gives us new forms of visual and spoken language and opens us to new ways of sensing and feeling. Learning these different forms of communication and perception is for Marcuse, the inevitable precursor to social action. Brookfield (2005, p. 200) notes that 'if there is any truth to Marcuse's argument regarding the power of aesthetic dimension', then 'adult education that concerns itself with liberating the senses through creative, artistic expression is potentially revolutionary'.

8.3 Drama as a Learning Medium

Drama may manifest itself in a variety of educational experiences and may be enacted in a variety of forms. This project is concerned with drama in education rather than with theatre or creative drama. Drama in education (DIE) emphasises the forging of meaning within collective "as if" fictional contexts encountered while participants are "in role". DIE is viewed primarily as a learning medium, where many skills and strategies used in theatre serve educational goals. Theatre emphasises performance; creative drama, the development of personal awareness, sensitivity, and self-confidence. What is important here to understand is that although these 'forms of drama experience share the same common elements of theatre: focus, metaphor, tension, symbol, contrast, role, time and space' (Bowell and Heap 2001, p. 1), their goals differ. Explaining the relationship between theatre and drama, Dorothy Heathcote, DIE's most important pioneer, has remarked (Wagner 1999, p. 147): 'the difference is that in theatre everything is contrived so that the audience gets the kicks. In classroom drama the participants get the kicks'. There is also considerable confusion between drama and simulation. What can be said here is that 'the educational philosophy underlying the simulation movement is rigorously behavioristic' and that 'the purely cognitive approach favored by simulation exponents ignores entirely or tries to minimize the affective dimension' (Morgan and Saxton 1987, p. 72). The latter is organically infused in DIE.

According to Morgan and Saxton (1987, p. 21), DIE "operates in two frames: the expressive frame (the outer manifestation) and the meaning frame (the inner understanding)". Apart from certain drama skills in handling of the form (for example, sustaining a role) there are a great number of other skills acquired through DIE: creative and imaginative powers; practical skills for communicating and expressing ideas, feelings and meaning; investigative, analytical, experimental and interpretative capabilities; aesthetic understanding and critical skills. These are generic life skills and overlap with those advocated in the field of adult education. All of the above are possible, because

> These three lengths of yarn: the culture of the fictional world, the substance of the material world and the reference to the world outside, are woven together with a fourth, that of the stance of those involved, to form a mutual cultural landscape that should hold us in relation to each other, so that we can explore how one ought to be (Hulson 2006, p. 10).

In DIE the educator 'needs to adopt a kind of quadripartite thinking in order to manage this complex, creative, educative process' (Bowell and Heap 2005, p. 64). In particular, the teacher requires the head of the playwright "needing to think about how to help the children craft the narrative so the story unfolds in a way that carries within it the learning"; the head of the director 'needing to steer the children to the learning within the narrative through the best dramatic performance structure' and the head of the actor 'needing to give a performance that engages and beguiles the children and supports and challenges them in the creation of their own roles (Bowell and Heap 2005, p. 64). Above all, the educator needs to hold all of the other thinking simultaneously, together with knowledge and understanding of the real context of the learners, classroom, school, community, culture, and curriculum. The teacher in drama also needs 'a critical awareness of herself as she operates in each function' (Bowell and Heap 2005, p. 65). This forms the so-called dimension of "the self-spectator" in DIE. Ultimately, the quadripartite thinking of the teacher elicits a reciprocal quadripartite response from the learners, which has also embedded within it the self-spectator. The activation of the "spectator-actor", "spectator-playwright", "spectator-director", and "spectator-teacher/learner" within teacher and learners, results in 'a spiral of creative exchange between them whereby both experience the power of drama' (Bowell and Heap 2005, p. 66).

Use of applied drama and theatre in health promotion is increasing, but knowledge about its effectiveness for health and well-being is still limited. Interventions aimed at effecting attitudinal and behavioral change through interactive methods, such as role-play, video games and group work, have led to self-reported reductions in risk behavior in adults (Wright and Walker 2006). According to a recent review (Joronen et al. 2008) on school-based drama interventions in health promotion for children and adolescents, most reviewed school-based drama or theatre plays succeed in the short term in increasing knowledge and positive attitudes related to health behavior among school children. It seems that the challenge is to find or develop a theory, which combines educational, drama and health theories with valid and reliable measurements to examine the effects of the intervention.

8.4 A Vygotskian Epistemology of DIE

During the past half-century, several efforts have been made to reform teacher education on the basis of an explicit theoretical paradigm. Among these efforts, the paradigm of competency-based teacher education of the late-1960s and 1970s, with the personal orientation to teaching and teacher education as its counterpart, and the paradigm based on reflection and inquiry of the late-1980s and 1990s, achieved the clearest recognition and the strongest influence on teacher-education practice (Zeichner 1983; Feiman-Nemser 1990). However, commentators agree that the impact of these paradigms has remained limited, and concerns have been expressed about the anti-theoretical bias of teacher education and its lack of an explicit epistemology (Ginsburg and Clift 1990).

A strong point of competency-based teacher education (Houston 1974; McKenzie et al. 1996; Valli and Rennert-Ariev 2002), is that it defines a public standard for teaching as a framework for teacher education, with explicit objectives and assessment criteria for effective performance in the daily practice of teaching. This paradigm has been criticized for reducing the teacher's role to that of a 'technician' or 'executive' (Borman 1990, p. 395; Valli and Rennert-Ariev 2002, p. 203), as it leaves little room for a personal interpretation in the light of either personal preferences or specific demands and conditions of the situation.

The paradigm consisting of a combination of the teacher as researcher (Stenhouse 1975) and the reflective practitioner (Schön 1983), is noteworthy in its emphasis that professional repertoires are not established once and for all and are not given from outside a practice, but have to be continually reaffirmed, or modified by questioning experiences in the light of standards of evaluation (Posner 1985; Tom 1985; Valli and Taylor 1987; Zeichner and Liston 1987; Richardson 1989; Calderhead and Gates 1993). A frequently signaled problem with this paradigm is that it refers to a formal procedure that may be applied to many different aspects of teaching but does not itself refer to any substantive image of teaching to which reflection and inquiry are to be addressed (Van Manen 1977; Hursh 1987).

Vygotsky gave his name to a tradition in social science currently termed cultural-historical psychology or cultural-historical activity theory. Chaiklin (2001, p. 21) defines cultural-historical psychology as 'the study of the development of psychological functions through social participation in socially-organized practices'. Vygotsky's ideas are influential among educationists (e.g. Davydov 1988; Moll 1990; Forman et al. 1993; Bruner 1996; John-Steiner and Mahn 1996; Wenger 1998) and his views have been acknowledged, and used fruitfully, in studies of teacher learning and professional development (see, Tharp and Gallimore's (1988) *Rousing Minds to Life* and Lave and Wenger's (1991) *Situated Learning*). However, the Vygotskian tradition has not been examined explicitly and consistently in order to devise a paradigm of teacher education.

The chief distinguishing mark of this perspective is that it does not set out from opposition between organism and environment (or individual and society), but from the idea of a unified system in which these two elements are joined together in a dialectical relationship, where humans are both shaping and shaped by their living conditions. Thus, a neo-Vygotskian perspective foreshadows a way of integrating the valuable elements of the teacher education paradigms, as outlined above, because it concentrates on the connections between individual functioning and development *and* the sociocultural practices in which individuals take part.

A first principle to be derived from the Vygotskian theoretical framework is that professional learning and development are best conceived and conditioned as an aspect of evolving participation in a social practice. Participation in an activity system involves both the performance of action and the assignment of meaning. Activity is fundamentally defined by the meanings it seeks to realize: the needs and motives it seeks to satisfy, and the goals it seeks to achieve (Huizen et al. 2005). Public meanings invested in the activity need to be explored by participants in relation to what makes participation personally meaningful to them (Zinchenko and Davydov 1985). In other words, they will have to orient themselves towards a public standard of teaching that reflects the values and goals in the cultural and political setting of the schooling in which they are engaging. This orientation should not lead them to be recruited into any existing ideology, but clarify and define their own allegiance and commitment to teaching as the core of their professional identity. Participation also involves forms of social interaction and co-operation in an institutional context, and the use of cultural resources (Gallimore and Tharp 1990; Lave and Wenger 1991). These situations are by themselves of a nature to encourage a negotiation of meaning between participants as a prerequisite for co-operation.

Another crucial element of the theory is that the personality is viewed as an integrative system, in which rational, volitional, and emotional aspects are welded together (Bozhovitch 1979; Chaiklin 2001, p. 241). This integrated view of personality development creates a clear distinction between the Vygotskian perspective as the basis for a paradigm of teacher education and the recent elaboration of a 'situational perspective' which is primarily based on the cognitive tradition (Putnam and Borko 2000).

Hence, competence is no longer the primary target of a teacher-education program, but is defined and developed in relation to an image of 'good' teaching that invites commitment to the ideal forms inspiring this activity, and to the identity of the actor as a participant in this activity (Wenger 1998). Reflection and inquiry are focused on the continuing harmonization of action and meaning. The Vygotskian perspective, while preserving the major strengths of earlier paradigms (the development of a personal orientation towards a standard of competence and reflective inquiry), holds promise for teacher education by offering the possibility for integration into a more comprehensive and theoretically more satisfying ensemble (Huizen et al. 2005).

A social constructionist approach might be seen to be close to DIE's epistemological orientation since the focus of the educator is on how social concerns are shaping learners who work in a group in order to give them some experience of the complex, competing demands that are made upon them in the social world. However, according to Roper and Davis (2000), both social constructionism and its current adversary of cognitivism, are considered inappropriate for DIE. The writers argue that

> Where cognitivism only knows the images, information and representations in minds, social constructionism only knows discourses, language and signifiers in the social world. Both are equally unable to acknowledge or give a role to material reality apart from reducing it to its own terms (Roper and Davis 2000, p. 225).

Trying to escape this mind over culture (and vice versa) dualism, which is advocated by the philosophical idealism of both these epistemologies, Roper and Davis (2000, p. 226), propose Vygotsky's (1978) dialectical materialist approach that seeks to place 'language and cultural tools in relationship to mind and material reality as a dialectical unity of opposites where none reduces to any of the others' but instead 'each undergoes change within the conditions of interlocking environment, species, socio-cultural history and individual development' (see also Wertsch 1985). Subsequently, whereas in the social constructionist epistemology the fictional world created by the learners and the teacher in DIE is a social construction of reality and 'the reality is in this created world', a Vygotskian approach would be underpinned by a different epistemology where the fictional world created in the drama would be 'related to an objective world existing independently of any consciousness' (Roper and Davis 2000, p. 230). Instead of the focus being on the issues involved in a fictional reality, DIE, following Vygotsky, attempts to relate these to the actual living content of the world, 'understood as existing prior to and co-existent with the invented world, thus forming a constant point of reference' (Roper and Davis 2000, p. 230).

In the following section, andragogy, self-direction, transformation theory and situated cognition are included as models that illustrate adult learning theories in sociocultural context. These theories encompass the complexity of adult learning through models that complement and support Vygotsky's sociocultural theory (1978) and support the need for practice-oriented models for planning and implementation of professional development.

8.5 Andragogy

Theories on cognitive development have been proposed to explain various aspects of cognitive, social, and emotional development to inform adult educators (Gardner 1983; Magolda 1992; Perry 1970; Silverman and Casazza 2000). Some researchers (Glickman et al. 2007; Gordon 2004; Loucks-Horsley et al. 2003) suggest that planning for effective professional development requires alignment of the professional development formats and approaches with the theories of adult learning. The earlier is andragogy.

In 1968, Malcolm Knowles proposed to the United States a European theory of adult learning that differentiated andragogy, "the art and science of helping adults learn" (Knowles 1980, p. 43), from pedagogy. Knowles based his theory of andragogy upon five assumptions that describe the adult learner. These included: (a) an independent spirit to direct his or her own learning, (b) the use of life experiences as a resource for learning, (c) a connection of learning needs to changing social roles, (d) a desire to apply new learning to relevant situations, and (e) intrinsic motivational factors that drive learning (Merriam 2001a, b). These principles relate directly to professional development as well as classroom learning situations. Since its introduction, andragogy has not been without its skeptics and detractors. Many have centered around the criticism that andragogical principles assume an autonomous learner who is free of social and contextual constraints and the original theory ignored the context in which learners live, work, and learn (Pratt 2002). More recent theorists have taken Knowles' (1980) view of adult education and altered it to include such aspects as cognitive processes, environment, context, culture, physiological processes, emotion, and other factors that impact learning, creating a richer field of theory that guides our understanding of how adults learn.

8.6 Self-Direction

Self-directed learning (SDL) is regarded by many as a key principle in adult education (see Garrison 1992, 1997; Knowles et al. 1998; Brookfield 2000; Brockett and Hiemstra 1991; Norton 2001; Boyer and Maher 2004). In its broadest sense, SDL describes a process in which individuals 'take the initiative, with or without the help of others, in diagnosing their learning needs, formulating learning goals, identifying human and material resources for learning, choosing and implementing appropriate learning strategies, and evaluating learning outcomes' (Knowles 1975, p. 18).

The HP standards also define a health – literate person as a responsible, productive citizen, an effective communicator, and a self – directed learner. One disposition related to this standard is the teacher who values critical thinking and self – directed learning as habits of the mind (Interstate New Teachers Assessment and Support Consortium 1995).

The main reason for the self-spectator to be embedded in the learners' quadripartite response is that to be engaged in drama is to be engaged in a process of "education for self-direction", as Dorothy Heathcote (1995) points out in the video *Pieces of Dorothy*. By using this exact term, Heathcote is creating a connection between drama in education and self-directed learning and particularly with Garrison's (1997) comprehensive model, which posits the challenge to adult educators of integrating metacognitive processes in self-directed learning.

Although 'the term self-direction has misled many into elevating the individual above the collective' (Candy 1991, p. 311), Garrison (1997) views self-directed learning from a 'collaborative constructivist' perspective, where cognitive and collaborative learning processes are integrated resulting in educational experiences that are both personally meaningful and socially worthwhile, leading to collaborative self-directed learning.

8.7 Situated Cognition and Communities of Practice

Situated cognition is another model of adult learning that is representative of the context-based adult learning framework. It is based on the notion that "learning in context is paying attention to the interaction and intersection among people, tools, and context within a learning situation" (Hansman 2001, p. 44). According to Hansman, it also is "incorporating the learners' developmental needs, ideas, and cultural context into the learning experience" (p. 44).

The 'Communities of Practice' model embodies the ideas of situated learning. All communities of practice (CoP) are characterized by the domain of knowledge they share, a community of people who care about the domain, and the shared practice they are developing to be used effectively within the domain (Wenger et al. 2002). The domain is the common ground or common identity of the group, whether it is related to education, healthcare, or computer imaging. The community is the social network of learning and inquiry should be its guiding force, as members seek to develop and extend their domain and practice through social interaction. Finally, the practice is the action put forth by the community to further the domain. All of these factors combine to form the community of practice concept. The characteristics of this shared association, as defined by Wenger (1998), are: (a) mutual engagement, (b) a joint enterprise, and (c) a shared repertoire.

While Wenger's work has been mainly with all types of organizations, including corporate work groups and health-care workers, his theories can be applied to the education environment in many ways. Many authors have discussed the use of communities of practice related to instructional design (Au 2002; Barton and Tusting 2005; Jawitz 2007; Keppell 2007; McLaughlin 2003; Sim 2006). These sources investigate CoP in various different aspects of adult learning, including online learning, post-graduate instruction, and teacher preparation. Research into the effective practices and organizational structures of professional learning (Cochran-Smith and Lytle 1990; Darling-Hammond 1998; Hargreaves 1994; McLaughlin 1998) have led to the understanding that "the old workshop delivery model for teachers must give way to vibrant and ongoing professional learning communities" (Lieberman and Wood 2002, p. 40). The literature on CoP related to professional development is an area that has developed just recently (Corley and Thorne 2005; Guskey and Huberman 1995; Hinson et al. 2005; Meyer 2005) and clearly there is room for more investigation in this area.

The promotion of health, equity and sustainability is most effective where the school uses its full organizational potential. Team work and cooperation in school are then key factors. Teacher training for comprehensive health promotion can possibly increase its sustainability by focusing on collaboration issues. Study results indicate that teacher collaboration is scarce and difficult to sustain (Stoll et al. 2006). If teacher collaboration is happening, it is more often informal collaboration as a result of personal initiatives than formal collaboration (Jourdan et al. 2008). On the basis of mainly sociological arguments, Edwards and colleagues (2002) outline collaborative strategies, as a major way of assisting teachers in their attempts to improve their teaching.

8.8 Transformative Learning and Critical Thinking

Aiming at better explaining how the assumptions under which a person operates are changed, through the process of learning, Mezirow (1978, 1990, 2000) presented his notion of "perspective transformation" to the literature on adult education, influenced by such writers as Habermas (1970, 1971), Freire (1970), and Gould (1978).

The theory holds the key to learning that is transformational in nature are experiences that cause the learner to become "...critically aware of one's own, cultural and psychological, tacit assumptions and expectations and those of others and assessing their relevance for making an interpretation" (Mezirow 2000, p. 4). Paramount to the success of transformative learning are critical self-reflection and reflective discourse. Both tenets have been recognized as important to creating change in participant beliefs and attitudes, which leads to higher levels of implementation of new learning in the work setting (Mezirow 1997, 1998; Baumgartner 2001; Guskey 2002; King and Lawler 2003; Merriam 2001a; McKeown and Beck 2004; Wlodkowski 2003; Ross-Gordon 2002).

The transformational paradigm of drama finds its echo in the pedagogy of Freire where education is seen 'as a liberating and humanistic task that views consciousness as intention towards the world' (Freire 1985, p. 115) and 'transformation is not just a question of methods and techniques but a different relationship to knowledge and to society' (Freire and Shor 1987, p. 107). There is a lived hybridity in the process of artistic acting, which may serve the intentions of a transformational pedagogy. Learners will be asked to take on roles and imagine themselves differently. Mezirow (2009, p. 96) also identifies perspectival learning as 'an essential dimension of transformative learning'. The much-discussed function of empathy in drama is a matter of taking another's perspective. In drama, an actor enters into the world of an Other and learns about the perspective of the Other. That assumed – or created – world of the Other is a medium for learning (Henry 2000).

In acting 'as-if' the world was otherwise, learners may be encouraged to discover that at personal, local, national and international levels 'they are free to negotiate, translate and therefore transform the problem of identities and the problem of the representation of identities' (Neelands 2004, p. 54). DIE's intention is to capture this flux of change. In its essential ephemerality and in the collision of realities-fictionalities between the stage and the social space, drama can create a sense of instability between what has been, what is and what might be. According to Mezirow (2009, p. 5), 'imagination of how things could be otherwise is central to the initiation of the transformative process'. Mezirow (1981) also focuses on the idea that learning is the process of making meaning from experiences. DIE provides knowledge which involves 'the cognitive, conative and the affective aspects of experience...functioning together' (Henry 2000, p. 58). Drama employs the mode of the 'self-spectator' in an attempt to help learners 'make sense of the layering of drama experience as it moves toward the possibility of some kind of self-transformation in the real context' (Bowell and Heap 2005, p. 67). In this sense, drama is in accord with transformative learning theory's 'metacognitive process of reassessing reasons supporting our problematic meaning perspectives' (Mezirow 2009, p. 95). Through dialogue and the presentation of issues as problems, learners acquire the ability to intervene in, rather than accept, reality, an intervention, which results from their critical consciousness of the situation. The investigation of learners' own meaningful themes establishes a climate of creativity and enthusiasm for risking experimentation, in which learners critically analyze reality and intervene as subjects in the historical process, understanding themselves not as spectators but as co-authors of action to transform the world. Neelands (1996, p. 29) writes that drama 'is a personally transforming cultural resource, one which makes the invisible influences of culture visible and discussible, and serves as a mirror of how we are made, and of who we might become'.

A key element of transformative learning theory is critical thinking and reflection that more specifically requires, first 'critical assessment of the sources, nature and consequences of our habits of mind' – and second, 'participating fully and freely in dialectical discourse to validate a best reflective judgement' (Mezirow 2009, p. 94).

The American *National Health Education Standards* (Joint Committee on NHES 1995) highlight the importance of critical thinking in the context of our field or discipline. The standards define a health – literate person as "a critical thinker and problem solver who uses decision making and goal setting in a health promotion context". The concept of health literacy (Joint Committee on National Health Education Standards 1995; Tappe and Galer-Unti 2001; American Association of Health Education 2008) is a good example of a cognitive approach in health education that employs both the skill of critical thinking and decision making in practice. Decision making is viewed as a systemic approach to education, designed to equip learners with skills that enable them to make self – satisfying decisions and appropriate choices based on up – to – date factual information. Underlying the concept of health promotion is the notion that to achieve good health persons must have some measure of control over the decisions and conditions they encounter over time and across circumstances. Decisions are the result of an interpretative process both in understanding what the needs are and what can and should to be done about them.

Constructivist theory suggests that learners construct and reconstruct information to learn (Brooks and Brooks 1993). These constructions evolve when learners actively gather, generate, process, and personalize health – related information rather than passively receive knowledge from teachers or health – related resources. Within constructivist theory, when teaching for understanding, health educators must *facilitate* the students to work with the content. As such, when learners are challenged to go beyond facts into constructing personal meaning and understandings about health, their behavioral outcomes may be enhanced and extended as well. Constructivist approaches try not to view the learner's behaviors as objects of analyses that can be manipulated and controlled. Instead, constructivism assesses student learning in the context of teaching and adapts the curriculum to address students' suppositions (Brooks and Brooks 1993).

Such an approach is commended by Stuckey (2009) who writes about the 'creative power' of the arts to deconstruct existing knowledge. Her adult learners initially demonstrated strong resistance to self-care and regulation in terms of their own health management, but her study highlights how a creative arts-based approach empowered the participants to 'challenge assumptions of learning and open the possibilities of what counts as knowledge', inviting them to explore 'multidimensional ways of learning about the world' and 'create space to allow them to make sense of and take control of their own health condition' (Stuckey 2009, p. 62).

Although both critical thinking and decision making are identified and advocated for, in the literature reviewed, discussion on how to proceed, with making changes in the way teachers are prepared for teaching critical thinking in the classroom, is extremely limited to nonexistent in sources reviewed by the author of this article. Ubbes, Black and Ausherman (2010) admit that, to date, health education scholars have not fully outlined how critical and creative thinking skills can be used as cognitive tools for studying our behavioral science and educational theories as parallel, complementary processes. When learners are encouraged to generate critical questions about self and others and to probe health content for underlying meanings and assumptions, they are better able to understand the 'hows' and 'whys' of health behavior. Meyers (1987) adds that few opportunities to learn about teaching critical thinking are available through professional development, disciplinary conferences and everyday collegial dialogue. Haas and Keeley (1998) describe faculty resistance to teaching for critical thinking in higher education as a common problem. According to the authors, although college instructors purport to acknowledge critical thinking as an educational outcome for their students, it is evident many faculty fail to make critical thinking a reality in their classrooms.

The social construction of reality is another focus of transformative learning theory and 'learning is regarded as a method by which this may be changed' (Jarvis 2004, p. 133). O'Toole (1992, p. 21) suggests that drama may be viewed as an activity 'inherently collective and processual, and thus both socially and individually developmental'. Drama and transformative learning theory share the focus on creating 'the foundation in insight and understanding essential for learning how to take effective social action in a democracy' (Mezirow 2009, p. 96).

Democracy is the first principle of the European Network of Health Promoting Schools, and one of the ten principles listed as fundamental to WHO's Health Promoting Schools framework (1996). Democratic learning is "a set of purposeful activities always building toward increasing student activity, choice, participation, connection, and contribution. It always aims for students individually and collectively to take on greater responsibility for their own learning" (Glickman 1999). In practice, democratic learning involves students actively working with real world problems, ideas, materials, and people as they learn skills and content. School experiences managed in relation to these principles are posed as a way to broaden and intensify students' involvement in what Habermas (1990) has termed the *lifeworld* of schools: cultural traditions, ceremonial rituals, participation in clubs, and teacher – student relationships; and the *systemsworld* of the school: programs of study, school governance. Katherine Weare (1998) emphasizes that democracy needs a balance between participation, warm relationships, clarity and autonomy. As Anderson and Ronson (2010) put it, learning about democracy through health promotion initiatives is an important way for students to feel like they are a part of their school's ethos and progress, and to enjoy the freedom to be deterministic and expressive about the meanings they are making relative to course content, personal experience and world events. The health literate citizen is concerned not only about knowledge and interpersonal relationships but social justice, equity, and involvement – Whose interests are being served? Who dominates? Who is silent? How do determinants of health interact with decisions for health? Why are people in upper income households half as likely to require hospital care as people living in low income households?

8.9 Meaning – Making and Social Change

Meaning – making is the construction of 'comprehension' from an individual's experience. This may be the discovery of completely new core constructs or the reframing of current ideas. It requires an engagement with people, places, ideas or things, to create an 'internal' space in which an energetic information exchange can occur. This is what enables the individual to grasp an understanding of the unity between their 'inner' and 'outer' worlds.

Drama is a medium for meaning making and a means of making inner meanings public through the use of symbols. Courtney (1995, p. 28) wrote: 'To think adequately (that is, to be intelligent) is to be skilful with the feelings inherent in symbols'. In his discussion of dramatic meaning O'Toole (1992, p. 44) recognises its 'evanescence' which makes it 'extremely difficult to control', since the meaning in drama 'is not laid down' but instead 'emerges'. O'Toole is at pains to find an inclusive approach. He clarifies his disagreement with New Criticism's notion of the absolute distinction between artist and audience, where the reader is to embrace the meaning of the author and thus this meaning is entire in itself and must not be socialised. He also contradicts Barthes's (1977) view that each individual has limitless capacity for renegotiating any artist's meanings. He posits:

> The processuality of meaning in drama is not without bound because the art form is collective and …a proportion of the meaning, which emerges is therefore shared – as social meaning, at once a part of and apart from the personal constructions of meaning which each individual negotiates within the experience (O'Toole 1992, p. 217).

Henry Giroux, in *The Abandoned Generation: Democracy Beyond the Culture of Fear* (2003), posits that learning is a meaning – making process. Accordingly, individuals consciously strive for meaning, to make sense of their environment in terms of past experience and their present state in an attempt to create order, resolve incongruities, and reconcile external realities with prior experience. Giroux emphasizes that knowledge is mainly acquired through social processes wherein culture, context, and community are logically and philosophically active. He pushes for an approach to education that engages students in not only critical analysis of the context but of the implications this knowledge has for how we think, behave and participate in our life experiences. Views about health that focus on what is happening in a person's life – such as level of income, relationships with close and intimate others, participation in social and civic life, employment stability, to name a few – are seen as determinants of health which impact the capacity one has to live to the fullest. Accordingly, school work must be linked to larger purposes such as creating more equitable and just public spheres within and outside educational institutions – exercising rights and entitlements – as a part of active citizenry. In order to be effective, therefore, health promoters must find their inspiration in not only understandings of health but of the fundamental purposes of school and beliefs about learning (p. 208).

Using health education as a tool to bring about social change is not a new concept. Grounded in the social movements in the 1960s and 1970s, health education has taken a fresh look at what social change means in the contemporary world. As one searches for the underlying causes of illness, one finds that many social inequities are contributing factors. Here, the role of the health educator is to work within the political arena demanding and legislating changes that eliminate health disparities (Auld and Dixon-Terry 1999; Leviton 2002; O' Rourke 2002; Pinzon-Perez 2003).

8.10 Emancipatory Learning

Health promotion as "the process of enabling people to increase control over, and improve, their health" (WHO 1986, p. 1), shares a similar ethos with empowerment as "the process by which people, organizations and communities gain mastery over their lives" (Rappaport 1981, p. 3). Health promoters often describe health promotion as a social movement, concerned with health "prerequisites" such as "peace, shelter, food, income, a stable ecosystem, social justice and equity" (WHO 1986, p. 1) and acknowledge the disempowering qualities of government bureaucracies by contending that health promotion belongs to the whole community. Community groups do organize and act in the name of health promotion. However, they are more likely to do so around specific issues involving welfare rights, pollution, housing, safety, or employment concerns.

Weare (1998) asserts that a key strategy for a health promoting school is to ensure that its organization, management structures and ethos are empowering and encourage participation. She goes on to stress that "empowerment aims to be genuinely democratic by ensuring that the action or process is done with, rather than to, people". Labonte (2010) also identified a tension between "power over" and "power with". He claims that *power over* tries to educate others to the educator's terms, his ways of viewing the world. *Power with* tries to find some common ground between what he knows, and how he talks about it, and what communities know, and how they talk about it.

Since the appearance of *Pedagogy of the Oppressed* (Freire 1970), learning to challenge and change existing systems has been a dominant issue in adult education. Freire emphasizes the need for a social critique of power, that is, of understanding the structural, particularly the ideological, forces of oppression and the need to link this theoretical understanding to a radical political practice. If we want students to analyze the social and political structures that constitute their lives, we have to begin with those that are socially and politically the closest. In effect, this means that teachers must enable students to understand what power they, the teachers, have over them; the strategies and tactics by which this power is operated; and, paradoxically, the strategies and tactics by which they could be empowered to take control of their own learning (see Roberts 2000).

Emancipatory adult education practices are often centered on 'helping students to see through and transgress traditional pedagogical practices – who decides, who assigns, who controls and who knows' (Inglis 1997). Emancipatory learning is about showing people how to read their lives and the family, groups, organisations and society in which they are involved in terms of a struggle for power. In contrast, empowerment can be understood as the process by which individuals or groups seek, by working within the existing system, to obtain greater economic, political, and social power. Empowerment is, thus, centered on creating self-confidence, self-expression, and an interest in learning (see Dew 1997). There is, then, 'a balance as well as a tension between individual transformative learning and emancipatory education. The former is a necessary, but not a sufficient, base for the latter' (Inglis 1997). Education for liberation and emancipation is a collective educational activity, which has as its goal social and political transformation. If personal development takes place, it does so within that context.

> Liberation is a social act. Liberating education is a social process of illumination. Even when you individually feel yourself free, if this feeling is not a social feeling, if you are not able to use your recent freedom to help others to be free by transforming the totality of society, then you are exercising only an individualist attitude towards empowerment and freedom (Freire and Shor 1987, p. 109).

The distinction between empowerment (working within the system) from emancipation (trying to change the system) is crucial, because it can help educators understand how a process that supposedly leads to increased or devolved power leads, in effect, to a more subtle form of incorporation.

8.11 A Foucauldian Perspective of Power

Perhaps the most famous postmodern effort to contest simple ideas of oppression comes at the beginning of Michel Foucault's (1977) *Discipline and Punish.* In this work, Foucault outlines many disciplinary technologies such as seating charts, forming lines, taking attendance, timetables, the distribution of individual bodies within a partitioned space, individuation and documentation, hierarchical observation, normalising judgments, and examination that are found in most public schools around the world. Contemporary poststructuralists drawing from the work of Foucault (1977, 1980) contend that students are socialised into hierarchies of normalisation by teachers who use techniques of discipline, surveillance, and regulation that penetrate into the smallest details of everyday life (see Olssen 1999; Weedon 1997; Fairclough 1995). Teachers have been encouraged to work with and in the postmodern to implement praxis-oriented emancipatory pedagogies in which they deconstruct classroom authority to see how it is constituted and constituting. Educational researchers have begun to fruitfully use the work of Michel Foucault to analyze and problematize many practices and structures of schooling (see for example, Walkerdine 1988; Popkewitz 1998; Butin 2001; Bevir 1999; Caughlan 2005; Schutz 2004; Gore 1999; Baptiste 2000; Brookfield 2005). Reconstituted authority relations should be participatory, dialogic, and pluralistic; antihierarchical; and conducive for decentering the centres of hegemonic power and knowledge (see Popkewitz and Brennan 1997; Gore 1999; Baptiste 2000; Mayo 2000; Butin 2001; Schutz 2004).

Foucault's argument (1980) that every new truth has the potential to create its own new "regime of truth," and that every form of "emancipatory" power relies on the counter power to which it is opposed, has been appropriated to make clear that progressive "enlightened" educational practices may ultimately lead to their own "distopias". The implications are that 'educational practices that may appear more democratic, participatory, or progressive may in fact be more effective forms of disciplinary power' (Brookfield 2005, p. 335). Elsewhere, Foucault (1982) examines even more complex forms of control and domination. He distinguishes between relatively crude disciplinary power, which is overtly imposed, and more subtle "pastoral" power, which operates more subversively and indirectly. Foucault locates the beginning of the movement of pastoral techniques from the religious to the secular in the eighteenth century. In the appropriation of pastoral techniques by secular institutions, the rhetoric of redemption is maintained and the agents of this power have multiplied to include police, social workers, health-care workers and teachers. They act to govern by 'structuring the possible field of action' of those in their care (Foucault 1982, p. 221). In the twentieth-century school, the language of psychology replaced the language of religion 'in facilitating particular ways of thought, self-reflection, and self-regulation' and the ensemble of pastoral techniques 'is particularly well suited for shaping the self-conduct of individuals' (Caughlan 2005, p. 15). From a postmodern perspective, in progressive school settings, the 'governing of the individual' occurs not through 'explicit defining of set procedures' as in more traditional settings, 'but through the deployments of reasoning' (Popkewitz 1998, p. 24). warns:

> Unlike traditional settings that tend to sanction divergences from a static norm, then, pastoral settings foster particular forms of creativity, often harnessing them to serve the (loosely coupled) systems in which participants are enmeshed. And because control in progressive classrooms is distributed throughout the environment instead of located in (apparently) identifiable figures or systems, it is extremely difficult for participants to detect or resist.

In fact, as many postmodern writers note (see for example, Popkewitz 1998; Bevir 1999; Butin 2001; Schutz 2004), multiple systems of control often work together without any clear locus of control. Although there is no single "leader" and there is no "center, only a flexible network of distributed roles and responsibilities, the activities of the group as a whole serve purposes set for them by the institution of school or the teacher" (Gee et al. 1997, p. 62). Instead of 'suppressing the agency of participants as more disciplinary approaches might, pastoral approaches co-opt the creativity of participants by recruiting their desires and motivations' (Schutz 2004, p. 17) and because control in progressive classrooms is distributed throughout the environment instead of located in (apparently) identifiable figures or systems, it is extremely difficult for participants to detect or resist Schutz (2004, p.15).

In the field of adult education, Brookfield (2005, p. 120) asserts that 'the possibility of converting "power over" learners into "power with" them continues to this day to exercise a hold on adult educators' imaginations'. He also admits (Brookfield 2005, p. 129) that for progressively inclined adult educators Foucault's analysis 'of how apparently emancipatory adult educational practices often contain oppressive dimensions is particularly disturbing'. Most adult educators often function as persuaders and organisers, using justifiable coercion, but choose not to acknowledge doing so (see Baptiste 2000) although 'sovereign power is easily detected and usually discredited by those within the field who see themselves as "true" adult educators dedicated to empowering learners in a respectful way' (Brookfield 2005, p. 144). If we are fated to exercise power, what we can draw from Foucault's work in building a theory of adult learning is 'the need to study how adults learn to recognize that they are themselves agents of power, perpetually channeling disciplinary power, but also possessing the capacity to subvert dominant power relations' (Brookfield 2005, p. 143).

8.12 Power Issues in Drama in Education

Giroux (2003) encourages progressive educators and their students to stand at the edge of society, to think beyond existing configurations of power in order to imagine the unthinkable in terms of how they live with dignity, justice, and freedom. Within acts of "moral imagination" students are posing problems that begin with "What if" and "Why not" (p. 146). Neelands (1996) identifies four modes of empowerment in drama, of which two can be directly linked to a process of changing understanding: personal empowerment and cultural empowerment. In defining DIE in the light of these modes, Neelands (1996, p. 29) writes that drama 'is a personally transforming cultural resource, one which makes the invisible influences of culture visible and discussible, and serves as a mirror of how we are made, and of who we might become'. O'Toole (1992, p. 45) posits:

> "Drama for empowerment" is quite a savage political battleground of right and left, with one set of exponents using drama to socialize children into the society in which they find themselves and the other using it to give them a critically active and activist orientation.

His position resonates with the distinction between empowerment and emancipation, as previously discussed. His warning for drama practitioners using drama in the emancipator framework, 'committed to assist in changing attitudes', of the danger of 'replacing one set of concealed power constraints with another' (O'Toole 1992, p. 62), is echoing Foucault's challenge. Many theorists in the field discuss the "colleagueness" between learners and teacher, generated by the co-creativity of drama. O'Toole (1992, p. 2) defines DIE as a process where the teacher is 'negotiating and renegotiating the elements of dramatic form in terms of the context and purposes of the participants'. He actually suggests that instead of drama being used as a teaching method where 'the fictional contexts are taken directly or indirectly from other curriculum subjects and thus only minimally negotiable' (O'Toole 1992, p. 57), drama's principal practitioners should be looking for 'a resolution in terms of empowering the students through giving them access to the negotiations of the playwriting process, and in fact to the elements of the art form itself' (O'Toole 1992, p. 63). Bowell and Heap (2005) point in the same direction. Elaborating on the teacher's quadripartite thinking as discussed earlier, they remark:

> the pupils need to be able to make a quadripartite response to what is happening within the drama so that they are in creative partnership with the teacher. The pupils learn how to contribute to the extension and deepening of the play they are in and to feel sufficiently empowered to initiate further developments of the narrative (Bowell and Heap 2005, p. 66).

Subsequently, in this multifaceted spiral of creative discourse, where learners provide feedback to the teacher, who in turn responds, changes in the shape and direction of the drama can be initiated by both learners and/or teacher. This situation creates challenges, 'not least because these functions are generally engaged simultaneously yet are driven by potentially different needs – the teacher being learning-objective driven and the pupils narrative driven' – but the teacher also recognizing that 'in order for the learning objective to be met, the narrative of the drama must create the imperative in which the learning can take place' (Bowell and Heap 2005, p. 63).

It seems that there is sufficient evidence in the literature to support drama as a dialogical, liberating form of education. Through dialogue and the presentation of issues as problems, learners acquire the ability to intervene in, rather than accept, reality, an intervention, which results from their critical consciousness of the situation. The investigation of students' own meaningful themes establishes a climate of creativity and enthusiasm for risking experimentation, in which learners critically analyze reality and intervene as subjects in the historical process, understanding themselves not as spectators but as co-authors of action to transform the world. The co-operation engendered in drama, another characteristic of dialogical action, which only occurs among subjects, what Neelands (2004, p. 24) calls the teacher – learner partnership in drama, provides the requirements for a truly co-operative undertaking in that learners are able to negotiate, decide what happens and when, and work with others towards a mutually satisfying goal, critically co-developing the session with the teacher. According to Freire and Shor (1987, p. 107) 'the liberating educator has to be very aware that transformation is not just a question of methods and techniques but a different relationship to knowledge and to society'. Drama, due to its ways of critically discovering knowledge and intention to transform the world, can therefore, be considered as a form of emancipatory education, in which 'the liberating educator is not doing something to the students but with them' (Freire and Shor 1987, p. 110).

8.13 Social Cognitive Theory and Self-Efficacy

The goal of any in-service training is behavioral – that teachers implement the curriculum as intended. Thus, it is logical to conceptualize the teacher training program as a behavior change intervention designed to affect the (implementation) behavior of independent individuals with a diverse range of interests, expertise, and experience, operating autonomously in a variety of classroom settings and under a variety of conditions. The modern psychology of behavior change and social learning theory, as its most prominent model (Bandura 1977b), recognizes that there are four prerequisites for the adoption of a behavior. The individual must (1) want to adopt the behavior (i.e., it must meet some internalized need or desire), (2) know what actions constitute the behavior, (3) have the tools to perform the behavior, and (4) have the ability and confidence to perform the behavior (self-efficacy).

Bandura's (1986, 1989b) social cognitive theory emphasizes the important role of human cognition and the interplay of personal, behavioral, and environmental influences on human growth. Bandura's (1989b) social cognitive theory recognizes that cognitive and personal factors, behavior, and the environment impact and interact with one another, a process he calls reciprocal determinism, and that learning and human adaptation occur within a social context. This causal model does not require nor imply that all sources are of equal strength, nor that they occur simultaneously. Within the triadic model, personal and behavioral "causation reflects the interaction between thought, affect and action… What people think, believe and feel, affects how they behave" (Bandura 1989b, p. 3). Finally, there is the interaction between behavior and environment. Throughout the processes of everyday life, our actions alter the environmental conditions, and the environment exerts influence on our actions. This bidirectional influence means "people are both products and producers of their environment" and evidently echoes a Vygotskian perspective. "They affect the nature of their experienced environment through selection and creation of situations" (Bandura 1989b, p. 4). In addition to learning from our own experiences, we can also learn from the experiences of others, either by social modeling or by vicarious experience.

Meanings in DIE are created for all "spect-actors" through the actual, fictional and symbolic uses of human presence in time and space. These may also be enhanced by the symbolic use of objects, light and sound. Creating a link between DIE and Social cognitive theory, Bandura (1989b) asserts that symbolic representation is a uniquely human skill, which provides people with the means for understanding the environment. It is through the use of symbols as representations in cognitive processes that people form meaning from interactions with the world and others. "Symbols serves as the vehicle for thought" (Bandura 1989b, p. 9). Without the ability to use symbols, humans would not be able to communicate with others; organize thoughts, experiences, or emotions into schema for making meaning; or test potential solutions to problems through thought rather than constant trial and error. As a bridge between the arts and social sciences, symbols function as widely shared public metaphors that interact with the felt-meanings of society Symbols therefore play a role in all aspects of social cognitive theory, including humans' ability to provide forethought to their actions. "A major function of thought is to enable people to predict the occurrence of events and to create the means for exercising control over those that affect their daily lives" (Bandura 1989a, p. 1176).

One important aspect of Bandura's theory is the concept of self-efficacy (1977a, 1978, 1997a). Throughout his career, Bandura attempted to define and develop the construct known today as self-efficacy, which is the belief "in one's capacity to organize and execute the courses of action required to produce given attainments" (Bandura 1997b, p. 3). Efficacy beliefs influence how resilient people are when dealing with obstacles, how long a person will persist in the face of failure or difficulty, and how much energy one will put forth on any given task (Bandura 1977b). It is important to note the distinction between self-efficacy, as defined here, and self-esteem. Self-esteem is a measure or judgment of self-worth, while self-efficacy is a measure or judgment of capability. Bandura (1977a, 1986) believed that there were four sources of self-efficacy information: (a) mastery experiences, (b) verbal persuasion, (c) vicarious experiences, and (d) emotional arousal. Much like the triadic model for social cognitive development, efficacy is influenced by the interaction of cognitive processes with motivational, affective and selection processes (Bandura 1997a). Affective or emotional factors, such as mood, anxiety, stress, and depression, can influence motivational processes. Cognitive processes alone, though, do not create beliefs of self-efficacy. Drama operates, through symbols, at the nexus of intelligence and emotion. The aesthetic response in drama involves cognition and emotion in a "thinkingly feeling" or "feelingly thinking" way (Bundy 2003, p. 174).

Despite the fact that self-efficacy and Bandura's theory is one of the most often used in health promotion (Glanz et al. 2008), its use in teacher education is limited (e.g. Howard et al. 2001; Shechtman et al. 2005). A recent review of training programs for teachers in the area of adolescent prevention (Shek and Wai 2008) urges researchers, for improved effectiveness of intervention programs targeting teachers, to base their interventions on social cognitive theory and especially the concept of self-efficacy.

8.14 Epilogue

This chapter explored the theoretical connections, between a drama in education based pedagogy, adult education and health promotion. The Vygotskian paradigm was proposed as a comprehensive epistemology for teacher education and a Foucauldian perspective was offered as a promising theoretical basis for the elaboration of such a paradigm. A drama in education based pedagogy was advocated to provide a useful lens for viewing teaching and learning in teacher education.

Engaging in curriculum integrated, multidisciplinary DIE processes will provide a more holistic understanding of health, particularly if they are cemented by peer group learning and the support of the wider school community. These types of cross-disciplinary approaches require the whole person's involvement, which in turn is what allows a deeper exploration of personal identity. Acknowledging the importance of an interdisciplinary approach necessitates the inclusion of the concept of 'interconnectedness'. Programs that acknowledge 'wholeness' also recognise the interconnection and overlaps of all of their parts and the affects of these amalgams. School ethos for example will directly impact on the success of health promotion programs, as both the implicit and explicit values of the school will either support or stall it. Every aspect of the school, including teacher training support and involvement, has an impact. The long-term sustainability of these programs is contingent on the tangible and sustainable support provided by school management representatives.

The primary agent for educational change continues to be the classroom teacher, and by providing effective opportunities for teachers to increase their knowledge and skills in content and pedagogy, as well as change their beliefs and attitudes towards instructional practices, strategies, and resources, professional development is in a position to be a powerful tool for increasing student achievement. However, much of the literature on professional development shows that these types of activities have only enjoyed moderate success at achieving measurable change in teaching practices and student outcomes (Abrami et al. 2004; Guskey 2002; McBride et al. 1994; Richardson 1989). Change within education has not come easily. Educational research that is conducted by university faculty and government agencies often appears to miss the mark of affecting its audience, perceived by teachers as only tangentially connected to their everyday concerns (Bryant et al. 2001; Guskey 2002; McBride et al. 1994). This disconnect creates a chasm between quality research-driven initiatives and those they are designed to assist. The goal of professional development must first be, then, to create an atmosphere of continuing learning for teachers, contextually situated and promoting involvement in professional development in meaningful experiences that will assist teachers and ultimately students in reaching new levels of success.

Providers of professional development, who utilize adult learning theory when planning educational opportunities, can better meet the needs of their target audience, teachers as adult learners. In order to bridge the gap between research and practice, professional development must encompass not only the needs of the delivery agency, but it must also meet the needs of participants (Hargreaves 1994). This can be accomplished in a multitude of ways, through various forms of content, structure, and mode that are consistent with the philosophical and theoretical frameworks of the particular reform initiative. Adult educators today, including providers of professional development, following the principles of andragogy to create a learner-centered environment, recognize the importance of valuing the adult learner's previous experience and knowledge, motivational factors, and need for relevant instruction when planning quality programs. Adult learning theories of self-directed learning, andragogy, transformative and emancipator learning, situated cognition and social cognitive theory, all emphasize a component of critical reflection, coupled with opportunities for feedback and follow-through, as crucial to effective learning for any adult. These theories emphasize the human need to strive to full potential throughout adulthood.

Educational theory needs further conceptualization in health promotion. The synergistic strengths of combined theories can convey the knowledge, issues, skills, perceptions, and values associated with health promotion. The combined pedagogical techniques and strategies of each theory need to be refined in order to contribute to an expanded vision of how to teach for creativity, critical thinking, and a desire for life-long learning – all mental habits that support healthy and sustainable societies.

A drama based pedagogy combined with the adult learning theories can provide a framework for teachers' professional development in the field of health promotion. This framework can enhance both adult and non-adult learners' literacy, investigating, critical thinking, problem-solving, decision-making and cooperation skills, by encouraging them to value the diversity of the world's people, cultures, and ecosystems and actively promote equity, justice, peace, the democratic process, and the protection of the environment in their own community and the world. In this process of active citizenship learners will be motivated and enabled to care about the physical, emotional and spiritual health of themselves and others, pursue healthy, hopeful, purposeful lives and meaningful relationships and make wise choices for a sustainable future, both personal and global.

Undoubtedly, there is a need for teacher education initiatives, which will employ a drama in education based pedagogy, in order to assess and evaluate its efficacy in practice.

References

Abrami, P. C., Poulsen, C., & Chambers, B. (2004). Teacher motivation to implement an educational innovation: Factors differentiating users and non-users of cooperative learning. *Educational Psychology, 24*(2), 201–216.

American Association of Health Education. (2008). Health literacy: A position statement of the American Association of Health Education (AAHE). http://www.aahperd.org/aahe/pdf_files/pos_pap/HealthLiteracy.pdf. Accessed 12 Jun 2013.

Anderson, A., & Ronson, B. (2010). Democracy – The first principle of health promoting schools. In J. M. Black, S. R. Furney, H. M. Graf, & A. E. Nolte (Eds.), *Philosophical foundations of health education* (pp. 207–225). San Francisco: Jossey-Bass.

Aprill, A., & Townsell, R. (2007). The arts as an occasion for collective adult learning as authentic community development. *New Directions for Adult and Continuing Education, 116*, 51–63.

Archer-Cunningham, K. (2007). Cultural arts education as community development: An innovative model of healing and transformation. *New Directions for Adult and Continuing Education, 116*, 25–36.

Au, K. H. (2002). Communities of practice: Engagement, imagination, and alignment in research on teacher education. *Journal of Teacher Education, 53*(3), 222–227.

Auld-Laine, M. E., & Dixon-Terry, E. (1999). The role of health education associations in advocacy. In J. M. Black, S. R. Furney, H. M. Graf, & A. E. Nolte (Eds.), *Philosophical foundations of health education* (pp. 311–318). San Francisco: Jossey-Bass.

Bandura, A. (1977a). Self-efficacy: Toward a unifying theory of behavioral change. *Psychological Review, 84*(2), 191–215.

Bandura, A. (1977b). *Social learning theory*. Englewood Cliffs: Prentice-Hall.

Bandura, A. (1978). Reflections on self-efficacy. *Behavioral Research and Therapy, 1*(1), 237–269.

Bandura, A. (1986). *Social foundations of thought and action: A social cognitive theory*. Englewood Cliffs: Prentice Hall.

Bandura, A. (1989a). Human agency in social cognitive theory. *American Psychologist, 44*(9), 1175–1184.

Bandura, A. (1989b). Social cognitive theory. In R. Vasta (Ed.), *Annals of child development* (Six theories of child development, Vol. 6, pp. 1–60). Greenwich: JAI Press.

Bandura, A. (1997a). Self-efficacy. *Harvard Mental Health Letter, 13*(9), 4.

Bandura, A. (1997b). *Self-efficacy: The exercise of control*. New York: Freeman and Company.

Baptiste, I. (2000). Beyond reason and personal integrity: Toward a pedagogy of coercive restraint. *Canadian Journal for the Study of Adult Education, 14*(1), 27–50.

Barone, T. E., & Eisner, E. (1997). Arts-based educational research. In R. M. Jaeger (Ed.), *Complementary methods for research in education* (pp. 22–45). Washington, DC: American Educational Research Association.

Barthes, R. (1977). *Image, music, text*. New York: Hill and Wang.

Barton, D., & Tusting, K. (2005). *Beyond communities of practice: Language, power and social context*. Cambridge/New York: Cambridge University Press.

Basch, C. E., Sliepcevich, E. M., Gold, R. S., Duncan, D. F., & Kolbe, L. J. (1985). Avoiding type III errors in health education programs evaluations: A case study. *Health Education Quarterly, 12*(4), 315–331.

Baumgartner, L. M. (2001). An update on transformational learning. In S. B. Merriam (Ed.), *The new update on adult learning theory* (pp. 15–24). San Francisco: Jossey-Bass.

Bevir, M. (1999). Foucault, power, and institutions. *Political Studies, 47*, 345–359.

Boal, A. (1979). *Theatre of the oppressed*. London: Pluto Press.

Borman, K. M. (1990). Foundations of education in teacher education. In W. R. Houston (Ed.), *Handbook of research on teacher education* (pp. 393–402). New York: Macmillan.

Botvin, G. J., Dusenbury, L., Baker, E., James-Ortiz, S., & Kerner, J. (1989). A skills approach to smoking prevention among Hispanic youth. *Journal of Behavioral Medicine, 12*, 279–296.

Botvin, G. J., Baker, E., Filazzola, A. D., & Botvin, E. M. (1990). A cognitive-behavioral approach to substance abuse prevention: One year follow-up. *Addictive Behavior, 15*(1), 47–63.

Bowell, P., & Heap, B. S. (2001). *Planning process drama*. London: David Fulton Publishers.

Bowell, P., & Heap, B. S. (2005). Drama on the run: A prelude to mapping the practice of process drama. *Journal of Aesthetic Education, 39*(4), 58–69. http://www.jstor.org/stable/3527392. Accessed 9 Sept 2009.

Boyer, N. R., & Maher, A. P. (2004). Constructing scaffolds for social online learning: Using self-directed frameworks with virtual groups. *International Journal of Self-Directed Learning, 1*(1), 26–38.

Boyle, B., Lamprianou, I., & Boyle, T. (2005). A longitudinal study of teacher change: What makes professional development effective? Report of the second year of the study. *School Effectiveness & School Improvement, 16*(1), 1–27.

Bozhovitch, L. I. (1979). Stages in the formation of the personality in ontogeny. *Soviet Psychology, 17*(3), 3–24.

Branagan, M. (2005). Environmental education, activism and the arts. *Convergence, 38*(4), 33–50.

Brockett, R. G., & Hiemstra, R. (1991). *Self-direction in adult learning: Perspectives in theory, research, and practice*. London/New York: Routledge.

Brookfield, D. S. (2000). Self-directed learning as a political idea. In G. A. Straka (Ed.), *Conceptions of self-directed learning: Theoretical and conceptual considerations* (pp. 9–22). New York: Waxmann.

Brookfield, D. S. (2005). *The power of critical theory: Liberating adult learning and teaching*. San Francisco: Jossey-Bass.

Brooks, J. G., & Brooks, M. G. (1993). *In search of understanding: The case for constructivist classrooms*. Alexandria: Association for Supervision and Curriculum Development.

Bruner, J. S. (1996). *The culture of education*. Cambridge: Harvard University Press.

Bryant, D. P., Linan-Thompson, S., Ugel, N., Hamff, A., & Hougen, M. (2001). The effects of professional development for middle school general and special education teachers in implementation of reading strategies in inclusive content area classes. *Learning Disability Quarterly, 24*, 251–264.

Bundy, P. (2003). Aesthetic engagement in the drama process. *Research in Drama Education, 8*(2), 171–181.

Butin, D. W. (2001). If this is resistance I would hate to see domination. *Educational Studies, 32*(2), 157–176.

Calderhead, J., & Gates, P. (1993). *Conceptualizing reflection in teacher development*. London: Falmer.

Cameron, H. (1991). Effect of inservice training on implementation of a health curriculum in Nova Scotia, Canada. *Journal of School Health, 61*, 131–135.

Candy, P. C. (1991). *Self-direction for life-long learning: A comprehensive guide to theory and practice*. San Francisco: Jossey-Bass.
Carroll, J. (2005). Developing effective teachers of mathematics: Factors contributing to development in mathematics education for primary school teachers. In P. Clarkson, A. Downton, D. Gronn, M. Horne, A. McDonough, R. Pierce, et al. (Eds.), *Building connections: Theory, research and practice*, Proceedings of the 28th Annual Conference of the Mathematics Education Research Group of Australasia (pp. 201–216). Sydney: MERGA.
Caughlan, S. (2005). Considering pastoral power: A commentary on Aaron Schutz's "rethinking domination and resistance: Challenging postmodernism". *Educational Researcher, 34*, 14–16.
Chaiklin, S. (2001). The institutionalisation of cultural-historical psychology as a multinational practice. In S. Chaiklin (Ed.), *The theory and practice of cultural-historical psychology* (pp. 15–34). Aarhus: Aarhus University Press.
Clover, D. E. (2006). Culture and antiracisms in adult education: An exploration of the contributions of arts-based learning. *Adult Education Quarterly, 57*(1), 46–61.
Clover, D. E. (2010). A contemporary review of feminist aesthetic practices in selective adult education journals and conference proceedings. *Adult Education Quarterly, 60*(3), 233–248.
Clover, D. E., & Stalker, J. (2007). *The arts and social justice: Re-crafting adult education and community cultural leadership*. Leicester: NIACE.
Cochran-Smith, M., & Lytle, S. (1990). Research on teaching and teacher research: The issues that divide. *Educational Researcher, 19*(2), 2–11.
Connell, D. B., Turner, R. R., & Mason, E. F. (1985). Summary of findings of the school health education evaluation: Health promotion effectiveness, implementation, and costs. *Journal of School Health, 55*, 316–321.
Corley, A., & Thorne, A. (2005). Reflecting on practice: A partnership approach. *Reflective Practice, 6*(2), 231–237.
Courtney, R. (1995). *Drama and feeling: An aesthetic theory*. Montreal: McGill–Queens University Press.
Darling-Hammond, L. (1998). Policy and change: Getting beyond the bureaucracy. In A. Hargreaves, A. Lieberman, M. Fullan, & D. Hopkins (Eds.), *The international handbook of educational change* (pp. 642–666). Dordrecht: Kluwer.
Davydov, V. V. (1988). Problems of developmental teaching. *Soviet Education, 30*, 8–10.
Dew, J. R. (1997). *Empowerment and democracy in the workplace: Applying adult education theory and practice for cultivating empowerment*. Westport: Quorum Books.
Dickinson, D. (2002). Learning through the arts. New Horizons for Learning. http://www.newhorizons.org/strategies/arts/dickinson_lmarts.htm. Accessed 22 Sept 2004.
Dirkx, J. M. (2001). The power of feelings: Emotion, imagination, and the construction of meaning in adult learning. In S. B. Merriam (Ed.), *The new update on adult learning theory*. San Francisco: Jossey-Bass.
Donnelly, A., Morgan, D. N., Deford, D. E., Files, J., Long, S., Mills, H., et al. (2005). Transformative professional development: Negotiating knowledge with an inquiry stance. *Language Arts, 82*(5), 336–346.
D'Onofrio, C. N. (1989). Making the case for cancer prevention in schools. *Journal of School Health, 59*, 225–231.
Dusenbury, L., & Falco, M. (1995). Eleven components of effective drug abuse prevention curricula. *Journal of School Health, 65*(10), 420–425.
Edwards, A., Gilroy, P., & Hartley, D. (2002). *Rethinking teacher education: Collaborative responses to uncertainty*. London: Routledge.
Eisner, E. W. (1995). What artistically crafted research can help us understand about schools. *Educational Theory, 45*(1), 1–6.
English, L. M., Fenwick, T. J., & Parsons, J. (2003). *Spirituality of adult education and training*. Malabar: Krieger.
Evans, R. I. (1988). Health promotion – Science or ideology? *Health Psychology, 7*, 203–219.
Fairclough, N. (1995). *Critical discourse analysis*. London: Longman.

Feiman-Nemser, S. (1990). Teacher preparation: Structural and conceptual alternatives. In W. R. Houston (Ed.), *Handbook of research on teacher education* (pp. 212–233). New York: Macmillan.
Finn, K. V., & Willert, H. J. (2006). Alcohol and drugs in schools: Teachers' reactions to the problem. *Phi Delta Kappa, 88*(1), 37–40.
Firth, G. R. (1977). Ten issues on staff development. *Educational Leadership, 35*(3), 215.
Flay, B. R., Miller, T. Q., Hedeker, D., Siddiquim, O., Britton, C. R., Brannon, R. B., Johnson, C. A., Hansen, W. B., Sussman, S., & Dent, C. (1995). The television, school and family smoking prevention and cessation project: VIII. Student outcomes and mediating variables. *Preventive Medicine, 24*, 29–40.
Florio-Ruane, S. (2002). More light: An argument for complexity in studies of teaching and teacher education. *Journal of Teacher Education, 53*(3), 205–215.
Forman, E. A., Minick, N., & Stone, C. A. (1993). *Contexts for learning: Sociocultural dynamics in children's development*. Oxford: Oxford University Press.
Foucault, M. (1977). *Discipline and punish: The birth of the prison*. New York: Vintage.
Foucault, M. (1980). *Power/knowledge: Selected interviews and other writings 1972–1977*. New York: Pantheon Books.
Foucault, M. (1982). How is power exercised? In H. Dreyfus & P. Rabinow (Eds.), *Michel Foucault: Beyond structuralism and hermeneutics* (pp. 216–226). Chicago: University of Chicago Press.
Freire, P. (1970). *Pedagogy of the oppressed*. New York: Herder and Herder.
Freire, P. (1985). *The politics of education*. London: Macmillan.
Freire, P., & Shor, I. (1987). *A pedagogy for liberation: Dialogues on transforming education*. South Hadley: Bergin & Garvey.
Gallimore, R., & Tharp, R. (1990). Teaching mind in society: Teaching, schooling, and literate discourse. In L. C. Moll (Ed.), *Vygotsky and education: Instructional implications and applications of sociohistorical psychology* (pp. 175–205). Cambridge: Cambridge University Press.
Gardner, H. (1983). *Frames of mind: The theory of multiple intelligences*. New York: Basic Books.
Garet, M. S., Porter, A. C., Desimone, L., Birman, B. F., & Yoon, K. S. (2001). What makes professional development effective? Results from a national sample of teachers. *American Educational Research Journal, 38*(4), 915–945.
Garrison, D. R. (1992). Critical thinking and self-directed learning in adult education: An analysis of responsibility and control issues. *Adult Education Quarterly, 42*(3), 136–148.
Garrison, D. R. (1997). Self-directed learning: Toward a comprehensive model. *Adult Education Quarterly, 48*(1), 18–33.
Gee, J. P., Hull, G., & Lankshear, C. (1997). *The new work order*. Boulder: Westview Press.
Ginsburg, M. B., & Clift, R. T. (1990). The hidden curriculum of preservice teacher education. In W. R. Houston (Ed.), *Handbook of research on teacher education* (pp. 450–465). New York: Macmillan.
Giroux, H. (2003). *The abandoned generation: Democracy beyond the culture of fear*. New York: Palgrave Macmillan.
Glanz, K. Rimer, K. B., & Viswanath, K. (2008). *Health behavior and health education: Theory, research and practice*. San Francisco: Jossey-Bass.
Glickman, C. D. (1999). *Revolutionizing America's schools*. San Francisco: Jossey-Bass.
Glickman, C. D., Gordon, S. P., & Ross-Gordon, J. M. (2007). *SuperVision and instructional leadership: A developmental approach*. Boston: Pearson.
Glynn, T. J. (1989). Essential elements of school-based smoking prevention programs. *Journal of School Health, 59*, 181–188.
Goleman, D. (1995). *Emotional intelligence: Why it can matter more than IQ*. New York: Bantam.
Gordon, S. P. (2004). *Professional development for school improvement: Empowering learning communities*. Boston: Pearson.
Gore, J. M. (1999). On the limits to empowerment through critical and feminist pedagogies. In D. Carlson & M. W. Apple (Eds.), *Power/knowledge/pedagogy* (pp. 271–288). Boulder: Westview Press.

Gould, R. L. (1978). *Transformations*. New York: Simon & Schuster.
Grace, A., & Wells, K. (2005). Out is in: Arts-informed community-based approach to social and cultural learning by and for queer young adults. In S. Mojab (Ed.), *Conference proceedings of the Canadian Association for the Study of Adult Education* (pp. 112–118). London: University of Western Ontario.
Grace, A. P., & Wells, K. (2007). Using Freirian pedagogy of just ire to inform critical social learning in arts-informed community education project for sexual minorities. *Adult Education Quarterly, 57*(2), 95–114.
Greene, M. (1995). *Releasing the imagination, essays on education, the arts and social change*. San Francisco: Jossey-Bass.
Guskey, T. R. (2002). Professional development and teacher change. *Teachers & Teaching, 8*(3/4), 381–391.
Guskey, T. R., & Huberman, M. (1995). *Professional development in education: New paradigms and practices*. New York: Teachers College Press, Columbia University.
Haas, P. F., & Keeley, S. M. (1998). Coping with faculty resistance to teaching critical thinking. *College Teaching, 46*(2), 63–68.
Habermas, J. (1970). *Toward a rational society*. Boston: Beacon.
Habermas, J. (1971). *Knowledge and human interests*. Boston: Beacon.
Habermas, J. (1990). *Moral Consciousness and Communicative Action* (trans: Lenhardt, C., & Nicholson, S. W.). Cambridge: MIT Press.
Hansman, C. A. (2001). Context-based adult learning. In S. B. Merriam (Ed.), *The new update on adult learning theory* (pp. 42–52). San Francisco: Jossey-Bass.
Hargreaves, A. (1994). *Changing teachers, changing times: Teachers' work and culture in the postmodern age*. New York: Teachers College Press.
Heathcote, D. (1995). *Pieces of Dorothy, [videotape]*. Newcastle: University of Newcastle upon Tyne.
Helitzer, D., Yoon, S., Wallerstein, N., & Garcia-Velarde, L. D. (2000). The role of process evaluation in the training of facilitators for an adolescent health education program. *Journal of School Health, 70*(4), 141–147.
Henry, M. (2000). Drama's ways of learning. *Research in Drama Education, 5*(1), 45–62.
Hinson, J. M., Laprairie, K. N., & Cundiff, J. M. (2005). One size does not fit all. *T.H.E. Journal, 32*(11), 26–30.
Houston, W. R. (1974). *Exploring competency based education*. Berkeley: McCutchan.
Howard, N. M., Horne, A. M., & Jolliff, D. (2001). Self-efficacy in a new training model for the prevention of bullying in schools. *Journal of Emotional Abuse, 2*(2/3), 181–191.
Huizen, P. V., Oers, V. B., & Wubbels, T. (2005). A Vygotskian perspective on teacher education. *Journal of Curriculum Studies, 37*(3), 267–290.
Hulson, M. (2006). *Schemes for classroom drama*. Stoke on Trent/Sterling: Trentham Books.
Hursh, D. (1987). *Becoming reflective teachers: Preservice teachers' understanding of school and society*. Paper presented at the meeting of the Midwest Educational Research Association, Chicago.
Inglis, T. (1997). Empowerment and emancipation. *Adult Education Quarterly, 48*(1), 3–18.
Interstate New Teachers Assessment and Support Consortium (INTASC). (1995). *Next steps: Moving toward performance – Based licensing in teaching*. Washington, DC: Council of Chief State School Officers.
Jarvis, P. (2004). *Adult education and lifelong learning: Theory and practice* (3rd ed.). London: Routledge Falmer.
Jawitz, J. (2007). New academics negotiating communities of practice: Learning to swim with the big fish. *Teaching in Higher Education, 12*(2), 185–197.
Johnson, C. C. (2006). Effective professional development and change in practice: Barriers science teachers encounter and implications for reform. *School Science & Mathematics, 106*(3), 150–161.
John-Steiner, V., & Mahn, H. (1996). Sociocultural approaches to learning and development: A Vygotskian framework. *Educational Psychologist, 31*(3–4), 191–206.

Joint Committee on National Health Education Standards. (1995). *National health education standards: Achieving health literacy*. Atlanta: American Cancer Society.
Joronen, K., Rankin, S. H., & Stedt-Kurki, P. (2008). School-based drama interventions in health promotion for children and adolescents: Systematic review. *Journal of Advanced Nursing, 63*(2), 116–131.
Jourdan, D., Samdal, O., Diagne, F., & Carvalho, G. S. (2008). The future of health promotion in schools goes through the strengthening of teacher training at a global level. *Promotion and Education, 15*(3), 36–38.
Kazemek, F. E., & Rigg, P. (1997). '…the sense of soul…goes hand in hand with an aesthetic response': Art in adult literacy…. *Adult Basic Education, 7*(3), 131.
Keppell, M. J. (2007). *Instructional design: Case studies in communities of practice*. Hershey: Information Science Publishing.
King, K. P., & Lawler, P. A. (2003). Trends and issues in the professional development of teachers of adults. *New Directions for Adult & Continuing Education, 98*, 5.
Kinloch, V. (2007). Youth representations of community, art, and struggle in Harlem. *New Directions for Adult and Continuing Education, 116*, 37–49.
Knowles, M. S. (1975). *Self-directed learning. A guide for learners and teachers*. New York: Association Press.
Knowles, M. S. (1980). *The modern practice of adult education: From pedagogy to andragogy* (2nd ed.). Chicago: Follett.
Knowles, J. G., & Cole, A. (2008). *Handbook of the arts in qualitative research*. Thousand Oaks: Sage.
Knowles, M. S., Holton, E. F., & Swanson, R. A. (1998). *The adult learner: The definitive classic in adult education and human resource development*. Houston: Gulf.
Labonte, R. (2010). Health promotion and empowerment: Reflections on professional practice. In J. M. Black, S. R. Furney, H. M. Graf, & A. E. Nolte (Eds.), *Philosophical foundations of health education* (pp. 179–195). San Francisco: Jossey-Bass.
Lave, J., & Wenger, E. (1991). *Situated learning: Legitimate peripheral participation*. New York: Cambridge University Press.
Lawrence, R. (2005). Artistic ways of knowing: Expanded opportunities/or teaching and learning. In *New directions for adult and continuing education, 107*. San Francisco: Jossey-Bass.
Levenson-Gingiss, P., & Hamilton, R. (1989). Evaluation of training effects on teacher attitudes and concerns prior to implementing a human sexuality education program. *Journal of School Health, 59*, 156–160.
Leviton, D. (2002). Potential untapped: Health education and health promotion as a means to peace. *International Electronic Journal of Health Education, 1*, 12–26.
Lieberman, A., & Wood, D. (2002). Untangling the threads: Networks, community and teacher learning in the National Writing Project. *Teachers & Teaching, 8*(3/4), 295–302.
Loucks-Horsley, S., Love, N., Stiles, K. E., Mundry, S., & Hewson, P. (2003). *Designing professional development for teachers of science and mathematics*. Thousand Oaks: Corwin.
Magolda, M. B. (1992). *Knowing and reasoning in college: Gender-related patterns in students' intellectual development*. San Francisco: Jossey-Bass.
Manning, C., Verenikina, I., & Brown, I. (2010). Learning with the arts: What opportunities are there for work-related adult learning? *Journal of Vocational Education and Training, 62*(3), 209–224.
Marcuse, H. (1969). *An essay on Liberation*. Boston: Beacon.
Mayo, C. (2000). The uses of Foucault. *Educational Theory, 50*(1), 103–116.
McBride, R. E., Reed, J. L., & Dollar, J. E. (1994). Teacher attitudes toward staff development: A symbolic relationship at best. *Journal of Staff Development, 15*(2), 36–41.
McCaul, K. D., & Glasgow, R. E. (1985). Preventing adolescent smoking: What have we learned about treatment construct validity? *Health Psychology, 4*, 361–387.
McCormick, L. K., Steckler, A. B., & McLeroy, K. R. (1995). Diffusion of innovations in schools: A study of adoption and implementation of school-based tobacco prevention curricula. *American Journal of Health Promotion, 9*, 210–219.

McGraw, S. A., Stone, E. J., Osganian, S. K., Elder, F. P., Perry, C. L., Johnson, C. C., Parcel, G. S., Webber, L. S., & Luepker, R. V. (1994). Design of process evaluation within the Child and Adolescent Trial for Cardiovascular Health (CATCH). *Health Education Quarterly, 2*(suppl.), S5–S26.

McKenzie, P., Mitchell, P., & Oliver, P. (1996). *Competence and accountability in education.* Aldershot: Arena.

McKeown, M. G., & Beck, I. L. (2004). Transforming knowledge into professional development resources. *Elementary School Journal, 104*(5), 391–408.

McLaughlin, M. W. (1998). Listening and learning from the field: Tales of policy implementation and situated practice. In A. Hargreaves, A. Lieberman, M. Fullan, & D. Hopkins (Eds.), *The international handbook of educational change* (pp. 70–84). Dordrecht: Kluwer.

McLaughlin, T. H. (2003). Teaching as a practice and a community of practice: The limits of commonality and the demands of diversity. *Journal of Philosophy of Education, 37*(2), 339–352.

Mclaughlin, T. F., & Vacha, E. F. (1993). Substance abuse prevention in the schools: Roles for the school counselor. *Elementary School Guidance & Counseling, 28*(2), 124–132.

Merriam, S. B. (2001a). Andragogy and self-directed learning: Pillars of adult learning theory. In S. B. Merriam (Ed.), *New directions for adult and continuing education* (pp. 3–14). San Francisco: Jossey-Bass.

Merriam, S. B. (2001b). *The new update on adult learning theory.* San Francisco: Jossey-Bass.

Merriam, S. B., Caffarella, R. S., & Baumgartner, L. M. (2007). *Learning in adulthood: A comprehensive guide* (3rd ed.). San Francisco: Jossey-Bass.

Meyer, R. (2005). Taking action for meaningful professional development. *Language Arts, 82*(5), 376.

Meyers, C. (1987). *Teaching students to think critically: A guide for faculty in all disciplines.* San Francisco: Jossey-Bass.

Mezirow, J. (1978). Perspective transformation. *Adult Education, 28,* 100–110.

Mezirow, J. (1981). A critical theory of adult learning and education. *Adult Education, 32*(1), 3–24.

Mezirow, J. (Ed.). (1990). *Fostering critical reflection in adulthood: A guide to transformative and emancipatory learning.* San Francisco: Jossey-Bass.

Mezirow, J. (1997). Transformative learning: Theory to practice. *New Directions for Adult and Continuing Education, 74,* 5.

Mezirow, J. (1998). On critical reflection. *Adult Education Quarterly, 48*(3), 185.

Mezirow, J. (2000). *Learning as transformation: Critical perspectives on a theory in progress* (1st ed.). San Francisco: Jossey-Bass.

Mezirow, J. (2009). An overview on transformative learning. In K. Illeris (Ed.), *Contemporary theories of learning: Learning theorists…in their own words* (pp. 90–105). London/New York: Taylor & Francis Routledge.

Moll, L. C. (1990). Introduction. In L. C. Moll (Ed.), *Vygotsky and education: Instructional implications and applications of sociohistorical psychology* (pp. 1–27). Cambridge: Cambridge University Press.

Morgan, N., & Saxton, J. (1987). *Teaching drama. A mind of many wonders.* London: Hutchinson.

Neelands, J. (1996). Agendas of change, renewal and difference. In J. O'Toole & K. Donelan (Eds.), *Drama, culture and empowerment* (pp. 20–29). Brisbane: IDEA Publications.

Neelands, J. (2004). Miracles are happening: Beyond the rhetoric of transformation in the Western traditions of drama education. *Research in Drama Education, 9*(1), 47–56.

Norton, M. (2001). Learning in the borderlands: Perspectives and possibilities in community-based learning. In M. C. Taylor (Ed.), *Adult literacy now!* (pp. 3–16). Toronto: Culture Concepts Inc.

Olson, C. M., Devine, C. M., & Frongillo, E. A. (1993). Dissemination and use of a school-based nutrition education program for secondary school students. *Journal of School Health, 63,* 343–348.

Olssen, M. (1999). *Michael Foucault: Materialism and education.* Westpoint: Bergin and Harvey.

O' Rourke, T. (2002). Health care reform: Insights for health educators. *American Journal of Health Education, 33*(5), 297–300.

O'Toole, J. (1992). *The process of drama: Negotiating art and meaning*. London/New York: Routledge.

Parcel, G. S. (1984). Theoretical models for application in school health education research. *Journal of School Health, 54*, 39–49.

Patterson, J. L., & Czajkowski, T. J. (1979). Implementation: Neglected phase in curriculum change. *Educational Leadership, 37*, 204–206.

Perry, W. G. J. (1970). *Forms of intellectual and ethical development in the college years*. Troy: Holt, Rinehart, and Winston.

Perry, C. L., William, C. L., Veblen-Mortenson, S., Toomey, T. L., Komro, K. A., Anstine, P. S., McGovern, P. G., Finnegan, J. R., Forster, J. L., Wagenaar, A. C., & Wolfson, M. (1996). Project Northland: Outcomes of a community wide alcohol use prevention program during early adolescence. *American Journal of Public Health, 86*, 956–965.

Pinzon-Perez, H. (2003). Lessons from developing countries: Health education in the global village. *American Journal of Health Education, 34*(2), 101–103.

Popkewitz, T. S. (1998). *Struggling for the soul: The politics of schooling and the construction of the teacher*. New York: Teachers College Press.

Popkewitz, T., & Brennan, M. (1997). Restructuring of social and political theory in education: Foucault and a social epistemology of school practices. *Educational Theory, 47*, 287–313.

Posner, G. J. (1985). *Field experience: A guide to reflective teaching*. New York: Longman.

Pratt, D. D. (2002). Good teaching: One size fits all? *New Directions for Adult and Continuing Education, 93*, 5.

Putnam, R. T., & Borko, H. (2000). What do new views of knowledge and thinking have to say about research on teacher learning? *Educational Researcher, 29*(1), 4–15.

Quant, P. L. (1967). Rural studies & the newsom child. *Hertfordshire Rural Studies Association Journal*, April 1967, p. 12.

Rademaker, L. L. (2003). Community involvement in arts education: A case study. *Arts Education Policy Review, 105*(1), 13–24.

Rappaport, J. (1981). In praise of paradox: A social policy of empowerment over prevention. *American Journal of Community Psychology, 9*, 1–25.

Read, D., & Stoll, W. (2010). Healthy behavior: The implications of a holistic paradigm of thinking through bodymind research. In J. M. Black, S. R. Furney, H. M. Graf, & A. E. Nolte (Eds.), *Philosophical foundations of health education* (pp. 145–167). San Francisco: Jossey-Bass.

Richardson, V. (1989). The evolution of reflective teaching and teacher education. In R. T. Clift, W. R. Houston, & M. C. Pugach (Eds.), *Encouraging reflective practice: An analysis of issues and programs* (pp. 3–19). New York: Teachers College Press.

Roberts, P. (2000). *Education, literacy and humanisation. Exploring the work of Paulo Freire*. Westport: Bergin and Harvey.

Rohrbach, L. A., Graham, J. W., & Hansen, W. B. (1993). Diffusion of a school-based substance abuse prevention program: Predictors of program implementation. *Preventive Medicine, 22*, 237–260.

Roper, B., & Davis, D. (2000). Howard Gardner: Knowledge, learning and development in drama and arts education. *Research in Drama Education, 5*(2), 217–233.

Ross, J. G., Luepker, R. V., Nelson, G. D., Saavedra, P., & Hubbard, B. M. (1991). Teenage health teaching modules: Impact of teacher training on implementation. *Journal of School Health, 61*, 31–33.

Ross-Gordon, J. M. (2002). Effective teaching of adults: Themes and conclusions. *New Directions for Adult and Continuing Education, 93*, 85.

Scher, A. (2007). Can the arts change the world? The transformative power of community arts. *New Directions for Adult and Continuing Education, 116*, 3–11.

Schon, D. A. (1983). *The reflective practitioner: How professionals think in action*. New York: Basic Books.

Schutz, A. (2004). Rethinking domination and resistance: Challenging postmodernism. *Educational Researcher, 33*(1), 15–23.

Schwartz, S., & Carpenter, K. M. (1999). The right answer for the wrong question: Consequences of type III error for public health research. *American Journal of Public Health, 89*(8), 1175–1180.

Shechtman, Z., Levy, M., & Leichtentritt, J. (2005). Impact of life skills training on teachers' perceived environment and self-efficacy. *The Journal of Educational Research, 98*(3), 144–154.

Shek, D. T. L., & Wai, C. L. Y. (2008). Training workers implementing adolescent prevention and positive youth development programs: What have we learned from the literature? *Adolescence, 43*(172), 823–845.

Silverman, S., & Casazza, M. (2000). *Learning and development: Making connections to enhance teaching*. San Francisco: Jossey-Bass.

Sim, C. (2006). Preparing for professional experiences: Incorporating pre service teachers as 'communities of practice'. *Teaching & Teacher Education, 22*(1), 77–83.

Smith, D. W., McCormick, L. K., Steckler, A. B., & McLeroy, K. R. (1993). Teachers' use of health curricula: Implementation of growing healthy, project SMART, and the THTM. *Journal of School Health, 63*, 349–354.

Smith, D. W., Steckler, A. B., McCormick, L. K., & McLeroy, K. R. (1995). Lessons learned about disseminating health curricula to schools. *Journal of Health Education, 26*(1), 37–43.

Sobol, D. F., Rohrbach, L. A., Dent, C. W., Gleason, L., Brannon, B. R., Johnson, C. A., & Flay, B. R. (1989). The integrity of smoking prevention curriculum delivery. *Health Education Research, 4*, 59–67.

Song, Y. I. K. (2009). Community participatory ecological art and education. *International Journal of Art and Design Education, 28*(1), 4–13.

Stenhouse, L. (1975). *An introduction to curriculum research and development*. London: Heinemann.

Stoll, L., Bolam, R., McMahon, A., Wallace, M., & Thomas, S. (2006). Professional learning communities: A review of the literature. *Journal of Educational Change, 7*, 221–258.

Stone, E. J., Osganian, S. K., McKinlay, S. M., Wu, M. C., Webber, L. S., Luepker, R. V., Perry, C. L., Parcel, G. S., & Elder, J. P. (1996). Operational design and quality control in the CATCH multicenter trial. *Preventive Medicine, 25*, 384–399.

Stuckey, H. L. (2009). Creative expression as a way of knowing in diabetes adult health education. *Adult Education Quarterly, 60*(1), 46–64.

Supovitz, J. A., & Turner, H. M. (2000). The effects of professional development on science teaching practices and classroom culture. *Journal of Research in Science Teaching, 37*(9), 963–980.

Sussman, S. (1991). Curriculum development in school-based prevention research. *Health Education Research, 6*(3), 339–351.

Sussman, S., Dent, C. W., Burton, D., Stacy, A. W., & Flay, B. R. (1995). *Developing school-based tobacco use prevention and cessation programs*. Thousand Oaks: Sage.

Taggart, V. S., Bush, P. J., Zuckerman, A. E., & Theiss, P. K. (1990). A process evaluation of the District of Columbia "Know Your Body" project. *Journal of School Health, 60*, 60–66.

Tappe, M. K., & Galer-Unti, R. A. (2001). Health educator's role in promoting health literacy and advocacy for the 21st century. *Journal of School Health, 71*(10), 477–482.

Tappe, M. K., Galer-Unti, R. A., & Bailey, K. C. (1995). Long-term implementation of the teenage health teaching modules by trained teachers: A case study. *Journal of School Health, 65*, 411–415.

Tharp, R. G., & Gallimore, R. (1988). *Rousing minds to life: Teaching, learning, and schooling in social context*. Cambridge: Cambridge University Press.

Thompson, J. (2002). *Bread and roses: Arts, culture and lifelong learning*. Leicester: National Institute for Adult and Continuing Education.

Tisdell, E. J. (2003). *Exploring spirituality and culture in adult and higher education*. San Francisco: Jossey-Bass.

Tom, A. (1985). Inquiring into inquiry-oriented teacher education. *Journal of Teacher Education, 36*(5), 35–44.

Tortu, S., & Botvin, G. J. (1989). School-based smoking prevention: The teacher training process. *Preventive Medicine, 18*, 280–289.

Trounstine, J. (2007). Texts as teachers: Shakespeare behind bars and changing lives through literature. *New Directions for Adult and Continuing Education, 116*, 65–77.

Ubbes, A. J., Black, M. J., & Ausherman, A. J. (2010). Teaching for understanding in health education: The role of critical and creative thinking skills within constructivism theory. In J. M. Black, S. R. Furney, H. M. Graf, & A. E. Nolte (Eds.), *Philosophical foundations of health education* (pp. 95–107). San Francisco: Jossey-Bass.

Valli, L., & Rennert-Ariev, P. (2002). New standards and assessments? Curriculum transformation in teacher education. *Journal of Curriculum Studies, 34*(2), 201–225.

Valli, L., & Taylor, N. (1987). *Reflective teaching: Preferred characteristics for a content and process model*. Paper prepared for the Reflective Inquiry Conference, Houston. Department of Curriculum and Instruction, University of Maryland.

Van Manen, M. (1977). Linking ways of knowing with ways of being practical. *Curriculum Inquiry, 6*(3), 205–228.

Vygotsky, L. S. (1978). In M. Cole, V. John-Steiner, S. Scribner, & E. Souberman (Eds.), *Mind in society: The development of higher psychological processes*. Cambridge, MA: Harvard University Press.

Wade, D. T. (2001). Research into the black box of rehabilitation: The risks of a type III error. *Clinical Rehabilitation, 15*, 1–4.

Wagner, B. J. (1999). *Dorothy Heathcote: Drama as a learning medium*. Portsmouth: Heinemann.

Walkerdine, V. (1988). *The mastery of reason: Cognitive development and the mastery of reason*. London: Routledge.

Weare, K. (1998). The health promoting school: An overview of concepts, principles and strategies and the evidence for their effectiveness. In European Network of Health Promoting Schools, *First workshop on practice of evaluation of the Health Promoting School* (pp. 9–18). http://www.schoolsforhealth.eu/upload/pubs/FirstWorkshoponpracticeofevaluationoftheHPS.pdf. Accessed 2 May 2012.

Weedon, C. (1997). *Feminist practice and post-structuralist theory*. Oxford: Blackwell.

Wenger, E. (1998). *Communities of practice: Learning, meaning, and identity*. New York: Cambridge University Press.

Wenger, E., McDermott, R., & Snyder, W. M. (2002). *Cultivating communities of practice: A guide to managing knowledge*. Boston: Harvard Business School Press.

Wertsch, J. V. (1985). *Vygotsky and the social formation of mind*. London: Harvard University Press.

Wesley, S. (2007). Multi-cultural diversity: Learning through the arts. *New Directions for Adult and Continuing Education, 116*, 13–23.

Willinsky, J. (2005). Just say know? Schooling the knowledge society. *Educational Theory, 55*(1), 97–112.

Wlodkowski, R. J. (2003). Fostering motivation in professional development programs. *New Directions for Adult and Continuing Education, 98*, 39.

World Health Organization. (1986). *Ottawa charter for health promotion*. http://www.who.int/hpr/NPH/docs/ottawa_charter_hp.pdf. Accessed 14 May 2012.

World Health Organization. (1996). *Promoting health through schools – The World Health Organization's Global School Health Initiative*. Geneva: World Health Organization.

Wright, N. M. J., & Walker, J. (2006). Homeless and drug use – A narrative systematic review of interventions to promote sexual health. *AIDS Care, 18*, 467–478.

Yorks, L., & Kasl, E. (2002). Toward a theory and practice for whole-person learning: Reconceptualizing experience and the role of affect. *Adult Education Quarterly, 5*(3), 176–192.

Zeichner, K. M. (1983). Alternative paradigms of teacher education. *Journal of Teacher Education, 34*(3), 3–9.

Zeichner, K. M., & Liston, D. P. (1987). Teaching student teachers to reflect. *Harvard Educational Review, 57*(1), 23–48.

Zinchenko, V. P., & Davydov, V. V. (1985). Foreword. In J. V. Wertsch (Ed.), *Vygotsky and the social formation of mind*. Cambridge: Cambridge University Press.

Chapter 9
Professional Competencies Within School Health Promotion – Between Standards and Professional Judgment

Monica Carlsson

Abstract The purpose of the study is to critically explore the formulations of competencies and standards in the European project "Developing Competencies and Professional Standards for Health Promotion Capacity Building in Europe", and to discuss them in relation to school health promotion. The analysis shows that 'a production logic' and economic values are emphasized in the motivation of the project and in the knowledge base underpinning the competency-framework. The discussion of the responsiveness of the formulations in relation to school health promotion points out that there are matches between these formulations, and essential values and approaches in school health promotion. However, by underemphasizing the potential of education and learning, and reducing changes at individual and group level to behavioral change, the formulations of competencies and standards are not in concert with essential values and approaches in school health promotion, and the usefulness of the formulations impaired for professionals in this field. Issues related to the use of competency-based standards within the field of education, are addressed in a concluding discussion.

Keywords Professional competences • School health promotion

9.1 Introduction

This chapter explores the formulations of competencies and standards in the European project "Developing Competencies and Professional Standards for Health Promotion Capacity Building in Europe", in short the CompHP project (Dempsey et al. 2010, 2011b; Speller et al. 2012). The aim of the CompHP project is to develop

M. Carlsson, Ph.D. (✉)
Department of Education, Aarhus University, Tuborgvej 164, 2400 Copenhagen, Denmark
e-mail: monica@edu.au.dk

V. Simovska and Patricia Mannix-McNamara (eds.), *Schools for Health and Sustainability: Theory, Research and Practice*, DOI 10.1007/978-94-017-9171-7_9,

competency-based standards and a pan-European accreditation system for health promotion practice, education and training, at the level of graduate or post graduate qualifications in health promotion or related disciplines, including health education and education (Dempsey et al. 2011b: 3). The competencies term in the project is defined as *"a combination of the essential knowledge, abilities, skills and values necessary for the practice of health promotion"*, and nine domains of core competencies are identified: enable change, advocate for health, mediate through partnership, communication, leadership, assessment, planning, implementation, evaluation and research (ibid).

Concepts of health promotion and professional competencies related to health promotion practice are complex and multidimensional, as they will have fluctuant meanings in different sociocultural, historical and political contexts and be embedded in different theories and practices. Health promotion has been defined in the following manner: *health promotion = health education x healthy public policy* (Tones and Tilford 2001). The understanding that health promotion includes both health education, based on strategies at the individual, group and community levels, and healthy public policy, based on strategies targeting the population level, and that different competencies are needed for working with strategies at the different levels, seems to be widely accepted in health promotion theory (see e.g. Green and Tones 2010). The ways in which competencies and competence development are formulated differ between contexts and are closely related to rationales and values in these contexts (Rychen and Salganik 2003; Dall'Alba and Sandberg 2006; Carlsson 2006). Health promoting school formulations seems to invoke mainly democratic and eco-holistic value discourses by highlighting competencies needed for: linking health and education issues and systems, creating safe and supportive school environments, upholding social justice and equity, and working with student participation and empowerment approaches (Young and Williams 1989).

One the one hand, professional competencies requirements within the school system are highly specialized and well described, e.g. the stipulation of competencies in teacher education acts. On the other hand, there has traditionally been room for professional interpretation and judgment in educational practices, which indicates that educational legislation has been flexible enough not to limit the professional autonomy and decision-making abilities of the practitioners. Criticism of a competency and standards approach in education argues that such an approach is based on an engineering (or a mechanistic) model of education, with a tendency to undervalue professional judgement and experience, and disregard educational values and principles (Elliot 2004; Biesta 2010). This chapter is driven by an interest to elucidate this critique by exploring ruptures and matches between the formulations of the competencies and standards in the CompHP project, and the values and approaches prevalent in school health promotion.

Standards are a form of social regulation, filling the gaps between legal regulations (e.g. education acts), and professional norms and actions (Timmermans and Epstein 2010). The CompHP project's aim of developing competency-based standards can be seen in the context of the EU Commission's agreement to develop flexible governance tools such as indicators and standards. These can be characterized as "soft law" tools; i.e., not legally binding, but based on reflexivity and persuasion, and intended to provoke institutions into reflection and self-criticism of their own practice (Moos 2009). Research on governance in the educational system points out, that the values in, and rationale for, developments of educational standards and indicators in projects funded by the EU may not always be explicated (Young 2007; Moos 2009). A critical exploration and discussion of the underlying discourses on values and rationales in the EU funded CompHP project's competency-based standards is therefore essential.

9.1.1 Aims and Methods in the CompHP Project

The aim of the literature review in the CompHP project is to provide an overview of the international and European literature published on the development of competencies frameworks for health promotion, with reference to work in related fields (Dempsey et al. 2010). The results of the review informs a draft of the core competency framework (Dempsey et al. 2011a), and this framework is the base for the development of the CompHP professional standards. The development of the competency framework and standards took place through a process including a survey on the draft standards, focus group with experts, and consultation with practitioners, academics, policy makers and employers (Speller et al. 2012). The standards are designed to form the basis for the CompHP Pan-European accreditation framework for health promotion, and are for that purpose aligned to the standards in the European qualifications framework (Speller et al. 2012).

The main data sources in the literature review are identified through a search performed mainly in medical and public health databases and include reports, articles and other information sources, published in English in 1980–2009. With these data sources, the boundaries of the framework must be considered – the resulting competency framework will be more closely aligned to requirements in medical and public health practice, than to those within the field of education. Another key concern related to the data sources in the review is that they include few references to theories and empirical research on competencies and competence development, and therefore the competencies framework, as pointed out in the review, is not based on evidence concerning competencies for health promotion.

9.1.2 Analytical Framework in the Study

The purpose of the study is to critically explore the formulations of competencies and standards in the CompHP project, to discuss them in relation to school health promotion, and to address issues related to the use of competency-based standards within the field of education, and the role of education in health promotion. The following questions guide the study:

- Which values and approaches are included and emphasized in the CompHP competencies and standards, and which are excluded and underemphasized?
- Are the formulations of the CompHP competencies and standards responsive to essential values and approaches in school health promotion?

The analysis of the first question will focus on the formulations of the value-base and the knowledge-base underpinning all nine core competencies within the CompHP project, and on the 'enable change' competency domain, which is the domain that best corresponds with the qualification and socialization objectives of the school system. Based on this analysis the second question will be addressed, where the aim is to identify ruptures and matches between the formulations of the competencies and standards in the CompHP project, and essential values and approaches in school health promotion.

The analysis of the CompHP formulations is inspired by a critical discourse analysis approach to qualitative inquiry. Critical discourse analysis does not pretend to be objective or neutral: instead it is upfront in both providing a positioned perspective, and revealing positions (Porter 2007). It offers the possibility of heightened reflexivity enabling us to use or resist the positions that have been revealed in the analysis (Cheek 2004: 1149).

The analysis of the CompHP formulations is based on two sets of discursive positions, found meaningful in relation to the formulations: The first is 'a logic of production vis-à-vis a logic of development', juxtaposing economic and democratic values, and used in analyses of competence development in workplace learning (Ellström 2006; Ellström and Kock 2009). The second is 'systems-centered health promotion vis-à-vis people-centered health promotion', used in the definition of health promotion as including both healthy public policy strategies and health education (Tones and Tilford 2001).

A logic of production and economic values is prominent in OECDs initiative in relation to competence development, referring to 'national competence accounts' (DeSeCo 2002), and in the development of EU higher education policy, including the European qualifications framework, where the European Commission discourses on 'measurable' educational outcomes are in the foreground (Keeling 2006). *A logic of development* and democratic values is prominent in the Ottawa Charter for health promotion (WHO 1986), referring to empowerment, participation, equality and justice. The Bangkok Charter (WHO 2005) highlights issues focusing on the investment needed to meet the health challenges of globalization (Potvin and Jones 2011), and hereby introduces an economic discourse of efficiency in health promotion.

Systems-centered health promotion approaches foreground policy and functions, and emphasize organizational development and leadership as key strategies (see e.g. Baric 1993; Grossman and Scala 1993). Robertson (1998) points out that the systems-centered approach represents a movement towards a population health discourse in health promotion. *People-centered health promotion* is a term coined by the Canadian researchers Raeburn and Rootman (1989), foregrounding individuals, groups and communities as political actors and subjects in health promotion. It emphasizes the existence of different interests in health promotion, and the resultant need for strategies and methods that can handle diversity and plurality (see e.g. Labonte 1995; Labonte et al. 1999).

I have delimited the discourse analysis to focus on the approximately twenty pages in the reports that presents the competency framework and the standards (Dempsey et al. 2011a; Speller et al. 2012). The remaining parts of the three reports in the CompHP project provided information about the context in which the texts were developed: The literature review provided generic insight into the practice of developing competency-frameworks, and the reports on the competency framework and standards offered insight into the foundation for, as well as the practice and processes in, the CompHP project. Space does not permit a full account of the results of the analysis, but the results will be used to illustrate the two main analytical positionings of discourses elaborated above.

9.2 A Logic of Production Vis-à-Vis a Logic of Development

The ComHP competencies framework and standards are underpinned by the concepts, principles and ethical values of health promotion as defined by the Ottawa Charter and subsequent charters (including the Bangkok Charter) and declarations (Speller et al. 2012: 13). Thus we can expect to identify discourses from both charters in the CompHP formulations. The section below focuses on the motivation of the ComHP project, the selection of core competencies, and on the formulations of values in the competency framework. Values not referring directly to democratic and economic discourses are not discussed here (e.g. values that invoke moral discourses related to health and medical practices, such as 'being honest about what health promotion can achieve', and 'causing no harm').

9.2.1 A Project Argued from the Production Logic

The production logic and the economic discourses in the standards are strengthened by the alignment to the standards in the European qualifications framework, motivated by the demands for work force flexibility and efficient operations in the workplace. The reference to this framework is made explicit in the following description of standards, as drawing together perspectives on *"what a learner needs to learn to be effective in employment"*, and *"what the learner has learnt and is competent to carry out in employment."* (Speller et al. 2012: 9). The production logic is furthermore explicit in the reference to the growing workforce within the field as motivation for the ComHP project: *"[...] health promotion is an evolving field in Europe with a diverse and growing workforce [...] there is a need for core competencies [...] to unify and strengthen health promotion workforce capacity across*

Europe." (Dempsey et al. 2011b: 2). In arguing for the necessity of establishing core competencies, these are understood as a way of coordinating practice from a distance in order to ensure consistency across locations.

The selection of the nine core competencies in the framework indicates that a balance between formulations that respectively invoke democratic or economic discourses is essential in the CompHP project (Dempsey et al. 2011a). Three of the competencies – 'enable change', 'advocate for health', 'mediate' – are directly linked to the democratic values and principles of participation and empowerment in the Ottawa Charter. Four of the competencies – 'leadership', 'assessment', 'planning' and 'implementation' – invoke a management discourse about effective mechanisms for governance, which is a key discourse in the Bangkok Charter. (The last two competencies – 'communication' and 'evaluation and research'- are not explicated on a level that makes it possible to link them to either democratic or economic discourses.)

9.2.2 A Value-Base Characterized by Navigating Between Democratic and Economic Value Discourses

The introduction to the value base in the competencies framework refers to the democratic discourses of equity, justice, and autonomy: *"Ethical values and principles for health promotion include a belief in equity and social justice, respect for the autonomy and choice of both individuals and groups, and collaborative and consultative ways of working"* (Dempsey et al. 2011a: 8). The philosophical foundation of this formulation is a belief that social change is brought about by an emancipation of the subject. References to a democratic value discourse are made in four of the values underpinning the framework: 'rights', 'diversity and inequalities', 'social injustice and needs', and 'collaboration, partnership, and empowerment'. These formulations are underpinned by the new health promotion discourse of the Ottawa Charter, recognizing the structural and social determinants of health, and that health cannot be separated from broader societal goals.

The formulation of the value *"Seeking the best available information and evidence needed to implement effective policies and programs"* (Dempsey et al. 2011a: 8) invokes a logic of production by emphasizing effectiveness. The phrase 'best available information and evidence' is closely connected to the politics of evidence discourse in health promotion (Carlsson and Simovska 2009), a discourse which presupposes that evidence is quantifiable. The value *"Being accountable for the quality of one's own practice"* (Dempsey et al. 2011a: 8) invokes a technical and economic discourse by proposing new public management solutions in health promotion. Values are contingent; i.e., they take on a different meaning in different contexts, and their interpretation depends on the disposition and prerequisites that the reader brings with her or him. If the phrase *'best available information and evidence'* is read without knowledge of the evidence discourse in health promotion, the meaning could be somewhat different for the reader. Twenty years ago, the value *"Being accountable for the quality of one's own practice"* would most likely have been read as a reference to a professional accountability discourse, and not as a reference to a new public management discourse. The meaning and interpretation of values are in other words influenced by movements and developments within the field of health promotion.

One of the more complex and contingent values in the value base in the competencies framework is '*sustainable development and sustainable health promotion action*'. The Ottawa Charter emphasizes ecological sustainability, holism and interdependence (Robertson and Minkler 1994), which is in harmony with Dryzek's (1997) description of sustainable development as a holistic view of development integrating environmental, economic and social considerations. However, in health promotion literature the concept of sustainable development mainly seems to be understood as an outcome related to the implementation and effectiveness of programs: Something that *"appeared at the end of a program like a gold star"* where the challenge lays in *"sustaining a programme content"* (St. Leger 2005: 317–318). The question is if the concept in the CompHP value formulation is to be understood as a holistic view of development, linking to a democratic discourse, or as an outcome related to the effectiveness of programs, linking to an economic discourse? The concept is not defined in the CompHP reports, however the formulation of the competency statement *"Facilitate the development and sustainability of coalitions and networks for health promotion action"* in the 'mediate domain' of the competencies framework (Dempsey et al. 2011b: 5) indicates an understanding of sustainable development as an outcome of coalitions and networks for health promotion action.

9.3 Systems-Centered Health Promotion Vis-à-Vis People-Centered Health Promotion

The distinction between systems-centered and people-centered health promotion can be further elaborated by Porter's distinction between technocracy and socio-ecology in her analysis of discourses in the Ottawa and Bangkok Charters (Porter 2007). The first applies to technical discourses, while the second covers discourses emphasizing diversity and interdependence. The focus in the analysis below is on the formulations in the 'enable change domain' in the competency framework and standards, and in the summary of knowledge requirements across the nine core competencies.

9.3.1 From Catalyzing Change and Empowerment to Capacity Building

The enable change domain is described as follows in the introduction to the domain in the standards: *"Enable individuals, groups, communities and organizations to build capacity for health promotion action to improve health and reduce health inequalities."* (Speller et al. 2012: 16). Capacity building is in the glossary of the competencies framework defined as the development of knowledge, skills, commitment, structures, systems and leadership necessary to *enable effective health* (Dempsey et al. 2011b: 13). While the introductory formulation uses the term 'enable individuals', and invokes a discourse emphasizing interdependence between individuals, groups, communities and organizations, the glossary formulation 'enable effective health' seems to invoke a more technical discourse, glossing over diversity in people and contexts.

Half of the terms that define capacity building in the glossary refer to people-centered health promotion concerns (knowledge, skills, commitment), and half to systems-centered concerns (structures, systems and leadership). One could say that this formulation provides an excellent example of the authors' attempts to navigate between the two. However, in the summary of knowledge requirements across the nine core competencies, the following knowledge is emphasized: program planning and management; human and financial resource management; effective leadership; evidence for effective health promotion; and quality assurance and monitoring. That is, knowledge of change management in systems-centered health promotion approaches is emphasized, while knowledge of change in relation to people-centered health promotion is underemphasized (Dempsey et al. 2011b: 13–17).

The inspiration of "The Galway Consensus Conference Statement" is apparent in the competencies framework: Eight of the nine domains of core competencies in the CompHP framework are formulated in the Galway statement (the ninth core competency is communication). The first competency domain in the Galway Consensus Conference Statement is also about enabling change, although the main paragraph emphasizes 'catalyzing change' and 'empowering individuals and communities', and not capacity building: *"Catalyzing change – enabling change and empowering individuals and communities to improve their health"* (Dempsey et al. 2010). In a critical discussion of the competencies in the Galway Consensus Conference Statement, Green and Tones (2010: 54) points out that the political aspects of health promotion, such as empowerment, are understated. One could say that is even more the case in the CompHP competency framework and standards, since 'the enable change' formulation doesn't refer to empowerment but to building capacity. The definition of capacity building in the competencies framework only implicitly refers to education and learning in the formulation 'development of knowledge, skills and commitment', and the competency formulation does not have the critical and transformative connotations of the 'catalyzing change' formulation in the Galway statement. Education, as Porter (2007) points out in the analysis of the Bangkok Charter, can be reduced to a technical function, and with the CompHP formulation, education in health promotion runs the risk of being perceived as such.

9.3.2 Narrowing the Scope and Potentials of the Enable Change Domain

The formulation of the five competencies in the enable change domain in the competencies framework, navigates between systems-centered and people-centered health promotion discourses: One the one hand there are formulations such as 'cross-sector collaboration' and 'collaboration with stakeholders to reorient health and other services', and on the other hand formulations that seem to invoke people-centered health promotion discourses: 'approaches that support empowerment, participation, partnership and equality', and 'development of personal skills that will maintain and improve health'.

However, in the standard related to the specific knowledge, skills and performance criteria required for enabling change, it is apparent that a only a few knowledge, skills and performance criteria are formulated in relation to people-centered health promotion. The knowledge related to this is 'knowledge of behavioral change techniques', the skill 'behavioral change techniques', and the performance criteria is: *"Use appropriate behavioral change techniques for specified individuals or groups to facilitate development of personal skills to maintain or improve health, and develop the capacity of others to support behavioral change"* (Speller et al. 2012: 16). Firstly, these formulations narrow the scope of changes at individual level to behavioral change, and thereby also the scope of health education in health promotion. Secondly, the formulations offer very little inspiration in relation to developing qualifications for professionals who mainly works with individual change in health promotion (e.g. nurses involved in daily health promotion activities such as health interviews).

The 'enable change' competency domain in the standard includes knowledge about communication processes and information technology, but not knowledge about learning approaches and processes. Learning is only referred to in the introduction to the standards, in a section describing the professional and ethical basis of health promotion practice: *"being aware of different learning approaches and preferences"* (Speller et al. 2012: 15). The lack of interest in learning in the competency framework and standards is of general concern in relation to the quality of the framework: The development of competencies in health promotion takes place trough learning, and therefore knowledge about learning prerequisites, approaches, processes and outcomes, and knowledge about contextual factors and mechanisms that influence these outcomes and processes should be in focus in an 'enable change' competency domain.

Transformative skills formulated as 'reflecting on own behavior and practice', 'recognizing the need for, and making use of, opportunities for own and others' development' and 'objectively and constructively reviewing the effectiveness of own area of work' (ibid.), are referred to only in the introduction to the standards – and not integrated in the nine core competencies in the standards. Multilevel entry level programmes within health promotion create some difficulty in developing a post-graduate curriculum, since the students will have different learning prerequisites shaped by different graduate programs and work experiences. In order to meet this challenge, the following transformative learning principles were integrated in a curriculum model for a MA in Health Promotion that was based on a competency-based framework from the Australian Health Promotion Association (Madsen and Bell 2012): the centrality of experience, critical reflections, reflective dialogue, and self-reflective strategies. Transformative learning principles and skills are closely related to professional expertise and judgment, and the underemphasizing of these in the

CompHP standards supports the critique (by Hammersley 2004 and others) that competence-based standards tend to undervalue professional experience and judgment. The omission of 'critical' in the 'reflecting' formulation in the CompHP standards, and the individualistic discourse in the phrasing of 'own behaviour' and 'own area of work', strengthens the tendency that Green and Tones points out in relation to the Galway statement, namely of understating political aspects of health promotion.

9.3.3 Summary of Analysis

The logic of production and economic values are emphasized in the motivation for the CompHP project. The selection of the core competencies, and the value base of the competencies framework, indicates that a balance between formulations that invoke democratic and economic discourses is intended: References to both democratic values (e.g. respect for diversity, addressing inequalities and social injustice) and economic values (e.g. efficiency, accountability) are made. However, references to traditional values in educational practice, such as professional duty and judgement, are not included.

Systems-centered and people-centered health promotion concerns are balanced in the introduction to the 'enable change' domain in the competency framework, and in the formulations of the five competencies in the domain. However, the standards are mainly related to systems-centered health promotion concerns, and the few people-centered formulations narrow the scope of changes at individual level to behavioral change. Learning approaches and transformative skills are only mentioned in the introduction to the standards, not in the standards themselves, and thereby the potential of education and learning in health promotion change is underemphasized.

Based on the summary of what is included/excluded and emphasized/underemphasized, an assumption about whom is addressed in the competency framework and standards (and who not), can be put forward: The CompHP competency framework and standards mainly addresses professionals who work with population health in leading and management roles, which lead to that the usefulness of the formulations is impaired for professionals working with people face to face, as in school health promotion and health education.

9.4 School Health Promotion and the CompHP Formulations

The discussion will point to ruptures and matches between central values and approaches in school health promotion on the one hand, and the formulations in the CompHP project on the other hand. On the basis of this discussion, the question of whether the CompHP formulations are responsive to essential values and approaches in school health promotion will be addressed.

9.4.1 Values in School Health Promotion and the CompHP Formulations

The basic values underpinning the health promoting schools approach in Europe are: equity; active participation of students; development of students' action competence; importance of the social and physical environment of the school; and integration of health promotion policies as part of school development (WHO 2002). With the references to equity, students' participation, and action competence, these formulations invoke the democratic value discourse from the Ottawa Charter, which the value-base in the CompHP competency framework also draws upon. In the terms of reference for Schools for Health in Europe (SHE), it is stated that a strategy founded on the health promoting schools approach helps school communities to: manage health and social issues; enhance student learning; improve school effectiveness (SHE Network 2007). With the terms 'manage health' and 'improve school effectiveness', the technical and economic value discourses identified in the value-base in the CompHP competency framework, are also to be found in health promotion school policy documents.

Although a theoretical distinction between different values in health promotion policy can be made, health promotion in practice is based on a multitude of values that are in constant interaction with each other (Gregg and O'Hara 2007). Schools simultaneously answer to a range of core accountabilities and values which are difficult to combine, creating numerous dilemmas for the practitioners, including: a marketplace accountability that focuses on efficiency and competition; a bureaucratic accountability that focuses on outcomes and indicators, and a professional accountability that focuses on professional expertise and judgment (Moos 2009). The references to economic values of efficiency, and the technical discourses of quality assurance and evidence-based practice in the CompHP formulations, match the focus on narrow and measureable learning outcomes in schools, and on evidence-based practice as a mean to achieve these outcomes.

In teacher training for health education, navigating between school and health concerns and issues is seen as essential, requiring the development of critical reflection and judgment (Jourdan 2011). Research on implementation and outcomes of health promoting school projects indicates that the interplay between professional values in educational practices in schools and the values in the implemented projects substantially influences both the implementation and project outcomes (Carlsson and Simovska 2012; Clarke et al. 2010). Knowledge about and skills in coping with these different values are developed in situational analysis, which is promoting the importance of professional judgment as a core value in school health promotion. Elliot (2004) points out that the focus on competence brings with it an interest in an outcomes-based education, characterized by a need to justify professional practices as effective and efficient means of producing desirable outputs. Professional judgment hereby loses its position as a core value, and becomes a means to reach values related to marketplace and bureaucratic accountabilities. In accordance with this, it's no surprise that references to professional expertise and judgment are underemphasized in the CompHP competencies framework and standards.

9.4.2 Approaches in School Health Promotion and the CompHP Formulations

Collaborative and participatory approaches in teaching and learning are emphasized in school health promotion (Simovska et al. 2006), based on the values underpinning the health promoting schools approach in Europe (WHO 2002). The formulation of the democratic value discourse in the introduction to the value-base of the CompHP competency framework – *"respect for the autonomy and choice of both individuals and groups, and collaborative and consultative ways of working"* – match these approaches. There is furthermore a reference to collaborative and participatory approaches in the description of the 'enable change' domain in the framework; however, in the standards related to the domain, these approaches are linked to systems-centered health promotion concerns (e.g. 'cross-sector collaboration'), and not to the teaching and learning approaches in focus in school health promotion.

The understanding of the settings approach in school health promotion is in accordance with the Ottawa Charter, where it is described as an approach that shifts the focus from individual behavior change towards community action in everyday life settings (WHO 1986). In the formulations of the 'enable change domain' in the CompHP standards, the scope and potentials of changes at individual and group level is narrowed down to behavioral change. Thus the CompHP formulations on approaches don't match the understanding of the settings approach that has been of such vital importance in the health promoting school. The Ottawa Charter underlines holism, sustainability and interdependency between different agents, perspectives and interests, which is illustrated in the holistic models of the health promoting school (see e.g. Parsons et al. 1997, in: Green and Tones 2010: 451). The systems-centered and people-centered health promotion concerns are balanced in the introduction to the 'enable change' domain in the Comp HP competency framework, and thus match the eco-holistic models of the health promoting school. However, the standards related to this domain mainly refer to systems-centered health promotion concerns (e.g. effective leadership, financial resource management), and therefore differ from the health promoting school models, where pupils, teachers, parents, and community partners are involved in creating safe and supportive school environments. As Dooris (2009) points out, a standards approach, with its precise and unambiguous criteria for educational outcomes, might be a challenge in relation to encompassing the complexity of the settings approach.

'Skills-based health education' (described in Aldinger and Whitman 2005) and 'action-oriented health education' (described in Jensen 1997) illustrate the span in educational philosophy, principles and approaches within school health promotion: From utilitarian to critical philosophy, from coping to empowerment, and from intentions of adopting healthy behaviors to shaping social change (Carlsson 2012). Transformative health education discourses as 'action-oriented health education' are based on approaches involving learners in the formulation of health problems and action-possibilities related to solutions of these problems, and providing opportunities for real-life experiences with initiating and carrying out change (Carlsson and Simovska 2012). It casts children and youths as agents in health promoting changes (Simovska and Carlsson 2012), while the CompHP formulations mainly casts managers and organizations as agents. By downplaying empowerment and related discourses such as critical action and critical thinking, and by formulating people-centred change strategies in terms of behavioral change techniques, the formulations in the CompHP framework and standards provide a better match with 'skills-based health education' approaches, than transformative health education approaches.

9.4.3 The CompHP Formulations and Their Responsiveness to School Health Promotion

Agents in different settings and situations will interpret formulations of frameworks and standards in different ways, depending on the disposition and prerequisites that the reader brings with him/her in the reading. As a head of studies of 15 professionally oriented Master of Education (MEd) programs, including a MEd in Health Promotion and Education, I have worked with competencies frameworks for educational programs within the humanities, as well as with the standards in the Danish qualifications framework, and found them useful as tools for reflexivity in development and accreditation processes (Carlsson and Adriansen 2010). In comparison to the CompHP standards, the standards in the Danish qualifications framework formulations are relatively open and pliable, constructed in order to be elaborated in line with the purpose and content of the different programs, and with educational values and methodologies. Being situated in the field of education and humanities/social science, it is perhaps no surprise that I find the CompHP standard formulations not quite in concert with this field (while recognizing that a practitioner within a public health domain will interpret the formulations differently).

The key positive aspects of a competency- and standards based approach described in the CompHP review includes: the usefulness of a shared vocabulary for defining boundaries for competencies in professional practices and in curriculum development; that it can stimulate communication and reflexivity in and between professional practices involved with health promotion; and inspire development of programs and curricula. With the emphasis of participatory, collaborative and settings approaches in school health promotion, and the intention to integrate health promotion policies as part of school development, a shared vocabulary and communication and reflexivity in and between professional practices in school health promotion are most certainly of value.

The usefulness of the CompHP formulations as a shared language for competencies and standards in professional practices within school health promotion is impaired by the CompHP project's overemphasis on systems-centered health promotion and public health work descriptors, and underemphasis on people-centered health promotion and descriptors meaningful for professionals working with people in face to face health promotion and education. Although there are matches between values and approaches in the CompHP project and in school health promotion, they often take on a different meaning when used in relation to either system-centred or people-centred health promotion. By downplaying professional judgment, and values and approaches linked to people-centred health promotion; by underemphasizing the potential of education and learning; and by reducing changes at individual and group level to behavioral change, the CompHP formulations cannot be considered responsive to the essential values and approaches in school health promotion. Furthermore, and as pointed out in the summary of the analysis, neither do the CompHP formulations seem to address professionals in school health promotion.

In continuance of the discussion above of values and approaches in school health promotion, I would like to emphasize two perspectives that should be considered in relation to efforts to strengthen professional competencies within school health promotion:

- If we want to address professional dilemmas related to the demands of simultaneously having to answer to different accountabilities and values in schools, competencies in navigating between the concerns based on professional values and approaches in educational practices in schools and the values and approaches in school health promotion, are essential. Here it is especially important to consider the links between health related, and educational concerns and aims, as well as the structural and cultural barriers for working with health in schools.
- Educational policy tasks schools with addressing a range of different issues, including health and sustainability issues, and integrating them in curricula. If we want to address issues related to sustainable development in school health promotion, an understanding of sustainable development as a holistic view of development that implies learning is more meaningful than understanding it as an outcome related to the effectiveness of programs. Since the approaches in working with health and sustainability issues in schools are similar, both encompassing a holistic view, democratic processes, and transformative learning principles (Cooke et al. 2010), the prospects for linking the two seems promising.

9.5 Issues in Using Competency-Based Standards Within the Field of Education

The contextual and value-bounded nature of educational competencies, and the need for situational analysis in decision making, gives professional judgement a crucial role in practice development in education (Eraut 2004). The more complex the practice field is, such as the fields of education and school health promotion, the less likely it is that competency-based standards will offer formulations that are meaningful in concrete contexts and situations. Pre-conceptualized competencies in competency-based standards therefore need to be further developed by professionals in concrete contexts and situations. Or as a colleague said on reading curricular guidelines, 'attack them as you would attack a mountain top in the Tour de France'!

Competency-based standards can be used as tools for regulating and directing social agents and behavior, or they can be used as tools that guide social agents and behavior by promoting reflexivity and learning. Lessons from countries that have developed accreditation systems on the basis of national qualification frameworks, and that have applied a rule based approach, indicates that it can lead to unwanted implementation effects such as resistance against change, and development of a hidden curriculum (Young 2007; Allais 2007). However, a guideline-based implementation approach can also lead to resistance strategies, as in the implementation of the Danish qualification framework in educational programs within humanities, where teachers referred to the process of transforming competency description from the framework to descriptions of learning objectives in the curriculum guidelines, as 'bullshit-bingo', and as 'a style exercise that doesn't fulfill its purpose' (Sarauw 2011: 210).

Competency-based standards offered by the CompHP project, or by other agents, can potentially inspire a development of local standards for school health promotion (on national, regional or school level): These in turn can stimulate communication and reflexivity and coordinate actions in and between professional practices involved with school health promotion, and through processes of reflexivity and learning, they might lead to the transformation of educational curricula. The review in the CompHP project emphasizes that views are divided concerning the extent to which core competencies should be linked to quality assurance or formal accreditation mechanisms (Dempsey et al. 2010, 113). In the discussion of whether standards should be used as tools for regulation or guidance, I side with the latter view, and, as such, it concerns me if the CompHP standards are to be used as a basis for an accreditation mechanism (as is the intention in the CompHP project), since this means the standards will be implemented as rules.

The CompHP competency framework and standards were developed through a deliberative process, including a survey on the draft standards, a focus group with experts, and consultation with practitioners, academics, policy makers and employers. The question of who these experts are that have been given a voice in the development process is discussed in a paper on the CompHP project (Battel-Kirk et al. 2009), and in the review in the CompHP project it is mentioned, that the competency- and standards based approach can be criticised *"as potentially reinforcing conventional discourses about professional, norms, behaviours and attitudes, and perpetuating existing domains of professional legitimacy"* (Dempsey et al. 2010: 95). The CompHP development processes must have shaped the formulations of the framework and standards in a number of ways, and it would have been relevant to describe these in the CompHP project, in order to shed light on which concrete issues were raised and by whom.

Standards, as a technology in high stakes accountability, can, as the discussion of issues in competency-based standards above has indicated, leave little room for reflective practice. However, if the standards are used as a guiding tool, they will not to hinder processes of reflexivity, learning, and adjustments of competencies formulations and standards, or to impede or interfere with professional judgment in health promotion. Studies of the implementation of the standards in the European qualifications framework for higher education point out that, even within culturally similar regions such as the Nordic countries, implementation has differed substantially (Young 2007), and not only between countries but also between institutions (Sarauw 2011). If the CompHP standards are implemented within school health promotion education and practice, it will be interesting to explore how they are operationalized and used in different fields of health promotion, and follow the transformation processes at different levels.

9.5.1 The Role of Education in School Health Promotion

In the start of the 1990s, health promotion was criticized for emphasizing strategies on the level of individuals – it was phrased as 'half a holism' (Mathiesen 1993), referring to the visions of ecological sustainability, holism and interdependency emphasized in the Ottawa Charter. Health promotion seems to be in the process of developing another 'half a holism' – this time emphasizing public health strategies, that invokes bureaucratic and marketplace accountabilities; capacity building (instead of empowerment); change management (instead of catalysing change); effective leadership (instead of individuals and groups participating in change processes); evidence for effective health promotion (instead of professional judgment); and quality assurance and monitoring (instead of trust in professional judgment).

The Ottawa Charter expanded the scope and goals of health promotion practice: it placed political aspects of health promotion, such as empowerment, in the foreground; it cast people in the role of social agents participating in and taking decisions about health matters; and it tasked health promotion with supporting this with education, including ensuring learning opportunities (Porter 2007). As argued in the analysis, the lack of interest for learning in the competency framework and standards in the CompHP project is of general concern: Development of competencies in health promotion takes place through learning, and therefore knowledge about learning approaches, processes and outcomes, and about contextual factors and mechanisms that influence these outcomes and processes, should be prioritized in competency-based standards within health promotion. Most importantly, competency-based standards within school health promotion should be considered in close relation to not only values in health promotion, but to the purpose of education.

In relation to this, I would like to rephrase Biesta's (2010) challenge to education as a challenge to school health promotion: It constitutes a threat to the wider democratic deliberations of the aims and the conduct of health promotion in schools if questions about the purpose of education lose their relevance, and focus shifts instead to technical questions concerning how we can make health promotion processes more manageable, accountable and effective. As Green and Tones (2010) point out in the discussion of the Galway statement, health education within a critical empowerment approach has an important role to play in health promotion, because of its insistence on the core health promotion values and principles of empowerment and participation. The empowerment focus, transformative education approaches based on involving learners in the formulation of health problems and action-possibilities for solving these problems, and transformative learning principles emphasizing critical reflection, thinking and action, can ensure that education is not reduced to a technical function in school health promotion.

References

Aldinger, C., & Whitman, C. V. (2005). Skills for health: Skills-based health education to teach life skills. In S. Clift & B. B. Jensen (Eds.), *The health promoting school: International advances in theory, evaluation and practice* (pp. 151–172). Copenhagen: Danish University of Education Press.

Allais, S. M. (2007). Why the South African NQF failed: Lessons for countries wanting to introduce national qualification frameworks. *European Journal of Education, 42*(4), 523–547.

Baric, L. (1993). The settings approach – Implication for policy and strategy. *Journal of the Institute of Education, 21*(1), 17–24.

Battel-Kirk, B., Barry, M. M., Taub, A., & Lysoby, L. (2009). A review of the international literature on health promotion competencies: Identifying frameworks and core competencies. *Global Health Promotion, 16*(12), 12–20.

Biesta, G. J. J. (2010). Why 'what works' still won't work. From evidence-based education to value-based education. *Studies in Philosophy and Education, 29*(5), 491–503.

Carlsson, M. (2006). Between the vocational and the academic world – Boundary learning and navigation at the master programs in education. In I. Christensen & T. Fristrup (Eds.), *Universitetspædagogiske refleksioner* [Reflections on higher education pedagogy]. Copenhagen: Danish University of Education Press.

Carlsson, M. (2012). Sundhedspædagogiske perspektiver på læring og undervisning [Health education perspectives on learning and teaching]. In V. Simovska & J. M. Jensen (Eds.), *Sundhedspædagogik i sundhedsfremme* [Health education in Health promotion]. Copenhagen: Gad Publisher.

Carlsson, M., & Adriansen, H. K. (2010). *Masteruddannelsernes kompetenceprofil* [The competence profile in the master of education programs]. DUE, unpublished development project report. Copenhagen: DUE.

Carlsson, M., & Simovska, V. (2009). Fokus på evidens indenfor sundhedsundervisning og sundhedsfremme [Focus on evidence within health education and health promotion]. In M. Carlsson, V. Simovska, & B. B. Jensen (Eds.), *Health education and health promotion – Theory, research and practice*. Copenhagen: Unipress.

Carlsson, M., & Simovska, V. (2012). Exploring learning outcomes of school-based health promotion – A multiple case study. *Health Education Research, 27*(3), 437–447.

Cheek, J. (2004). At the margins? Discourse analysis and qualitative research. *Qualitative Health Research, 14*(8), 1140–1150.

Clarke, A. M., O'Sullivan, M., & Barry, M. M. (2010). Context matters in programme implementation. *Health Education, 110*, 273–294.

Cooke, S. M., Davis, J. M., Blashki, G., & Best, A. (2010). Healthy children, healthy planet: The case for transformative education in schools and early childhood education. *International Public Health Journal, 2*(4), 561–570.

Dall'Alba, G., & Sandberg, J. (2006). Unveiling professional development: A critical review of stage models. *Review of Educational Research, 76*(3), 383–412, American Educational Research Association.

Dempsey, C., Barry, M., & Battel-Kirk, B. (2010). *Literature review developing competencies for health promotion*. Galway: Health Promotion Research Centre, National University of Ireland.

Dempsey, C., Battel-Kirk, B., & Barry, M. M. (2011a). *The CompHP core competencies framework for health promotion handbook*. Paris: International Union of Health Promotion and Education (IUHPE).

Dempsey, C., Battel-Kirk, B., & Barry, M. M. (2011b). *The CompHP core competencies framework for health promotion handbook. Short version*. Paris: International Union of Health Promotion and Education (IUHPE).

DeSeCo. (2002). *Definition and selection of competences (deseco): Theoretical and conceptual foundations*. www.portalstat.admin.ch/deseco/deseco_strategy_paper_final.pdf

Dooris, M. (2009). Sundhedsfremmende settings: Teori, politik og praksis (Health promoting settings: Theory, politics, and practice). In M. Carlsson, V. Simovska, & B. B. Jensen (Eds.), *Sundhedspædagogik og sundhedsfremme. Teori, forskning og praksis* (Health Education and Health promotion – theory, research and practice). Copenhagen: Unipress.

Dryzek, J. S. (1997). *The politics of the earth: Environmental discourses*. Oxford: Oxford University Press.

Elliot, J. (2004). Making evidence-based practice educational. In G. Thomas & R. Pring (Eds.), *Evidence-based practice in education*. Buckingham: Open University Press.

Ellström, P.-. E. (2006). Two logics of learning. In E. Antonacopoulou, P. Jarvis, V. Andersen, B. Elkjær, & S. Høyrup (Eds.), *Learning, working and living mapping the terrain of working life learning*. London: Palgrave Macmillan.

Ellström, P.-. E., & Kock, H. (2009). Competence development in the workplace: Concepts, strategies, and effects. In K. Illeris (Ed.), *International perspectives on competence development*. London: Routledge.

Eraut, M. (2004). Practice-based evidence. In G. Thomas & R. Pring (Eds.), *Evidence-based practice in education*. Buckingham: Open University Press.

Green, J., & Tones, K. (2010). *Health promotion: Planning and strategies*. London: Sage publications.

Gregg, J., & O'Hara, L. (2007). Values in principles evident in current health promotion practice. *Health Promotion Journal of Australia, 18*(1), 7–11.

Grossman, R., & Scala, K. (1993). *Health promotion and organizational development: Developing setting for health*. Copenhagen: WHO Regional Office for Europe.

Hammersley, M. (2004). Some questions about research and evidence-based practice in education. In G. Thomas & R. Pring (Eds.), *Evidence-based practice in education*. Buckingham: Open University Press.

Jensen, B. B. (1997). A case of two paradigms within health education. *Health Education Research, 12*(4), 419–428.

Jourdan, D. (2011). *Health education in schools. The challenge of teacher training*. Saint-Denis: Inpes, coll. Santé en action.

Keeling, R. (2006). The Bologna process and the Lisbon Research Agenda: The European Commission's expanding role in higher education discourse. *European Journal of Education, 41*(2), 203–223, Oxford: Blackwell Publ. Ltd.

Labonte, R. (1995). Population health and health promotion: What do they have to say to each other? *Canadian Journal of Public Health, 86*, 165–168.

Labonte, R., Feather, J., & Hills, M. (1999). A story/dialogue method for health promotion knowledge development and evaluation. *Health Education Research, 14*(1), 39–50.

Madsen, W., & Bell, T. (2012). Using health promotion competencies for curriculum development in higher education. *Global Health Promotion, 19*(1), 43–49.

Mathiesen, J. (1993). Sundhedsformidling, Sundhedsfremme, Forebyggelse. *Sundt og Socialt*, nr. 2 s. 4–8.

Moos, L. (2009). A general context for new social technologies. *Nordisk Pedagogik, 29*, 79–92.

Porter, C. (2007). Ottawa to Bangkok: Changing health promotion discourse. *Health Promotion International, 22*(1), 72–79, Oxford University Press.

Potvin, L., & Jones, C. M. (2011). Twenty-five years after the Ottawa Charter: The critical role of health promotion for public health. *Canadian Journal of Public Health, 102*(4), 244–248.

Raeburn, J., & Rootman, I. (1989). Towards an expanded health field concept: Conceptual and research issues in a new era of health promotion. *Health Promotion, 3*, 383–392.

Robertson, M. (1998). Critical reflections on the politics of need: Implications for public health. *Social Science and Medicine, 47*, 1419–1430.

Robertson, M., & Minkler, M. (1994). New health promotion movement: A critical examination. *Health Education & Behavior, 21*, 295–312.

Rychen, D. S., & Salganik, L. H. (Eds.). (2003). *Key competences for a successful life and a well functioning society*. Cambridge/Göttingen: Hogrefe & Huber.

Sarauw, L. L. (2011). *Kompetencebegrebet og andre stiløvelser*. Ph.d.-afhandling. Københavns Univerisitet. Afd. For Pædagogik.

SHE Network. (2007). *Terms of reference SHE network.* Identified July 15, 2012, at http://www.schoolsforhealth.eu

Simovska, V., & Carlsson, M. (2012). Health-promoting changes with children as agents: Findings from a multiple case study research. *Health Education, 112*(3), 292–304.

Simovska, V., Jensen, B. B., Carlsson, M., & Albeck, C. (2006). *Towards a healthy and balanced growing up.* Methodological guidebook, http://www.shapeupeurope.net

Speller, V., Parish, R., Davidson, H., Zilnyk, A., & CompHP Project Parterns. (2012). *The CompHP professional standards for health promotion handbook.* Paris: IUHPE.

St. Leger. (2005). Questioning sustainability in health promotion projects and programs, *Health Promotion International, 20*(4), 317–319, Oxford: Oxford University Press.

Timmermans, S., & Epstein, S. (2010). A world full of standards but not a standard world: Toward a sociology of standardization. *Annual Review of Sociology, 36*, 69–89.

Tones, K., & Tilford, S. (2001). *Health promotion: Effectiveness, efficiency and equity* (3rd ed.). Cheltenham: Nelson Thornes.

World Health Organization. (1986). *The Ottawa charter for health promotion*. Geneve: WHO.

World Health Organization. (2002). *The Egmond Agenda. A tool to help establish and develop health promotion in schools and related sectors across Europe.* Egmond aan Zee, the Netherlands, 25–27 September 2002/ENHPS Technical Secretariat. Copenhagen: WHO Regional Office for Europe.

World Health Organization. (2005). *The Bangkok charter for health promotion in a globalized world.* Geneva: WHO.

Young, M. (2007). Qualification frameworks: Some conceptual issues. *European Journal of Education, 42*(4), 445–457.

Young, I., & Williams, T., (Eds.). (1989). *The healthy school: A Scottish Health Education Group*. Word Health Organization report. Edinburgh.

Chapter 10
Health Education in the Context of Performance Driven Education: Challenges and Opportunities

Sharon Moynihan, Jennifer Hennessy, and Patricia Mannix-McNamara

Abstract The education of the whole person is promoted as the central premise of the Irish education system. However, the practice of schooling at post primary level diverges significantly from the espoused ethos of holistic and personal development, with measurement, accountability and performativity taking centre stage. The significant influence of a consumerist agenda in education has meant that in the drive to educate for exam success, the development of affective education has been severely challenged. This has placed significant pressure on schools to treat the cognitive and affective as mutually exclusive and to teach the cognate discipline alone. In this context, a narrowed edition of the subject becomes prioritised to the detriment of the promotion of students' health and well-being.

The subject Social Personal and Health Education (SPHE) is dedicated to the promotion of the health and well-being of students. Ireland adopted this specific curricular approach to suit the already dominant discrete curricular structure of Irish schooling. The subject of SPHE and the health promoting schools initiative has much to offer in terms of student development, yet both have experienced significant challenges in terms of parity of esteem. This chapter will provide a critical analysis of the role of education and of the health promoting school within school systems where pedagogical practice is driven by a different value system, that of exam performativity and knowledge reproduction. The chapter will also make recommendations in terms of the development of a sustainable, affective curriculum and health promoting schools initiative.

S. Moynihan, B.Ed. (✉) • J. Hennessy, Ph.D.
Research Centre for Education and Professional Practice,
Faculty of Education and Health Sciences, University of Limerick, Limerick, Ireland
e-mail: Sharon.Moynihan@ul.ie; Jennifer.Hennessy@ul.ie

Patricia Mannix-McNamara, Ph.D.
Faculty of Education and Health Sciences, University of Limerick, Limerick, Ireland
e-mail: Patricia.M.McNamara@ul.ie

V. Simovska and Patricia Mannix-McNamara (eds.), *Schools for Health and Sustainability: Theory, Research and Practice*, DOI 10.1007/978-94-017-9171-7_10,

Keywords Health education • Exam performativity • Affective education • Curriculum • Student health and well-being • Social and personal development

10.1 History of Post Primary Health Education in Ireland

The post primary education system in Ireland comprises two distinct phases; the junior cycle, catering for students aged 12–15 years and the senior cycle, catering for students aged between 15 and 18 years. Upon completion of the three year junior cycle programme of study, a summative examination, called the Junior Certificate Examination, is undertaken by all students. The successful completion of the two year senior cycle programme is assessed primarily through a summative examination entitled the Leaving Certificate Examination.

In 2000, the Department of Education and Science approved the recommended syllabus for the subject Social, Personal, Health Education (SPHE) at junior cycle. At this time all schools were issued with a department circular mandating the introduction of the subject on a phased basis over three years in all post primary schools from September 2000 (Department of Education and Science 2000b). The concept of introducing personal education into schools was not a new one, having up to this point been addressed on an *ad hoc* basis. Personal education had become popularised in the previous two decades, housed under the auspices of pastoral care programmes in schools. School children's exposure to pastoral care and personal education often occurred in the formally timetabled religion classes. Programmes of Life Skills education also became quite popular in schools. These were offered at the discretion of school management however, and were dependent upon individual teacher interest and motivation. This notwithstanding, the increasing engagement with the personal and social education of school children, even in this *ad hoc* style, signaled the beginning of a sea change in educational provision that now increasingly recognised the need for attention to the holistic education of young people. The impetus for the provision of personal and social education was influenced by the changing nature of Irish society and consequently the increasingly diverse needs of school children (SPHE Support Service 2005). Of concern immediately prior to the introduction of SPHE, was the growing problem of substance misuse. The earlier maturation of young people and the information communications revolution meant children were more exposed to global messages and values often detrimental to their well-being. In terms of substance misuse, the programme entitled "On My Own Two Feet" was introduced in 1993 as a pilot initiative in order to raise awareness of substance abuse among post primary school students (Department of Education and Science 1994). It was successful in its aim and led to the recognition of the potential for health specific curriculum. Prior to the implementation of SPHE, a specific module on Relationships and Sexuality Education (RSE) was developed and introduced into schools in 1995 (SPHE Support Service 2005) in order to address the challenges of sexuality education for young people. While controversial at the time, drawing significant media attention and quite some resistance, the introduction of RSE has proved a fruitful addition

(Maycock et al. 2007) to the school curriculum and is now subsumed into SPHE, functioning as a core component of the SPHE, curriculum in Irish schools.

The Education Act of 1998 acknowledges the responsibility of schools to provide holistic education and to promote the moral, spiritual, social and personal development of students. While personal and social education has been relatively neglected in terms of curriculum development (Fullan 2001), the development and implementation of this suite of personal and social education in Ireland is important in terms of the balance of educational provision for young people. It was always intended that SPHE and RSE be implemented in the context of a Health Promoting School (HPS). The Irish Network of Health Promoting Schools was established in 1995 and its vision was clear. It sought to ensure that the implementation of these initiatives would occur in a context of supportive schools which were health promoting. However, the success of the health promoting schools initiative has left much to be desired in Irish schools.

10.2 Social, Personal and Health Education

In September 2003, educational policy decreed that SPHE become a compulsory subject in all schools with a formal place allocated on the school timetable as part of the junior cycle core curriculum (Department of Education and Science 2003). The aims of SPHE are closely aligned with the central aim of Irish education, namely the holistic development of the student, with specific focus placed on the development of the emotional, social, moral, intellectual, creative, critical and spiritual dimensions (Department of Education and Science 2000a). Clearly, health education is valued by policy makers in the adoption of a specific curriculum in schools however; its implementation has been tokenistic to some degree. It is noteworthy, that SPHE was not made compulsory for senior cycle students (aged 15–18), given that this is a time in a student's life that is often emotionally challenging and intellectually draining. Senior cycle is a distinctly pressurised time for students as they are facing into high stakes examinations which are primarily used for the purpose of matriculation in Ireland. The absence of health education in senior cycle means that the school curriculum offers little protected time for students to discuss issues that are pertinent to their own lives or to promote awareness of students' physical and emotional well-being. While it was intended that a senior cycle curriculum would be drafted and implemented quickly following the junior cycle edition, worryingly, over a decade later, while a draft curriculum has been developed by the National Council for Curriculum and Assessment (NCCA), it has progressed no further, suggesting some potential ambivalence in terms of the subject.

The introduction of SPHE gave coherence and structure to health education in schools (Nic Gabhainn et al. 2007). The development of a comprehensive syllabus recognised the importance of the school as a health promoting setting while enhancing educational attainment. The framework for the implementation of SPHE, in keeping with the ethos of the health promoting school, was designed as a partnership initiative between the Department of Health and Children and the Department of Education and Science, in which both contributed to the establishment of the SPHE Support Service. This partnership between both departments resulted in the development of an SPHE regional support team which consisted of Regional Development Officers from the education setting and Health Promotion Officers from the health setting. This support service was established in order to assist teachers and principals in the implementation of the new curriculum. The National Health Promotion Strategy 2000–2005 provided the national policy context for this support service (SPHE Support Service 2005).

The subject SPHE has much to offer the education of young people. The curriculum aims are clearly set out and aspire to; "provide students with a unique opportunity to develop the skills and competence to learn about themselves and to care for themselves and others and to make informed decisions about their health, personal lives and social development" (Department of Education and Science 2000a: 3). The specific aims of SPHE are outlined in the Guidelines for Teachers document as (a) the enablement of students to develop personal and social skills; (b) the promotion of self-esteem and self-confidence; (c) the enablement of students to develop a framework for responsible decision making; (d) the provision of opportunities for reflection and discussion and (e) the promotion of physical, mental and emotional health and well-being (Department of Education and Science 2001: 5). The aims of the curriculum are broad ranging and tailored to build students' self-awareness and decision making capacities in order to enable them to address some of the challenges of modern life. Complex issues such as substance use, sexuality, bullying, stress and suicide among others are given space for discussion and debate in SPHE. The SPHE curriculum by its very nature is spiral in design which means that each module is taught in year one of the junior cycle and revisited each year in growing depth. The themes of the modules are specifically prescribed and are entitled; Belonging and Integration; Self-Management; Communication Skills; Physical Health; Friendship; Emotional Health; Relationships and Sexuality; Influences and Decision Making; Substance Use and Personal Safety.

10.3 The Link Between Educational and Health Outcomes

SPHE has also a lot to offer in terms of contribution to educational outcomes. It has been reported that the learning experiences provided for by the SPHE class may have a positive contribution to all learning success, as a student with high self-esteem will be better able to avail of that which school has to offer (Mohammadi et al. 2010; SPHE Support Service 2005). The link between better health and improved educational outcomes is established in the literature (Basch 2011; Stewart-Brown 2006; Wolford Symons et al. 1997). It is argued that healthier students are better learners (Basch 2011). The literature suggests that health status and achievement are inextricably linked (Jourdan et al. 2010) and points to the fact that adolescents experience difficulty in learning when their health is not optimum (Freudenberg and Ruglis

2007; Carnegie Council on Adolescent Development 1995). It is also noted that high-quality educational standards and experience influence the development of a healthier population (Jourdan et al. 2010; Young 2008).

Investment in the health and well-being of school children not only has immediate positive implications for student learning but it is also a critical determinant of health across the life span (Fiscella and Kitzman 2009). Therein, inequality in one contributes to inequality in the other. Moreover, level of educational attainment is asserted to have an impact upon health because it influences factors such as future occupational status, and income (Cutler and Lleras-Muney 2006). Lower educational attainment is negatively correlated with earlier onset of chronic disease conditions (Fleishman 2005) as well as a range of risk behaviours, including smoking, poor diet, poor physical activity, early sexual activity, teenage pregnancy and crime. Accordingly, Fiscella and Kitzman (2009: 1074) and Lleras-Muney (2005) claim that the relationship between education and health appears at least to be partly causal both in the short and long term. Clearly, given the importance of health and well-being in schooling, it appears imperative that schools attempt to achieve balance between the health education of students and their concurrent cognitive development. Education has, without doubt, the power to positively impact on economic prosperity, and also a nation's health outcomes (St. Leger et al. 2007). Therefore, investing in health education has significant merit. Arguably, investing social resources in schools is one of the most powerful ways to shape the lives of young people (Basch 2011). It is then somewhat surprising that while the discourse of health has, to some degree, permeated schooling; it has not done so in a sustained and prioritised manner. It remains that in the majority of education systems, health is perceived as the "add on" to the school curriculum. The primary aim of the school remains focused on academic achievement with its core business being teaching and learning (Mannix McNamara 2012).

Concurrent to the myriad and continuously increasing number of complex health issues confronting current school children, schools continue to grapple with enormous pressure to improve academic skills. Local school leaders and stakeholders either remain unconvinced that improving student health represents a means to achieving improved academic outcomes (Wolford Symons et al. 1997) or they have not been adequately educated as to the reciprocal link between health and educational outcomes. Yet, schools continue to experience considerable pressure to respond to health crises especially those that manifest amongst youth and while schools, are not a panacea for societal health issues; they can play a key role in the promotion of health, particularly when there is partnership between health and education.

There is little doubt that healthy children are in a better position to learn; however, curriculum in isolation is not enough to compensate for the adverse health factors that many students experience (American Cancer Society 1992). It is a dubious assumption to hold that children in any given school environment represent a homogenous population. Within even a single class a huge disparity may occur in terms of socio economic status; health and cognitive ability. A 'one size fits all' implementation may not meet the needs of all school children but certainly commitment to health, and in particular, to a comprehensive implementation agenda of health promotion in schools is a significant step in the right direction. The health promoting schools movement continues to gain a strong foothold globally as it is recognised that a whole school approach to health promotion contributes to improved learning outcomes and reduces health risk behaviours (IUHPE 2008; Kann et al. 2007). The challenge then, is the implementation of the health promoting schools approach in a sustained and effective manner in this type of system. The answer lies in the recognition of the inter-reliant and clearly reciprocal nature of both systems.

Given the significant pressures of performativity in schools, opting to concentrate on health education within curriculum only, while having its strengths as an approach, also brings with it significant challenges. The evidence is clear that whole school approaches are necessary in order for viable and sustainable implementation (Stewart-Brown 2006). The Irish experience demonstrates the potential of strong curriculum development however, implementing a discrete curriculum has created problems in terms of parity of esteem and teacher engagement. Whole school approaches that embed the promotion of health and well-being in the ethos and practice of the school are essential. The health promoting school places the promotion of health and well-being in schools at the heart of education. Without doubt SPHE has an important contribution to make to the health gain of children and to better educational outcomes, while also making cross-curricular links with other subjects, and more specifically in developing a HPS (SPHE Support Service 2005).

In order to ensure the successful implementation of SPHE, the Department of Education and Science proposed that the subject of SPHE should be embedded within a healthy whole school ethos. However, the development of a national and sustainable health promoting schools network has lagged significantly in Ireland. Many factors have influenced this limitation, not least of all the lack of an established relationship between SPHE and the health promoting schools initiative. It has been recognised that the ambiguity in terms of the link between SPHE and health promoting schools has been problematic (Burtenshaw 2003). While the policy vision for SPHE advocated that, SPHE should not become an isolated subject but rather draw on cross curricular links, this has not been forthcoming, not least of all due to the limited awareness of the content of SPHE among teachers who do not teach it and who are consequently challenged in terms of making subject comparisons and cross-curricular links.

By placing the subject on the school curriculum, it seems clear that the Department of Education and Skills is prioritising health education in Irish schools. However, lack of consistent implementation across the whole school population suggests some ambivalence as to its relevance for senior cycle students in schools. Implementing a new and discrete curriculum, somewhat at odds to the dominant drive in education (that of cognitive achievement), is a challenging and problematic endeavour. The subject of SPHE was a significant and courageous step in Irish education in terms of placing holistic education and health education firmly on the school agenda. However, very quickly some challenges began to impinge on the effectiveness of its implementation, not least of these being the challenges linked to the performativity agenda. The issues that now surround the implementation of

SPHE in Irish schools suggest the need for more careful attention to educational culture and practice prior to subject implementation.

10.4 Educational Purpose

Increasingly, in Ireland tensions are emerging between the pressures to educate students to perform well in the terminal exam and the need to pay attention to the personal and social needs of students. The dominance of the terminal exam has resulted in a 'teaching to the exam' ideology (Hennessy and Mannix McNamara 2011) that has consequently marginalised health education in schools. A problematic consequence of the exam focus in schools has meant that the tendency to teach for knowledge reproduction or 'teaching what to think' has emerged strongly in school pedagogy. In effect, the space to develop critical thinking and the broader liberal education agenda has become marginalised. Such deterministic education is contrary to the values of health education and to health promoting schools, resulting in divergent agendas competing for space on the school timetable. Pedagogical directives which advocate the need to "teach us rather how to think, than what to think" (Beattie as cited in Denbow 2004: 19) has unfortunately become relegated to education rhetoric. Without a doubt the school is a formidable terrain for 'epistemic apprenticeship' (Claxton 2009) and holds the possibility to develop confident, inquiring and critical minds. It is important that this potential is broad ranging in terms of intellectual and affective development.

Decision making about the development and purpose of education, encompasses judgements about that which is perceived to be educationally desirable and valued (Hume 1739), and this is the cause of much contention. While the rhetoric of current education policy is laudable in its emphasis on life-long learning and student oriented pedagogy, critical analysis of the practice of education yields a different and somewhat antithetical focus wherein the virtues of the former ideal are sacrificed in the name of expedience, standardisation and performance. The sizeable chasm between rhetoric and reality as well as policy and practice harks 'troubled times' for Irish teachers (Gleeson and Knights 2006) who attempt to negotiate the inherent challenges of a system divided, yet even more so for health education teachers whose subject is considered by many as antithetical to the main priority of academic examinable subjects. Teachers consequently experience tensions between their educational beliefs and their pedagogical practice.

As Irish teachers are credited with 'legendary' autonomy (OECD 2009) in terms of their teaching, one might assume that the capacity to integrate health education would be easier in this context however, the increased testing agenda has created what Lynch (1985) terms a 'say/do dichotomy' in Irish life. The development of SPHE as an academic subject has proven consistent in this respect. The performativity agenda is not isolated to the Irish experience yet the response by many Irish teachers has unfortunately been little other than a nod of acquiescence. It is unsurprising that in education systems where exam performativity dominates, teachers' beliefs about education quickly become "part of an older, increasingly displaced discourse" (Ball 2003: 223). Consequently, teachers' commitment to students' affective development often languishes (McNess et al. 2003). Health education and health promotion in schools in effect become collateral damage, disappearing from teachers' daily priorities. For many teachers then, provision of health education is a complex challenge, involving not only pedagogical proficiency but also the strength to withstand competing pressures borne from the dominant cultures of exam performativity. The lack of SPHE at senior cycle evidences the impact of the dominance of exam priorities where health education has been unsuccessful in gaining a place on the timetable.

The terminal examination of post primary education entitled the Leaving Certificate examination, acts as a 'towering presence' on the educational landscape (NCCA 2002: 45). As the leaving certificate examination forms the primary basis for the allocation of places into university in Ireland, many students tend to conflate the Leaving Certificate with the number of points achieved in the exam rather than their learning for each individual subject (Hyland 2011).

The Irish education system as outlined, celebrates high levels of public esteem where a strong degree of public trust exists in teachers (Teaching Council of Ireland 2010). Yet, this notwithstanding, there are significant struggles both at the levels of practice and ideology, which often hinder the potential for the more engaged educational encounters and the potential for broader, holistic and more liberal educational experiences within the classroom. This may account for the sizeable numbers of teachers who report teaching to be a 'difficult' job (Teaching Council of Ireland 2010), a projection which is intensified in light of the findings of the Teaching and Learning International Study (TALIS), (OECD 2009) in which teachers' average job satisfaction was found to be somewhat lower in Ireland than in other comparison countries. Broad liberal education is struggling for its very survival in modern schools where instrumentalism and technocratic rationality dominate (Smyth and Shacklock 1998). National policy espouses broader aims, for example according to the National Council for Curriculum and Assessment (NCCA) in Ireland, "The general aim of education is to contribute towards the development of all aspects of the individual, including the aesthetic, creative, cultural, emotional, intellectual, moral, physical, political, religious, social and spiritual development, for personal and family life, for living in the community, and for leisure" (NCCA 2012). However, critics of the Irish education system argue that education as a process has thus become means-end technicism, where schooling which terminates in examination, feeds the needs of industry and business (O'Brien 2008). It is little wonder then that teachers struggle to mediate the holistic needs of their students while balancing the teaching of content for the exam.

The relegation of personal and social education appears a global phenomenon with notable educationalists such as Noddings (2003) suggesting that education has been, and continues to be, deeply mired in a form of Puritanism that excludes the pursuit of happiness and care in the attempt to enhance performance and raise standards of achievement. It is little wonder then that the dominating effect of both the Junior Certificate and Leaving Certificate exams on teaching and learning practices gives rise to a pernicious teaching to the test ideology (NCCA 2010). In this context it is particularly difficult for health curriculum and for the health promoting schools initiative to make significant inroads in embedding themselves in school practice. Irish education policy makers need to pay attention to the cautionary proviso that lack of attention to affective development results in negation of the provision of a holistic education, leading to a compromised vision of democratic education (LeBlanc et al. 2009). There is currently an unbalanced and harmful overemphasis on academic measurable outcomes to the neglect of the other elements that are necessary to participation in a humane and democratic society. Such reductionism poses significant concerns for the teacher of SPHE.

The Irish education system is dominated by academic curricula in which knowledge is rigidly stratified, resulting in subjects constantly competing for timetable allocation, status and resources. As a result they have essentially become 'balkanised' from one another (Hargreaves 1994) leaving teachers in the difficult position of having to fight for the status of their own subject. It is widely accepted that education is not just about academic performance and that developing students' knowledge of their social context, personal intelligence and their health and wellbeing is critical to quality of life. Yet the dominance of the academic performance serves to frequently obscure the broader education agenda. Indeed Quant sums it up eloquently in the challenge

> True education is not for every man the scrap of paper he leaves school with. Dare we as teachers admit this? Dare we risk our existence by forcibly expressing our views on this? Are we to rely on exams for all to prove ourselves worthy of the kindly eye of the state? (Quant 1967, cited in Goodson 1983)

10.5 Subject Specific Challenges

New subjects are not born easily (Trant 2007) and often many obstacles are faced by principals and teachers in terms of allocating time and commitment to a new subject. The introduction of any subject is a lengthy and challenging process. According to Fullan (2001: 69) "educational change is technically simple and socially complex". The implementation of new subjects is not a discrete event but a difficult process (Lawton 1996), as was the case with the subject of SPHE.

10.5.1 Leadership

The main challenges faced by principals when initially implementing SPHE were curriculum overload, lack of co-ordination and planning time and inadequate staff training in SPHE (Geary and Mannix McNamara 2003). A new subject needs to be valued and esteemed by the school leadership if it is to have any chance of successful engagement by teachers and students. In the case of SPHE this became somewhat problematic as school principals became an obstacle to its implementation, evidencing low participation rates at management briefings on the subject (Geary and Mannix

McNamara 2003). This in turn had an impact on school commitment to the subject. This is not to apportion significant blame on to school principals, as many would argue the subject was imposed upon schools without adequate prior consultation with principals.

Whatever the reason for such reticence to engage with SPHE, the reality has been that SPHE has become the poor relation in an already overcrowded school timetable. It has also meant a distinct lack of coherence in terms of which teachers are designated to teach SPHE. Teachers are asked to teach SPHE, a subject like any other, without teacher education on its content or pedagogy. Guidance on the teaching of SPHE is provided by the SPHE Support Service and consists of a module of 40 h in-service. It is clearly not enough to engender parity of esteem for the subject or teacher competency. School principals generally select the teacher for SPHE (Geary and Mannix McNamara 2003) and the evidence suggests that SPHE teachers rotate all too frequently. Teachers are often designated to teach SHPE dependent on the spaces on their timetable once the other subjects have been fully allocated. This has significant implications for the implementation of the subject, in terms of teacher suitability, teacher competency and commitment. If teachers do not choose to teach the subject themselves, they may not be invested in its success nor committed to its value. Consequently this may have an adverse effect on student experiences of the subject (Mannix McNamara et al. 2012).

Clearly if school leadership, and in particular if school principals, do not have adequate knowledge of the subject it exacerbates the challenges of top down reform and poses distinct challenges for the status of SPHE among the staff and students. Good leadership is thus seen as a key factor for successful implementation of health education and health promotion in schools (Grieg Viig et al. 2012). Fullan (2001: 115) argues that the teacher is instrumental in curriculum change because "educational change depends on what teachers do and think – it's as simple and complex as that" (Fullan 2001: 115). The same is true of principals, of school management and indeed of education policy makers.

10.5.2 SPHE and Parity of Esteem

Subjects are constantly competing for timetable allocation, status and resources. As a result a distinct hierarchy has emerged in terms of subject status and esteem in schools. This is not a phenomenon particular to Ireland; subject hierarchy is evident in educational institutions globally. Almost 30 years have passed since Goodson reported this hierarchy of knowledge, advocating that "High status in the secondary school curriculum is reserved for abstract theoretical knowledge divorced from the working world of industry and the everyday world of the learner" (Goodson 1983: 202).

'Academic subjects' and written examinations have become closely entwined (Goodson 1983). The academic tradition is centrally exam focused with an emphasis on theoretical knowledge. If a subject is to be formally assessed it carries with it the guarantee of high status (Goodson 1983). SPHE does not sit well with this dominant ideology about what counts as legitimate and valuable knowledge. SPHE, by its nature is affective in orientation. It is concerned with personal and social learning

Fig. 10.1 Effective implementation of SPHE

and is, in fact situated learning rather than abstract theorising. It is an applied subject and is not formally assessed. From its inception SPHE has struggled to gain esteem in schools and subsequent research on its implementation appears to suggest that little has changed in this regard (Nic Gabhainn et al. 2007).

In a system that appears to place value on performativity and assessment, teachers are challenged to gain support for the subject, while competing with examination classes for resources and timetable allocation. Lack of esteem for SPHE has been credited with a consequent lack of teacher involvement in the subject (Burtenshaw 2003). The evidence points to teachers regarding SPHE as less important than the examination subjects (Nic Gabhainn et al. 2007), with many expressing the view that SPHE took valuable time away from the 'more important subjects' (p. 26). Value is further diminished when teachers have a negative attitude towards the subject and when principals do not view SPHE as a big priority (Nic Gabhainn et al. 2007).

10.5.3 Effective Implementation of SPHE

As has been outlined above, it is clear that SPHE has faced many challenges since its inception including lacking esteem and value among its stakeholders. The model Fig. 10.1 outlines the factors needed for successful implementation of SPHE in order to increase the worth of the subject among principals, teachers, parents and students alike. These include; the provision of the subject in both the junior and senior cycle; the option of gaining an SPHE qualification during initial teacher education while also exposing all student teachers to SPHE during pre-service education. Other factors which are important to the successful implementation of SPHE include, covering all topics on the syllabus and adherence given to the spiral nature of the curriculum; implementing SPHE as part of an overall Health Promoting Schools framework; school management prioritising SPHE and recognition given that SPHE positively contributes to educational outcomes. All of these factors would raise the status of the subject and contribute to effective implementation of SPHE.

10.6 Fostering Sustainable Implementation

10.6.1 Student Participation

Research indicates that the majority of students find the subject SPHE interesting and relevant to them (Nic Gabhainn et al. 2007). Recently in Ireland young people were consulted about how they value the Irish education system. When asked to identify the skills and subjects deemed as essential in the Irish education system, Irish students highlighted the value of SPHE, noting that they would like more SPHE classes and advocating that SPHE needs to be compulsory across the junior and senior cycle curriculum. However, they also made a strong case for the need for SPHE to be restructured and improved (Roe 2011). Listening to the voices of students is essential for subjects like SPHE as the core agenda of affective education, such as SPHE, focuses on responsiveness to students' needs. As key stakeholders, they need to be engaged in this process. Student participation empowers young people to influence and change their lives (Barnekow et al. 2006). Students need to be consulted on the subject so that they have a vested interest in its implementation and if they are to believe in its worth and value on the curriculum.

10.6.2 Teacher Competency Development

For the successful implementation of SPHE, where the subject is esteemed and meets the needs of students and teachers alike, teachers need more support and education. A key outcome of SPHE is the development of student health literacy. The literature indicates that teachers need to be equipped with the requisite competencies and skills in order to support students to attain optimal levels of health literacy. Health literacy is defined by Nutbeam (1998) as "the achievement of a level of knowledge, personal skills and confidence to take action to improve personal and community health by changing lifestyles and living conditions" (p. 10). Paakkari and Paakkari (2012) propose that healthy literacy is composed of five distinct components; theoretical knowledge, practical knowledge, critical thinking, self-awareness and citizenship. This requires competent teachers with the specific knowledge and skills to develop these attributes in students. Indeed St. Leger (2001: 204) identifies the importance of schools and of national policy in prioritising this agenda when he writes: "Increased and widespread empowerment of students through health literacy concepts is possible, but only if there is a will to support schools in their efforts." The political will to support the development of health education is necessary but it needs to be met by teachers and schools who are committed to health education for their students. It is not acceptable to impose a mid-career expectation on teachers to now become health education teachers as the key to ensuring that teachers are committed to the holistic education of student's lies in their professional formation.

Teachers are at the centre of an education system that is a constant state of change. Previous generations of teachers possibly required an education for stability, however Kress (2000: 133) makes the case that the coming era requires an "education for instability." Teachers need to be flexible and responsive to students needs in a manner that was not required as much previously. Consequently, teachers require strong, professional competencies (Selvi 2010). Health promotion effectiveness is dependent upon a workforce that is equipped with core, flexible and adaptable skills (Barry 2008). There should be a constant emphasis on teachers' competencies in research and analysis (Selvi 2010). Initial teacher training should focus on understanding and developing teachers' competencies (Selvi 2010). SPHE needs professional teachers who are equipped with competence in this regard. In the same way that St. Leger (2001) argues the development of critical health literacy requires teachers to be competent in developing advocacy and social change skills and to be knowledgeable around major health issues; SPHE requires similar competency development for teachers.

The use of competencies as a model for teacher education has been prompted in the discourse on teacher quality in light of new student demands, the changing knowledge needed by teachers and the balance between accountability and autonomy (Day 2002). Therefore, research suggests that there is a need for universities to play a key role in training health promotion workforce (Shilton 2009). Competency development was considered essential to the 'professionalisation' of health promotion in Australia (Howat et al. 2000), and is increasingly gaining momentum elsewhere, largely thanks to the global efforts of Barry et al. (2009). Focusing on the development of teacher competencies is an urgent necessity in order to effect more successful and sustained implementation of health education in schools, whether it is through the health promoting schools agenda or via specific curriculum.

10.6.3 Fostering Pedagogical Competence

SPHE differs from other subjects on the curriculum because the pedagogical structure demands experiential pedagogies. Teachers have identified that the transition between their traditional pedagogies and this new expectation of facilitation and experiential learning as their greatest challenge (Geary and Mannix McNamara 2003). It is unsurprising that teachers find this transition difficult when they have had limited exposure to experiential pedagogies. In order for the implementation and delivery of SPHE to be successful, quality teaching is a vital ingredient (Nic Gabhainn et al. 2007). There is undoubtedly a need for a national qualification in SPHE at undergraduate level. This would contribute to raising the status of the subject (Burtenshaw 2003: 12).

Teacher training is pivotal in the development of health education and health promotion in schools (Jourdan et al. 2008). Universities have a key role in supporting health curriculum in schools by providing pre-service teachers with the adequate training. Some teachers are currently teaching SPHE without any form of training (Burtenshaw 2003). Teachers are positive about the quality and benefits of the 40 h professional development offered by the SPHE support service (Millar 2003), but are clear that is it not enough. This professional development acts as in-service for qualified teachers, so in effect, this means that teachers attending SPHE professional development may already be teaching SPHE and subsequently attending training. This has significant implications for SPHE competence and esteem.

The national policy is that "every teacher is a teacher of SPHE" (Department of Education and Science 2000a: 6), yet presently, there is no nationally mandated university based programme for post primary SPHE teachers. Exposure to health education and health promotion is critical in initial teacher education in order for sustained implementation so that all teachers perceive they have a role in the affective development of students (Jourdan et al. 2010). Currently pre-service teacher training in SPHE is conducted on an *ad hoc* basis and is varied among teacher education providers, with some offering little more than an overview of SPHE (Lyons 2008) and others offering no exposure at all. Ironically, in terms of the national policy that all teachers are teachers of SPHE, significant numbers of pre-service teachers may not intend to engage with the subject. Mannix McNamara et al. (2012) reported that the majority (three quarters of respondents) of the final year teacher education students do not intend to teach the subject. Reasons cited included limited knowledge of SPHE, lack of exposure during initial teacher education, less than positive post primary experiences of the subject and general lack of interest in health education. If national policy states that all teachers are teachers of SPHE then health education should be placed on the curriculum for initial teacher education (Mannix McNamara et al. 2012). This would assist in engendering more openness and esteem for the subject, which are necessary requirements for sustained and optimal implementation. Clearly some attention is needed for the promotion of the role of the teacher as an educator of the whole person.

10.7 SPHE Within a Health Promoting School

Schools are recognised as important settings for health promotion (Mukoma and Flisher 2004; St. Leger et al. 2010) because of the wide audience they encapsulate and also the length of time that children remain in school. Schools can have an influence on students' decision making and perception of health, as many of the behaviours that have a profound effect on health status, such as physical activity levels and dietary choices are established during the schooling years of young people (Mohammadi et al. 2010). Knowledge alone is insufficient to empower people to make health promoting lifestyle choices and behaviour changes (Kischuk et al. 1990; Bellew and Wayne 1991; Klepp et al. 1994). Teaching health education may increase knowledge but it does not necessarily change behaviour (Lister-Sharp et al. 1999). To address this issue the didactic approach has been replaced by a more holistic model of teaching health in schools, in the hope that an all-encompassing approach will help to change people's destructive behaviour in relation to their health. This perspective has led to the development of the settings approach to school health promotion.

> Schools health programmes that co-ordinate the delivery of education and health services and promote a healthy environment could become one of the most efficient means available for almost every nation in the world to improve significantly the well-being of its people. Consequently, such programmes could become a critical means of improving the condition of humankind globally (WHO 1997: 9).

Implementing a discrete health education subject that does not sit within a broader agenda of the promotion of health is problematic. It was always the intention that SPHE be part of a larger school agenda of the promotion of health and well-being (Department of Education and Science 2000a). However, the practice in this regard has left a lot to be desired. The lack of a consistent and coherent national health promoting schools network implementation has hampered SPHE in schools. An effective school health education programme is one that combines health education with other health-promoting initiatives in the school, and involves parents and families and the community (Seffrin 1990; Young 1993; Denman 1994). This approach signals a movement away from the discrete health education class as the only place of information in relation to health. Multi-faceted approaches are more effective than classroom only or single intervention approaches in achieving health and educational outcomes (Stewart-Brown 2006; Moon et al. 1999). Health promoting schools are the ideal context within which to implement health education curriculum (Stewart-Brown 2006).

The World Health Organisation have defined a health promoting school as one that is "constantly strengthening its capacity to be a healthy setting for living, learning and working by focusing on all the conditions that effect health" (WHO 2012: 1). It is a working model focused on process, which in terms of the dominant ideology of exams and performativity may appear somewhat antithetical. The health promoting school provides much needed relief from the performativity pressures that young students now contend with.

The aims of the HPS approach are numerous and wide ranging. They include development of a broad and holistic view of health, and the support of students through the provision tools which will enable them to make healthy choices. In particular, it seeks to provide a healthier environment, using interactive (in the case of SPHE experiential) teaching methods, building communication and seeking partners in the community (Jensen and Simovska 2002). There are clearly a wide range of benefits of the HPS approach for students, teachers, parents, communities and society as a whole. The HPS is considered one of the most powerful approaches to promoting health, empowerment and action competence, which is the ability to take action and generate change, in schools (Clift and Jensen 2005).

Using the HPS approach should improve not only health outcomes but educational outcomes also (Whitman and Aldinger 2009; Stewart-Brown 2006; SHE 2009). It is lack of recognition of the positive impact of health education and health promoting schools on the educational outcomes of students by those in education that hampers teacher commitment to its promotion. If education policy makers recognised this benefit they would perhaps pay closer attention to its successful and sustained implementation in schools. The purposes of health promoting schools are to enhance learning outcomes and to facilitate action for health by developing knowledge and skills (IUHPE 2008). The evidence points to the fact that whole school implementation is more likely to be effective in terms of a range of outcomes than a classroom based approach (St. Leger et al. 2007). Whole school approaches that demonstrate coherence between policies and practices that promote and support social inclusion and commitment to education, facilitates improved learning outcomes, increases emotional well-being and reduces health risk behaviours (Lister-Sharp et al. 1999; Young and Currie 2009).

Teachers' active participation in HPS is dependent on many factors including their perspectives on the HPS approach (Adamson et al. 2006; Barnard et al. 2009). Even within the education sector, teachers have varying perspectives on how they see their role in health promotion which is very much dependent on their subject but also their own epistemologies of teaching (Jourdan et al. 2010). There is a need to achieve a common understanding of the concept and goals of the HPS approach as well as recognition of the positive impact on health promotion in schools on educational outcomes in order for coherent implementation and for stronger teacher commitment. Increased professional development of teachers is a necessary and significant first step. As St. Leger (2004: 408) advocates "Let us rethink school health away from kits and projects to solve problems and use the school as an on-going setting where health is created, supportive environments are built, partnerships made and many skills are learned".

SPHE and the health promoting schools approach are not 'add on', nor are they another problem that needs to be addressed within schools. They provide a framework that schools can use to address many of their own educational goals and use them as the medium through which these goals can be achieved. The Health Promoting School approach is a sustainable approach to health education in schools, not dependent on a single subject teacher but rather includes the whole school community. Health education and the promotion of health in schools is a complex and worthwhile endeavour, one which we cannot assume that teachers will automatically have the skills to effect.

Funding This work has been funded by the Irish Research Council.

References

Adamson, G., Mcaleavy, G., Donegan, T., & Shevlin, M. (2006). Teachers' perceptions of health education practice in Northern Ireland: Reported differences between policy and non-policy holding schools. *Health Promotion International, 21*(2), 113–120.
American Cancer Society. (1992). *National action plan for comprehensive school health education* (pp. 4–7). Atlanta: American Cancer Society.
Ball, S. J. (2003). The teacher's soul and the terrors of performativity. *Journal of Education Policy, 18*(2), 215–228.
Barnard, M., Becker, E., Creegan, C., Day, N., Devitt, K., Fuller, E., Lee, L., Neil, H., Purdon, S., & Ranns, H. (2009). *Evaluation of the National Healthy Schools Programme: Interim report*. London: National Centre for Social Research.
Barnekow, V., Buijs, G., Clift, S., Jensen, B. B., Paulus, P., Rivett, D., & Young, I. (2006). *Health promoting schools: A resource for developing indicators*. Copenhagen: WHO Regional Office for Europe.
Barry, M. (2008). Capacity building for the future of health promotion. *Promotion & Education, 15*(4), 56–58.
Barry, M., Allegrante, J., Lamarre, M. C., Auld, M., & Taub, A. (2009). The Galway Consensus Conference: International collaboration on the development of core competencies for health promotion & health education. *Global Health Promotion, 16*(2), 5–11.
Basch, C. (2011). Healthier students are better learners: High- quality, strategically planned, and effectively coordinated school health programs must be a fundamental mission of schools to help close the achievement gap. *Journal of School Health, 81*(10), 650–662.
Bellew, B., & Wayne, D. (1991). Prevention of smoking among school children: A review of research, and recommendations. *Health Education Journal, 50*(1), 3–8.
Burtenshaw, R. (2003). *Report from the review of social, personal and health education at junior cycle by the SPHE support service post-primary*. http://www.sphe.ie/review2.pdf. Accessed 21 May 2012.
Carnegie Council on Adolescent Development. (1995). *Great transitions: Preparing adolescents for a new century* (pp. 22–23). New York: Carnegie Corporation of New York.
Claxton, G. (2009). *What's the point of schooling?* Oxford: One World Publications.
Clift, S., & Jensen, B. B. (2005). *The health promoting school: International advances in theory, evaluation and practice*. Copenhagen: Danish University of Education Press.
Cutler, D., & Lleras-Muney, A. (2006). *Education and health: Evaluating theories and evidence* (NBER Working Paper No. 12352). http://ssrn.com/abstract_913315. Accessed 30 May 2012.
Day, C. (2002). School reform and transitions in teacher professionalism and identity. *International Journal of Educational Research, 37*, 677–692.
Denbow, J. (2004). *Proudly teach words to teach by*. Victoria: Trafford Publishing.
Denman, S. (1994). Do schools provide an opportunity for meeting the health of the nation targets? *Journal of Public Health Medicine, 16*(2), 219–224.
Department of Education and Science. (1994). *On my own two feet: Educational resource materials for the prevention of substance abuse*. Dublin: Government Publications.
Department of Education and Science. (2000a). *Social personal and health education*. Dublin: Government Publications.
Department of Education and Science. (2000b). *Social, personal and health education: Circular M22/00*. Dublin: Department of Education and Science.
Department of Education and Science. (2001). *Social, personal and health education: Guidelines for teachers*. Dublin: Government Publications.
Department of Education and Science. (2003). *Social personal and health education: Circular M11/03*. Dublin: Department of Education and Science.
Fiscella, K., & Kitzman, H. (2009). Disparities in academic achievement and health: The intersection of child education and health policy. *Pediatrics, 123*(3), 1073–1080.

Fleishman, J. (2005). *Methodology report #15: Demographic and clinical variations in health status*. Rockville: Agency for Healthcare Research and Quality.
Freudenberg, N., & Ruglis, J. (2007). Reframing school dropout as a public health issue. *Preventing Chronic Disease, 4*(4), 1–7.
Fullan, M. (2001). *The new meaning of educational change*. New York: Teachers College Press.
Geary, T., & Mannix McNamara, P. (2003). *Implementation of social, personal & health education at junior cycle: National Survey Report*. Limerick: University of Limerick.
Gleeson, D., & Knights, D. (2006). Challenging dualism: Public professionalism in 'Troubled' Times. *Sociology, 40*(2), 277–295.
Goodson, I. (1983). *School subjects and curriculum change*. Kent: Croom Helm Ltd.
Grieg Viig, N., Fosse, E., Samdal, O., & Wold, B. (2012). Leading and supporting the implementation of the Norwegian network of health promoting schools. *Scandinavian Journal of Educational Research, 56*(6), 671–684.
Hargreaves, A. (1994). *Changing teachers, changing times*. London: Cassell.
Hennessy, J., & Mannix McNamara, P. (2011). Packaging poetry? Pupils' perspectives of their learning experience within the post-primary poetry classroom. *English in Education, 45*(3), 206–223.
Howat, P., Maycock, B., Jackson, L., Lower, T., Cross, D., Collins, J., & van Asselt, K. (2000). Development of competency based university health promotion courses. *Promotion & Education, 7*(1), 7–33.
Hume, D. (1739). *A treatise on human nature*. London: J. M. Dent and Sons Ltd.
Hyland, A. (2011). *Entry to higher education in Ireland in the 21st century*. Dublin: National Council for Curriculum and Assessment (NCCA) and the Higher Education Authority (HEA).
International Union for Health Promotion and Education. (2008). *Achieving health promoting schools: Guidelines for promoting health in schools*. http://www.iuhpe.org/uploaded/Publications/Books_Reports/HPS_GuidelinesII_2009_English.pdf. Accessed 30 Sept 2011.
Jensen, B. B., & Simovska, V. (Eds.). (2002). *Models of health promoting schools in Europe*. Copenhagen: WHO Regional Office for Europe.
Jourdan, D., Samdal, O., Diagne, F., & Carvalho, G. (2008). The future of health promotion in schools goes through the strengthening of teacher training at a global level. *Global Health Promotion, 15*(3), 36–38.
Jourdan, D., Mannix McNamara, P., Simar, C., Geary, T., & Pommier, J. (2010). Factors influencing the contribution of staff to health education in schools. *Health Education Research, 25*(4), 519–530.
Kann, L., Telljohann, S. K., & Wooley, S. F. (2007). Health education: Results from the school health policies and programs study 2006. *Journal of School Health, 77*(8), 408–434.
Kishchuk, N., O' Loughlin, J., Sylvain, P., Masson, P., & Sacks-Silver, G. (1990). Illuminating negative results in evaluation of smoking education programmes. *Journal of School Health, 60*, 448–451.
Klepp, K., Oygard, L., Tell Grethe, S., & Vellar Odd, D. (1994). Twelve year follow up of a school-based health education programme. *European Journal of Public Health, 4*, 195–200.
Kress, G. (2000). A curriculum for the future. *Cambridge Journal of Education, 30*(1), 133–145.
Lawton, D. (1996). *Beyond the national curriculum: Teacher professionalism & empowerment*. Sevenoaks: Hodder & Stoughton.
LeBlanc, P. R., Gallavan, N. P., & The Association of Teacher Educators. (2009). *Affective teacher education*. Lanham: Rowman & Littlefield.
Lister-Sharp, D., Chapman, S., Stewart-Brown, S., & Sowden, A. (1999). Health promoting schools and health promotion in schools: Two systematic reviews. *Health Technology Assessment, 3*(22), 1–207.
Lleras-Muney, A. (2005). The relationship between education and adult mortality in the United States. *Review of Economic Studies, 72*(250), 89–221.
Lynch, K. (1985). Ideology, interests and Irish Education: Ciarán Benson talks to Tom Garvin, Kathleen Lynch and Billy Roche. *The Crane Bag, 9*(2), 12–24.

Lyons, M. (2008). *National council for curriculum and assessment report for the National Parents Council*. www.npcpp.ie/fileadmin/files/reports/080408_NCCA_report.pdf. Accessed 4 Feb 2011.
Mannix McNamara, P. (2012). So what is really worth fighting for in our schools: the role of the teacher in challenging educational times, *TUI Annual Congress Journal*, Waterford: TU.
Mannix McNamara, P., Moynihan, S., Jourdan, D., & Lynch, R. (2012). Pre-service teachers' experience of and attitudes to teaching SPHE in Ireland. *Health Education, 112*(3), 199–216.
Maycock, P., Kitching, K., & Morgan, M. (2007). *Relationships and sexuality education in the context of social personal and health education: An assessment of the challenges to full implementation of the programme in post primary schools*. Dublin: Department of Education and Science.
McNess, E., Broadfoot, P., & Osborn, M. (2003). Is the effective compromising the affective? *British Educational Research Journal, 29*(2), 243–257.
Millar, D. (2003) *Review of records of in-service training for teachers of SPHE: February 2001 to April 2003*. http://www.sphe.ie/downloads/SPHE%20REVIEW/REVIEW%20-%20 INSERV%20RECORD%202001-3%20D%20MILLAR.pdf. Accessed 9 Sept 2011.
Mohammadi, N. K., Rowling, L., & Nutbeam, D. (2010). Acknowledging educational perspectives on health promoting schools. *Health Education, 110*(4), 240–251.
Moon, A. M., Mullee, M. A., Rogers, L., Thompson, R. L., Speller, V., & Roderick, P. (1999). Helping schools to become health promoting environments – An evaluation of the Wessex Healthy Schools Award. *Health Promotion International, 14*(2), 111–122.
Mukoma, W., & Flisher, A. J. (2004). Evaluations of health promoting schools: A review of nine studies. *Health Promotion International, 19*(3), 357–368.
NCCA. (2002). *Developing senior cycle education: consultative paper on issues and options*. Dublin: National Council for Curriculum and Assessment.
NCCA. (2010). *Innovation and identity: Ideas for a new junior cycle*. Dublin: National Council for Curriculum and Assessment.
NCCA. (2012). *A rationale for Social, Personal and Health Education (SPHE)*. http://www.curriculumonline.ie/en/PostPrimary_Curriculum/Junior_Cycle_Curriculum/Junior_Certificate_Subjects/Social,_Personal_and_Health_Education/SPHE_Syllabus/. Accessed 12 June 2012.
Nic Gabhainn, S., O'Higgins, S., Galvin, M., Kennedy, C., & Barry, M. (2007). *The implementation of SPHE at post primary level: A case study approach*. (Executive summary: Report to social, personal & health education support service). Galway: Department of Health Promotion.
Noddings, N. (2003). *Happiness and education*. Cambridge: Cambridge University Press.
Nutbeam, D. (1998). Health promotion glossary. *Health Promotion International, 13*(4), 349–363.
O'Brien, M. (2008). *Well-being and post-primary schooling: A review of the literature and research*. Dublin: National Council for Curriculum and Assessment.
OECD. (2009). *Creating effective teaching and learning environments: First results from TALIS*. Paris: Organisation for Economic Co-operation and Development.
Paakkari, L., & Paakkari, O. (2012). Health literacy as a learning outcome in schools. *Health Education, 112*(2), 133–152.
Quant, P. L. (1967). Rural studies & the newsom child. *Hertfordshire Rural Studies Association Journal*, April 1967, p. 12.
Roe, S. (2011). *Report of a consultation with young people on reform of the junior cycle*. Dublin: Government Publications.
Seffrin, J. R. (1990). The comprehensive school health education curriculum: Closing the gap between state-of-the-art and state-of the-practice. *Journal of School Health, 60*(4), 151–156.
Selvi, K. (2010). Teachers' competencies. *Cultura International Journal of Philosophy of Culture and Axiology, 3*(1), 167–175.
SHE. (2009). *Vilnius Resolution: Better schools through health*. Woerden: NIGZ. http://www.schoolsforhealth.eu/upload/Vilnius_resolution.pdf. Accessed 10 Jan 2012.
Shilton, T. (2009). Health promotion competencies: Providing a road map for health promotion to assume a prominent role in global health. *Global Health Promotion, 16*(2), 42–46.

Smyth, J., & Shacklock, G. (1998). *Re-making teaching: Ideology, policy and practice*. London: Routledge.

SPHE Support Service. (2005). *The SPHE story – An example of incremental change in the school setting*. http://www.sphe.ie/downloads/res_eval/SPHE_Story.pdf. Accessed 8 Sept 2011.

St. Leger, L. (2001). Schools, healthy literacy and public health: Challenges and possibilities. *Health Promotion International, 16*(2), 197–205.

St. Leger, L. (2004). What's the place of schools in promoting health? Are we too optimistic? *Health Promotion International, 19*(4), 405–408.

St. Leger, L., Kolbe, L. J., Lee, A., McCall, D., & Young, I. (2007). School health promotion: Achievements, challenges and priorities. In D. V. McQueen & K. Jones (Eds.), *Perspectives on health promotion effectiveness* (pp. 107–124). Saint-Denis: Springer.

St. Leger, L., Young, I., Blanchard C., & Perry, M. (2010). *Promoting health in schools: From evidence to action.* An International Union for Health Promotion and Education (IUHPE) Publication. http://www.iuhpe.org/index.html?page=516&lang=en#sh_advevid. Accessed 30 Sept 2011.

Stewart-Brown, S. (2006). *What is the evidence on school health promotion in improving health or preventing disease and, specifically, what is the effectiveness of the health promoting school approach?* WHO Regional Office for Europe, p. 26. www.euro.who.int/document/e88185.pdf. Accessed 2 Sept 2011.

Teaching Council of Ireland. (2010). *Evaluation of public attitudes to the teaching profession: Summary of findings from iReach market research.* Maynooth: Teaching Council.

Trant, A. (2007). *Curriculum matters in Ireland.* Dublin: Blackhall Publishing.

Whitman, C., & Aldinger, C. (Eds.). (2009). *Case studies in global school health promotion.* New York: Springer.

Wolford Symons, C., Cinelli, B., James, T., & Grof, P. (1997). Bridging student health risks and academic achievement through comprehensive school health programs. *Journal of School Health, 67*(6), 220–227.

World Health Organisation. (1997). Promoting health through schools: Report of a WHO Expert Committee on Comprehensive School Health Education and Promotion. *World Health Organization Technical Report Series, 870*(i–vi), 1–93.

World Health Organisation. (2012). *Health promoting schools.* http://www.who.int/school_youth_health/gshi/hps/en/print.html. Accessed May 27 2012

Young, I. (1993). Health promoting schools: Healthy eating policies in schools – An evaluation of the effects on pupils' knowledge, attitudes and behaviour. *Health Education Journal, 52*(1), 3–9.

Young, I. (2008). *Promoting health in schools: From evidence to action.* Paris: IUHPE.

Young, I., & Currie, C. (2009). The HBSC study in Scotland: Can the study influence policy and practice in schools? *International Journal of Public Health, 54*(2), 271–277.

Part III
Examining Implementation

Chapter 11
Implementation Strategies to Promote and Sustain Health and Learning in School

Oddrun Samdal and Louise Rowling

Abstract A health promoting school approach involves a complex dynamic of group behaviours and system changes within the school by staff and students, in collaboration with external stakeholders. To ensure efficient implementation of the health promoting school approach there is, however, a need to identify how the approach can best be implemented, which so far has not been sufficiently addressed within the evaluation research. This chapter will carefully enunciate implementation components that will allow practitioners to understand the function of each component, and present theory based guidelines so each component can be utilised with fidelity. This approach differs from implementation guidelines for pre-packaged programs, which fail to provide guidance for a whole school approach. A systematic literature review has identified eight implementation components for health promoting schools: (1) Preparing and planning for school development; (2) Policy and institutional anchoring; (3) Professional development and learning; (4) Leadership and management practices; (5) Relational and organisational support context; (6) Student participation; (7) Partnership and networking and (8) Sustainability. The eight components identified overlap closely with implementation components for school improvement which has as the end-point improved school effectiveness and student learning: (1) a coherent instructional guidance system, (2) the professional capacity of its faculty, (3) strong parent-community-school ties, (4) a student-centred learning climate, and (5) leadership that drives change. With this background, the chapter will discuss the common core of promoting health and learning in school and demonstrate how similar implementation processes may be applied to achieve an efficient and sustainable change process for

---

O. Samdal, Ph.D. (✉)
Department of Health Promotion and Development, University of Bergen, Bergen, Norway
e-mail: Oddrun.Samdal@uib.no

L. Rowling
Faculty of Education and Social Work, University of Sydney, Sydney, NSW, Australia
e-mail: louise.rowling@sydney.edu.au

V. Simovska and Patricia Mannix-McNamara (eds.), *Schools for Health and Sustainability: Theory, Research and Practice*, DOI 10.1007/978-94-017-9171-7_11,

both health and learning. The chapter concludes with an examination of possible contributions of organisational and complex adaptive systems theories and empirical findings to future health promoting schools work.

Keywords Health promoting schools • Health promotion • Implementation • School change • School effectiveness • Student learning

11.1 Introduction

As most young people attend school throughout their childhood and adolescence, schools may provide an ideal setting for enhancing students' general health, health-related behaviours, and subjective well-being. This has been recognised in the Schools for Health action in Europe (www.schoolsforhealth.eu), building on the World Health Organisation (WHO) Health promoting schools in Europe and other parts of the world. Schools, however, do have competing priorities. For many, achieving academic success, and maintaining the engagement and discipline of students, is more important than meeting health-related behavioural objectives (Samdal et al. 1998; Valois et al. 2011). Individual schools and the education sector as a whole have from time to time perceived to be 'exploited' by the health sector in their mission of teaching individual life skills to promote health-related behavioural change and the well-being of students. To achieve progress in advancing both educational and health goals, greater attention needs to be given to a more holistic orientation, which is the intent of the health promoting schools approach. Health promoting schools build on the principle that characteristics of the school environment may have an impact on both academic achievement and school commitment, as well as an impact on health-related behaviours and subjective well-being (Samdal 2008).

A supportive school environment may be considered a resource for the development of academic achievement, health-enhancing behaviours, general health and subjective well-being, while a non-supportive school environment may constitute a risk (Samdal and Torsheim 2012). Hoyle and colleagues (2010) call for a refocus, from getting support for health programs to "finding the niche of the health promotion process in on-going school improvement efforts" (p. 165). From this perspective the ultimate health promoting school aim of increased subjective well-being and behaviours conducive to health is not only an end-point but also a premise for educational aims and educators. Recently a WHO report (Suhrcke and de Paz Nieves 2011) highlighted this need to shift perspective from seeing improved health as a product of education to seeing it as a factor that could determine educational outcomes. Subjective well-being may in this perspective be seen as an important prerequisite for learning and academic achievement in school (Basch 2010). Further, academic coping may bolster adolescents' self-esteem, which again may positively feed back into subjective well-being (Mortimore 2007). Educational concerns of problem behaviours and alienation from school on the other hand tend to cluster with health-compromising behaviours (Samdal et al. 1998). By preventing development of health-compromising behaviours such as smoking and early use of alcohol, students may be redirected from following a course of action that is detached from and in opposition to school values. Likewise from their review Suhrcke and de Paz Nieves (2011) found a negative correlation between risky health behaviours and (ill) health conditions on the one hand and on the other hand, education as measured through both educational achievement and academic performance. Basch (2010: 76) in a report on educational equity identified the impact of ill health on students:

> If their ability to concentrate, use memory, and make decisions is impeded by ill-nourishment or sedentary lifestyle, if they are distracted by negative feelings, it will be more difficult for them to learn and succeed in school. If their relationships at school with peers and teachers are negative, they will be less likely to be connected with and engaged in school, and therefore less motivated and able to learn……Urban minority youth are disproportionately affected by educationally relevant health disparities.

However Suhrcke and de Paz Nieves (2011) noted that despite this connection being established there was a gap in the research on the link between good health and educational achievement and academic performance.

11.2 The Evolving Nexus Between Implementation of Health Promoting Schools and School Improvement Processes

Over a decade ago Lister-Sharp and colleagues (1999) proposed that there was a need in systematic reviews of health promoting schools to capture evidence on the perspectives of people's lives and that theoretical frameworks and literature be included in judging which research be included in the review. These proposals acknowledge that the actions of people involved in implementation of health promoting schools are a factor that: may influence outcomes, vary between individuals, and is theoretically and conceptually important to recognise and assess. The implementation actions are principally applied by teachers and other school personnel within a disciplinary context of educational conceptual theories and frameworks. The closer the conceptual match between health promoting schools action and teacher's professional disciplinary orientations, the better the implementation is likely to be. However, it is only in the last few years that the importance of utilizing educational research about school change to promote teaching and learning for academic achievement combined with improving health outcomes, has been accepted as a key focus area. Prior to this, much health promoting schools implementation was designed from a health behaviour change perspective without any link or motivation to improve student academic performance in school (Valois et al. 2011). This chapter argues for a focus on the interaction between promotion and sustainability of health and learning, by demonstrating how we can connect quality health promoting schools implementation to educational priorities, and school improvement.

The gradual shift in approaches that has followed from the introduction of the health promoting schools approach, now involves research building on the premises and acknowledgement of: educational aims; school as an institution in constant change; and the development of integrating the principles of empowerment and participation in health promotion, to simultaneously promote both health and learning. Davò-Blanes and La Parra (2013) reviewed the role of student participation and identified that it increased student motivation and self-confidence through their sense of engagement, as well as increased health knowledge.

There is a need to better understand how the principles of health promotion and educational research interests coalesce and can be implemented in health promoting schools. This chapter draws connections between education and health research articulating the links and commonalities in desired outcomes. It has been argued that the traditional model of health curriculum and health services has thwarted the shift to a closer linking of educational and health goals and outcomes (Valois et al. 2011), a perspective that has a wider and joint focus. Specific areas for action and outcomes have been identified to address the education to health link and the health to education link (Basch 2010).

Bryk (2010) has identified five essential supports for school improvement: (1) a coherent instructional guidance system, (2) the professional capacity of its faculty, (3) strong parent-community-school ties, (4) a student-centred learning climate, and (5) leadership that drives change. Bryk's support elements for school improvement to a large extent overlap with core components identified by Samdal and Rowling (2011, 2013) in their meta-analysis of health promoting schools implementation: (1) Preparing and planning for school development; (2) Policy and institutional anchoring; (3) Professional development and learning; (4) Leadership and management practices; (5) Relational and organisational support context; (6) Student participation; (7) Partnership and networking; and (8) Sustainability. In all the reviews of implementation of school improvement and health promoting schools the school leadership is seen as a key feature for successful changes processes (Samdal and Rowling 2011, 2013). In the following sections, the bringing together and the overlap of the components needed for both school improvement and health promoting schools, will be addressed to demonstrate how implementation of health promoting schools also may contribute to school improvement.

11.2.1 Leadership

Many health researchers cite the critical role of the principal in the success of health promoting schools. It is the educational literature that can assist the health sector to articulate this component of implementation to achieve health outcomes. For example the principal's involvement embeds health in the school improvement plan with targets, specific goals and objectives related to healthy schools (Valois et al. 2011). The role of the leadership is to align the health promoting schools initiative with the educational visions and aims of the school by identifying how both health promotion and educational goals can be met by the same strategies. For the school principal leading the implementation process is frequently considered more vital for its success than facilitating the needed practical arrangements through management and administrative actions, although both aspects of the leadership role are needed (Daft 1999; Fullan 2005; Larsen and Samdal 2008). Fullan (2005) highlights that the required balance of leadership strategies and actions include: building understanding and commitment to the change processes; creating strong cohesion and relationships in staff; developing relevant knowledge; application of sound and joint decision making; and effective human resource management (Fullan 2005). Adding to Fullan's general principles for school leadership, a Norwegian study of the implementation of health promoting schools found that more systematic educational implementation approaches were likely to occur if the leadership had a strong focus on pedagogical development and general school development combined with the health promoting school perspective (Samdal et al. 2010).

The present approach typically involves attribution of leadership as a sole focus on the principal as leader, their role and their part in the process. There are other forms of leadership that could be equally beneficial. Distributed leadership (Spillane 2006) acknowledges the complexity of schools and the increasing and varying demands being experienced. Along with leadership imbued in a principal, there can be leadership as an organisational quality (Spillane 2006), a leadership structure where school personnel are active participants, collegiality is valued and joint ownership and responsibility are present. The concept of distributed leadership aligns closely with health promotion principles of participation, empowerment and ownership, characteristic of quality implementation of health promoting schools (Rowling and Samdal 2013). Research has identified that distributed leadership is vital for whole school innovation and is essential to achieve systemic change (Rowling 2009) and teacher efficacy (Inchley et al. 2007).

11.2.2 Preparing and Planning for School Development

Over the years implementation of health promoting schools has increasingly acknowledged the importance of a whole school approach aiming at organisational change processes rather than primarily individual health behaviour change. This approach is consistent with the educational interest of school development as the processes simultaneously can contribute to increased school effectiveness and student achievements as well as improved health and well-being (Valois et al. 2011). An initial motivation for implementing the health promoting school principles should therefore be easy to stimulate. In the longer term planning process motivation is key to nurture the joint focus of promoting both educational effectiveness and health, to further stimulate the prioritising of implementation of the principles of health promoting schools. Schools are familiar with planning of actions to improve health, but frequently it is from experience in a limited time frame of change through for instance, a project (Hubbard 2009) and primarily building on their immediate ideas for action. Conversely, health promoting schools represent a long term change process, and in order to make sure that long term goals can be achieved, a systematic planning process is needed. The planning component may be seen to parallel Bryk's (2010) emphasis on the underpinning of a coherent instructional guidance system as the key to school improvement. Bryk in his approach primarily addresses the need for clear instructions for teachers and students with regard to student learning. The basic principle of these instructions is to provide a clear rationale and structure for which content should be focussed on and which actions should be performed to achieve school improvement and effectiveness. The aim of the planning phase is parallel to that proposed by Bryk, namely to identify and agree on key actions to promote health and learning (Samdal and Rowling 2013).

The first step of the planning phase is to stimulate alignment and readiness for implementation by sensitising the leadership and school staff to the aim of the change process and how it can be best performed to achieve the desired change (Elias et al. 2003; Flaspohler et al. 2008; Sabatier 1997; Stith et al. 2006; Weiner et al. 2009). Readiness is found to be particularly stimulated by goal commitment and collective efficacy, through establishing shared values and beliefs that the suggested change process is of importance, to develop an organisational climate conducive to student learning and academic achievement, a key aim of the educational setting (Bandura 1998; Weiner et al. 2009). Further readiness involves competence building and organisational facilitation, for example through time set aside for the implementation process and balancing timing of the initiation of the new change process with other on-going commitments in the school (Flaspohler et al. 2008).

Schools need to be guided to spend time on the planning process. When organisational readiness is achieved the next step would be an analysis of needs for change, followed by reflections on how to meet the needs by building on theories and empirical findings. In this process, programme theory or programme planning models such as Program Logic (Kellogg Foundation 2004) or Green and Kreuter's (2005) PRECEDE model may prove useful. Such models provide step-by-step guidance on how to perform a needs analysis as well as identifying actions that can be powerful in changing organisational and individual aspects of importance to meet the identified needs. In the identification of actions it is relevant and important to search for existing programmes and interventions already proven effective to achieve the desired organisational or individual change, thereby ensuring good use of time and resources. When no previous programme exists the school may need to develop actions themselves, applying theory and empirical findings that have identified useful mechanisms to achieve the desired change. Further, it is crucial to systematically plan the implementation of the identified actions by addressing the need for time and resource allocation, competence development and involvement of all stakeholders. These issues will be addressed in the following description of other components.

A key part of the planning phase is also to make a plan of evaluating and monitoring change and development. Identification of change should ideally build on a baseline survey, identifying the frequency of the students' and staffs' behaviour, perceptions or knowledge that need to change. Moreover, it is of importance to assess students' and staffs' appraisal of the activities as it indicates their motivation and commitment to continue participation in the activities. The planning of the implementation process should also include actions that maximise sustainability. Such actions relate to for instance long term anchoring of the initiative in policy plans and continued resource allocation for professional development (Larsen and Samdal 2008; Sabatier 1997).

A well-structured and efficient planning process is recommended, performed by a coordination committee with representatives from all relevant stakeholders in the school (staff, students, parents, and others) (Denman 1999; Firth et al. 2008). For effective communication with the leadership, which is core to resource allocation and priority of the actions in the daily life of the school, the principal or deputy is suggested as a key representative in the committee (Samdal et al. 2010). The committee should oversee the different phases of the planning process and make sure that all stakeholders have an opportunity to influence the decisions taken, as this will stimulate commitment and motivation to participate in the implementation of the agreed actions.

11.2.3 Policy and Institutional Anchoring

Over the last two to three decades increased formalisation of school practices has taken place. Most schools today therefore develop policy or action plans for each school year to agree on and communicate their priorities for school improvement to parents, local authorities and other stakeholders (Valois et al. 2011). The focus of the improvement is primarily on how to improve student learning, and as part of that, also improving the learning environment, i.e. the ethos of the school. Such anchoring of actions for school improvement is found to be core also for the implementation of health promoting schools (Bond et al. 2001; Deschesnes et al. 2003; Hoyle et al. 2008; Samdal 2008; Samdal et al. 2010; Samdal and Rowling 2011; Samdal and Rowling 2013). Policies articulate vision, mandate, decisions and agreed actions, and when summarised in a written document they provide a clear strategy and commitment for action (Bond et al. 2001; Hoyle et al. 2008). A written policy thus helps a school to maintain focus, secure continuity and commit existing as well as new teachers, to agreed priorities at the school (Samdal et al. 2010). The written policy can also be seen to coincide with Bryk's (2010) system for coherent instructional guidance, where the system approach implies that a clear structure for what to do and how to do it commits the teaching staff in their contribution to school improvement.

Related to the planning phase, a key function of the policy or improvement plan is to develop ownership among staff and stakeholders for the aims and objectives listed in the plan (Deschesnes et al. 2003; McBride et al. 1999). To achieve ownership it is therefore important that both staff and stakeholders have been part of identifying and deciding on the aims and objectives of the plan and thereby future actions. This is best achieved through a consultative process spearheaded by the school leadership. In this way the staff and stakeholders are more likely to be aligned and motivated to implement actions to achieve the agreed objectives. Further, the consultation process will also take into account the competencies in the staff and among stakeholders ensuring that what is agreed is realistic and achievable (Heward et al. 2007; Hopkins and Jackson 2003). A written policy will also ensure that priority is given from the leadership through curriculum planning and resource allocation, which will help facilitate the implementation of the suggested actions.

11.2.4 Professional Development and Professional Learning

The school personnel are key initiators of change and their competence and buy-in to the change process are considered vital to the outcomes in both implementation of health promoting schools (Samdal 2008) and in general school improvement (Bryk 2010). Adding to teachers' competencies and capacities, the role of teachers' concerns in change and innovation has long been recognised in educational research. Professional development needs to address their concerns about how an innovation will impact on their classroom and their students. In particular the role of teachers under the right conditions is crucial (Fullan and Stiegelbauer 1991). However, health education research has been more interested in distal outcomes of training of teachers to implement a particular curriculum/program with fidelity that could then be correlated with health outcomes, and has thereby focussed training of specific competencies. Both elements, i.e. concerns and competencies, need to exist to achieve the desired outcomes.

Professional development in combination with school based professional learning is crucial to ensure teacher capacity for participation through building their skills, competencies, and efficacy. The aim of this developmental process is to assist teachers to apply and adjust their actions to varying situations by having capacity to evaluate what actions are appropriate given available constraints and resources (Jourdan 2011). Such action in context is considered professional learning.

Application of both professional development and professional learning is important for effective capacity building. Professional development frequently takes place outside the school premises, typically where staff from two or more schools meet for joint training and networking. Professional development content is normally identified and decided upon by people from above (a top down approach) and the focus is primarily on knowledge and competency development. Professional learning has its base in the school and the staff's perceived need for competence development (a bottom up approach), and focuses on knowledge, attitudes, skills, aspirations and behaviour (Easton 2008). Professional learning actions include group based conversations analysing practice and adequate management of concrete situations, shadowing colleagues, coaching from principal or colleagues, mentoring, and planning individually or in groups (Easton 2008). A popular approach to coaching and mentoring is by adopting the role of being a critical friend (Butler et al. 2011). Effective performance of such a role involves being helpful and supportive without pressuris-ing; and to help and challenge colleagues to meet agreed criteria by ensuring accom-plishment of certain standards (Barnard et al. 2009).

11.2.5 Student Participation

Stakeholder participation and in particular student participation represents the core of health promotion principles and the health promoting schools (Barnekow et al. 2006; Buijs 2009; Inchley et al. 2007; Simovska 2007). Two sets of theoretical concepts are offered as a way of framing student participation. These are: the demo-cratic approach to health promoting schools and the sociocultural perspective (Simovska 2008). The former has a strong focus on individual and collective empowerment. It was particularly dominant in the European approach to health promoting schools, whereas globally this principle was not as great a focus in health promoting schools action, particularly because of the political sensitivity in emerg-ing societies in Asia and the South West Pacific where democratic movements were not so influential. The second theoretical framework, the socio-cultural approach to student involvement, focuses on the influences of such factors as the diversity of economies, ethnic groups and education systems. Different perspectives on the stu-dent's role in learning may be a result of a hierarchical or collaborative culture in a country's education system that flows down to school and teaching environments. This observation pinpoints the importance of cultural context at national level as well as local school level, when implementing actions in a given setting, in this case, the student's role in the school setting (Samdal and Rowling 2013).

Student involvement is considered core for school effectiveness and improvement, as identified by Bryk's (2010) support component for a student-centred learning climate. In line with relational pedagogy student participation can be seen as a means to achieve motivation for both health and learning (Boyd et al. 2006). During the three recent decades, classroom pedagogy has increasingly moved into a student-centred approach, allowing students to influence what and how they are going to learn. Student-active approaches also involve engagement in the governance and decision-making in the school. Such governance frequently builds on training and opportunity for student leadership and active participation (Holdsworth and Blanchard 2006). This student-centred pedagogy underpins the empowerment and capacity building of students, which are corner stones of health promotion.

In this sense, students who experience that they are allowed to influence deci-sions and learning strategies in school are likely to develop stronger intrinsic moti-vation for school which again may positively influence both their academic achievement and overall well-being (Danielsen et al. 2009; Reeve and Jang 2006; Wierenga 2002). Student participation is also a goal in itself as development of autonomy and self-efficacy is key to educational training and aims, in that it pre-pares students for their future role of asserting and performing active citizenship

(Jensen and Simovska 2005; Larson 2000; Stefanou et al. 2004). Simovska (2008) stresses that active participation is critical for students' general development. Through active involvement and decision-making in their learning process, students learn how to work with and understand complexities of their lives and also experience personal meaningful learning. A recent review (Griebler et al. 2012) identified the personal effects of participation as involving a range of skills, competences and knowledge including: increased communication skills, improved collaboration skills, improved decision-making and problem solving skills, increased learning capacity and learning research skills. These are all educational outcomes that have been a result of health projects and programs in schools. Additionally student participation produced some effects on the school as an organization. These included: better acceptance of and compliance with rules, improved school engagement, change in or new school policies and changes in or new infrastructure. These findings support this chapter's theme of the nexus between learning and health outcomes.

11.2.6 Relational and Organisational Support Context

Research highlights the importance of organisational and contextual/relational support for implementing health-promoting schools (Rowling and Samdal 2011; Samdal and Rowling 2011; Simovska and Carlsson 2011). Similarly educational research identifies climate and culture (Boyd et al. 2006; Hargreaves et al. 2001), and organisational capacity (Flaspohler et al. 2008; Hopkins and Jackson 2003) as key to achieve school development and change. The focus on the support context addresses structures, strategies and practices that stimulate smooth and efficient implementation of actions and activities (Weiner et al. 2009). In this regard the concepts of climate and culture involve relational support (Bandura 1998) whereas organisational structures, including timetabling, physical environment and fiscal resources provide organisational support (Leithwood et al. 2007).

The coordination committee and the leadership have vital roles to play in identifying actions and structures to support the development of a stimulating relational and organisational support context. As outlined above the leadership fulfils this role through their management actions of: allocating resources for professional development and facilitating structures through timetabling and regulation of staff collaboration. Moreover, in their role they are responsible for developing readiness, alignment and motivation for implementing health promoting school actions as well as student learning (Bond et al. 2001; Elias et al. 2003; Sabatier 1997). An efficient tool in such development is to establish discussion and consultative groups in the staff that meet regularly to share and analyse their experiences (Kallestad and Olweus 2003). Similarly, student cohesion through a positive and well-functioning social climate at school is vital to students' life satisfaction as well as for their academic achievement (Danielsen et al. 2009; Samdal et al. 1998, 1999). Parents may also, if invited by the school, and when they perceive that their inputs are valued, represent a key support for schools in the implementation of change processes (Cuttance and Stokes 2000). Support functions may involve building readiness among their children, partaking in concrete actions or events that focus on building better school-community communication, or in fund-raising activities giving schools more resources for their implementation process.

Finally, as commented earlier, it is important to be aware that implementation operates in a political national and regional context (Samdal and Rowling 2013). This context is vital in relation to the expectations of schools and for the overall resource allocation for new initiatives. In Portugal a nationally comprehensive professional development and school level implementation of health promoting schools was abruptly stopped with a change to a more conservative government and the impact of the financial crisis (Gaspar de Matos et al. 2013).

11.2.7 Partnerships and Networking

Partnerships are a key underpinning principle of a settings approach. Samdal and Rowling (2011) in their narrative synthesis of empirical work that focused on implementation of health promoting schools, categorised seven out of the eight key sources as including this implementation component. The WHO Jakarta Declaration on Leading Health Promotion (WHO 1997) highlighted partnerships, identifying their positive outcomes as sharing of expertise, skills and resources.

From an education and school change perspective, it is essential that partnerships exist between health personnel, parents and school staff. Schools need to 'own' change processes (Reynolds et al. 2000) so common goals need to include a focus on this element of the change process. A similar understanding to that of school personnel is required by partnering sectors. Some health staff prefer this shared approach, linking their technical expertise in a coordinated way (Marx and Northrop 2000). Functional partnerships facilitate this mutual commitment and a shared vision (Deschesnes et al. 2010). Genuine collaboration with community stakeholders can be achieved by the principal, who due to their standing in the community, can draw on a range of school community members, parents, businesses, and local agencies with greater authority than other school staff (Valois et al. 2011).

A key strategy of the health sector has been offering funding to schools. These resources, usually in the form of one off grants have had impacts on practice, policy and attitudes (as represented through dialogue/language). In terms of practice, grants have 'bought' staff time, for dialogue with health personnel and opportunities to think and act horizontally linking with other school and curriculum change initiatives (McBride et al. 1995). But grants have also contributed to project thinking and project funding that can hamper whole school change (Hubbard 2009).

Previous research has shown that partnerships have been difficult to develop in health promoting schools (Samdal et al. 2010). Effective collaboration models demand complementary collaborative approaches between school and community, enhancing each other's work and evolving into comprehensive, integrated approaches (Center for Mental Health and Schools 2008). Shared aims and commitment, sharing of power and mutuality are key components of such a collaborative approach (Cuttance and Stokes 2000). Clearly defined roles between the participants, the institutional infrastructure and anchoring are also vital (Center for Mental Health and Schools 2008; Deschesnes et al. 2010). Systematic approaches to building trust, is another key principle to stimulate partnership development (Boot et al. 2010).

Also educational research has underpinned the benefit of functional partnerships. Here lateral capacity building across schools has been highlighted as this provides an excellent opportunity where principals and teacher leaders collaborate with other schools, to learn from and contribute to school improvement (Fullan 2005). Bryk (2010) in his identification of support components for school improvement, highlights strong parent-community-school ties as key to student learning and achievement. Here parents' roles in supporting and communicating the school's aims and strategies for student learning, is considered crucial to the individual student's achievement and progress. Further, institutions in the local community are seen to represent and provide important support for values and behaviours motivating learning and priority to scholastic performance.

11.3 Sustainability for Health Promoting Schools

As pinpointed earlier, schools are often recipients of projects and therefore can be driven by project thinking and project funding, initiating a specific initiative and keeping its focus for a limited period of time (Hubbard 2009). This means that when the funding is finished, a new priority takes over and the previous one is frequently forgotten. School effectiveness research is becoming increasingly aware of the importance of addressing systemic change and that such organisational change processes take time and need to be nurtured over years and decades. Similarly, Hoyle and colleagues (2010) argue that only when the health sector moves away from their project and topic oriented thinking and addresses health promotion based on educational strategies, can health promotion be achieved in school. This is the aim of the health promoting schools approach. As such there is no end point to the process, but a continuous focus on how the school climate and school organisation can be improved to stimulate both learning and well-being of both students and staff. This continuous process is identified in the sustainability component.

The driving force of the sustainability of the change processes in school is the principal (Bryk 2010; Fullan and Hargreaves 1992; Fullan 2001; Larsen and Samdal 2008). The focus on sustainability is kept through continuous follow-up of all the identified implementation components, i.e. sustainability needs to be planned from the start and maintained throughout a minimum of a decade to ensure lasting organisational change (Green and Kreuter 2005). In order to maintain staff and stakeholder's motivation to keep up their efforts to achieve change, it is important to evaluate progress, to know that the activities are helping the school to meet its objectives, or if not, provide guidance for higher intensity of activities or other types of actions.

Sustainability is cross cutting through all the components and needs to be initiated at the beginning of the change process. Facilitation of sustainability includes long term anchoring of the initiative in a policy plan, continued resource allocation for professional development, monitoring and evaluation of actions and progress (Larsen and Samdal 2008; Sabatier 1997). If successful implementation has been achieved, the school is ready to consider a new cycle of identifying priorities, while at the same time maintaining actions that were initiated in the first cycle. At this point the school has a lot of experience to build on, and may thus be able to more quickly initiate new actions.

Continued priority of professional learning and development is vital to maintain commitment and increase capacities in staff to perform agreed actions. Such a focus will also integrate new staff in alignment processes necessary for their involvement in the change process. Evaluation of the sustainability in terms of how teachers and school leadership change the way they work to implement health promoting schools can be monitored as indicators of systemic change (Easton 2008). Similarly, changed behaviour and increased mutual trust in partnerships should also be monitored to identify any needs for change in the sustainability phase.

11.4 Interactions Between the Components

This chapter has been arguing for the need to strengthen the coming together of health and education sectors. This will involve a merging of conceptualisations and perspectives in these two sectors. For example, Bryk (2010) underpins the interrelationship between the support components for school improvement, aligning them to ingredients in a cake, where all contribute to the result and stimulate the role of the other components. A similar claim has been made by Samdal and Rowling (2011) for the implementation components for health promoting schools.

The interaction can be seen for example in the key role of the leadership in stimulating and facilitating the implementation of all the components. Similarly, the planning component involves planning and preparing implementation of all components. Likewise the pivotal role of partnerships in implementation emerges when other components are explored. For example for sustainability, partnerships can stimulate actions and help institutionalise change for health outcomes that link with educational priorities. Organisational support from the health sector in the form of grants creates the conditions for empowerment and training of staff and activation or enhancement of relationships. The partnership is critical so the grants meet the priorities of the school at a time and in a form that is functional for personnel involved. Additionally relational support in the form of harmonious relationships between school personnel/parents/health staff is needed to achieve expectations of a joint project. The personal connection of school personnel with individuals in outside agencies helps build trust and confidence.

Another feature of partnerships is networking, a powerful means for promoting commitment, change and innovation. Networks are organizational forms that provide for collective learning processes and can thus reduce uncertainty in the implementation of innovation (Tsouros et al. 1998: 16). Informal teacher exchange of innovation, resources and targeted materials across schools is commonplace. This practice can be utilised and enhanced in a more formalised manner to stimulate collaboration and commitment in staff both within and between schools.

11.5 Discussion

The previous presentation of components for health promoting school implementation has highlighted the complexity of implementation in creating health promoting schools that attend to the priorities of both the education and health sectors. A recent development in health promotion is the examination of complexity theory (Resnicow and Page 2008; Shiell et al. 2008), complex adaptive systems (Keshavarz et al. 2010) and organisational systems (Dür 2013). The explanatory power of these perspectives has been identified as offering assistance to understanding the elements and processes in the interaction between health and education as operationalised in health promoting schools.

In a study conducted in Australia, Keshavarz and colleagues (2010) explored whether health promoting schools exhibit characteristics of complex adaptive systems. At its core, a complex adaptive system comprises a population of diverse rules-based agents located in multilevel and interconnected systems in a network shape. Agents in complex adaptive systems are often numerous, dynamic, autonomous, highly interactive, receptive to learning and adaptive. Agents of complex adaptive systems act in ways that are based on a combination of their knowledge, experience, feed-back from the environment, local values and rules.

The researchers aimed to explore the use of complex adaptive systems theory in developing understanding of the differential impact of health promoting schools implementation. Diversity among agents and between schools is a key observation in understanding why some evidence-informed polices and practice do not yield similar outcomes in some schools. Lack of acknowledgement of this diversity among schools, external agencies, including health workers and health agencies, results in creating unrealistic expectations of what schools might achieve and sustain in relation to pre-defined health goals (Keshavarz et al. 2010).

Complex adaptive systems are characterised by rules. These rules include the formal organisational rules, but are also formed in response to more informal school characteristics, described as the school "ethos". Teachers follow rules, both formal and informal, in their teaching and interpersonal interactions with colleagues, students and parents (Keshavarz et al. 2010). "Rules" were interpreted by individual agents, to create school-specific "rules" producing actions by agents that were adapted to more informal school characteristics and prevailing social norms and practices. The existence of informal rules may confound validity in interpretation of research results, making imperative for monitoring individual school implementation decision-making.

In summary, schools' collective behaviours were found to be dependent on the rules, interactions, information, values, context, time, and other systems' action, as well as availability of resources. Hence, schools' general behaviour was identified to be the result of the interplay of multiple factors, and is accordingly an emergent phenomenon that is not easily or fully predictable. Keshavarz and colleagues' (2010) description of complexity in implementing health promoting schools is depicted by Dür (2013) in even more detail and clearly parallels Bryk's (2010) components of school improvement. The focus on rules and collective behaviours may be seen to overlap with Bryk's focus on the instructional guidance system, other system actions include interaction with parents, and availability of resources taps the competence level of the school staff.

The analyses of the components of school development (Bryk 2010; Hopkins and Jackson 2003) and implementation of health promoting schools (Samdal and Rowling 2013), show that the two change processes work towards similar aims and use overlapping means. A common core seems to be the building of organisational capacity (Hopkins and Jackson 2003). This concept is typically identified as being composed by four components: (1) contribution, (2) alignment, (3) support, and (4) shared values. Contributions from staff, students and stakeholders (e.g. parents and health personnel) are necessary for action to occur. They also represent a basic principle of both health promotion and learning in letting those who are to perform and perhaps also be target of the change process, influence priorities of objectives, aims and actions. This influence will stimulate motivation to participate and contribute, and thereby impact on the outcome of the intervention through participants' persistency of actions over time. Looking at Bryk's (2010) school improvement components the student-centred climate may be seen to fall in this category. Similarly the student participation component from health promoting school implementation (Samdal and Rowling 2011) may be seen to cover the contribution part of organisational capacity.

Alignment may, in Bryk's components of school improvement seem to be captured by a coherent instructional guidance system in that this aims to have teachers agree to, and work towards the same aims using similar strategies for teaching and interaction with students. Among the components for implementation of health promoting schools *Preparing and planning for school development* and *Policy and institutional anchoring* may be seen to represent approaches that align staff, students and stakeholders to work toward agreed aims and objectives.

The support dimension of organisational capacity may in Bryk's categorisations be represented by the specific support received by parents and the local community in the focus on building strong ties between the school and local partners. This is reflected in the health promoting school implementation the components of *Relational and organisational support context* and *Partnership and networking.* They clearly articulate the relevance of building strong support both within the school organisation and with relevant external partners. Finally, shared values in the organisation capacity concept may, in Bryk's components, be seen to be represented by the professional capacity of its faculty, and similarly by *Professional development and learning* in the implementation of health promoting schools. This focus on teacher training and development across both school improvement and health promoting school implementation has a joint mission in building shared values as well as competencies to take part in the change process.

The importance of the school leadership for successful school improvement and implementation of health promoting school is not highlighted as an explicit component of organisational capacity. It may however be seen as a prerequisite for the development of organisational capacity in the process of building readiness for change and development. Further, only the components for health promoting school implementation specifically identify focus on sustainability as key to the change process. Harris and Lambert (2003) have however observed that schools that systematically have built and developed their capacity to implement change are more likely to sustain improvement over time. Thus the common core of building organisational capacity in both school improvement and implementation of health promoting schools underpins that implementation sustaining actions for both health and learning in school. However inclusion of the term 'sustainability' may suggest that health promotion is more in focus than is learning. To better communicate its overall school improvement goal in terms of promoting well-being and development of staff and students through organisational changes in school, the health promoting school implementation in the future should consider to more explicitly highlight the links between components of school improvement and health promotion, in pinpointing health promotion as both a prerequisite and catalyst of academic learning.

11.6 Conclusion and Suggestions for Future Research

A health promoting school approach involves a complex dynamic of group behaviours and system changes within the school by staff and students spearheaded by the school leadership, and in collaboration with external stakeholders. Research on school improvement and implementation of health promoting schools shows that the two change processes of promoting health and learning in school share a common core of developing organisational capacity that stimulates first readiness for change, then implementation competencies and finally sustainability.

Building on the theoretical claims that health promotion stimulates student learning, future research should assess in more detail if the implementation of health promoting schools also has positive empirical impact on student learning and school improvement in addition to health and well-being outcomes. The recent review on student participation (Griebler et al. 2012) is an excellent example of identifying the impact of this component. However, currently the research on implementation of health promoting schools seems to have provided more specificity about implementation components than about school improvement, particularly with regard to processes to ensure anchoring and alignment. A first important step could therefore be to test if the eight components for implementation of health promoting schools identified in the meta-analysis conducted by Samdal and Rowling (2011) are applicable also for school improvement processes. In this regard employment of organisational systems and complex adaptive systems theory and empirical findings (Dür 2013; Keshavarz et al. 2010) may be useful by aiming at improved understanding of the complexity of schools, and their functioning as complex adaptive systems; increased acknowledgement and understanding of the diversity between schools, how this affects health and learning outcomes; and how to stimulate more effective communication and collaboration between schools and the health sector, and between schools and parents regarding health and learning and how these interact.

References

Bandura, A. (1998). Health promotion from the perspective of social cognitive theory. *Psychology & Health, 13*(4), 623–649. doi:10.1080/08870449808407422.

Barnard, M., Becker, E., Creegan, C., Day, N., Devitt, K., Fuller, E., & Ranns, H. (2009). Evaluation of the National Healthy Schools Programme. Interim report. Department of Health. London: National Centre Social Research

Barnekow, V., Buijs, G., Clift, S., Jensen, B. B., Paulus, P., Rivett, D., & Young, I. (2006). *Health-promoting schools: A resource for developing indicators*. Copenhagen: World Health Organization.

Basch, C. E. (2010). *Healthier students are better learners: A missing link in school reforms to close the achievement gap*. A Research Initiative of the Campaign for Educational Equity Teachers College, Columbia University. Equity Matters: Research Review No. 6. http://www.equitycampaign.org/i/a/document/12557_EquityMattersVol6_Web03082010.pdf. Accessed 10 Nov 2011.

Bond, L., Glover, S., Godfrey, C., Butler, H., & Patton, G. C. (2001). Building capacity for system-level change in schools: Lessons from the Gatehouse Project. *Health Education & Behavior, 28*(3), 368–383. doi:10.1177/109019810102800310.

Boot, N., van Assama, P., Hesdal, B., & de Vries, N. (2010). Professional assistance in implementing school health policies. *Health Education, 110*(4), 294–308.

Boyd, R., MacNeill, N., & Sullivan, G. (2006). Relational pedagogy: Putting balance back into students' learning. *Curriculum Leadership, 4*(13), 1–2.

Bryk, A. S. (2010). Organizing schools for improvement. *Phi Delta Kappa, 91*(7), 23–30.

Buijs, G. J. (2009). Better schools through health: Networking for health promoting schools in Europe. *European Journal of Education, 44*(4), 507–520. doi:10.1111/j.1465-3435.2009.01410.x.

Butler, H., Krelle, A., Seal, I., Trafford, L., Sarah, D., Hargreaves, J., & Bond, L. (2011). *The critical friend. Facilitating change and wellbeing in school communities*. Camberwell: ACER Press.

Center for Mental Health and Schools. (2008). *Working Collaboratively: From School-Based Teams to School Community Higher Education Connections*. Los Angeles: Center for Mental Health and Schools. Available at: http://smhp.psych.ucla.edu/pdfdocs/worktogether/worktogether.pdf.

Cuttance, P., & Stokes, S. (2000). *Reporting on student and school achievement*. A research report prepared for the Commonwealth Department of Education, Training and Youth Affairs, Canberra. Retrieved from http://www.dest.gov.au/sectors/school_education/publications_resources/other_publications/reporting_on_student_and_school_achievement.htm

Daft, R. L. (1999). *Leadership: Theory and practice*. Fort Worth: Dryden Press.

Danielsen, A. G., Samdal, O., Hetland, J., & Wold, B. (2009). School-related social support and students' perceived life satisfaction. *The Journal of Educational Research, 102*(4), 303–318.

Davò-Blanes, M. C., & La Parra, D. (2013). Children as agents of their own health: Exploratory analysis of child discourse in Spain. *Health Promotion International, 28*(3), 367–377.

Denman, S. (1999). Health promoting schools in England – A way forward in development. *Journal of Public Health Medicine, 21*(2), 215–220.

Deschesnes, M., Martin, C., & Hill, A. J. (2003). Comprehensive approaches to school health promotion: How to achieve broader implementation? *Health Promotion International, 18*(4), 387–396. doi:10.1093/heapro/dag410.

Deschesnes, M., Couturier, Y., Laberge, S., & Capeau, L. (2010). How divergent conceptions among health and education stakeholders influence the dissemination of healthy schools in Quebec. *Health Promotion International, 25*(4), 435–443. doi:10.1093/heapro/daq040.

Dür, W. (2013). Applying system theory of organizational change to health promotion intervention in schools and their problems. In O. Samdal & L. Rowling (Eds.), *The implementation of health promoting schools: Exploring the theories of what, why and how*. London: Routledge.

Easton, L. B. (2008). From professional development to professional learning. *Phi Delta Kappa, 89*(10), 755–759.

Elias, M. J., Zins, J. E., Graczyk, P. A., & Weissberg, R. P. (2003). Implementation sustainability and scaling up of social emotional and academic interventions in public schools. *School Psychology Review, 3*, 303–319.

Firth, N., Butler, H., Drew, S., Krell, A., Sheffield, J., Patton, G. C., & Bond, L. (2008). Implementing multilevel programs and approaches that address student wellbeing and connectedness: Factoring in the needs of the school. *Advances in School Mental Health Promotion, 1*(4), 14–24.

Flaspohler, P., Duffy, J., Wandersman, A., Stillman, L., & Maras, M. A. (2008). Unpacking prevention capacity: An intersection of research-to-practice models and community-centered models. *American Journal of Community Psychology, 41*(3–4), 182–196.

Fullan, M. (2001). *Leading in a culture of change*. San Francisco: Jossey-Bass.

Fullan, M. (2005). *Leadership and sustainability. System thinkers in action*. Thousand Oaks: Corwin Press.

Fullan, M., & Hargreaves, A. (1992). *Teacher development and educational change*. London: Falmer Press.

Fullan, M., & Stiegelbauer, S. M. (1991). *The new meaning of educational change*. New York: Teachers College Press.

Gaspar de Matos, M., Sampaio, D., Baptista, I., & Equipa Aventura Social, FMH/UTL & CMDTl. (2013). Adolescents' health education and promotion in Portugal: A case study of planning for sustainable practice. In O. Samdal & L. Rowling (Eds.), *The implementation of health promoting schools: Exploring the theories of what, why and how*. London: Routledge.

Green, L. W., & Kreuter, M. W. (2005). *Health promotion planning: An educational and ecological approach* (4th ed.). Mountain View: Mayfield Publications.

Griebler, U., Rojatz, D., Simovska, V., & Forster, R. (2012). *Evidence for the effectiveness of student participation in designing, planning, implementing and evaluating school health promotion. A systematic review* (Working Paper No. 12). LBIHPR. http://pure.au.dk/portal/files/44474041/Participation_review_of_evidence.pdf

Hargreaves, A., Earl, L., Moore, S., & Manning, S. (2001). *Learning to change. Teaching beyond subjects and standards*. San Francisco: Jossey-Bass Inc.

Harris, A., & Lambert, L. (2003). *Building leadership capacity for school improvement*. Maidenhead: Open University Press.

Heward, S., Hutchins, C., & Keleher, H. (2007). Organizational change – Key to capacity building and effective health promotion. *Health Promotion International, 22*(2), 170–178. doi:10.1093/heapro/dam011.

Holdsworth, R., & Blanchard, M. (2006). Unheard voices: Themes emerging from studies of the views about school engagement of young people with high support needs in the area of mental health. *Australian Journal of Guidance and Counselling, 16*(1), 14–28.

Hopkins, D., & Jackson, D. (2003). Building the capacity for leading and learning. In A. Harris (Ed.), *Effective leadership for school improvement*. New York: Teachers College Press.

Hoyle, T. B., Samek, B. B., & Valois, R. B. (2008). Building capacity for the continuous improvement of health-promoting schools. *Journal of School Health, 78*(1), 1–8.
Hoyle, T. B., Bartee, R. T., & Allensworth, D. D. (2010). Applying the process of health promotion in schools: A commentary. *Journal of School Health, 80*(4), 163–166.
Hubbard, R. (2009). Tinkering change vs. system change. *Phi Delta Kappa, 90*(10), 745–747.
Inchley, J., Muldoon, J., & Currie, C. (2007). Becoming a health promoting school: Evaluating the process of effective implementation in Scotland. *Health Promotion International, 22*(1), 65–71. doi:10.1093/heapro/dal059.
Jensen, B. B., & Simovska, V. (2005). Involving students in learning and health promotion processes – Clarifying what? How? and why? *Promotion & Education, 12*(3–4), 150–156.
Jourdan, D. (2011). *Health education in schools. The challenges of teacher training*. Saint-Denis: INPES.
Kallestad, J. H., & Olweus, D. (2003). Predicting teachers' and schools' implementation of the Olweus bullying prevention program: A multilevel study. *Prevention and Treatment, 6*(1), Article 21. doi:10.1037/1522-3736.6.1.621a.
Kellogg Foundation. (2004). *Logic model development guide*. Retrieved from http://www.wkkf.org
Keshavarz, N., Nutbeam, D., Rowling, L., & Khavarpour, F. (2010). Schools as social complex adaptive systems: A new way to understand the challenges of introducing the health promoting schools concept. *Social Science and Medicine, 70*(10), 1467–1474.
Larsen, T., & Samdal, O. (2008). Facilitating the implementation and sustainability of second step. *Scandinavian Journal of Educational Research, 52*(2), 187–204. doi:10.1080/00313830801915820.
Larson, R. (2000). Towards a psychology of positive youth development. *American Psychologist, 55*(1), 170–183.
Leithwood, K., Mascall, B., Strauss, T., Sacks, R., Memon, N., & Yashkina, A. (2007). Distributing leadership to make schools smarter: Taking the ego out of the system. *Leadership and Policy in Schools, 6*(1), 37–67. doi:10.1080/15700760601091267.
Lister-Sharp, D., Chapman, S., Stewart-Brown, S., & Soden, A. (1999). Health promoting schools and health promotion in school: Two systematic reviews. *Health Technology Assessment, 3*(22), 1–207.
Marx, E., & Northrop, D. (2000). Partnerships to keep students healthy. *Educational Leadership, 57*(6), 22–24.
McBride, N., Midford, R., & Cameron, I. (1999). An empirical model for school health promotion: The Western Australian school health project model. *Health Promotion International, 14*(1), 17–25. doi:10.1093/heapro/14.1.17.
McBride, N., Cameron, I., Midford, R., & James, R. (1995). Facilitating health promotion in Western Australian schools: Key factors for success. *Health Promotion Journal of Australia, 5*(1), 11–16.
Mortimore, P. (2007). *The road to improvement: Reflections on school effectiveness*. Lisse: Swets & Zeitlinger Publishers.
Reeve, J., & Jang, H. (2006). What teachers say and do to support students' autonomy during learning activities. *Journal of Educational Psychology, 98*, 209–218.
Resnicow, K., & Page, S. E. (2008). Embracing chaos and complexity: A quantum change for public health. *American Journal of Public Health, 98*(8), 1382–1389.
Reynolds, D., Teddlie, C., Hopkins, D., & Stringfield, A. (2000). Linking school effectiveness and school improvement. In C. Teddlie & D. Reynolds (Eds.), *The international handbook of school effectiveness research* (pp. 206–231). New York: Falmer Press.
Rowling, L. (2009). Strengthening "school" in school mental health promotion. *Health Education, 109*(4), 357–368.
Rowling, L., & Samdal, O. (2011). Filling the black box of implementation for health promoting schools. *Health Education, 111*(5), 347–362.
Rowling, L., & Samdal, O. (2013). Theory and empirically based implementation of elements in components. In O. Samdal & L. Rowling (Eds.), *The implementation of health promoting schools: Exploring the theories of what, why and how*. London: Routledge.

Sabatier, P. A. (1997). Top–down and bottom–up approaches to implementation research. In M. Hill (Ed.), *The policy process: A reader* (2nd ed., pp. 272–295). London: Prentice Hall.

Samdal, O. (2008). School health promotion. In K. Heggenhougen & S. Quah (Eds.), *International encyclopedia of public health* (Vol. 5, pp. 653–661). San Diego: Academic.

Samdal, O., & Rowling, L. (2011). Theoretical base for implementation components of health promoting schools. *Health Education, 111*(5), 367–390.

Samdal, O., & Rowling, L. (2013). Theory based components for implementation of health promoting schools. In O. Samdal & L. Rowling (Eds.), *The implementation of health promoting schools: Exploring the theories of what, why and how*. London: Routledge.

Samdal, O., & Torsheim, T. (2012). School as a resource or risk to students' subjective health and well-being. In B. Wold & O. Samdal (Eds.), *An ecological perspective on health promotion: systems, settings and social processes* (pp. 48–59). London: Bentham.

Samdal, O., Nutbeam, D., Wold, B., & Kannas, L. (1998). Achieving health and educational goals through schools: A study of the importance of school climate and students' satisfaction with school. *Health Education Research, 13*(3), 383–397.

Samdal, O., Wold, B., & Bronis, M. (1999). The relationship between students' perceptions of the school environment, their satisfaction with school and perceived academic achievement: An international study. *School Effectiveness and School Improvement, 10*(3), 296–320.

Samdal, O., Viig, N. G., & Wold, B. (2010). Health promotion integrated into school policy and practice: Experiences of implementation in the Norwegian network of health promoting schools. *Journal of Child and Adolescent Psychology, 2*, 43–72.

Shiell, A., Hawe, P., & Gold, L. (2008). Complex interventions or complex systems? Implications for health economic evaluation. *British Medical Journal, 336*, 1281–1283.

Simovska, V. (2007). The changing meanings of participation in school based health education and health promotion. *Health Education Research, 22*(6), 864–878.

Simovska, V. (2008). Learning in and as participation. In A. Reid, B. B. Jensen, J. Nikel, & V. Simovska (Eds.), *Participation and learning perspectives in education and environment, health and sustainability* (pp. 61–80). New York: Springer.

Simovska, V., & Carlsson, M. (2011). Health promoting changes with children as agents: Findings from multiple case study research. *Health Education, 112*(3), 292–304.

Spillane, J. (2006). *Distributed leadership*. San Francisco: Jossey-Bass.

Stefanou, C. R., Perencevich, K. C., DiCintio, M., & Turner, J. C. (2004). Supporting autonomy in the classroom: Ways teachers encourage student decision making and ownership. *Educational Psychologist, 39*, 97–110.

Stith, S., Pruitt, I., Dees, J., Fronce, M., Green, N., Som, A., & Linkh, D. (2006). Implementing community-based prevention programming: A review of the literature. *The Journal of Primary Prevention, 27*(6), 599–617. doi:10.1007/s10935-006-0062-8.

Suhrcke, M., & de Paz Nieves, C. (2011). *The impact of health and health behaviours on educational outcomes in high income countries: A review of the evidence*. Copenhagen: WHO Regional Office for Europe.

Tsouros, A., Dowling, G., Thompson, J., & Dooris, M. (1998). *Health promoting universities: Concept, experience and framework for action*. Copenhagen: World Health Organisation, European Regional Office.

Valois, R. F., Slade, S., & Ashford, E. (2011). *The healthy school communities model. Aligning health and education in the school setting*. ASCD. http://www.ascd.org/ascd/pdf/siteascd/publications/aligning-health-education.pdf. Accessed 10 Nov 2011.

Weiner, B. J., Lewis, M. A., & Linnan, L. A. (2009). Using organization theory to understand the determinants of effective implementation of worksite health promotion programs. *Health Education Research, 24*(2), 292–305. doi:10.1093/her/cyn019.

WHO. (1997). *Jakarta declaration on leading health promotion into the 21st century*. http://www.who.int/healthpromotion/conferences/previous/jakarta/declaration/en/. Accessed 17 Aug 2012.

Wierenga, A. (2002). Losing and finding the plot: Storying and the value of listening to young people. *Scottish Journal of Youth Issues, 4*, 9–30.

Chapter 12
Healthy Principals – Healthy Schools? A Neglected Perspective to School Health Promotion

Kevin Dadaczynski and Peter Paulus

School principals have been widely neglected in school health promotion. Due to societal changes and important school reforms, the job requirements of this occupational group have also dramatically changed. Against this background, available research results on principals' health are presented. Moreover, two additional perspectives on the role of school principals in school health promotion are discussed in more detail in this chapter. While one perspective deals with the question as to whether and how principals affect teachers' health, the other illuminates the influence of principals on the implementation and overall success of health promoting activities in the school. The chapter closes with a discussion of the findings and recommendations for future research and practice.

12.1 Introduction

> *Wanted: A miracle worker who can do more with less, pacify rival groups, endure chronic second-guessing, tolerate low levels of support, process large volumes of paper and work double shifts (75 nights a year). He or she will have carte blanche to innovate, but cannot spend much money, replace any personnel, or upset any constituency*
>
> (Evans 1995).

Despite the progress school health promotion has made in recent years, principals and their roles have scarcely been examined in theory and practice. This restraint is somewhat surprising, given that the primary responsibility for most

K. Dadaczynski, Ph.D. (✉)
Centre for Applied Health Science, Leuphana University of Lueneburg, Lueneburg, Germany
e-mail: dadaczynski@leuphana.de

P. Paulus, Ph.D.
Institute of Psychology, Leuphana University of Lueneburg, Lueneburg, Germany
e-mail: paulus@uni.leuphana.de

V. Simovska and Patricia Mannix-McNamara (eds.), *Schools for Health and Sustainability: Theory, Research and Practice*, DOI 10.1007/978-94-017-9171-7_12,

school matters lies with the school administration. Thus, it is not surprising that a wealth of educational research has shown that school leaders do make a difference in school effectiveness and school improvement (e.g. Huber 1999; Scheerens and Bosker 1997; Townsend 2007). The same can be expected with regard to the success of school health promotion activities. But what exactly is the role school principals play in school health promotion? Slowly emerging evidence indicates that principals as 'gatekeepers' to school innovations have significant influence on whether or not a school will become and remain a healthy organization (e.g. Dadaczynski 2012; Rowling and Samdal 2011; Samdal and Rowling 2011; Viig et al. 2011). Within all phases of the school health development process, from its beginning to the end, principals are responsible for e.g. building and maintaining high motivation (e.g. through vision building), supporting their school staff in developing the skills needed for successful change, coordinating the processes and activities, and encouraging the school staff to sustain new practices and activities.

Additionally, at least two more perspectives in the area of principalship and school health can be distinguished. Carr (1994), who surveyed 94 Australian principals, found out that approximately one-third of all respondents had a high level of anxiety (31.9 %) and/or depression (31.9 %). In summarizing his findings, Carr titled his article '*Warning, principalship can be hazardous to your health*'. While in the past teachers and pupils received most of the attention in health and epidemiological research, we know very little about the health of principals and how they deal with their everyday demands. Hence, this perspective is interested in better illuminating the health status of principals, and in identifying factors which are associated with principals' health. With regard to the second perspective, Landsmann (1978) used a similar title saying '*Warning to principals: you may be hazardous to your teachers' health*'. This seemingly minor shift reflects the results of a large study which identified three major areas of health concerns in 9.000 teachers: stress and tension, the physical environment of the school, and diet and exercise – each of which can be significantly influenced by the principal. Hence, the second perspective asks whether and how principals affect the health and wellbeing of teachers.

It is safe to assume that these three perspectives (1) principals' health, (2) influence of principals on teachers' health, and (3) influence of principals on health promoting school activities) are highly interconnected. So it is quite conceivable that principals with low wellbeing and a high degree of e.g. psychosomatic complaints feel unable to initiate and support health promotion activities. Furthermore, in their study Harazd et al. (2009) found that wellbeing of principals is substantially associated with the wellbeing of teachers (particularly in primary schools) (r = .49). Based on this, it seems to be important to consider these perspectives jointly, and not isolation from each other. But before discussing these three perspectives in more detail, we want to start with a short description of the current job situation of school principals, which results from the socio-political and educational environment and recent changes.

12.2 School Principals in a Changing Working Context

To better understand the importance of school principals and their function, it is important to consider the broader context within which the school is operating. As emphasized by Huber (2004) each school is embedded in a community, in a particular educational system, which in turn is highly influenced by the society as a whole and the pace with which it is changing.

Amongst others, societal changes include e.g. growing multi-cultural diversity and a diversity of family patterns (e.g. patchwork or single parent families), all of which pose many challenges for schools. Moreover, changing gender roles (e.g. increased employment rate of women) result in an increased importance of schools as the primary educating authorities. In Germany as well as in other European countries, reconciling the demands of family and employment has been one main argument in favor of expanding of all-day-schools (Hagemann et al. 2011).

In addition to these broader societal changes, a number of school reforms and developments need to be mentioned. Firstly, schools all over Europe are faced with the challenge that information in knowledge-based societies is getting more complex and specific, with a progressively shorter half-life. At the same time, schools also need to teach subject-independent skills that will enable young people to cope with everyday demands and independently shape their lives. Hence, schools are faced with the need to adopt a broader understanding of education, encompassing the training of cognitive skills and the promotion of personality development. A further emerging trend worldwide concerns a tendency towards decentralization in some areas, while at the same time centralization in other school domains simultaneously increases. Decentralization is the result of rigid system-based management processes which often ignore specific school-based needs and conditions (Cheng 2002). Hence, decentralization represents an attempt to support the autonomy and accountability of schools by transferring decision power from a system-wide level to the school itself (Huber 2004). Paradoxically, there is also a parallel tendency towards more influence from the system level through e.g. national standardized tests and curricula as well as national school inspections (ibid).

Although there are many additional school reforms than those described (e.g. inclusion as a main challenge for European schools but also technological changes), it becomes clear that all of these changes have a large influence on the school and the way it is managed. Schools and principals in particular need to react to these new demands by flexibly and continuously adapting their educational practice to societal and individual needs. In the following sections, we describe some tasks and functions of principals and forms of stress that may arise from their work.

12.2.1 Tasks and Functions of School Principals

Based on the changing working context, the roles and tasks of school principals have also changed dramatically in many countries during recent years. As argued by Huber (2004, 2007), it is impossible to define a generally accepted ‘role’ of school

Table 12.1 Range of tasks of school principals (Huber 2007)

Working with people inside the school	Working with people outside the school	Managing resources
As an important 'change agent' in developing the school	As 'homo politicus' who act diplomatically with political acumen	As an administrator and organiser, as manager of the organisation
As personnel developer, who is responsible for the advanced training of school staff and for cooperation within the school	As a representative of the school in public	As an architect and facility manager responsible for building maintenance, renovation and expansion
As a 'people person', i.e. as a trusted contact person for teachers, pupils and parents	As mediator, as liaison between internal and external interests	As financial professional and entrepreneur
As a teacher with teaching duties	As contact person, sometimes also as target	
As a role model inside and outside the classroom		

principals. Rather, school leadership tasks have many different aspects that can be best described as a colorful bunch of different tasks that can be grouped in several categories. As shown in Table 12.1, school principals are responsible for working with people inside and outside the school.

In this context, school principals need to act as initiators and facilitators of organizational change processes (e.g. development of the curriculum and education programs), but also as staff developers who promote the professional and academic growth of their school staff (e.g. though meetings, staff evaluation, in-service training). Furthermore, it must be taken into account that school principals have their own teaching duties which differ significantly among European countries and types of schools (principals from smaller schools typically have more teaching duties). Outside the school itself principals need to represent the school within the community and establish/maintain partnerships with community organizations. They need to act as 'homo politicus' in that they learn to behave diplomatically with great political sensitivity. Moreover, school principals have to demonstrate the ability to mediate between different interests and expectations (e.g. parents, school authorities). Finally, principals also need to manage resources, such as the school budget as well as the facilities.

In addition to this there have been a number of further attempts to systemize the tasks and responsibilities of school principals (Leithwood and Day 2007; Pont et al. 2008). Based on a whole array of educational leadership studies, Leithwood (1994, see also Leithwood and Day 2007) developed four broad categories of successful school leadership practices: *building vision and setting direction, understanding and developing people, designing the organization, and managing the teaching and learning program.* These categories do not only give information on what needs to be done but also on how things need to be done in order to enhance school quality. Therefore these categories are strongly associated with leadership behavior (see 12.1.5).

Building vision and setting direction as the first category encompasses efforts which aim to develop a high motivation among the school staff. Successful school principals have the ability to illustrate the necessity and importance of change processes to all school members (e.g. by emphasizing the benefits). Only if the goals are shared by all members, change processes (such as becoming a health promoting school) have a high chance of being sustainable.

Understanding and developing people as the second basic school leadership practice refers to the competencies and skills that are necessary to realize change processes within the school. Successful school principals not only provide the necessary knowledge and skills, but also establish the conditions that enable the skills to be applied in different situations and contexts. This requires that the skills and desires but also limitations of all school members are known and taken seriously.

Designing the organization as the third category includes practices which focus on the development of conditions and infrastructure that allow staff to translate their motivation into the school practice. This requires a school principal who promotes a culture of cooperation and of productive relationships within and outside of the school.

Finally, the last category refers to the core business of schools which is to *manage the teaching and learning program.* Successful principals translate their visions and ideas into the school practice by using efficient and effective management techniques. This includes e.g. schedules and plans to implement the school curriculum but also to support teachers with regard to teaching issues.

In summary, principalship is associated with numerous tasks and responsibilities which require a variety of qualifications and skills. In light of this multitude of work demands it is not surprising that Evans (1995) describes school principals as '*miracle workers*' (see quotation at the beginning of this article). In contrast to a traditional understanding school principals are not just teachers with additional administrative functions (in the sense of 'primus inter pares'). Rather, they are educational leaders comparable with entrepreneurs of small and medium-sized enterprises. The question that arises here is how school principals perceive their working conditions respectively what kind of stress arises from the principalship.

12.2.2 Work Related Stress of School Principals

Most research in the field of work-related stress of school principals has been conducted in the 1980s to 1990s and at the beginning of the twenty-first century. Surprisingly, although a number of school reforms have taken place during recent years a declining interest in work-related stress of school principals can be ascertained.

With regard to the frequency of high or very high stress that is perceived during the work, there is great variability ranging from 21 % (Borg and Riding 1993) over 43 % (Phillips et al. 2007), to 70 % (Darmody and Smyth 2011). Much of this

Table 12.2 Work related stress of school principals

Stress category	Sources
Relationship problems (e.g. teachers, administrative staff, parents, pupils)	Borg and Riding (1993), Carr (1994), Chaplain (2001), Cooper and Kelly (1993), Darmody and Smyth (2011), Friedman (2001), Gmelch and Swent (1981), Mackler (1996), Williamson and Campbell (1987) and Whitaker (1996)
Workload	Borg and Riding (1993), Cooper and Kelly (1993), Gmelch and Swent (1981), Friedman (2001) and Mackler (1996)
Time management (e.g. frequent interruptions, meetings)	Gmelch and Swent (1981), Williamson and Campbell (1987) and Whitaker (1996)
Lack of support (e.g. from education authority, teachers)	Borg and Riding (1993), Chaplain (2001), Darmody and Smyth (2011) and Whitaker (1996)
Financial management/lack of resources	Borg and Riding (1993), Chaplain (2001) and Williamson and Campbell (1987)
External contact/cooperation	Cooper and Kelly (1993) and Hüfner (2010)
State/federal rules & policies	Carr (1994), Chaplain (2001), Friedman (2001) and Gmelch and Swent (1981)
Role conflict & ambiguity	Gmelch and Swent (1981), Mackler (1996) and Whitaker (1996)
Lack of recognition & respect	Mackler (1996) and Whitaker (1996)
Other working conditions (e.g. salary, responsibilities, teaching duties)	Borg and Riding (1993), Darmody and Smyth (2011) and Hüfner (2010)

difference can be explained by the cultural, societal and political differences in the countries examined but also by methodological inconsistencies (e.g. definition and measurement of stress, type of sample, sample size, recruitment).

The results of the studies reveal comprehensive compilations of stressors, which can be grouped in a number of stress categories (see Table 12.2). In his study, Carr (1994) identified three major sources of stress: feeling a lack of support from the education authority, coping with heavy work demands and difficulty in interpersonal relationships with other teachers. Gmelch and Swent (1981) found 35 stressors which could be assigned to five groups: administrative stress (e.g. time, meetings, workload), administrative responsibility (supervision, evaluation, gaining public support), interpersonal relations (conflicts between parents and school, among teachers), intrapersonal conflicts (conflict between performance and expectations) and role expectations (expectation of self and the various publics). In a further study Gmelch and Swent asked 243 US school principals the degree to which each of the 35 situations induces stress. Among the stressors most often mentioned belong complying with state and federal rules and policies, interpersonal conflicts or frequent interruptions from work. Similarly, on the basis of a study of 820 primary and secondary school principals from Israel four stress factors emerged: problems with parents (e.g. pressure from parents, rude behavior), work overload (e.g. meetings, overburdening instructions by the Ministry), teacher problems (unsatisfactory functioning, disobedient teachers), and problems with administrative and technical team (uncooperative and incompetent staff) (Friedman 2001). Using the Administrative Stress Index (ASI), Williamson and Campbell (1987) found four stressors in 243 US high school principals: management of time, relations with superiors, relations with subordinates, and matters of finance. Interestingly, school principals of large high schools perceived more stress due to time management issues whereas principals from smaller schools more often suffered from difficulties with subordinates.

Further differences could be found in a study of Maltese school principals, which yielded four stress categories: lack of support and resolving conflicts, inadequate resources, workload, and work conditions and responsibilities (Borg and Riding 1993). Inadequate resources were assessed as most stressful, followed by workload and work conditions and responsibilities. With regard to sex, male principals report more stress due to inadequate resources and work conditions and responsibilities than their female colleagues. When compared with primary schools, principals from secondary schools more often report to be stressed from work conditions, responsibilities, inadequate resources, lack of support, and resolving conflicts.

12.3 The Health of School Principals

As described in the previous section, the demands placed on school principals are very complex and can cause a great deal of strain in a number of ways. From health research, it is widely known that stress can negatively affect the health of an individual if internal or external resources are not available to deal appropriately with the demands (Antonovsky 1987; Lazarus and Folkman 1984). Although school principals have been widely neglected in health research in the past, there have been a few studies which are summarized in the following sections. Here we distinguish between short-term health outcomes (e.g. wellbeing, satisfaction, and sub-threshold mental and physical health complaints) and long-term health outcomes (e.g. burnout, sick leave, early retirement due to illness).

12.3.1 Short-Term Health Outcomes of School Principals

Unfortunately, there is a clear lack of studies which examine positive health outcomes such as wellbeing or satisfaction. We are aware of only two studies from Germany and the Switzerland which used positive measures of health. In their study, Harazd et al. (2009) compared the wellbeing of principals and teachers by using the WHO-II wellbeing index. When compared to teachers, school principals had a significant higher wellbeing ($M=3.56$ vs. 3.09, $d=.62$). Within the Swiss study, three quarter of the principals examined (n = 85) reported a high to very high vitality, confidence and energy (Landert 2009). On the other hand, more than 30 % of the same sample had problems relaxing after work.

With regard to health complaints there are a number of studies from different countries. Dewa and colleagues (2009) e.g. used the SF-12 health survey to assess the physical and mental health of Canadian principals and vice-principals from a large school district in Ontario (n = 108). Compared with norm-based standards about one quarter of respondents had a low physical health status and almost half had a low mental health status. There were significant associations between health status and job satisfaction, i.e. principals with lower mental health status were less satisfied with their salary, supervision, contingent rewards or communication.

Moreover, various studies have been conducted which compared different occupational groups. Phillips et al. (2008) examined 290 school principals from the UK concerning their physical complaints (e.g. headache, insomnia) and mental complaints (e.g. constant irritability). Compared to a general population of workers group and a group of managers, school principals had worse physical and mental health outcomes. In particular, mental health complaints were higher for female and primary school principals than for male and secondary school principals. These results were confirmed by a series of German studies that compared the risk of having psychosomatic complaints (e.g. nervousness, exhaustion, irritability) over 65 occupational groups (Hasselhorn and Nübling 2004; Hasselhorn 2009). While in the first study of 1999 school principals had the fifth-highest risk for poor mental health (OR: 1.9), 6 years later they ranked highest in this area. However, divergent findings come from a study conducted in the UK (Johnsen et al. 2005). In comparing 26 occupations, school principals ranked under the norm in health complaints, suggesting a better than average physical and mental health for this occupational group.

Finally, there are indications of an association between leadership behaviour and the mental health of principals. In a German study with 860 principals, Warwas (2009) identified five different leadership groups. Principals with a generalist orientation (high degree of leadership, collegiality and administration) and those with a focus on administration had worse mental health outcomes and more time-related stress, whereas principals with a focus on leadership and those with high orientation on collegiality had better mental health outcomes.

12.3.2 Long-Term Health Outcomes of School Principals

Although the current state of data on principals' health appears to be insufficient, there are some studies with a focus on burnout among this occupational group. It is worth noting that research regarding burnout syndrome is not unproblematic. Firstly, despite – or perhaps because – of its lay-scientific usage there is no common definition of burnout. Secondly, as a result of terminological imprecision, there are varying inclusion and exclusion criteria and various ways to operationalize burnout. Thirdly, there are no clear cut-off values that tell us when exactly a clinical relevance exists. Despite these problems the majority of research on teachers' health focused on burnout has relied mostly on the Maslach Burnout Inventory (MBI, Maslach and Jackson 1981). The MBI consists of three subscales: emotional exhaustion (chronic state of depletion and fatigue), personal accomplishment (decreased feeling of competence and successful achievement at work), and depersonalization (cynical stance towards the clients such as teachers, pupils, parents).

Available research on burnout among school principals has produced partially mixed findings, making clear conclusions impossible. In their study, Harazd et al. (2009) e.g. examined the extent of emotional exhaustion of teachers and their school principals. Compared to teachers, school principals were significantly less emotionally exhausted ($M=2.48$ vs. 2.01, $d=.87$). Whitaker (1992) presents results from a study of 107 US school principals that show a high level of exhaustion and depersonalization for 13 % of respondents. More current data comes from a study with 228 primary school principals from the southwest of North America (Combs et al. 2009). About 27 % reported moderate levels of burnout, whereas 9 % suffered from high levels of burnout. Interestingly, gender, age and years of experience in education were not associated with principals' burnout.

Along with studies on the prevalence of burnout, there is also research about the determinants of burnout. In a follow-up study, Whitaker (1996) interviewed those school principals with high levels of emotional exhaustion and depersonalization. The causes identified in this study generally correspond with the stressors described in section 12.1.3. School principals with high levels of burnout reported a high degree of relationship problems, constant interruptions from work, a large number of meetings and too much paperwork. In another study, Gmelch and Gates (1998) found moderate to high correlations between emotional exhaustion of principals and level of stress ($r=.57$), task-based stress ($r=.51$) and conflict-mediating stress ($r=.41$). Moreover, Devos et al. (2007) indicate that symptoms of burnout are associated with self-efficacy. In other words: the higher the confidence in one's own abilities, the lower the emotional exhaustion, and depersonalization and the higher the personal accomplishment of school principals.

Other long-term health outcomes are the number of sick days, as well as early retirement due to illness. Unfortunately, there is no data available that provides information on the number of sick days of school principals. However, findings from the economic sector suggest that this indicator is of limited use to describe the health status of executives. In a German study of the health status of executives from the industrial and service sectors, respondents reported only a low number of sick days during the last 12 months ($M=4.8$) (Wilde et al. 2009). The picture changed after considering the number of days respondents went to work despite having a sickness ("presenteeism"), which was more than 8 days. As concluded by the authors, executives are not less frequently ill, but attend work more often while being sick. It can be assumed that admitting health problems and weaknesses does not fit the professional role and external expectations of executives. Moreover, external circumstances (e.g. work demand) may not permit absence from work. To what extent this can be applied to school principals, however, remains unknown and requires further research.

Finally, with regard to early retirement due to illness there are first indications from a German study, which assessed all medical examinations of school principals from 1997 to 1999 in the Bavarian region (Weber et al. 2005). Results indicate that 84 % of the school principals examined by health authorities ($n=342$) were unable to work due to health problems (median age: 58 years). In 45 % of all cases, psychosomatic disorders such as depressive disorders and exhaustion syndromes were the main reasons for early retirement.

While this section focused on the health situation of school principals, we now turn to the second perspective, which deals with the question, as to whether and how school principals and their behavior positively or negatively affect the health of teachers.

12.4 The Influence of School Principals on Teachers' Health

Initial findings from a German study suggest that a potential association between principals' behavior and the health of teachers may depend on the concept of health a school principal has. In this study, Harazd et al. (2009, p. 94) asked 32 school principals about their understanding of teachers' health. Two contrasting cases show significant differences:

- *"You've got me totally stumped. Don't know. Someone who is not sick and reasonably resilient. I'm not a doctor. No idea, don't know."* (male principal)
- *"A healthy teacher is someone who is able to pursue his/her job with joy and who is free to address the children. That's for me a healthy teacher, who experiences his/her job as satisfactory. Someone who not only lives for the job, but also has other interests that satisfies him/her."* (female principal)

While the male principal shows a narrow understanding of health, which is defined as the absence of illness, the female respondent has a holistic concept of health that is oriented on personal fulfillment and satisfaction of needs. On the basis of these results, it can be assumed that school principals with a narrow understanding of health will be less sensitive and active regarding the health of teachers, whereas principals with a broader concept of health will be more interested in the health of their teachers. It is difficult to generalize about a gender effect from these contrasting cases alone. However, our own research with 860 school principals showed evidence that female principals are more sensitive towards mental health-related problems of their teachers and pupils than their male colleagues (Dadaczynski and Paulus 2011).

In examining possible mechanisms through which school principals affect the health of teachers, two factors seem to be important: social support and relationship quality as well as the leadership style of school principals.

Looking at the first mechanism, on the basis of his extensive Potsdam study of teachers' health (n = 17.000), Schaarschmidt (2005) found that the extent of mental health problems and number of sick days of teachers were associated with the level of social support by the school principal. That means that, the more social support that was provided by the principal, the less teachers reported mental complaints and sick days. More detailed information about the impact of specific forms of social support can be found in a study with 385 special and 313 general education teachers from Virginia (Littrell et al. 1994). Firstly, this study suggests a gap between the importance teachers attach to all dimensions of social support (emotional, appraisal, informational, instrumental) and the extent of support that is perceived by the teachers. Secondly, emotional support from the principal (e.g. through positive regard, comfort, and understanding) is perceived as most important and more often received. Thirdly, regression analyses showed that emotional and informational support predicted teachers' job satisfaction. This result is of particular relevance, because as we know from meta-analyses (e.g. Faragher et al. 2005), perceived job satisfaction is associated with a number of health outcomes (particularly burnout, depression, anxiety, self-esteem). Moreover, teachers who perceived more emotional support from the school principals reported fewer health problems. Further evidence comes from Nelson et al. (2001), who examined predictors of occupational stress among 415 teachers of students with emotional and behavioral disorders (EBD). Multiple regression analyses revealed that strong relationships with the principal, a high degree of decision power, good relationships with colleagues, and the ability to work with children exhibiting externalizing behaviour contributed to lower levels of occupational stress.

Closely associated with the first mechanism is the question as to whether, and how, leadership behavior affects the health of teachers. A commonly used typology in leadership research and practice has been suggested by Burns (1978), who distinguishes between transactional and transformational leaders. In short, transactional leadership is characterized by an exchange of outcomes that is of value for both, the superior and the subordinate. Typically, transactional leaders explain to their employees what they expect from them, and what the employee will receive as reward or compensation if the outcomes fulfill the requirements. Thus, this leadership style is based on mutual dependence and extrinsic motivation. In contrast, transformational leadership is more focused on intrinsic needs and motivation, by stimulating positive changes (e.g. an orientation towards organizational goals) among the followers. Transformational leaders achieve this aim in different ways, e.g. by inspiring and intellectually stimulating their employees; by demonstrating strong personal characteristics (e.g. self-confidence); or by articulating shared goals and visions (Bass 1990). This conceptualization has been further elaborated by Bernard M. Bass, who suggests that good leaders can show elements of both leadership styles. Furthermore, he introduced a third leadership style, which is characterized by the avoidance of leading (i.e. laissez-faire leadership) (Bass 1985). Available research on school leadership indicates some significant associations with aspects related to teachers' health. In their meta-analysis of unpublished research, Leithwood and Sun (2012) identified high to moderate effects of transformational leadership on teachers' internal state (weighted mean $r = .57$). Among all teacher variables, transformational leadership had the strongest influence on job satisfaction (.76) or school commitment (.70). These results could be confirmed by a German study, which examined the association between principals' leadership and health related variables in 2.400 teachers (Harazd and van Ophuysen 2011). While transformational and transactional leadership were positively associated with job satisfaction ($r = .50/r = .38$) and commitment ($r = .65/r = .52$), the opposite was found for laissez-faire leadership ($r = -.40/r = -.51$). Moreover, moderate to weak positive associations could be found between wellbeing and transformational leadership ($r = .32$) and transactional leadership ($r = .25$), and negative associations with laissez-faire leadership ($r = -.20$).

Table 12.3 Direct salutogentic leadership (Harazd et al. 2009, p. 127)

Comprehensibility	Transparent decision-making processes
	Promoting information flow
	Clarity of job tasks
	Understandable explanations
Manageability	Match person and job tasks
	Promote self-esteem through feedback and recognition
	Consider individual strengths
	Create optimized organisational structures
	Enable cooperation
	Promote material exchange
Meaningfulness	Act in a goal-oriented way
	Convey visions
	Engage in common goal-setting
	Explain action/provide goal clarity

Finally, based on Antonovsky's sense of coherence, Harazd et al. (2009) proposed a salutogentic leadership style, which can be understood as the ability to promote teachers' sense of comprehensibility, manageability, and meaningfulness. This salutogentic leadership style is expressed in interpersonal situations (direct salutogentic leadership, Table 12.3) as well as on an organizational level. Hierarchical Linear Modeling (HLM) indicates that about 13 % of teachers stress experience can be explained through this style of leadership.

12.5 The Influence of School Principals on School Health Promotion Activities

A large body of studies on educational research indicates that school principals have a strong influence on the success of educational change processes and the effectiveness of the school. School principals are of key importance in initiating school innovations, but also in successfully implementing and anchoring these innovations (Fullan et al. 1980). The same can be assumed with regard to activities on school health promotion, which also present very complex innovations and change processes. Unfortunately, research on school health promotion and prevention has mostly focused on effectiveness by neglecting questions regarding implementation. On the basis of a comprehensive review of over 1.200 prevention programs, Durlak (1997) concludes that only 5 % reported data on implementation. It is only in recent years that research on implementation has gained more attention (e.g. Barry et al. 2005; Fagan and Mihalic 2003). In the next sections, we present some first findings on school principals and their influence on school health promotion activities. Here we differentiate between programs of school health promotion and the holistic health promoting school approach. As opposed to the health promoting school (HPS), a program is understood as a time-bounded activity that is mostly focused on a specific behavioral topic (e.g. bullying) and target group (e.g. primary school pupils).

12.5.1 The Influence of School Principals on Programmes of School Health Promotion

In their study, Rohrbach et al. (1993) examined determinants of program implementation of the *A*dolescent *A*lcohol *P*revention *T*rial (AAPT), an 8-year efficacy trial which aimed at the prevention of substance use among fifth- and seventh grade pupils in Los Angeles. Among other implementation factors (e.g. teacher training), a principal intervention consisting of a 30-min one-to-one meeting about the intervention and its effectiveness was tested. Compared with schools without the principal intervention, schools with trained school principals implemented a greater amount of the program (70 % vs. 49 %).

Further information can be found in the evaluation of the *P*romoting *A*lternative *TH*inking *S*trategies curriculum (PATHS), a delinquency prevention program (Kam et al. 2003). Evaluation results of three schools revealed that high principal support and a high quality of implementation contributed to greater reductions of pupils' aggression, behavioral dysregulation, and significant gains in emotional competencies.

Fagan and Mihalic (2003) reported the results of a process evaluation of the Life Skills Training (LST) drug prevention program from 292 schools and approximately 130.000 pupils. Based on classroom observations and reports of different parties (e.g. teachers, administration, LST instructor), findings show that strong support by the school principal facilitated the implementation of the LST program. Successful schools had an active principal who motivated their teachers to attend the training, attended at the teacher training workshop themselves, observed and even taught program lessons, and informed the teachers about the implementation progress. However, unlike Rohrbach et al. (1993) and Kam et al. (2003), no significant association between principals' support and implementation could be found (see also Mihalic et al. 2008). The authors explain this with the limitations of the measures used and the LST local coordinator who played a very active role in the implementation process.

In a Dutch study, Leurs et al. (2007) examined factors associated with high numbers of health promotion activities addressed by 180 primary school teachers during the previous 12 months (e.g. social skills, diet, mental health). Compared to less active teachers, teachers who addressed more than two health issues per year perceived less disadvantage (e.g. time, responsibility), had a higher self-efficacy, taught higher classes, and perceived more staff support (especially from the school administration).

Finally, Larsen and Samdal (2008) employed personal interviews with principals and teachers of four primary schools to examine the principals' role in implementing and sustaining Second Step, a social skill program that is widely disseminated in

Norwegian primary schools. Results revealed great variance between the schools. While principals from less successful schools exclusively used management strategies (e.g. resource planning), a principal from a school that successfully implemented Second Step combined management and leadership strategies. This included e.g. the communication of common goals and visions, but also the development of a collaborative culture among the teachers. Additionally, the principal from this school spent a significant amount of time preparing the program realization by collaboratively developing an own implementation model. As suggested in previous research (Viig and Wold 2005), program implementation was more successful when anchored at both the top and the bottom levels of the school. Furthermore, principals from successful schools maintain a focus on the program through e.g. giving feedback and supporting teachers, constantly reminding teachers of the need to continue with the program, or familiarizing new teachers with the program.

12.5.2 The Influence of School Principals on the Health Promoting School

Unfortunately, research on the principals' role within the HPS is limited. Based on their narrative review of studies published between 1995 and 2010, Samdal and Rowling (2011), Rowling and Samdal (2011) identified eight components for successful implementation of a health promoting school. One of these components (leadership and management practices) emphasizes school leadership as crucial for implementation success. In accordance with the findings from Larsen and Samdal (2008), a balance of management and leadership strategies is needed in order to enhance implementation quality. Leadership tasks include e.g. to establish readiness for change within the school community, to support teachers and to act as a role model within the change process. Management strategies, on the other hand, encompass the establishment of structures that facilitate organizational change (e.g. resource allocation, time tables). Given the variety of tasks, it has been suggested that leadership should be to distribute on many shoulders, and not just the principal's. This avoids overwhelming school principals, supports school staff participation, and strengthens the quality of relationships within the school (ibid).

In addition to this review, there are a number of studies which are based on many years of experience with the HPS approach. In their study, Inchley et al. (2006) identified four major issues for successful implementation of the HPS in Scotland. While the availability of resources seems a less decisive factor, a sense of ownership and empowerment, leadership and management, collaboration, and integration were most important. Especially in schools in which the principal took the lead, the project gained more status and acceptance. Moreover, principals' participation helped the project to ensure access to resources and establish cooperation with partners outside the school setting.

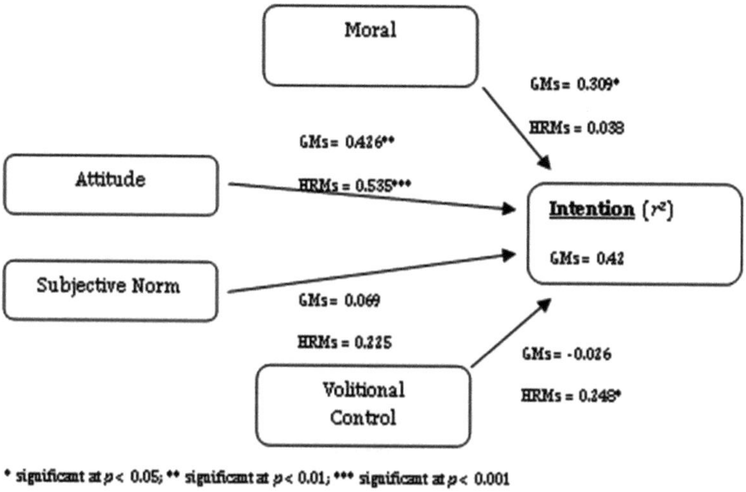

Fig. 12.1 Intention for WHP of general managers (GMs) and human resource managers (HRMs). Downey and Sharp (2007)

These results can be confirmed by the extensive experiences from the Norwegian Network of HPS (Viig et al. 2011). In cases in which the teacher acted as program leader, some problems emerged due to a lack of authority among colleagues and staff. As argued by the authors, this result contradicts one of the core pillars of health promotion, which is based on a bottom-up approach. Rather, as suggested by Larsen and Samdal (2008), a combination of bottom-up and top-down seems more appropriate in implementing the HPS. Further results from the Norwegian Network show that a positive attitude towards the HPS, as well as the ability to motivate the school staff, were important characteristics of principals within the implementation process (Viig and Wold 2005). These factors not only promote the implementation of the project, but also enhance its sustainability. They contributed to the anchoring of the HPS by formalizing activities and fostering external collaboration and networking (Tjomsland et al. 2009).

Outside the school setting, interesting evidence comes from Downey and Sharp (2007), who examined factors that explain general managers' (GMs) and human resource managers' (HRMs) intention to allocate resources to workplace health promotion. Based on the Theory of Planned Behavior the results show that GMs' and HRMs' attitudes (e.g. perceived likelihood of outcomes resulting from Work Health promotion (WHP), desirability of outcomes of WHP) significantly influences their intention, while subjective norm (i.e. importance of approval of the behavior by a referent group, e.g. supervisors) had no influence (see Fig. 12.1). Moreover, GMs were motivated by their moral responsibility (i.e. feeling of personal moral obligation) whereas volitional control (i.e. individuals' perception of control over resource allocation on WHP) was found to be significant only for HRMs. In summary, all included antecedents accounted for 42–59 % of variance of GMs/HRMs intention for WHP.

12.6 Summary and Discussion

School health promotion often fails to take the important step from a project status towards an established school practice. On the basis of the available research summarized in this chapter, we believe that school principals and their different roles are one missing piece in explaining why some schools become and remain a healthy organization whereas others do not.

Although school principals and their influence on school quality have been widely examined in educational research, there is a clear lack of discussion within school health promotion research and practice. Often this discussion is reduced to the claim that school principals in their overall responsibility of the school also need to promote the health of their teachers and pupils. Although we generally agree with this demand, it is often overlooked that the ability to support school health promotion will strongly depend on the health and wellbeing of the school principals themselves. Unfortunately, specific research about the health of this occupational group is still in its infancy. However, there is a lot to suggest that health concerns of school principals have increased over the last years. Much of this can be explained by the working context, which changed rapidly due to social development and school reforms. In particular, the worldwide trend towards more decentralization caused an increase of responsibilities and work tasks of school principals, which in turn may result in more work-related stress and worse health outcomes.

When summarizing the research results concerning the health of principals it should be taken into account that the comparability of the studies is limited due to methodological, cultural and political differences. However, there is some evidence suggesting that compared to physical health problems, school principals more frequently suffer from mental health problems such as psychosomatic complaints, anxiety, depression or symptoms of burnout. Moreover, it seems that female and primary school principals have worse mental health outcomes than male and secondary school principals. One explanation for the school type effect could be that principals from smaller schools such as primary schools typically have less administrative support (e.g. secretariat), a smaller school management team, more teaching duties, and closer ties with pupils and teachers that make them more vulnerable to relationship problems. With regard to other occupational groups, research findings are mixed. With the exception of one study from the UK, it seems that compared to non-school occupations, school principals have a poorer health status. However, compared to teachers, results also revealed that principals have a higher wellbeing and are less often affected by emotional exhaustion. It may be argued here that school principals have more resources at work (e.g. autonomy, room to maneuver) that 'buffer' experiences of work-related stress. Unfortunately, there is a clear lack of studies that examine the extent of resources and their effect on the health of school principals.

To summarize the second perspective of this chapter, available research results suggest that school principals have a significant influence on the health of their teachers. School teachers who perceive more support (especially emotional support) from, and good relations with, their principals report fewer health problems, including fewer sick days. Furthermore, compared to an absence of leadership (laissez-faire leadership), leadership behaviour that is focused on intrinsic needs and motivation, e.g. by articulating shared goals and visions or inspiring and cultivating intellectual stimulation seems to be more favourable in terms of teachers' wellbeing, job satisfaction and school commitment. It might be assumed that the ability to provide (emotional) support and to show health promoting leadership behavior (e.g. direct salutogentic leadership) depends on self-management strategies and the health of the principals. Unfortunately, we found no studies on this issue and, hence, can only hypothesize about these relationships. Clearly, future research is warranted. We did find, however, some first indications that principals' conceptualizations regarding their own health could be a possible third variable in explaining the influence of school principals on teachers' health.

Finally, available research findings suggest that school principals are of vital importance for the successful implementation and sustainability of programs on school health promotion and the health promoting school. Successful principals combine leadership (e.g. communication of goals and vision) and management strategies (e.g. resource allocation) to support health promotion activities in their school. Moreover, they build an effective communication strategy, develop a specific implementation plan, attend training workshops, maintain their support over time (e.g. remind teachers to continue health promoting activities, familiarize new teachers with health promotion), or formalize activities (e.g. include activities in the school curriculum). As argued by Larsen and Samdal (2008), it is not the presence of all these aspects that contributes to the success of school health promotion activities. Rather, it is the way these aspects are addressed by the principals' management and leadership style. Interestingly, contrary to the principles of health promotion, the health promoting school approach seems more effective when based on a simultaneous bottom-up and top-down approach. This also implies that leadership should be broadly distributed, including teachers, parents and pupils. Here again the question emerges as to whether and how the health status of school principals affects the ability to support activities on school health promotion. Finally, as visible through the brevity of chapter 12.1.6.2, the number of available publications examining the influence of school principals on the health promoting school is very limited. To stimulate further research, findings from other areas (such as workplace health promotion) should be used. The study from Downey and Sharp (2007) is such an example, showing that the Theory of Planned Behaviour is useful in predicting managers' intentions towards workplace health promotion. Whether these results can be transferred to the school setting is currently being examined in a study conducted by the authors.

12.7 Conclusions and Future Perspectives

In this chapter, we tried to illuminate the health of principals and their role in school health promotion. Based on the findings presented, we think it is time to widen our thinking of the health promoting school approach by including school principals and their role within this approach more explicitly in research and practice. Specifically, the following aspects should receive more attention in future research and practice. Firstly, more research on principals' health and its determinants is needed. To ensure comparability, research on principals' health should be transnational and based on the same methodology. Research instruments should encompass short-term and long-term health indicators as well as positive (i.e. resource oriented) measures of health. Secondly, health promotion in schools should explicitly aim at school principals. To ensure this, specific evidence based interventions that aim to strengthen the health of principals need to be developed, implemented and integrated in the health promoting school approach. Thirdly, to ensure success and sustainability, each activity in school health promotion needs to support school principals in their role as school leaders and managers (e.g. through specific trainings, material, supervision, communication tools). Furthermore, to enhance implementation research, activities on school health promotion should evaluate the extent to which school principals influence its success. Fourthly, health promotion and prevention should be an inherent part of initial and continuing professional training of school principals. This should encompass developing the capability for health promoting self-leadership, but also a greater sensitivity for health-related issues within the school (e.g. relationship between health and education or leadership and teachers' health) or the capacity to support health promotion activities. Finally, we need more research about the connections between the different perspectives presented here. This includes e.g. the question as to whether and how the health status of school principals influences the ability to support health promotion in the school or the relationship quality with teachers, non-teaching staff and pupils.

References

Antonovsky, A. (1987). *Unraveling the mystery of health: How people manage stress and stay well*. San Francisco: Jossey-Bass.

Barry, M. M., Domitrovich, C., & Lara, M. A. (2005). The implementation of mental health promotion programmes. *Promotion & Education, 12*(Suppl. 2), 30–36.

Bass, B. M. (1985). *Leadership and performance beyond expectations*. New York: Free Press.

Bass, B. M. (1990). From transactional to transformational leadership: Learning to share the vision. *Organizational Dynamics, 18*, 19–31.

Borg, M. G., & Riding, R. J. (1993). Occupational stress and job satisfaction among school administrators. *Journal of Educational Administration, 31*, 4–21.

Burns, J. M. (1978). *Leadership*. New York: Harper & Row.

Carr, A. (1994). Anxiety and depression among school principals – Warning, principalship can be hazardous to your health. *Journal of Educational Administration, 32*, 18–34.

Chaplain, R. P. (2001). Stress and job satisfaction among primary headteachers. A question of balance? *Educational Management & Administration, 29*, 197–215.

Cheng, Y. C. (2002). The changing context of school leadership. In K. Leithwood & P. Hallinger (Eds.), *Second international handbook of leadership and administration* (pp. 103–132). Dordrecht: Kluwer Academic Publishing.

Combs, J., Edmonson, S. L., & Jackson, S. H. (2009). Burnout among elementary school principals. *AASA Journal of Scholarship and Practice, 5*, 10–15.

Cooper, C. L., & Kelly, M. (1993). Occupational stress in head teachers: A national UK study. *British Journal of Educational Psychology, 63*, 130–143.

Dadaczynski, K. (2012). Die Rolle der Schulleitung in der guten gesunden Schule. In DAK-Gesundheit & N. R. W. Unfallkasse (Eds.), *Handbuch Lehrergesundheit – Impulse für die Entwicklung guter gesunder Schulen* (pp. 197–228). Köln: Carl Link.

Dadaczynski, K., & Paulus, P. (2011). Psychische Gesundheit aus Sicht von Schulleitungen: Erste Ergebnisse einer internationalen Onlinestudie für Deutschland. *Psychologie in Erziehung und Unterricht, 58*, 306–318.

Darmody, M., & Smyth, E. (2011). *Job satisfaction and occupational stress among primary school teachers and school principals in Ireland. A report compiled by the ESRI on behalf of the teaching council.* http://tinyurl.com/7ldcyq2. Accessed 14 Mar 2012.

Devos, G., Bouckenooghe, D., Engels, N., Hottin, G., & Aelterman, A. (2007). An assessment of well-being of principals in Flemish primary schools. *Journal of Educational Administration, 45*, 33–61.

Dewa, S. C., Dermer, S. W., Chau, N., Lowrey, S., Mawson, S., & Bell, J. (2009). Examination of factors associated with the mental health status of principals. *Work, 33*, 439–448.

Downey, A. M., & Sharp, D. J. (2007). Why do managers allocate resources to workplace health promotion programmes in countries with national health coverage. *Health Promotion International, 22*, 102–111.

Durlak, J. A. (1997). *Successful prevention programs for children and adolescents.* New York: Plenum.

Evans, R. (1995). Getting real about leadership. *Education Week, 14*, 29:36.

Fagan, A. A., & Mihalic, S. (2003). Strategies for enhancing the adoption of school-based prevention programs: Lessons learned from the blueprints for violence prevention replications of the life skills training program. *Journal of Community Psychology, 31*, 235–253.

Faragher, E. B., Cass, M., & Cooper, C. L. (2005). The relationship between job satisfaction and health: A meta-analysis. *Occupational and Environmental Medicine, 62*, 105–112.

Friedman, I. A. (2001). Burnout in school principals: Role related antecedents. *Social Psychology of Education, 5*, 229–251.

Fullan, M. G., Miles, M. B., & Taylor, G. (1980). Organizational development in schools: The state of the art. *Review of Educational Research, 50*, 121–183.

Gmelch, W. H., & Gates, G. (1998). The impact of personal, professional and organizational characteristics on administration burnout. *Journal of Educational Administration, 36*, 146–159.

Gmelch, W. H., & Swent, B. (1981). Stress and the principalship: Strategies for self-improvement and growth. *NASSP Bulletin, 65*, 16–19.

Hagemann, K., Jarausch, K. H., & Allemann-Ghionda, C. (Eds.). (2011). *Children, families, and states: Time policies of childcare, preschool and primary education in Europe.* Oxford: Berghahn Books.

Harazd, B., & Ophuysen van, S. (2011). Transformationale Führung in Schulen. Der Einsatz des „Multifactor Leadership Questionnaire“(MLQ 5 x Short). *Journal for Educational Research Online, 3*, 141–167.

Harazd, B., Gieske, M., & Rolff, H.-G. (2009). *Gesundheitsmanagement in der Schule. Lehrergesundheit als neue Aufgabe der Schulleitung*. Köln: Wolters Kluwer.

Hasselhorn, H.-M. (2009). Arbeitsbedingte psychische Erschöpfung bei Lehrern im Vergleich mit anderen Berufsgruppen. In U. Kliebisch & R. Melofski (Eds.), *LehrerGesundheit. Anregungen für die Praxis* (pp. 73–82). Baltmannsweiler: Schneider Verlag Hohengehren.

Hasselhorn, H.-M., & Nübling, M. (2004). Arbeitsbedingte psychische Erschöpfung bei Erwerbstätigen in Deutschland. *Arbeitsmedizin - Sozialmedizin - Umweltmedizin, 39*, 568–576.

Huber, S. G. (1999). Was macht Schule wirksam? *Schul-Management, 30*, 10–17.

Huber, S. G. (2004). School leadership and leadership development – Adjusting leadership theories and development programs to values and the core purpose of school. *Journal of Educational Administration, 42*, 669–684.

Huber, S. G. (2007). Schulleiter/-innen – multifunktionale Wunderwesen oder professionelle Teamarbeiter? Anforderungen an Schulleitung. *SchVw Spezial, 3*, 32–35.

Hüfner, G. (2010). *Befragung von Schulleitungen 2010 an bayrische grund-, Haupt- und Förderschulen*. München: Bayrischer Lehrerinnen- und Lehrerverband.
Inchley, J., Muldoon, J., & Currie, C. (2006). Becoming a health promoting school: Evaluating the process of effective implementation in Scotland. *Health Promotion International, 22*, 65–71.
Johnsen, S., Cooper, C., Cartwright, S., Donald, I., Taylor, P., & Millet, C. (2005). The experience of work-related stress across occupations. *Journal of Managerial Psychology, 20*, 178–187.
Kam, C.-M., Greenberg, M. T., & Walls, C. T. (2003). Examining the role of implementation quality in school-based prevention using the PATHS Curriculum. *Prevention Sciences, 4*, 55–63.
Landert Partner. (2009). *Arbeitssituation der Schulleitungen im Kanton Thurgau. Evaluation.* Zürich: Landert Partner.
Landsmann, L. (1978). Warning to principals: You may be hazardous to your teachers' health. *National Elementary Principal, 57*, 69–72.
Larsen, T., & Samdal, O. (2008). Facilitating the implementation and sustainability of Second Step. *Scandinavian Journal of Educational Research, 52*, 187–204.
Lazarus, R. S., & Folkman, S. (1984). *Stress, appraisal, and coping*. New York: Springer.
Leithwood, K. (1994). Leadership for school restructuring. *Educational Administration Quarterly, 30*(4), 498–518.
Leithwood, K., & Day, C. (2007). Starting with what we know. In C. Day & K. Leithwood (Eds.), *Successful principal leadership in times of change* (pp. 1–15). New York: Springer.
Leithwood, K., & Sun, J. (2012). The nature and effects of transformational school leadership: A meta-analytic review of unpublished research. *Educational Administration Quarterly*. doi:10.1177/0013161X11436268.
Leurs, M., Bessems, K., Schaalma, H. P., & de Vries, H. (2007). Focus points for school health promotion improvements in Dutch primary schools. *Health Education Research, 22*, 58–69.
Littrell, P. C., Billingsley, B. S., & Cross, L. H. (1994). The effects of principal support on special and general educators' stress, job satisfaction, school commitment, health, and intent to stay in teaching. *Remedial and Special Education, 15*, 297–310.
Mackler, J. (1996). A survivor's guide to the principalship: Overcoming the challenges. *NASSP Bulletin, 80*, 84–89.
Maslach, C., & Jackson, S. E. (1981). *Maslach burnout inventory ("Human Services Survey")*. Palo Alto: Consulting Psychologists Press.
Mihalic, S. F., Fagan, A. F., & Argamoso, S. (2008). Implementing the LifeSkills training drug prevention program: Factors related to implementation fidelity. *Implementation Science, 3*, 5.
Nelson, J. R., Maculan, A., Roberts, M. L., & Ohlund, B. J. (2001). Sources of stress for teachers of students with emotional and behavioral disorders. *Journal of Emotional and Behavioral Disorders, 9*, 123–130.
Phillips, S. J., Sen, D., & McNamee, R. (2007). Prevalence and causes of self-reported work related stress in head teachers. *Occupational Medicine, 57*, 367–376.
Phillips, S. J., Sen, D., & McNamee, R. (2008). Risk factors for work-related stress and health in head teachers. *Occupational Medicine, 58*, 584–586.
Pont B., Nusche D., & Moormann H. (2008). *Improving school leadership. Volume 1: Policy and practice*. www.sourceoecd.org/education/9789264044678. Accessed 11 July 2012.
Rohrbach, L. A., Graham, J. W., & Hansen, W. B. (1993). Diffusion of a school-based substance abuse prevention program: Predictors of program implementation. *Preventive Medicine, 22*, 237–260.
Rowling, L., & Samdal, O. (2011). Filling the black box of implementation for health-promoting schools. *Health Education, 111*, 347–366.
Samdal, O., & Rowling, L. (2011). Theoretical and empirical base for implementation components of health-promoting schools. *Health Education, 111*, 367–390.
Schaarschmidt, U. (Ed.). (2005). *Halbtagsjobber? Psychische Gesundheit von Lehrerinnen und Lehrern. Analyse eines veränderungsbedürftigen Zustands (2. Auflage)*. Weinheim: Beltz.

Scheerens, J., & Bosker, R. J. (1997). *The foundations of educational effectiveness*. Oxford: Elsevier Science Ltd.

Tjomsland, H. E., Larsen, T. M. B., Viig, N. G., & Wold, B. (2009). A fourteen year follow-up study of health promoting schools in Norway: Principals' perceptions of conditions influencing sustainability. *The Open Educational Journal, 2*, 54–64.

Townsend, T. (Ed.). (2007). *International handbook of school effectiveness and improvement*. Dordrecht: Springer.

Viig, N. G., & Wold, B. (2005). Facilitating teachers' participation in school-based health promotion – A qualitative study. *Scandinavian Journal of Educational Research, 49*, 83–109.

Viig, N. G., Fosse, E., Samdal, O., & Wold, B. (2011). Leading and supporting the implementation of the Norwegian Network of Health Promoting Schools. *Scandinavian Journal of Educational Research*. doi:10.1080/00313831.2011.621139.

Warwas, J. (2009). Berufliches Selbstverständnis und Beanspruchung in der Schulleitung. *Zeitschrift für Erziehungswissenschaften, 12*, 475–498.

Weber, A., Weltle, D., & Lederer, P. (2005). Ill health and early retirement among school principals in Bavaria. *International Archives of Occupational and Environmental Health, 78*, 325–331.

Whitaker, K. S. (1992). Principal burnout and personality type: Do relationships exist? *The Record in Educational Administration and Supervision, 14*, 87–95.

Whitaker, K. S. (1996). Exploring causes of principal burnout. *Journal of Educational Administration, 34*, 60–71.

Wilde, B., Bahamondes-Pavez, C., Hinrichs, S., & Schüpbach, H. (2009). Gesundheit und Arbeitsbedingungen von Führungskräften auf der unteren und mittleren Hierarchieebene – Konsequenzen neuer Steuerungsformen. In GfA (Ed.), *Arbeit, Beschäftigungsfähigkeit und Produktivität im 21. Jahrhundert* (pp. 351–354). Dortmund: GfA Press.

Williamson, J., & Campbell, L. (1987). Stress in the principalship: What causes It? *NASSP Bulletin, 71*, 109–112.

Chapter 13
Three Approaches to School Health Education as a Means to Higher Levels of Health Literacy

Leena Paakkari

Abstract It is widely acknowledged that school health education is an important means for supporting the development of pupils' health literacy. A growing number of papers have described or suggested a variety of classroom-based and whole-school practices for developing health literacy. However, few of these papers have systematically addressed the various ways of approaching health education in schools or sought to analyze how these approaches differ from each other. This paper aims to do this. It does so by representing three approaches to school health education: the *facts and skills approach*, the *individual thinking approach*, and the *personal growth and citizenship* approach. The approaches differ in complexity. They can be used in planning for learning experiences aimed at supporting the development of higher levels of health literacy. Furthermore, they can be used in teacher training when the aim is to help teacher trainees to become aware of their current ways of seeing school health education, and the differences that may exist between their understanding and more complex forms of understanding.

Keywords Health education • Schools • Health literacy • Learning

13.1 Introduction

It is widely acknowledged that school health education is an important means for supporting the development of pupils' health literacy (see Benham-Deal et al. 2010; Hubbard and Rainey 2007; Nutbeam 2000; St Leger 2001). In schools, health education is often linked to classroom-based teaching of health issues, which aims at to promote the learning of health literacy through instruction and formative assessment (Benham-Deal and Hodges n.d.). However, if we consider health education to be any

L. Paakkari, Ph.D. (✉)
Faculty of Sport and Health Sciences, University of Jyväskylä, Jyväskylä, Finland
e-mail: leena.paakkari@jyu.fi

V. Simovska and Patricia Mannix-McNamara (eds.), *Schools for Health and Sustainability: Theory, Research and Practice*, DOI 10.1007/978-94-017-9171-7_13,

intentional activity aimed at developing health literacy among pupils, we have to recognize that such learning takes place also within the broader school context. A whole-school approach, such as a health-promoting school initiative, will include classroom-based education but will also expand its focus towards "changing the social and physical environment of the school, and creating links with the wider community" (Stewart-Brown 2006, p. 4). This is done to strengthen the capacity of the schools "as a healthy setting for living, learning and working" (World Health Organization 1998, p. 2). The improvement of health literacy is one of the main goals of modern school health programs along with the other goals, in particular the improvement of health behaviors, educational achievement and social outcomes (Kolbe 2005). Taking into account that the school is first and foremost an educational institution with educational goals, it can be argued that health literacy is the fundamental goal of a whole-school approach (see Kolbe 2002; see also Benham-Deal and Hodges n.d.).

Health literacy has been defined in varying ways, but most definitions focus on people's abilities to be or become empowered to take care of their own health (Kickbusch 2008; Wu et al. 2010), and also the health of others (St Leger 2001; see also Nutbeam 2000, 2008). This implies that health literacy is more than the individual's "capacity to obtain, process and understand basic health information and services needed to make appropriate health decisions" (Ratzan and Parker 2000, p. vi). It is also the capacity to change living conditions so as to contribute to better health for oneself and others (Nutbeam 1998; see also Abel 2007). This broader perspective is also what schools aim at when promoting health literacy among pupils: to educate pupils to be critical and active citizens who will be able to seek, evaluate, and construct knowledge, and in addition be able to make ethically-responsible decisions and take actions that are beneficial not only to themselves, but also to other people and the community (Paakkari and Paakkari 2012). With this goal in view, Paakkari and Paakkari (2012) have suggested that as a desired educational outcome in schools health literacy is made up of five core components: *theoretical knowledge*, *practical knowledge*, *critical thinking*, *self-awareness*, and *citizenship*. They argue that in the best possible situation, all the components will be intertwined with each other, and that the role of the schools is to develop all of them. However, not all educational practices support the acquisition of all the competences mentioned above.

There have been an increasing number of papers describing and suggesting a variety of classroom-based and whole-school practices for developing health literacy (e.g. Marx et al. 2007; Nutbeam 2000; St Leger 2001; Tappe and Galer-Unti 2001; see also Brey et al. 2008). However, for the most part these papers have not attempted to give a systematic description of the various ways of approaching health education in schools, or to examine how these approaches might differ from each other in terms of (for example) their learning objectives, the role of the pupils, and the role of the teacher (but cf. Nutbeam 2000; Paakkari 2012; St Leger 2001). This paper aims to address this gap in the field. It does so by presenting three approaches to school health education—a *facts and skills approach*, an *individual thinking approach*, and a *personal growth and responsibility approach*. These differ from each other in their complexity, and more specifically in terms of the aims, the teacher's role, the pupils' role, the nature of that knowledge that will be dealt with during the teaching-learning events, and the nature of the reflection that pupils are expected to carry out.

The approaches listed here were built into the context of health education (as a school subject) within the PhD thesis of Paakkari (2012), in which the three approaches were developed. Thus Paakkari encompassed the conceptions of teacher trainees relating to (i) health education as a school subject, (ii) its teaching, and (iii) its learning. The conceptions were examined within a phenomenographic research tradition, within which the conceptions represent qualitatively varying ways of seeing something. Using empirical and logical evidence, the conceptions can be put into a hierarchy based on how comprehensive, sophisticated, or advanced they are (see Marton and Booth 1997). The *facts and skills approach* represents the least comprehensive, sophisticated or advanced view of seeing health education, whereas the *personal growth and responsibility approach* represents the most comprehensive, sophisticated or advanced view of what health education is and does. Although the approaches were examined in the context of health education as a school subject, they can be used to describe different ways of seeing school health education in general. In fact, in this paper the term 'health education' should not to be limited to one particular school subject, or even merely to education taking place in the classroom; rather it should be seen as any activity that is intended to promote the learning of health issues in a school context. Hence, the approaches in question also serve as a framework for examining the teaching and learning of health literacy, located within an overall health-promoting school context. Below, the approaches will be described in more detail, as will the various components of health literacy supported by each approach.

Figure 13.1 summarizes the essential aspects of each of the approaches mentioned above. The aspects in question represent the critical differences between the various approaches, and at the same time they highlight what teachers should discern and focus on when aiming to direct attention and activity towards a particular approach.

13.2 Three Approaches to Health Education

13.2.1 The Facts and Skills Approach

The *facts and skills approach* to health education covers an understanding of health education in which the aim is to promote pupils' factual and practical knowledge capital. Alternatively, one could say that the aim is to support factual and practical knowledge as the core components of health literacy.

Theoretical knowledge could be described as involving all-round education on (or factual knowledge about) health matters (see Bereiter and Scardamalia 1993; Tynjälä 2008a, b). The acquisition of such knowledge serves as a basis for other

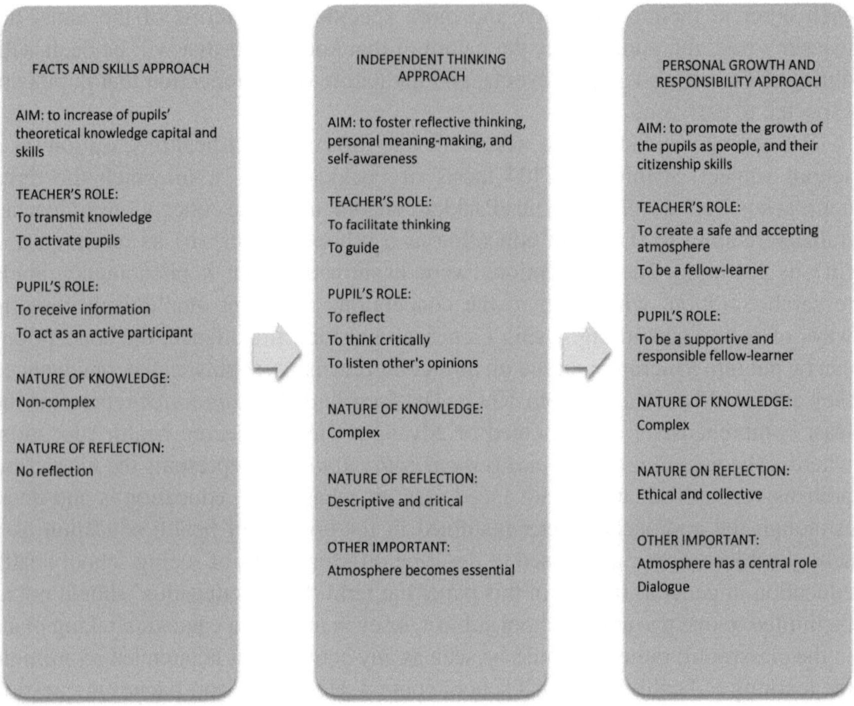

Fig. 13.1 Summary of the critical aspects of the three approaches to health education (Paakkari 2012)

components, since it may, for instance, help pupils to understand a phenomenon more deeply, and to create links between separate health issues (Paakkari and Paakkari 2012). Practical knowledge on the other hand—also referred to as skills or procedural knowledge (Bereiter and Scardamalia 1993; Tynjälä 2008a)—includes the basic health-related skills that pupils should acquire in order to behave in a way that enhances their health; these might include for example the ability to brush their teeth properly, follow safety traffic regulations, and give first aid. In this approach, the skills that pupils learn are the kind that will mainly be necessary or important for pupils later in their lives.

The facts and skills approach thus involves two modes of seeing pupils' learning: (i) learning as gaining and reproducing acquired health knowledge (cf. Paakkari et al. 2011), and (ii) learning as the application of health knowledge. At the same time, the pupil's role expands from that of being merely a knowledge recipient to that of an active participant—since the acquisition of skills does after all call for practical exercises. However, since the skills are for later use, i.e. to be taken and used on some indefinite future occasion, learning is not truly situated within the pupils' experiences, and pupils are not supported so that they truly reflect on the knowledge from the point of view of their own lives. In fact, the role of reflection during the teaching-learning situation is not emphasized. Thus, it could be said that this approach regards knowledge as non-complex or non-problematic, that is, as something that can be said to be simply right or wrong. This way of seeing can be related to the argument that people who regard knowledge as certain and unchangeable, and teachers as experts who exert authority, tend to hold traditional (as opposite to constructivist) conceptions of teaching and learning (Aypay 2010; Chan and Elliott 2004; Otting et al. 2010). According to Tsai (2007) and Muis and Foy (2010), teachers who understand knowledge as something certain and non-complex (thus taking a positivist view) tend to pursue teacher-focused practices. Moreover, within the positivist perspective, health as a concept may be looked at from various dimensions (physical, social, and psychological), but is not treated as something personal and perceived (in terms of the concept of "perceived health").

In accordance with this perspective, the role of the teacher is to keep a firm grip of the teaching and learning. Teaching could be described as the transmission of knowledge when the role of the teacher is mainly to deliver the health knowledge he or she regards as important to the pupils. St Leger (2001) has argued that purely providing knowledge to pupils, in line with a top-down approach to health education, only rarely results in sustainable changes to health behavior. It is true that when the focus of the teaching moves to supporting the active processing of knowledge through application, the role of the teacher expands to cover aspects such as that of inspirer and tutor. However, even if the teacher does inspire pupils for example to seek out information, this will tend to take place within the reading options the teacher has chosen. Begoray et al. (2009) have demonstrated how pupils perceive such teaching as encouraging dependence on the teacher rather than independent acting and thinking. In so far as the participation of the pupils is supported only within the framework of what the teacher has planned and what he or she aims to achieve (with ready-planned outcomes), we could see such participation as no more than token (Simovska 2000, 2004).

It is likely that teachers possessing an understanding of knowledge as something certain and non-complex will support the development of pupils' understanding of knowledge and knowing in a corresponding direction (cf. Sormunen 2004). It is therefore fully to be expected that the facts and skills approach will help pupils to gain knowledge of many health matters (factual knowledge) and will develop a number of health skills (practical knowledge). However, pupils may not gain critical insights into health issues (critical thinking) or understand how separate issues are linked to their own lives, and they may be uncertain as to whether a certain form of health-promoting behavior actually suits them personally (self-awareness). For example, pupils may be unable to evaluate whether they themselves wish to smoke, or whether they smoke only because of pressure from their peers.

When the aim is to support the learning of health literacy in terms of factual knowledge, the teaching-learning events are usually built around teacher- or expert-led lectures, bulleted lists of items, or other activities that focus on helping pupils to receive and recall information from books, posters, or a teacher, and on promoting lower level thinking skills such as listing, naming and describing (see Biggs and Tang 2011). Support for practical knowledge calls for learning-by-doing activities (e.g. practicing first aid skills) or other means of advancing pupils' application skills.

13.2.2 The Independent Thinking Approach

The *independent thinking approach* focuses on promoting pupils' critical and reflective thinking. Now, the aim is to support pupils so that they understand health issues from their own personal perspectives and create personal meanings. Thus, this approach highlights the focus on what a pupil learns *through* health content—not what a pupil learns *about* health content as was the case in the previous approach. At the same time, the focus on enhancing the development of health literacy expands to cover critical thinking and self-awareness as the core components of health literacy (see Paakkari and Paakkari 2012).

Critical thinking as an ability enables pupils to distinguish the conditions that promote health from those that do the opposite (see Abel 2007). In addition, critical thinking enables pupils to identify and work out causal relationships, to set out firm arguments, to assess assumptions, to judge the credibility of arguments, and ultimately, to make sound health decisions after pondering various options (see Fisher 2001). ten Dam and Volman (2004) add that critical thinking enables people to contribute to society "in a critical and aware manner" (pp. 370–371).

Self-awareness as a core component of health literacy involves the ability to inquire into and evaluate one's own thoughts, feelings, and behavior (Grant et al. 2002). In addition, it involves the ability to inquire into one's character, with its strengths and weaknesses (World Health Organization 1994, p. 2). Self-awareness as a health literacy component may focus on the self in general, or on the self as a learner (metacognitive knowledge, also called self-regulatory knowledge; see Bereiter and Scardamalia 1993). The self-in-general aspect is particularly important, since it promotes personal meaning-making in health issues and awareness of the often tacit (implicit) routines of daily living; it also serves as a precondition for interpersonal relationships and communication, and for the development of empathy (World Health Organization 1994, p. 2). However, in this approach the reflection focuses on the self and not on others (as is the case in the following approach). The self-as-learner aspect enables pupils to set learning goals for themselves, to adopt various strategies to reach these goals, to monitor the learning process, to manage time, and to attribute reasons for what has happened in their learning (see Zimmerman 2002).

The *independent thinking approach* to health education includes two ways of seeing learning: (i) emphasizing the pupils' ability *to develop* their own meanings, and (ii) emphasizing the widening of horizons, that is, the *transformation* of thinking. Central to both modes of learning is the need for reflective thinking. However, the nature of the reflection differs in each case. When the focus is on developing meanings, it is expected that the pupils will demonstrate descriptive reflection—asking *how* and *what* questions (How does this health matter relate to my life? What do I think about this particular topic? What are my current ways of behaving in a health-promoting manner?). In fact, questions like these relate to the larger question of "Who am I?" In asking this question, the pupils mainly draw a picture of themselves (their ways of thinking and behaving), but do not yet critically evaluate the assumptions underlying these thoughts and behaviors. At the same time, a transformation in thinking requires pupils to be able to show critical reflection, that is, to ask *why* questions: Why do I think or behave as I do? Why should I make this decision and not any other? Furthermore, critical thinking here moves from merely assessing health issues from various perspectives towards assessing those issues from the point of view of one's own values, and "daring" to express a differing point of view (ten Dam and Volman 2004).

Along with the above, the role of the pupils will expand (from what it was in the facts and skills approach) to cover aspects of, first, the *reflective*, and later the *critical* thinker. Within the learning process, the opinions of others play a crucial role. When the focus is on developing meanings, the perspectives of others enable one to become aware of one's own perspectives and those of others. Hence, it becomes accepted that there may be different ways of seeing the same thing. However, as the focus moves to seeing learning as a widening of horizons or as understanding something in a new way, the perspectives of others serve as an essential means for changing one's own ways of thinking. It is important that these perspectives are evaluated carefully; in other words, that they are not merely accepted without question, and this questioning clearly calls for critical thinking skills. Here, we may see elements of the ideas of ten Dam and Volman (2004), suggesting that the development of critical thinking is a *social* process. Moreover, if the pupils learn to separate their own hopes and wishes from those of their parents and friends, and if they succeed in balancing the expectations of others with their own aspirations—or as Baxter Magolda (2001) puts it "letting go of external control and beginning to replace it with one's internal voice" (p. 94)—they will become able to think critically and define their own values.

Self-reflective and critical thinking is important because the knowledge that is handled in the teaching and learning of health issues is complex, problematic, and uncertain. In this case the knowledge is not just something that exists in books, but also something that exists in oneself and in others. Thus, "certainty" (in whatever degree) will not be evaluated only against scientific truth, but also against personal "truth." At the same time, people's health will be understood as personal, perceived, and relational, and not merely as something to be diagnosed. What is best for one person's health is not necessarily best for another person.

The focus on supporting pupils' individual and independent thinking requires teachers to challenge pupils to think critically, to ponder matters from their own personal perspectives, and to create their own points of view. Moreover, teachers should encourage pupils to construct arguments for their opinions. Tones (2005) argues that if education does not encourage people to think critically in order to make sound decisions, it will be nothing more than instruction, training, or simple brainstorming. Support for the expression of one's own thoughts calls for an environment that is tolerant and safe, in other words, an environment that allows pupils to have misconceptions, to make mistakes, and to learn from them (see Biggs and Tang 2011, p. 27). As compared to the previous approach, the teacher is seen as loosening his/her grip, allowing personal interpretations, situating the issue (the object of learning) within pupils' personal experiences, and validating pupils as knowers (see Baxter Magolda 1999). This view is in line with the findings of

Begoray et al. (2009), who reported that the pupils themselves showed a desire to personalize information and develop their own meanings; receiving overly generalized health information was perceived as insufficient. If one accepts the link between the teacher's and the pupils' epistemic beliefs, it can be said that the independent thinking approach to health education may ultimately help pupils to "think for themselves and to discover their own distinctive voice" (van Rossum and Hamer 2010; see Baxter Magolda 2001).

The independent thinking approach calls for teaching practices that are clearly different from those in the previous approach, since the learning of critical thinking and self-awareness as core components of health literacy cannot be advanced merely by continuing with lower level thinking skills. The development of critical thinking calls for practices that focus on higher level thinking skills such as analyzing, contrasting, prioritizing, and rating. These could involve activities such as problem solving, debating, or evaluation of arguments. The promotion of self-awareness calls for practices in which pupils relate issues to their own lives (for example though compiling portfolios and sleeping/eating diaries, with accompanying critical reflection), and evaluate their own learning. The learning of critical thinking and self-awareness requires both appropriate prior knowledge and accessible new knowledge (cf. Biggs and Tang 2011, p. 27), in other words, both the having and gaining of new factual knowledge.

13.2.3 The Personal Growth and Responsibility Approach

The *personal growth and responsibility approach* to school health education represents the most complex approach to health education. The approach combines an understanding of teaching and learning that sees health education as aiming to promote the growth of the pupils as people, while at the same time promoting their citizenship skills. In this context, the focus of the teaching and learning is on supporting pupils' *growth through learning about themselves, others, and the world*, and *progress towards shared meanings through growing with others in dialogue.* The aims are connected with developing pupils' citizenship as a core component of health literacy (see Paakkari and Paakkari 2012). In this context, citizenship is understood as the pupils' abilities to act in an ethically-responsible way and to take social responsibility. It is essential for pupils to gain these abilities if they are to promote community-level health (Abel 2007). Through having citizenship, they may be able to work for personal and community health (Benham-Deal and Hodges n.d.), influencing the policies and environments that affect their own and others' health (World Health Organization 1997). Paakkari and Paakkari (2012) put the matter thus:

> This highlights the importance of students being able to understand their rights and responsibilities, and also to be aware of the effects of their thoughts and actions on other people and the world at large. The point is that students should be able to consider health matters beyond their own perspective: to think of what other people or we (as a group or as a society) regard as important, what could be done to improve their or our health and well-being. In

> other words, students should become health literate about themselves in relation to others, understanding the perspectives of others and of the collective (p. 139).

This approach to teaching contains two modes of seeing learning, namely *learning as personal growth*, and *learning as collective meaning-making*. Growth refers to growing as a person through seeing something in a different way, through being a supportive fellow-learner, and through dialogue with others in a learning community. Thus, the role of the pupils opens up not only towards the content to be learned or towards one's self (as was the case in the previous approach) but also towards others (see Hoveid and Hoveid 2008): pupils are seen as having the capacity to be supportive fellow-learners and responsible members of the community. This calls for ethical reflection, that is, an ability to consider "the appropriateness of a variety of health-related practices […] and to show empathy when trying to understand something from the point of view of others" (Paakkari and Paakkari 2012, p. 145). However, ethical reflection seeks not merely self-understanding, but also a collective or shared meaning in the context of reflective discussion (Kwak 2007), which in turn calls for collective reflection viewed as meaning-negotiation within a dialogue (see De Lawter and Sosin 2000).

Collective reflection requires a genuine dialogue, a discussion in which all members may share their own ideas and experiences, learn about differences, and create a collective understanding. Lodge (2005) argues that dialogue is about engaging with others in building a shared narrative, one that ends up at a point that the individual would not have reached alone. Here, we may indeed see the elements of viewing knowledge also as something that is socially-constructed—as compared to the previous approach in which knowledge was developed individually. Dialogue may support the participation of pupils in decision-making in the present and not merely the development of citizenship skills for the future (Barrow 2010). Hence, we may see elements of genuine participation (as compared to token participation), which could be described here as something that focuses on processes of knowing and meaning-making in dialogue, within the contexts that the pupils are part of (Simovska 2000, 2004). Bereiter and Scardamalia (2010) for their part talk about genuine knowledge-creation using the principle of "real ideas, authentic problems." The possibility of participating and taking control, that is, of "sharing power" (Simovska and Jensen 2009), may promote pupils' self-esteem and self-efficacy (King and Occleston 1998), thus supporting growth as a person.

As the focus of the teaching-learning event moves towards supporting personal growth, the teacher becomes viewed as building a context (atmosphere) for learning that could be characterized as safe, tolerant, and accepting (cf. Pigozzi 2006). Though such an atmosphere was already seen as essential in the previous approach, here it has an accentuated role. Such an atmosphere permits a feeling of being respected, an increase in confidence, and the expression of personal (and sometimes uncertain) opinions and experiences. It supports critical thinking and reflection, and also genuine dialogue, allowing the evolution of supportive interaction between *all* the members of a class. At the same time as the focus of the teaching-learning moves to collective meaning-making, the teacher's role expands to include aspects of being a *reflective fellow-learner*. After all, the teacher along with his/her pupils forms a learning community in which the teacher can learn from the pupils as well.

The personal growth and responsibility approach highlights how a strong internal voice (related to personal growth) allows an equal relationship with peers and others, and how the internal voice is no longer merely in the background when one is discussing with others (cf. Baxter Magolda 2001). From the perspective of health literacy, these aspects are particularly important in developing *collective* health, that is, in deciding what is good for *us*—not for me or you. Now, within this process, the pupils may act as equal partners, having equally valid thoughts, without being "consumed" by others. Yet in these manifestations one is also allowed to choose a differing stance, that is, to follow one's own independent path (see Baxter Magolda 2001). Thus, the question is not merely one of a sense of competence, but rather one of having a sense of authority and agency (see Johnston et al. 2001).

The personal growth and responsibility approach calls for teaching and learning activities that direct pupils' thinking towards others and the broader context. Some of these could be role plays and dramas that would enable pupils to become aware of how others might think or feel in a given situation, thus increasing their ethical thinking skills and empathy for others, along with the ability to take into account the perspectives of others (World Health Organization 2003, p. 16). In addition, panel discussions and cases can help pupils to develop their ethical reasoning and collective thinking skills by looking at health topics from different angles, as can also tasks such as "diamond-ranking procedures" (Lakin et al. 2004, in Lakin and Littledyke 2008), in which pupils are asked to collectively rank or prioritize issues. However, when we consider the teaching practices that may truly support the development of pupils' citizenship as a component of health literacy, we may clearly see that the learning should also happen in "extended" classrooms, both within and outside classrooms and schools. Lakin and Littledyke (2008) have reported a good example of the sort of learning process that truly captures the elements of the personal growth and responsibility approach. They describe a project in which the aim was to "develop children's critical understanding of food production and consumption issues, with supporting school practices for a healthy diet maintained by ecologically sustainable and ethically sound food production" (p. 254). Here, the pupils were involved not merely in deciding the content of their school meals, but also in the growing, tending, harvesting, and preparing of vegetables for the meals.

13.3 Conclusions

If we accept that the aim of school health education is to promote pupils' health literacy, the approaches presented above can be examined according to how far they actually achieve this aim. Health literacy as a learning outcome means that the pupil becomes capable of evaluating, understanding, and developing health messages from the perspective of his/her own life. Moreover, health literacy enables pupils to become aware of themselves, other people, and the wider context, to make ethically sound

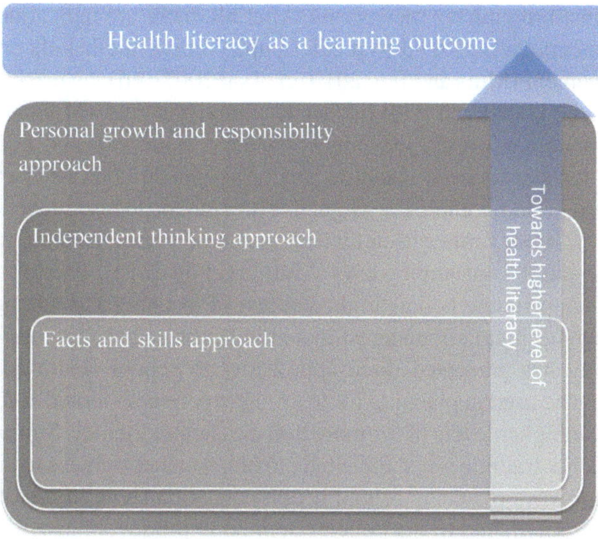

Fig. 13.2 Inclusive approaches to school health education as a means towards a higher level of health literacy (Paakkari 2012)

health decisions, and to work on and change their surroundings so that they influence not only their own, but also the health possibilities of others (Paakkari and Paakkari 2012). Figure 13.2 depicts how moving from a less advanced approach to health education (the facts and skills approach) towards a more advanced approach (the personal growth and responsibility approach) supports the achievement of a higher level (or more holistic form) of health literacy among pupils through the development of all the core components mentioned above. The hierarchical structure of the approaches is consistent with the notion that a more advanced approach(es) may include the aspects of previous approach(es) but not the other way round. Thus, when we emphasize personal growth and responsibility we are likely to include aspects that promote not only citizenship, but also the other components of health literacy. By contrast, a focus merely on facts and skills may overlook aspects from the other two approaches, with resulting gains in factual and practical knowledge, but not in critical thinking, self-awareness, or citizenship. It is in this sense that moving from the teaching of facts and skills towards an approach based on personal growth and citizenship is likely to promote a higher level of (or more holistic) health literacy.

One of the most critical aspects differentiating the least complex approach from more complex approaches is the way knowledge and knowing are understood (thus relating to the realm of epistemic belief, see Hofer and Pintrich 1997). While the facts and skills approach is consistent with an understanding of knowledge as something factual, discrete, and absolute, the approaches further up in the hierarchy move towards a view of knowledge as contextual, relative, and contingent (see Hofer and Pintrich 1997). Hence, it can be argued that when teachers are planning various teaching-learning sessions they should be aware of the kind of knowledge they will be instilling during the lessons. After all, there are indications that (i) the teacher's epistemic beliefs tend to predict pupils' epistemic beliefs and eventually pupils' achievement (Muis and Foy 2010; cf. Tsai 2007); (ii) the teacher's epistemic views tend to be associated with his/her teaching practices (Muis and Foy 2010; Tsai 2007)—in other words, viewing knowledge as certain and non-complex is aligned with teacher-focused teaching practices, and in contrast, viewing knowledge as uncertain and complex is aligned with student-focused teaching practices (cf. Tsai 2007). It has further been argued that these actual teaching practices tend to be linked to pupils' epistemic views (Muis and Foy 2010). The role of the teachers' and pupils' epistemic beliefs in the context of teaching and learning for health appears to have been greatly underestimated in health science literature.

This chapter has presented three approaches to school health education. The approaches can be used in planning for learning experiences aimed at supporting the development of higher levels of (or more holistic) health literacy. Similarly, they can be used in teacher training when the aim is to help teacher trainees to become aware of their current ways of seeing school health education (both in relation what happens in the classroom and to the policies applied in the school as a whole) and further, the differences that may exist between their own understanding and more complex forms of understanding. For a teacher trainer, getting to know how teacher trainees see health education (or the teaching and learning of health issues in schools) and how trainees' understanding differs from other ways of understanding health education, serves as an important resource for bringing about purposeful learning among the teacher trainees (see Lo et al. 2004). If teacher trainees for example, consider knowledge to be non-complex and view the pupils' role as merely that of recipients of knowledge, then the teacher trainer should try to create learning situations in which trainees are supported in such a way as to see the complexity of the knowledge, and to create activities that support pupils in being able to deal with problematic aspects of knowledge. In these cases, merely giving information (to trainees or to pupils) will be inadequate. Similarly, teacher trainees should be helped to see that this shift in understanding (and in teaching practices) is needed if they wish to develop higher level thinking skills among their pupils—skills that at the same time involve and contribute to the development of critical thinking, self-awareness, and citizenship as health literacy components, along with the necessary components of factual and practical knowledge. In this way, the approaches—including the conceptual structure in which they are located—can serve as a pedagogical tool for teacher trainers.

In future, it will be important to examine how pupils and their teachers see knowledge and knowing in health issues, and how the beliefs of the pupils develop throughout their school years. Similarly, it will be crucial to study whether these conceptions are related to each other, and to the teachers' teaching practices. A move towards pondering these epistemic beliefs brings into focus the fact that without a shift towards seeing knowledge as complex, a holistic or high-level form of health literacy is unlikely to develop among the pupils. This shift is all the more necessary in times when the managing of large amounts of knowledge in the mode of reasoning, analyzing, weighing evidence, and solving problems has become a paramount component of twenty-first century skills (Wagner 2008).

References

Abel, T. (2007). Cultural capital in health promotion. In D. V. McQueen & I. Kickbusch (Eds.), *Health and modernity – The role of theory in health promotion* (pp. 43–73). New York: Springer.

Aypay, A. (2010). Teacher education students' epistemological beliefs and their conceptions about teaching and learning. *Procedia Social and Behavioral Sciences, 2*, 2599–2604.

Barrow, W. (2010). Dialogic, participation and the potential for philosophy for children. *Thinking Skills and Creativity, 5*, 61–69.

Baxter Magolda, M. B. (1999). *Creating contexts for learning and self-authorship – Constructive-developmental pedagogy*. Nashville: Vanderbilt University Press.

Baxter Magolda, M. B. (2001). *Making their own way – Narratives for transforming higher education to promote self-development*. Sterling: Stylus Publishing.

Begoray, D. L., Wharf-Higgins, J., & MacDonald, M. (2009). High school health curriculum and health literacy: Canadian students' voices. *Global Health Promotion, 16*(4), 35–42.

Benham-Deal, T. B., & Hodges, B. (n.d.). Role of 21st century schools in promoting health literacy. National Education Association Health Information Network. Retrieved from http://crisisguide.neahin.org/healthliteracy/Images/BenhamDeal-Hodges%20Paper.pdf

Benham-Deal, T., Jenkins, J. M., Deal, L. O., & Byra, A. (2010). The impact of professional development to infuse health and reading in elementary schools. *American Journal of Health Education, 41*(3), 155–166.

Bereiter, C., & Scardamalia, M. (1993). *Surpassing ourselves—An inquiry into the nature and implications of expertise*. Chicago: Open Court.

Bereiter, C., & Scardamalia, M. (2010). Can children really create knowledge? *Canadian Journal of Learning and Technology, 36*(1). Retrieved from http://www.cjlt.ca/index.php/cjlt/article/view/585/289

Biggs, J., & Tang, C. (2011). *Teaching for quality learning at university* (4th ed.). Berkshire: McGraw-Hill Education.

Brey, R. A., Clark, S. E., & Wantz, M. S. (2008). This is your future: A case study approach to foster health literacy. *Journal of School Health, 78*(6), 351–355.

Chan, K.-W., & Elliott, R. G. (2004). Relational analysis of personal epistemology and conceptions about teaching and learning. *Teaching and Teacher Education, 20*(8), 817–831.

De Lawter, K., & Sosin, A. (2000, April 24–26). *A self-study in teacher education: Collective reflection as negotiated meaning*. Paper presented at the annual meeting of the American Educational Research Association, New Orleans.

Fisher, A. (2001). *Critical thinking—An introduction*. Cambridge: Cambridge University Press.

Grant, A. M., Franklin, J., & Langford, P. (2002). The self-reflection and insight scale: A new measure of private self-consciousness. *Social Behavior and Personality, 30*(8), 821–836.

Hofer, B. K., & Pintrich, P. R. (1997). The development of epistemological theories: Beliefs about knowledge and knowing and their relation to learning. *Review of Educational Research, 67*(1), 88–140.

Hoveid, H., & Hoveid, M. H. (2008). Teachers' identity, self, and the process of learning. *Studies in Philosophy and Education, 27*, 125–136.

Hubbard, B., & Rainey, J. (2007). Health literacy instruction and evaluation among secondary school students. *American Journal of Health Education, 48*(6), 332–337.

Johnston, P., Woodside-Jiron, H., & Day, J. (2001). Teaching and learning literate epistemologies. *Journal of Educational Psychology, 93*(1), 223–233.

Kickbusch, I. (2008). Health literacy: An essential skill for the twenty first century. *Health Education, 108*(2), 101–104.

King, P., & Occleston, P. (1998). Shared learning in action: Children can make a difference. *Health Education, 98*(3), 100–106.

Kolbe, L. (2002). Education reform and the goals of modern school health programs. *The State Education Standard, 3*(4), 4–11.

Kolbe, L. (2005). A framework for school health programs in the 21st century. *Journal of School Health, 75*(6), 226–228.

Kwak, D.-J. (2007). Re-conceptualizing critical thinking for moral education in culturally plural societies. *Educational Philosophy and Theory, 39*(4), 460–470.

Lakin, L., & Littledyke, M. (2008). Health promoting schools: Integrated practices to develop critical thinking and healthy lifestyles through farming, growing and health eating. *International Journal of Consumer Studies, 32*, 253–259.

Lo, M. L., Marton, F., Pang, M. F., & Pong, W. Y. (2004). Toward a pedagogy of learning. In F. Marton & A. B. M. Tsui (Eds.), *Classroom discourse and the space of learning* (pp. 189–225). Mahwah: Lawrence Erlbaum Associates.

Lodge, C. (2005). From hearing voices to engaging in dialogue: Problematising participation in school improvement. *Journal of Educational Change, 6*, 125–146.

Marton, F., & Booth, S. (1997). *Learning and awareness*. Mahwah: Lawrence Erlbaum Associates.

Marx, E., Hudson, N., Deal, T. B., Pateman, B., & Middleton, K. (2007). Promoting health literacy through the health education assessment project. *Journal of School Health, 77*(4), 157–163.

Muis, K. R., & Foy, M. J. (2010). The effects of teachers' beliefs on elementary students' beliefs, motivation, and achievement. In L. D. Bendixen & F. C. Feucht (Eds.), *Personal epistemology in the classroom – Theory, research, and Implications for practice* (pp. 435–469). Cambridge: Cambridge University Press.

Nutbeam, D. (1998). Health promotion glossary 1. *Health Promotion International, 13*(4), 349–364.

Nutbeam, D. (2000). Health literacy as a public health goal: A challenge for contemporary health education and communication strategies into the 21st century. *Health Promotion International, 15*(3), 259–267.

Nutbeam, D. (2008). The evolving concept of health literacy. *Social Science & Medicine, 67*, 2072–2078.

Otting, H., Zwaal, W., Tempelaar, D., & Gijselaers, W. (2010). The structural relationship between students' epistemological beliefs and conceptions of teaching and learning. *Studies in Higher Education, 35*(7), 741–760.

Paakkari, L. (2012). *Widening horizons: A phenomenographic study of student teachers' conceptions of health education and its teaching and learning* (Studies in sport, physical education and health 179). Jyväskylä: University of Jyväskylä.

Paakkari, L., & Paakkari, O. (2012). Health literacy as a learning outcome in schools. *Health Education, 112*(2), 133–152.

Paakkari, L., Tynjälä, P., & Kannas, L. (2011). Critical aspects of student teachers' conceptions of learning. *Learning and Instruction, 21*(6), 941–948.

Pigozzi, M. J. (2006). A UNESCO view of global citizenship education. *Educational Review, 58*(1), 1–4.

Ratzan, S. C., & Parker, R. M. (2000). Introduction. In C. R. Selden, M. Zorn, S. C. Ratzan, & R. M. Parker (Eds.), *Current bibliographies in medicine: Health literacy, National Library of Medicine* (pp. v–vi). Bethesda: National Institutes of Health, U.S. Department of Health and Human Services.

Simovska, V. (2000). Exploring student participation within health education and health promoting schools. In B. B. Jensen, K. Schnack, & V. Simovska (Eds.), *Critical environmental and health education* (pp. 29–43). Copenhagen: The Danish University of Education.

Simovska, V. (2004). Student participation: a democratic education perspective — Experience from the health-promoting schools in Macedonia. *Health Education Research, 19*(2), 198–207.

Simovska, V., & Jensen, B. B. (2009). *Conceptualizing participation – The health of children and young people*. Copenhagen: World Health Organization.

Sormunen, K. (2004). *Seitsemäsluokkalaisten episteemiset näkemykset luonnontieteiden opiskelun yhteydessä* [Seventh-graders' epistemic views related to science lessons] (Publications in Education No. 95). Joensuu: University of Joensuu.

St Leger, L. (2001). Schools, health literacy and public health: Possibilities and challenges. *Health Promotion International, 16*(2), 197–205.
Stewart-Brown, S. (2006). *What is the evidence on school health promotion in improving health or preventing disease and, specifically, what is the effectiveness of the health promoting schools approach?* (Health Evidence Network Report) Copenhagen: WHO Regional Office for Europe. Retrieved from http://www.euro.who.int/document/e88185.pdf
Tappe, M. K., & Galer-Unti, R. A. (2001). Health educators' role in promoting health literacy and advocacy for the 21st century. *Journal of School Health, 71*(10), 477–482.
ten Dam, G., & Volman, M. (2004). Critical thinking as a citizenship competence: Teaching strategies. *Learning and Instruction, 14*, 359–379.
Tones, K. (2005). Health promoting in schools: The radical imperative. In S. Clift & B. B. Jensen (Eds.), *The health promoting school: International advances in theory, evaluation and practice* (pp. 23–40). Copenhagen: Danish University of Education Press.
Tsai, C.-C. (2007). Teachers' scientific epistemological views: The coherence with instruction and students' views. *Science Education, 91*(2), 222–243.
Tynjälä, P. (2008a). Connectivity and transformation in work-related learning – Theoretical foundations. In M.-L. Stenström & P. Tynjälä (Eds.), *Towards integration of work and learning: Strategies for connectivity and transformation* (pp. 11–37). Dordrecht: Springer.
Tynjälä, P. (2008b). Perspectives into learning at the workplace. *Educational Research Review, 3*(2), 130–154.
van Rossum, E. J., & Hamer, R. (2010). *The meaning of learning and knowing*. Rotterdam: Sense Publishers.
Wagner, T. (2008). *The global achievement gap: Why even our best schools don't teach the new survival skills our children need—And what we can do about it*. New York: Basic Books.
World Health Organization. (1994). *Life skills education for children and adolescents in schools – Introduction and guidelines to facilitate the development and implementation of life skills programmes*. Geneva: World Health Organization.
World Health Organization. (1997). *Promoting health through schools: Report of a WHO Expert Committee on comprehensive school health education and promotion* (WHO Technical Report Series 870). Geneva: World Health Organization.
World Health Organization. (1998). *Health-promoting schools: A healthy setting for living, learning and working*. Geneva: World Health Organization.
World Health Organization. (2003). *Skills for health. Skills-based health including life skills: An important component of child-friendly health promoting school* (Information Series on School Health, Document 9). Geneva: World Health Organization.
Wu, A. D., Begoray, D. L., Macdonald, M., Higgins, J. W., Frankish, J., Kwan, B., Fung, W., & Rootman, I. (2010). Developing and evaluating a relevant and feasible instrument for measuring health literacy of Canadian high school students. *Health Promotion International, 5*(4), 444–452.
Zimmerman, B. J. (2002). Becoming a self-regulated learner: An overview. *Theory Into Practice, 41*(2), 64–70.

Chapter 14
Promoting Health Education in a Context of Strong Social and Cultural Heterogeneity: The Case of Reunion Island

Maryvette Balcou-Debussche and Crane Rogers

Abstract Contrasting social and cultural backgrounds are a challenge for health promoting schools, given the importance of social and cultural influences on health-related behaviour, and the sensitive and personal nature of health topics addressed at school. This chapter discusses the importance of social context and cultural diversity in health promoting schools in the context of the "Health Passport" project conducted in Reunion Island. The Reunionese population presents unique social divisions and contrasting cultural backgrounds, and is faced with unusually acute poverty and public health problems for a European territory, particularly among specific social groups. The "Health Passport" project illustrates potential pitfalls and key factors in successfully implementing sustainable health promotion programmes in socially and culturally heterogeneous contexts. In only three years, these teaching aids have become central to local health education policy in schools, and are currently used by over 400 teachers and 3,000 pupils and their families. The development and the evaluation of these booklets are discussed here. Based on empirical research, the efficacy of the health passports as experienced by educational staff, pupils and families are discussed. A variety of considerations favoured involvement and adoption of the booklets, demonstrating the importance of an open, inclusive approach for sustainable health promotion in schools.

Keywords Health promoting schools • Cultural diversity • Family implication • Reunion Island • European outermost regions

---

M. Balcou-Debussche, Ph.D. (✉)
ESPé de La Réunion, LCF-Icare, Université de La Réunion, F-97715 Saint-Denis, France
e-mail: maryvette.balcou-debussche@univ-reunion.fr

C. Rogers
Université Lyon 1 ESPé de l'Académie de Lyon, Université de Lyon, F-69622 Lyon, France
e-mail: crane.rogers@univ-lyon1.fr

V. Simovska and Patricia Mannix-McNamara (eds.), *Schools for Health and Sustainability: Theory, Research and Practice*, DOI 10.1007/978-94-017-9171-7_14,

14.1 Introduction

14.1.1 Social Context, Cultural Diversity, and Health Promotion in Schools

According to Pierret (1984), health-related practices develop under multi-dimensional systems of social, symbolic and cultural influences through which social groups develop practices of reference. These practices of reference also vary according to codes of behaviour and representations developed and shared in various living situations such as at work, in school, or at home. Individual representations of health develop from personal experience and evolve throughout the course of life, under the influence of multiple context-related behaviours. Thus an individual's representations of health and health-related behaviours will tend to evolve in a manner that is compatible with that person's living environment and that makes sense to the individual empirically, rather than complying perfectly with the individual's best interest from a medical standpoint (Jodelet 2006).

Schools are one context in which children develop representations of health. What is taught in health education may be more or less in contradiction with the representations children develop under other influences in other contexts. Teachers do not necessarily give much consideration to representations that children develop outside of school. However, given the sensitive and personal nature of the topics addressed in health education, as well as the need to redefine the teaching methods, the objectives and content (Downie et al. 1996), effective teaching of health education should arguably take account of teachers' and childrens' representations of health (Jourdan et al. 2002, 2010; Berger et al. 2011). Analysing these representations and the contextual practices of reference to which these representations are linked within the community, can aid teachers when designing teaching strategies and material. In culturally diverse contexts it can be helpful to first assess the diversity of representations, and in particular the gap between teachers' and students' representations.

There is now substantial literature attesting to the contributions and to the diversity of approaches among health promoting schools in Europe and indeed globally (Clift and Jensen 2005; IUHPE 2008; St Leger et al. 2007; Buijs et al. 2009). Similarly, cultural competency has been shown to contribute to professional effectiveness particularly when targeting specific populations, behaviours, or environments such as schools (Pérez and Luquis 2008). Given the importance of social and cultural influences on health related behaviour, contrasting social and cultural backgrounds of students are a challenge for health promotion in schools, and are also of particular interest for health promoting schools research. In this regard, the outermost regions and overseas territories of the European Union (E.U.)[1] provide a range of especially interesting cases as they typically present a mix of very different

[1] i.e. the eight Outermost Regions and the 26 Overseas Countries and Territories of E.U. member states (Denmark, France, the Netherlands, Portugal, Spain and the United Kingdom).

cultures, where the predominance of mainland European culture is often less evident. These territories are located across the globe in every ocean, are mostly islands, and are mostly relics of Europe's colonial past. They typically present complex histories of migrations, slave trade, and development of Creole cultures. They also present socio-economic contrasts and health problems that are more severe than in the mainland E.U. member states they belong to. These territories would certainly benefit from further developing health promotion in schools and in general.

Little material has been published to date on health promotion practices in these territories, and little effort has been made to compile or collect consistent data or to conduct comparative studies. Even among the outermost regions, which are integrally part of E.U. member states, European and national data schemes are often incomplete or not applied, or are not comparable between countries. To provide some illustration, the French Outermost regions[2] can be compared to mainland France. In a recent study on social and health inequities in France (Trugeon et al. 2010), a classification of all French municipalities clearly distinguished the French outermost territories from the rest of France, on grounds of age structure (more youth and higher birth rates), of health (differences in dominant pathologies and causes of mortality, and higher mortality rates) as well as lower education, greater poverty, and higher unemployment. Full illiteracy of adults is over twice as high as in mainland France and unemployment is over 2.5 times that of mainland France whereas health care and social services are less developed (INED 2012).

14.1.2 The Context of the Reunion Island

The Reunion Island is situated in the south-western part of the Indian Ocean, approximately 800 km East of Madagascar and 210 km West of Mauritius Island. Its 820,000 habitants are confined to an island of only 2,511 km^2. Discovered by the Portuguese in 1504, the island remained independent until 1665, when it became a French colony until 1946 when its status changed to a French overseas department. In 1997 it became one of the eight outermost regions of the European Union.

The Reunion's Creole population was constituted over the centuries by successive waves of migrants and slaves arriving from many different countries such as India, China, Madagascar, Africa and the Comoros islands. This varied cultural background gave way to a pluralistic society that through to the 1950s remained an essentially traditional rural society. During the last half-century Reunion has shifted rapidly to a post-industrial serviced-based economy characterised by an expanding middle class, and the development of public services and modern communications, plunging Reunion into the information age and global exchange. This transformation has left a substantial part of the island's traditional populations behind however, particularly those on the lowest levels of the social scale, similarly to their once-enslaved ancestors (Wolff and Watin 2010). In 2010 nearly 37 % of Reunionese

[2] French Guiana, Guadeloupe, Martinique, and Reunion.

families were receiving welfare, and approximately 40 % of the population received Universal Health Coverage.[3] Unemployment in 2010 averaged 29.5 % but was much higher for the 18–25 year age group, which represents 60 % of the Reunionese workforce.[4] These changes have initiated and/or reinforced strong social contrasts and among other aspects have affected the way Reunionese people eat, exercise and care for themselves (Balcou-Debussche 2010). Moreover, the challenges in maintaining population health in this fragile socio-economic context are exacerbated by the problem that public health issues in Reunion are much more acute than in metropolitan France.

Children are particularly exposed to specific health problems. Children under 15 years of age totalled over 190,000 children in 2005 (i.e., 24 % of the islands total population compared to 18.5 % in France) and the proportion of this age group in the whole population is increasing twice as fast as in mainland France. A school health survey conducted in 2003–2004 with adolescents at the 9th grade level showed that 11.8 % were treated for asthma (7.5 % for France), and that 20.7 % were overweight, including 5.4 % obese (respectively 15.9 and 4.0 % for all of France).[5] Overweight and obesity among 6 year-old Reunionese children increased in less than 20 years by a factor of 2.2 for boys and 3.6 for girls. Smoking and drug use has worryingly increased: children now start using drugs early and tend to develop regular use of multiple drugs at young ages.[6] The most common pathologies including, asthma, obesity or addictions afflict both adults and teenagers (Catteau 2012) whereas access to medical and preventive facilities is limited in many parts of the island.[7] The main chronic diseases are mostly linked to unhealthy behaviours (particularly alcohol and endemic drug use, and high fat/sugar diets) that are anchored in the political and cultural history of the island as well as its current situation (Wolff and Watin 2010), making effective prevention difficult.

Compared to mainland France, Reunion has particular need for ambitious health promotion action. Education professionals clearly have a key role to play in developing health promotion in schools as schools are the only place where continuous action can be undertaken for all of the children at each age level, and thereby for the entire population on the long term. It is in this perspective that the "Health passport" project was initiated in 2008. This project aimed to establish a sustainable educational policy for the health of children and their families, based on a contextualised reference document, and implemented with the help of school personnel (Balcou-Debussche 2009).

[3] http://www.ars.ocean-indien.sante.fr/fileadmin/OceanIndien/Internet/Votre_ARS/Etudes_et_publications/Bulletins_Infos_Reunion/DOSSIER_STAT_Etat_de_sante_Octobre2010_.pdf

[4] http://www.insee.fr/fr/themes/document.asp?reg_id=24&ref_id=17943

[5] http://www.ors-reunion.org/IMG/file/tableaux_bord/sante1a14ans_2007.pdf

[6] http://www.ors-reunion.org/IMG/file/tableaux_bord/TB_addictions_2011_synthese_produit.pdf

[7] http://www.fnors.org/fnors/ors/travaux/addictions.pdf

14.2 The Health Passport Project

14.2.1 Project Origins

The health passport project arose from the desire of the Department of Education of the University of Reunion[8] and the Regional Board of Education of Reunion[9] to work together on health education at primary school level.[10] From the start, the general intent was to promote health in Reunionese schools by addressing the three components of health promotion as defined by Downie: education, prevention, and protection (Downie et al. 1996). According to Jourdan (2011), this can be seen as developing students' knowledge and understanding of health issues, their personal and social skills enabling them to act on health-related issues and a critical view of their environment. The project was also to be built upon priorities and competency building strategies set out in French legislation on education (Decree No. 2006-830 July 11, 2006)[11] which focuses on developing children's life skills pertaining to social integration and responsible behaviour as well as to learning. The project was also to take local health priorities into account, and to enable the participatory intervention research upon which this chapter is based.

Until 2008, few Reunionese primary schools had developed projects focusing on health promotion. At that time, the University education department did not propose courses centred explicitly on health education, whereas the Board of education focused its efforts in health prevention at secondary school level to cope with problems of alcohol and drug abuse, adolescent pregnancy, or psychological distress. Shifting the focus to pupils at the earliest levels of schooling, the University and the Board of education aimed to develop prevention before children were exposed to these pathologies. It was anticipated that this change of focus would also make it possible to develop a more sustainable long-term approach that could systematically involve all the pupils of the targeted age group. In this approach, health promotion and education are seen as central to personal development and to citizenship education (WHO 1997, 1999), and as means of "learning to live together" in a holistic and systemic framework of positive health, rather than a prescriptive or normative approach to health (Tones and Tilford 1994).

In 2008 a project coordinator was identified at University to establish a joint project within the frame of the 2009–2012 education plan (Balcou-Debussche 2009). Three priority topics were selected on the basis of national teaching

---

[8] ESPé- Ecole Supérieure du Professorat et de l'éducation, Université de La Réunion.

[9] In France, public education is administered by "Academie" at the regional level, headed by a "Recteur" who organizes teaching, manages staff and schools, and implements the policy defined by the Ministry of National Education.

[10] Reunion is divided into 23 school districts that total 512 public schools, only 26 private schools, 45 200 pupils in pre-schooling, 77 400 pupils in elementary school, and 6 434 employees.

[11] http://www.legifrance.gouv.fr/affichTexte.do?cidTexte=JORFTEXT000000818367&dateTexte=&categorieLien=id

programmes and curricula, while focusing on key local health issues among children: nutritional education, sex education, and preventing addictive behaviour. The goals for nutritional education were to improve the quality and balance of meals served in the school's cantina, while developing students' awareness of healthy diet and encouraging physical activity. Plans for sex education involved conducting at least three teaching sessions on sex education at all levels of schooling. Prevention of addictive behaviour comprised of informing pupils of the dangers of excessive behaviours or use of substances of all natures such as drugs, alcohol, tobacco, physical activity, internet and video. Emphasis on these topics was not exclusive but rather a starting point towards developing a comprehensive health promoting school approach: promoting student health and welfare, mainstreaming health in all school activities, involving parents and partners and developing awareness of social justice and equity, and promoting social integration and commitment for education to facilitate the learning outcomes and reduce the behaviours that represent health risks (WHO 1997, 1999; Moon et al. 1999; Lister-Sharp et al. 1999; St Leger et al. 2009). General guidelines for the project were built on the results already achieved in health promoting schools, particularly key aspects with regards to cultural diversity: active involvement of students in learning situations, cooperation and consultation between parents, students and teachers, development of trust, respect, collaboration and openness (Graham et al. 2001; Masters 2004; Muijs and Reynolds 2005). The approach developed not only reinforced health education action in primary schools; it also enabled health education training in initial and continuing education for teachers.[12]

14.2.2 Project Management

The project coordinator initiated two committees: an executive committee comprising representatives of the institutional partners, and a technical committee composed of professionals in education or in teacher training (inspectors, councillors, school medical personnel, professors, school directors, and researchers). Between 2008 and 2010, the executive committee met 4 times and the technical committee met approximately every month. As work progressed, the idea emerged of creating a common tool that could circulate between schools and the pupil's families. This tool was expected to link pupils, teachers, parents and anyone else involved in a child's formal or non-formal education, be it in school, at home, or elsewhere.

Research has shown that certain educational practices have a positive impact on learning in the field of health, while others may be ineffective or even counterproductive (St Leger and Nutbeam 1999), so the team chose to first conduct a pilot study in five schools. As teacher willingness-to-participate in health education is also a crucial to the success of such projects (Jourdan et al. 2002; Viig and Wold

[12] Since 2010 becoming a teacher in France requires a master's degree in education. At Reunion University, master's level students in education receive at least 18 h of coursework in health education.

2005), pilot sites were selected among schools that had already begun developing health education or at least had set the goal of developing health education. These five schools were located in different districts throughout the island, in different contexts. The smallest school accommodated 95 pupils at pre-elementary levels (under 6 years of age) in a rural setting, whereas the largest totalled 420 elementary level pupils in a suburban setting. However, all five schools were located in areas classified as "education priority zones" by the Board of education, and all of them are primarily occupied by economically fragile populations. The entire project involved 55 teachers and 1,137 pupils.

14.2.3 The Health Passport Booklets

The health passport booklets were created through a collaborative process involving the education personnel of the five participating schools, taking into account current legislation and educational directives (ministerial texts on monitoring student health and health education),[13] and were then ratified by the two project committees. During the 2008–2009 school year, the project coordinator helped school teams build on work that had already been accomplished in health education in the participating schools, so as to reinforce team dynamics in continuity with prior action. Families were also involved in the construction of a common tool that was to be used both in school and at home. Elected representatives of parents on school boards participated in planning the implementation in schools, and meetings open to all families were organized to present and to discuss the project. Families were also informed of the project by written messages sent home, and were encouraged to discuss the project at school, with teachers. By working closely with local participants in the five different schools, the technical committee was able to propose a draft of a booklet in June 2009 for evaluation. A second revised version of the booklet was tested in the 5 schools during the 2009–2010 school years. A third and final version was then drawn up and in June 2010 was designated the official health passport by the Direction of education.

The resulting health passport booklets take the form of two twelve-page booklets adapted to different age levels or "cycles" of the French educational system[14]: one booklet was for preschool and early elementary school cycles (under 9 years of age) and the other for the higher cycle of elementary education (ages 8 and up). The three priority themes (nutritional education, sex education, and preventing addictive behaviour) are treated specifically at each cycle level, with content adapted to the age and the reading abilities of the targeted age group. The booklets primarily use an experience-based approach to directly involve the children, regardless of the topic being treated. At the end of the cycle 3 booklet, there is a summary of the topics and

[13] https://eduscol.education.fr/cid47750/education-a-la-sante.html

[14] Cycle 1 or the "early learning cycle" (ages 2–5), Cycle 2 or the "fundamental learning cycle" (ages 5–8) and Cycle 3 or the "reinforced learning cycle".

psycho-social competencies that have been covered during their elementary schooling, which can be transferred to the second level school with the pupil is to pursue in. The purpose is not to evaluate the pupils (psycho-social competencies are difficult to measure objectively), but to provide a clear synthesis of the health education work accomplished by pupils and their teachers in continuity throughout their primary education.

Another important aspect of the booklets is that they refer to the regional context of Reunion Island, including reference to specific local traits: food products, the climate and environment of the island, the school and living environments of different parts of the island. Indeed many of the teaching materials used in overseas territories and outermost regions are essentially based on living conditions in the metropolitan countries these territories belong to, and may therefore be less meaningful or involving for pupils. The use of photographs provided by members of the technical committee largely contributed to adapting the booklets to local context.

The booklets are also intended to stimulate teacher awareness and involvement in health education and promotion. Indeed one of the goals of the University department of education, as partner in the project, was to contribute to training teachers in health education. The booklets are designed to encourage teachers to include cross-curricular health education topics in their teaching, so that as use of the booklets develops, more and more teachers will become more aware of health education methods and goals, and develop their personal experience in teaching health education. The use of these booklets in a given school is also expected to reinforce teamwork among staff in a collective effort spanning from early pre-elementary class levels to the end of elementary schooling. Thus in different ways, the booklet aims to the benefit the students and their families by developing health competency and encouraging discussion of health issues and behaviours at home, but also the teachers and the school by developing teacher awareness and involvement in health promotion while reinforcing exchange and collaboration between families and the school.

14.3 The Pilot Phase

14.3.1 Data and Methods

The analysis presented here is based on three complementary data sources gathered during the pilot phase of the health passport project, between 2008 and 2010.

First different types of documents attesting to the process of creating the health passport were collected and analysed. These documents include draft versions of the health passport booklets, meeting proceedings of both the executive committee (4 meetings) and the technical committee (20 meetings), and various working documents such as notes taken during encounters with teachers.

Secondly, in 2008 and 2009 more than 100 h of classroom observations were conducted by the project coordinator and 18 videos of 10–15 min were produced from these observations. These videos show teaching sessions at pre-elementary to upper elementary levels on various topics such as personal hygiene and washing hands, childbirth, addictions, the body's energy balance, the differences between boys and girls.

Finally four questionnaires were collected for different populations involved in the pilot phase of the project. Questionnaires were distributed to parents of the five participating schools (approximately 800 families). For cycle 2 level children, teachers were asked to complete questionnaires for their pupils (n=291), collecting individual opinions on pupil preferences and perceptions, and their account of family reactions to receiving the health passport. Participating cycle 3 level pupils (n=570) completed an anonymous questionnaire following the first session of health education using the health passport booklet. Finally the 55 teachers that participated in the project all completed a document reporting their preferences, their criticism and their suggested improvements for the booklet they had used.

Return rates for teachers and pupils were receptively 98 % and 97.5 %, as data was collected in schools by the project manager. Return rates for parents averaged 63.8 %, ranging from 55 to 75 % in the five pilot schools. Data for quantifiable items were processed for occurrences and frequencies. Open text data were analysed through a grounded theory approach (Charmaz 2002) to content analysis (Holsti 1969; Krippendorff 1980) in order to reduce the risk of misinterpretation.

14.3.2 Results

14.3.2.1 The Pupils

The responses of the cycle 3 pupils of the 5 schools (n=570) indicate that they discussed their booklet, often at length, and not only with immediate family but also with others (neighbours, more distant family). The booklet was shown to the mother (64 %), the father (50 %), to brothers (25 %), to sisters (14 %) and/or to other people (8 %). It was mostly shown in the evening (37 %), upon arriving from school (29 %) or when doing homework (29 %).

Pupils and their families most often spent a good amount of time looking over the booklet (45 %) at least once (35 %), twice (29 %), three times (17 %) or more (15 %). The vast majority of the pupils (85 %) found that this moment was a positive and important moment, namely because "*the booklet makes us talk about ourselves with our parents*" (Aymeric, CM1). The booklet was found to be interesting (63 %) was preferred to other school documents (58 %), and was found to stimulate interest in health issues (56 %).

The study was carried out with 861 pupils spanning the entire range of elementary school class levels. Results for cycle 2 level pupils (n=291) were quite similar to results for cycle 3 level pupils and are not detailed here.

14.3.2.2 The Parents

Questionnaires included both open questions and Likert-type scale items (from 1 to 5). Return rates ranged from 55 to 75 % depending on the school, attesting to the interest parents granted to the health passport booklets. For the scale items, responses "satisfactory" (4) and "very satisfactory" (5) always totalled more than 50 % for all five schools. For the question regarding the importance of being able to follow the evolution of their child, the total of responses 4 and 5 reached 80 %.

For open questions the most frequently used terms in responses were *child, school, parents, interesting, family, discuss, education*. Parents reported that they appreciated being able to follow the work done in school outside the "fundamental" disciplines (reading and mathematics). They discovered the degree of concern teachers have for their children's health, and were pleased to see that the schools made efforts to develop relations with the families regarding health issues. Many parents found the booklet to be a means upon which they could build on their own efforts to educate their child about health issues such as eating well, getting enough sleep, watching too much television… Parents also found the booklets created opportunities to address sensitive questions "*The booklet enables children to discover things that parents dare not discuss with them. It is better that children learn about these things from an outside person who knows better how to discuss these things with children. Children often listen better to others than to their parents*" (Alexa, mother of a child in school 5).

The high return rates obtained for parents presumably reflect a combination of factors. As in other places, results show that most parents strongly support the promotion of health in schools and show willingness to participate more actively (Moon et al. 1999). Here particular efforts were made involve parents early on in the project and on various occasions in participating schools. Also the booklets were designed to stimulate regular exchange between home and school. Furthermore, attention was given to fully take the diversity of perspectives into account, to recognizing the parents' expertise and legitimacy within the project. Thus parent participation in the evaluation was probably favoured not only by their interest in their children's health, but also because their remarks were truly taken into account as much as possible.

14.3.2.3 The Teachers

The teachers (n = 55) also favourably welcomed the health passport booklets. The booklets were seen both as a tool upon which to build school policy and practices regarding health, and as a precious means of initiating exchange on more delicate subjects.

Although the teachers found the booklets attractive, they did express some criticism and make suggestions. Some teachers felt other topics should be included such as discovery of the human body or dangerous games, while others felt that the booklets should be less centred on the children themselves. Teachers teaching cycle

1 level pupils (pre-elementary level) felt that there were too few illustrations adapted to that level and too much text.

Some teachers echoed the concerns of certain families regarding the idea that the booklets could be used to keep records on children and that this could become prejudicial to their privacy and right to medical confidentiality. Moreover, some teachers were concerned that the booklets could serve to monitor teacher's practices. This was seen as a sensitive issue, as persons who would not normally be in a situation to evaluate teachers' practices would have access to the booklets.

14.4 Reflection on the Health Passport and Its Development

14.4.1 Accounting for the Diversity of Views

While the health passport booklets were well received by most, it is important to take criticism into account and to analyse the motivations behind resistance to the project. The booklets position health education as both a public issue in school and a private issue within the family, and as such the booklets necessarily question the respective place and roles of these two socialising environments. Because of this potentially opposing situation, the technical committee went to great lengths to record and accommodate as best as possible the various reactions and suggestions of participants throughout the experimental phase.

Although few families criticized the booklets, those who did expressed very different arguments, some of which questioned both the importance of health education and who should be involved in it. Certain families reported that involving parents was not necessary, or that the booklet was just an additional task to be dealt with in homework "*Families don't necessarily want to spend their weekend on diet, sexuality, or whatever*" (Reine-May, mother of a cycle 2 child). In contrast others felt the booklets intruded on strictly private issues that should not be treated openly in school: "*Discussing breast growth or the feeling of love with such young children is out of place*" (Lucette, mother of a cycle 2 child) or "*I'll teach my child to protect himself and take care of himself – I don't need your phony stereotyped passport to do that*" (Alex, father of a child in cycle 3). A few parents even felt that health education was not a priority or not the business of schools: "*Doesn't the direction of education have other priorities? Shouldn't schools be focussing more on fundamental learning?*" Other remarks concerned aspects of the booklets and their use. Some families felt the booklets recorded personal information about their children that could later be used to violate their privacy. Other parents criticized the booklets in light of other priorities such as the deplorable state of the school's toilets or the school's cafeteria, or criticized the expense: "*It's a waste of public money*" (Julien, father of a cycle 3 pupil) or "*With all the talk about sustainable development, was it rational to produce a booklet with full colour and glossy paper*" (Christine, mother of a cycle 3 pupil).

This diversity of objections illustrates the difficulties teachers encounter when teaching health education. Not only must teachers prepare and teach health education in class, they should also communicate with families on the importance of health education and on the purpose of the work being conducted in school with their children. In socially and culturally diverse populations such as that of Reunion Island, teachers would be well advised to take account of the variety of perspectives on health that families with different traditions and cultural backgrounds may have. To the contrary, France has long favoured policies of cultural assimilation by reference to a norm of national culture (Tucci 2011), so for teachers of the French national education system, taking account of the diversity of cultural perspectives on topics they teach may not come easy. The European framework for intercultural dialogue and mutual understanding between cultural groups recommends approaches that are developed collectively and thereby integrate the diversity of views on a topic through mutual tolerance, rather than juxtaposing contrasting perspectives on a topic. Because the experimental phase of the health passport project obtained reactions from more than 800 adults (families, teachers, decision-makers,) and from over 1,000 pupils, the technical committee was in a position to coordinate a truly collective and truly secular intercultural educative process. This is less a matter of ideology but rather is one of practical and ethical necessity in that it facilitates the inclusion of all those who are willing to participate and who recognize the right of others to see the issues or practices at hand differently. As stated by Martine Abdallah-Pretceille "For all matters pertaining to culture in the anthropological sense, the hardest part is not to act but to understand" (Abdallah-Pretceille 2003, p. 1). It is therefore necessary to develop the competencies enabling the participants (members of the committee, teachers) in such a process to listen and to observe, to understand and to accept the plurality and the complexity of others. This implies learning to search for and to detect aspects of the identity of "the other" that may be "hidden" behind their more evident first appearance, which in turn includes identifying the strategies associated with that underlying identity (Augé and Colleyn 2004, p. 17).

When comparing versions 2 and 3 of the booklets, it is evident for example that the place of the family was substantially reinforced following the early experimentation. In version 2, most of the page surface is dedicated to classroom work (3/4 of each page) and all of the investigative activities were to be done in school, whereas families were simply to confirm they have seen the work done in school (bottom 1/4 of each page). In version 3, each topic begins with statement "At school and at home, I learn to…", i.e.: establishing collaboration on a more equal basis. This change of posture implies that teaching staff must learn to think as parents tend to think, and to incorporate the families' perspective into their teaching approach.

This example demonstrates the importance of conducting an experimental phase so as to synthesize the array of family perspectives on the subjects and on the teaching materials, and thereby ensure that the approach and the materials will be socially acceptable for the vast majority of the community. This also implies that the designers of teaching materials must be able to understand and accept the various objections made by participating families, and be able to summarize this wealth of diverging representations in a positive, inclusive manner. Failure to recognize and to acknowledge cultural relevance and the legitimacy of local or minority practices and views may transform this diversity into a learning obstacle for children from these social groups, and could even lead families to consciously oppose the teaching of particular topics or have a negative impact on relations between certain families and the school system.

14.4.2 The Relation to Self and to One's Intimacy

Health and health education are both public and private matters, given the intimate nature of certain health topics, as well as their relation to personal identity and to culture. Public health policies cannot simply impose norms on the private lives of individuals, for ethical reasons and because of the principle of freedom of choice, but also because normative approaches tend to have an alienating effect on individuals whose behaviours are outside the norm (Jourdan 2012). Throughout the process of developing the booklets, the coordinating team was confronted with the opposition between public and private aspects of health, and progressively abandoned what they had initially considered as a 'common perspective' on health. For example, rules of personal hygiene indicated in version 1 of the booklets stated that children should wash themselves from head to feet, which seemed to be a commonly shared and logical manner of proceeding. During the experimentation it became apparent that several families did not proceed in this manner, particularly families from India, China and Madagascar. It also became apparent that one could wash the parts of one's body in a different order. The team coordinating booklet development searched for scientific grounds that might justify favouring one way of proceeding over another, only to conclude that there were none. Thus the coordinating team assumed their practices of reference to be a common norm when in fact different practices of reference existed within the range of social contexts encompassed by the five experimental schools. This case illustrates that practices are socially and culturally anchored on the basis of habit and common sense more often than on the basis of scientific evidence, implying that one should not readily pass judgement on the practices of others, particularly in more personal domains like health education.

The team also had to reflect on the place of the subject, in the sense of self or "I", and on how to involve the subject (a child, a pupil) but without the subject feeling forced or judged, which could occur if pupils were required to explain their personal behaviour regarding health practices. This might lead the subject to not speak out about his/her feelings or health related behaviour for example. For written material, the compromise adopted was to provide opportunity for children to reveal their behaviour when useful, while assuring them there would not be any negative consequences. In version 2 of the booklets, involvement was commended, and children were encouraged to express themselves as "I". However, anyone coming in contact with the health passport would be able to read the child's views. In version 3, to avoid making sensitive personal information accessible to others, the more sensitive items of the booklets were redesigned to have children write their answers on detachable pieces of paper that teachers would not necessarily have access to. This solution makes it possible to develop reflection and discussion of sensitive issues without having the children make definitive inscriptions in the booklets. This arrangement is essential as the booklets are to accompany children for the entire duration of their schooling, but should not by any means make sensitive or intimate information about the children available to third parties. The coordinating team thus had to anticipate possible or likely responses that children could have, and design items so as to avoid having children provide information on specific situations that could be seen as a confession or an accusation. Indeed, if children were to write such declarations in their booklets, schools may be forced to take action, whereas the health passport booklets were by not intended for such a purpose. The coordinating team did make a point of reminding participating teachers about the organisations that provide assistance with problems such as abuse and other forms of mistreatment, and this information is now systematically included in teacher training.

There was also considerable debate on whether to include the concept of "intimate" in the booklets or not. Some felt that privacy should not be mentioned because it is too difficult for children to conceptualise, while other opposed that it is important that we cite the idea, to get children to start developing the construct. In the final version of the booklets, intimacy is included at all level including the preschool level. At cycle 2 level, children are to colour-in reactions to the following statements where green means they agree, orange means they don't fully agree, and red means they don't agree at all: "I think everyone's intimacy is respected", "I think it's clean and well-lighted, and I go there just like at home" and "I find everything I need to wash my hands". Other questions treat other aspects of life in school such the school cafeteria or free time in the school courtyard. In this way the teaching staff gets feedback about how children perceive different moments and environments at school, both in terms of material aspects and in relation to their body, to hygiene and to their intimacy.

14.4.3 Health Education, Families and Social Diversity

Health education is driven by schools, but the diversity of family characteristics and their perceptions of health related topics can give rise to an array of issues that may affect the success of health education for their children. Just as Reunionese society is complex and difficult to grasp in terms of its dynamics, attempts at defining the typical "Reunionese family" or types of Reunionese families are also particularly complicated. Family structure in Reunion has often been treated from the somewhat narrow perspective of ethnic background, clinging to a dated view of Reunionese society without giving full account of the major changes that have occurred in recent decades, namely the great increase of intermarrying between communities, the emancipation of women, and the important changes in family models and functions

(Wolff and Watin 2010). Obviously some traits of the traditional extended family model, largely based on a sense of obligation, still exist. However contemporary Reunionese families are more autonomous, more democratic and family relations are now more often based on affinity than on obligation, although this may still occur. Among recent trends, young couples express the desire to move away from where their families have always resided, childbirth out of wedlock has become more common, and recently the development of "parental competencies" has become a public issue. As the preponderance of traditional family views subsides, children are no longer identified solely in connection with their family, but also as belonging to a social group and as having their own individual life story (Wolff and Watin 2010).

Adapting booklet content to the needs of different family conditions was an important part of the health passport project. The first versions of the booklets included passages with levels of language that were too sophisticated for some families, thus unintentionally running the risk of excluding them. In the second stage of booklet development, focus on using photographs, concrete examples, and simply worded commentary gave more satisfactory results. In the final versions, families that had been through difficult times, that had weak educational backgrounds, or that were living under difficult socio-economic conditions learned from the health passport booklets and found them to be a useful resource. Out of the 346 families that received the study questionnaire, 261 completed it, yielding a response rate of over 75 %, for which most were living in particularly deprived neighbourhoods. This high response rate apparently attests to the great interest families took in the booklets, and/or to the families' desire or willingness to develop ties and exchange with the schools. Families agreed to the following statements almost unanimously (>98 %): "*It is important to talk about sex education at school*", "*It is interesting to be able to follow the evolution of my child from the first to the third cycle*", and "*The booklet makes discussing delicate subjects with my child easier*". One parent wrote the following comment "*I think the booklet is great. It allows children to discover things that parents dare not bring up with them, by respect for their child. It's better that children learn about these things with other adults that know better how to address these sensitive subjects than parents do. Sometimes children listen better to others that to their parents*" (Robert, father of a cycle 3 child). In one school, the teaching staff organized an "open school" day on health: the pupils invited their parents to come to their school to discover the work they had done on health during the academic year. Parents participated in great numbers and many got involved in sports activities organized as part of the event, which also seems to indicate that parents are interested by their children's schools and do participate actively when they are invited to do so and feel accepted as they are, as compared to their more reluctant reactions when they are more or less ordered to be present for more official procedures regarding problems with their children for example.

These different results encourage pursuing reflection on the school's role in developing health education, and on the importance of respecting the diversity of family culture and perspectives. Schools can be a particularly effective context in which to develop health education, even within particularly sensitive and underprivileged populations, especially if they work with the families and are careful about respecting the diversity of family viewpoints.

14.4.4 Norms, Evaluation and Social Control

During the process of developing the booklets, the technical committee discovered that the evaluation they had designed to be formative was sometimes perceived more as a normative diagnostic. For example one parent stated: "*Why have nominative booklets that will pursue the child and his family throughout his education? Is the goal to collect statistics?*" [...] "*This health passport is typical of our aseptic, guilt- and fear-driven society, in which the paternalistic state takes responsibility off citizens and treats them as children*" (Axel, 42, pupil parent). Evaluation is also sometimes seen as an institutional manoeuvre to control teacher practice or as an intrusion into individual teaching practice. In France the excessively normative evaluation of teaching practices has been described as a "societal calamity" (Dejours 1998). Some teachers refuse such evaluations, sometimes rightly so, while others tend to not readily cooperate in the collecting of quantitative measures of the work conducted in class, arguing that an excessive number of repeated evaluations discourages pupil participation and hampers the quality of collective work in the classroom. In some cases teachers have feared that actions in health education might be used to compile records on pupils that could later be used against their interests, be it in school or for other purposes.

Because of these misgivings, the approach to evaluation in the booklets was debated at length both in committee meetings and with the educational teams of participating schools. It was necessary to clarify the distinction between teaching in general and the teaching of cross-curricular, competency-building topics such as health education, as this distinction was not clear to all teachers. In effect, schooling standards do not apply to cross-curricular topics in the same manner as with traditional disciplinary teaching. Teaching cross-curricular competency-building topics builds in large part upon social practices and representations of reference, which can vary greatly within the population of a school district. Therefore cross-curricular education, and particularly health education, cannot be developed through a normative approach aiming to measure, grade, and even prescribe behaviours based on a single definition of what is best.

A more appropriate approach would be to favour mutual understanding without grading. In this view, diversity of practice ought to be considered as contributing to the wealth of the educational process rather than as dispersion from a central ideal response. Differences in practice or in perspective can be discussed in classes or with families, and used as a means of formulating explicit representations of these topics and sharing different views and habits. Through such exchange, participants with comparatively unhealthy behaviours may come to admit that they should perhaps change their behaviours, under the influence of a group they belong to, rather than as a dictate coming from some outside authoritative figure.

Health education is largely about developing competencies and in particular psycho-social competencies that the participants can then use to develop appropriate and effective behaviour in life and with health related issues (Berger 2004).

Developing competencies requires building on different types of knowledge (academic learning and analysis of representations), on the ability to act and put into practice what one has learned, and also on learning to adapt to context, which is more difficult to evaluate than formal learning. Simply studying facts and rules will not suffice if the purpose is to help children develop effective health behaviour (Perrenoud 2008). In real-life individuals must grasp complex dynamic situations and resolve problems on the basis of very different and often incomplete information. Such real-life situations do not conveniently comply with the rules of a particular discipline or model (Meirieu 1994). While evaluation of health education projects and methods is obviously useful, independent evaluation schemes do not easily account for the complexity of real-life situations, the influence of culture, or the importance of context. Participative approaches to designing and evaluating health education tools and programmes are better adapted to accounting for real-life complexity and cultural diversity. They also favour active participation of the target population and all types of stakeholders and thus the acceptation and appropriation of the project and at the same time are known to contribute to building empowerment, which is a key aspect of health promotion (Themessl-Huber et al. 2008).

In the end, the technical committee had to incorporate all of the aforementioned considerations on evaluation into the project in a constructive manner. This involved rewriting many passages keeping in mind the different possible interpretations from different social and cultural perspectives not only of children and their families, but also of educational staff. These different perspectives were taken into account based on the feedback acquired during the participatory evaluation process, and if necessary by further consultation with certain participants. Graphic design and presentation was also reengineered to eliminate aspects that resembled testing in disciplinary teaching. For example, all the tables and all True/False questions in version 2 were removed in version 3. In version 2, for each section of the booklets, knowledge and competencies were evaluated using a scale from 1 to 4 (from "non-satisfactory" to "very satisfactory"), and then these partial results were summarized in the backs of the booklets. In version 3, all indications of performance to be reported by the children or by adults were deleted. The overall synthesis at the back of the book was modified several times, to eventually become a sort of guide announcing the knowledge and skills to be developed, clearly summarizing the aspects and progressions to be developed from cycle 1 to cycle 3 for the three health education themes in the booklets (nutrition, hygiene and sexuality, and preventing excessive and addictive behaviours). Moreover, rather than evaluate the level of knowledge and competencies, this guide enables teachers to indicate which aspects of the booklet had actually been covered in a given year and to what point, so as to keep track of what had been done with the pupils rather than how well they did. In this sense the guide became a tool for measuring progress in health education for families and schools alike, but without the potential restrictions of traditional evaluation using grades (Mukoma and Flisher 2004).

14.4.5 New Questions Arising from the Post-experiment Diffusion of the Booklets

The results of the pilot phase summarized here point to a certain number of issues that should be kept in mind when developing actions in health education. Foremost is the understanding that our relationships to health, to education and thus to health education vary considerably from person to person. Health education is at the crossroads between medicine and education science, between personal and public, between intimate and collective, and between formal learning and the development of psycho-social competencies. The complexity of health education is such that developing actions in this field requires taking a great number of precautions, both in the design of tools and actions and in their implementation by educational staff and with families.

One of the main factors influencing the success of health education appears to be the training of the persons involved. The success of the health passport booklets does not imply that Reunion has a successful and durable health education process. Simply developing and distributing the booklets does not guarantee that teachers will use them, let alone guarantee that they will use them effectively. The booklets must be accompanied by training and coaching, although it is not always easy to convince decision makers or the teachers to be trained of this. The project is currently in a critical phase as budget restrictions have affected support for the project. As the number of schools requesting the booklets continues to increase, the project coordinator is unable to respond to all these requests. Given current restrictions, two solutions are being tested. In each educational district, one school director is selected to be the local representative of the health passport and to explain it to those who are concerned or interested. At the same time, 18 h of health education is taught in the 3rd semester of the master's degree preparing students to become teachers, including a presentation of the health passport booklets.

While the financing of this project initially came from the region, more and more municipal governments are now financing the implementation of the health passport booklets in their cities. Several cities have now contracted with the Direction of education to ensure that the project will be maintained in the years to come. Thus it appears that interest in health education has been substantially boosted through this experiment, but also that an exemplary approach involving not only the schools but also the families was necessary to draw this substantial attention to health education. However the on-going pursuit and generalisation of this approach remains somewhat uncertain, particularly regarding eventual updates or ensuring the necessary support to those implementing the method for their first time, which requires both training and a minimum of resources and personnel to accompany and inform about the project.

14.5 Conclusion

The health passport project provides a good example of several factors that contribute to the success of sustainable actions in health education, and also illustrates the potential of intervention research as opposed to outside observation in this domain. The researcher's direct involvement as project coordinator greatly contributed to the success of the project from an education perspective, and also favoured teacher and parent involvement, as well as enabling greater access and understanding of different individuals' perspectives on health and the project.

Foremost one should keep in mind that such projects target heterogeneous groups that have contrasting perspectives of learning, health, and their relationship to others and to the world. The success and sustainability of health education depends in part on respecting this diversity of perspectives, on being open to others and taking the time to understand one another and to make room for each other in a collective, involving process that all agree to pursue over the long-term, rather than imposing action on people that are not convinced by the proposed action or that do not find a place for themselves within that action.

The booklets themselves have been produced for a given context with local priorities and are not intended for use elsewhere, although they could serve as an example. Focusing the booklets on local practices and familiar nearby locations made the booklets more meaningful to local families. The approach taken to developing the booklets is of more general significance as it involved an iterative participatory process which is transferable, and that contributes to developing health education in more than one way.

During this experience, project designers were led to question their representations and practices, not only regarding health but also teaching, and to recognize that others may have very different but yet legitimate representations and practices, and that because of these differences others may have unforeseen reactions to what the project designers propose. The experimental phase allowed booklet designers to synthesize this diversity of perspectives and adapt the booklets to the target public. It was also a learning experience for teachers, coming to recognize that fact-based teaching is not the only aspect of health education, and that creating an open inclusive relationship with students and their families is also essential to health education (Berger 2004). Thus the approach does not only involve creating a teaching tool, but rather using the creation of the tool to generate a dynamic in which teachers learn and become directly involved in a more open approach to teaching. When reproducing the approach, there is a necessary process of exchange to be reconstructed by participants that will need to go through the deconstruction of their representations and the process of coming to understand the other people and groups involved and their representations of the topics at hand (Jourdan 2011) in a specific local context that may differ from the ones the booklets were initially developed in. Taking these contextual variables into account plays a key role in the progression of our understanding of the teaching and learning processes

(Tupin and Dolz 2008). As the Health Passport project now involves over 5,000 people (pupils, teachers, parents, and students), we continue to pursue our analysis, particularly in the schools that have started using the booklets following the experimental phase.

References

Abdallah-Pretceille, M. (2003). *Former en contexte hétérogène. Pour un humanisme du divers*. Paris: Anthropos.

Augé, M., & Colleyn, J.-P. (2004). *L'anthropologie*. Paris: Presses Universitaires de France, Coll. Que sais-je?.

Balcou-Debussche, M. (2009). Un passeport santé pour le premier degré. *Expressions, 32*, 33–48. http://www.reunion.iufm.fr/Recherche/Expressions/32.html. Accessed 27 Oct 2013.

Balcou-Debussche, M. (2010). Manger et se soigner: entre manques et excès. In E. Wolff & M. Watin (Eds.), *La Réunion: une société en mutation*. Paris: Anthropos-Économica Univers créoles 7.

Berger, D. (2004). La formation des acteurs de l'éducation à la santé: vers des praticiens réflexifs. In D. Jourdan (Ed.), *La Formation des acteurs de l'éducation à la santé en milieu scolaire* (pp. 375–394). Toulouse: Éditions universitaires du Sud.

Berger, D., Bernard, S., Wafo, F., Hrairi, S., Balcou, M., & Carvalho, G. S. (2011). Éducation à la santé et à la sexualité: qu'en pensent les enseignants? Etude comparative dans 15 pays. *Carrefours de l'éducation, 32*, 81–103.

Buijs, G., Jociutè, A., Paulus, P., & Simovska, V. (Eds.). (2009). *Better schools through health: Learning from practice*. Third European conference on health promoting schools, Vilnius, Lithuania, 15–17 June 2009. http://www.marketingsociale.net/download/vil_bp.pdf. Accessed 27 October 2013.

Catteau, C. (2012). Les inégalités sociales de santé: l'exemple de La Réunion. *Expressions, 36*, 17–34. http://www.reunion.iufm.fr/Recherche/Expressions/36/Catteau.pdf. Accessed 27 Oct 2013.

Charmaz, K. (2002). Qualitative interviewing and grounded theory analysis. In J. F. Gubrium & J. A. Holstein (Eds.), *Handbook of interview research: Context & method* (pp. 675–694). Thousand Oaks: Sage.

Clift, S., & Jensen, B. B. (Eds.). (2005). *The health promoting school: International advances in theory, evaluation and practice*. Copenhagen: Danish University of Education Press.

Dejours, C. (1998). *Souffrance en France*. Paris: Editions du Seuil.

Downie, R. S., Tannahill, C., & Tannahill, A. (1996). *Health promotion. Models and values*. Oxford: Oxford University Press.

Graham, C., Cagiltay, K., Lim, B., Craner, J., & Duffy, T. (2001, March/April). *Seven principles of effective teaching*. The Technology Source. http://technologysource.org/article/seven_principles_of_effective_teaching/. Accessed 27 Oct 2013.

Holsti, O. R. (1969). *Content analysis for the social sciences and humanities*. Reading: Addison-Wesley, Publishing Company.

INED. (2012). *Les tendances démographiques et migratoires dans les régions ultrapériphériques: quel impact sur leur cohésion économique, sociale et territoriale?* (Report INED CE-DGRP). ec.europa.eu/regional_policy/sources/docgener/studies/pdf/rup_migratory/rapport8rup_shs_2_fr.pdf. Accessed 27 Oct 2013.

IUHPE. (2008). *Achieving health promoting schools: Guidelines for promoting health in schools*. Saint-Denis: The International Union for Health Promotion and Education.

Jodelet, D. (2006). Culture et pratiques de santé. *Nouvelle revue de psychosociologie, 1*, 219–239.

Jourdan, D. (2011). *Health education in schools: The challenge of teacher training*. Saint-Denis: Inpes, coll. Santé en action. http://www.inpes.sante.fr/CFESBases/catalogue/pdf/1380.pdf. Accessed 27 Oct 2013.

Jourdan, D. (2012). *La santé publique au service du bien commun? Politiques et pratiques de prévention à l'épreuve du discernement éthique*. Paris: Editions de Santé, coll Balises.

Jourdan, D., Piec, I., Aublet-Cuvelier, B., Berger, D., Lejeune, M. L., Laquet-Riffaud, A., Geneix, C., & Glanddier, P. Y. (2002). Éducation à la santé à l'école: pratiques et représentations des enseignants du primaire. *Revue française de santé publique, 14*(4), 403–423.

Jourdan, D., Mannix McNamara, P., Simar, C., Geary, T., & Pommier, J. (2010). Factors influencing the contribution of staff to health education in schools. *Health Education Research, 25*(4), 519–530.

Krippendorff, K. (1980). *Content analysis: An introduction to its methodology* (2nd ed.). Thousand Oaks: Sage Publications.

Lister-Sharp, D., Chapman, S., Stewart-Brown, S., & Sowden, A. (1999). Health promoting schools and health promotion in schools: Two systematic reviews. *Health Technology Assessment, 3*(22), 1–207.

Masters, G. (2004, April 5). "Beyond political rhetoric: What makes a school good" *OnLine Opinion – e Journal of Social and Political Debate*. http://www.onlineopinion.com.au/view.asp?article=2100&page=0. Accessed 27 Oct 2013.

Meirieu, P. (1994). *L'envers du tableau. Quelle pédagogie pour quelle école?* Paris: ESF.

Moon, A., Mullee, M., Rogers, L., Thompson, R., Speller, V., & Roderick, P. (1999). Schools become health promoting: An evaluation of the Wessex Healthy Schools Award. *Health Promotion International, 14*(2), 111–122.

Muijs, D., & Reynolds, D. (2005). *Effective teaching: Evidence and practice*. London: Paul Chapman Publishing.

Mukoma, W., & Flisher, A. J. (2004). Evaluations of health promoting schools: A review of nine studies. *Health Promotion International, 19*(3), 357–368.

Pérez, M. A., & Luquis, R. R. (Eds.). (2008). *Cultural competence in health education and health promotion*. San Francisco: Jossey-Bass.

Perrenoud, P. (2008). *Construire des compétences dès l'école*. Paris: ESF.

Pierret, J. (1984). Les significations sociales de la santé. In M. Augé & C. Herzlich (Eds.), *Le sens du mal. Anthropologie, histoire, sociologie de la maladie* (pp. 217–256). Paris: Editions Archives Contemporaines.

St Leger, L., & Nutbeam, D. (1999). Evidence of effective health promotion in schools. In D. Boddy (Ed.), *The evidence of health promotion effectiveness: Shaping public health in a New Europe*. Brussels: European Union.

St Leger, L., Kolbe, L., Lee, A., McCall, D., & Young, I. (2007). School health promotion. Achievements, challenges and priorities. In D. V. McQueen & C. M. Jones (Eds.), *Global perspectives on health promotion effectiveness* (pp. 107–124). New York: Springer.

St Leger, L., Young, I., Blanchard, C., & Perry, M. (2009). *Promoting health in schools: From evidence to action*. Saint-Denis: International Union for Health Promotion and Education.

Themessl-Huber, M., Lazenbatt, A., & Taylor, J. (2008). Overcoming health inequalities: A participative evaluation framework fit for the task. *Perspectives in Public Health, 128*(3), 117–122.

Tones, K., & Tilford, S. (1994). *Health education, effectiveness, efficiency and equity*. London: Chapman and Hill.

Trugeon, A., Thomas, N., Michelot, F., & Lémery, B. (2010). *Inégalités socio-sanitaires en France, de la région au canton*. Issy-Les-Moulineaux: Elsevier Masson, Collection Abrégés.

Tucci, I. (2011). Immigration, intégration et diversité en France et en Allemagne. *Informations Sociales, 163*(1), 116–123.

Tupin, F., & Dolz, J. (2008). Du périmètre des situations d'enseignement-apprentissage. *Les Dossiers des sciences de l'éducation, 19*, 141–156.

Viig, N. G., & Wold, B. (2005). Facilitating teachers' participation in school-based health promotion – A qualitative study. *Scandinavian Journal of Educational Research, 49*(1), 83–109.

WHO. (1997). *Promoting health through school*. Geneva: WHO Editor.

WHO. (1999). *Preventing HIV/AIDS/STI and related discrimination: an important responsibility of health promoting schools* (WHO information series on school health). Geneva: WHO Editor.

Wolff, E., & Watin, M. (2010). *La Réunion: une société en mutation*. Paris: Anthropos-Économica, Univers créoles 7.

Chapter 15
Implementing Mental Health Promoting Schools

Aleisha M. Clarke and Margaret M. Barry

Abstract This chapter examines the promotion of children's positive mental health within the health promoting schools framework. The chapter provides an overview of implementation components critical to health promoting schools practice. Following this, research on the implementation of the Zippy's Friends emotional wellbeing programme for primary school children is presented. This study is provided as an example of a multi-method evaluation approach, combining the use of a clustered randomised control design with qualitative approaches documenting the process of implementation, and case studies exploring the contextual factors that influence the process and outcomes. Key findings from this evaluation study are presented and the implications for researching the implementation of interventions within a health promoting schools approach are considered.

Keywords Mental health promotion • Emotional wellbeing • School intervention • Implementation • Children

15.1 Introduction

The increasing emphasis on the importance of social and emotional wellbeing for children and young people, including the development of emotional intelligence (Salovey and Mayer 1990; Goleman 1996) and emotional literacy (Steiner and Perry 1997), has led to an emphasis on social and emotional learning within the educational system (Durlak et al. 2011). Schools have an important function in nurturing children's social and emotional development as well as their academic and cognitive development. Enhancing children's positive mental health and wellbeing

A.M. Clarke, Ph.D. (✉) • M.M. Barry, Ph.D.
Health Promotion Research Centre, National University of Ireland, Galway, Ireland
e-mail: aleisha.clarke@nuigalway.ie; margaret.barry@nuigalway.ie

V. Simovska and Patricia Mannix-McNamara (eds.), *Schools for Health and Sustainability: Theory, Research and Practice*, DOI 10.1007/978-94-017-9171-7_15,

improves their ability to learn and to achieve academically as well as their capacity to become responsible adults and citizens (Payton et al. 2008; Zins et al. 2004; Weissberg et al. 1991). Current conceptualisations of positive mental health emphasise the importance of both hedonic and eudaimonic dimensions of wellbeing, including subjective emotional wellbeing and positive psychological and social functioning (Keyes 2007; Kovess-Masfety et al. 2005). The school setting can provide a supportive environment for promoting children's positive mental health and wellbeing, fostering the development of the social and emotional skills and resources necessary for positive youth development.

This chapter examines the promotion of children's positive mental health within the health promoting schools framework and explores the factors that influence the implementation of sustainable change at a whole school level. Drawing on the findings from the evaluation of an emotional wellbeing programme for primary school children, this chapter addresses how children's emotional, social and mental wellbeing can be promoted and strengthened within a health promoting schools framework.

There is a substantive body of international evidence that interventions which promote young people's emotional and social wellbeing in schools, when implemented effectively, can produce long-term benefits for young people (Weare and Nind 2011; Durlak et al. 2011; Payton et al. 2008; Jane-Llopis et al. 2005; Wells et al. 2003; Greenberg et al. 2001; Harden et al. 2001; Durlak and Wells 1997; Lister-Sharp et al. 1999). A meta-analysis examining the impact of 213 universal school-based interventions involving 270,034 school children age 5–13 years (Durlak et al. 2011) reported that compared to students in the control groups, children participating in social and emotional learning programmes demonstrated improvements in multiple areas including: enhanced social and emotional skills (mean ES = 0.57); improved attitudes towards self, school and others (mean ES = 0.23); enhanced positive social behaviour (mean ES = 0.24); reduced conduct problems – misbehaviour and aggression (mean ES = 0.22); reduced emotional distress – stress and depression (mean ES = 0.24). The review also found that in addition to improving students' social and emotional skills, these programmes significantly improved children's academic performance (mean ES = 0.27) yielding an average gain on achievement test scores of 11–17 percentile points.

Similarly, in a review of 52 systematic reviews of mental health interventions in schools, Weare and Nind (2011) concluded that interventions had wide-ranging beneficial effects on children and young people, on classrooms, families and communities and on a range of mental health, social, emotional and educational outcomes. Positive findings also emerge from a review of the evidence from low and middle income countries concerning the impact of school-based interventions on the mental health and well-being of young people living in poverty (Barry et al. 2013). Key characteristics of effective mental health interventions identified in these reviews include; a focus on positive mental health as opposed to prevention of mental disorder, teaching competence enhancement skills; starting early with the youngest children and continuing through the school grades. Reviewers of the evidence to date (Weare and Nind 2011; Barry and Jenkins 2007; Tennant et al. 2007;

Jane-Llopis et al. 2005; Browne et al. 2004; Wells et al. 2003; Lister-Sharp et al. 1999) conclude that taking a whole school approach, which embraces changes to the school environment as well as the curriculum, and involving parents, families and the local community, is more likely to be effective, resulting in enduring positive change. Taking a whole school approach is in keeping with the health promoting schools initiative with its focus on creating supportive environments for learning and development.

Health promoting schools offer a useful framework for strengthening the school's capacity as a mental health promoting setting for living, learning and working. This requires adopting a comprehensive approach with the use of coordinated strategies aiming to bring about change at the level of the individual, the classroom and the school in the context of the wider community. To date, however, the promotion of positive mental health has not received widespread attention as part of the health promoting schools approach. Weare and Markham (2005) state that effective work to promote mental health will not happen by chance. There is a need for explicit, coordinated strategies and programmes, based on sound research evidence and assessment of their effectiveness. Currently, the bulk of the evidence concerning the impact of school-based mental health promotion interventions is drawn from highly structured programmes delivered in a systematic way lending themselves to outcome research based on traditional randomised controlled trial (RCT) designs. This contrasts with the health promoting schools approach, which focuses on the process of effecting systems change at the environment level, rather than delivering discrete interventions.

Whilst recommending the need for mental health to be promoted within a whole school approach, Weare and Nind's (2011) review pointed to the lack of robust evaluations generating 'hard outcomes' from whole-school programmes. The type of whole school approach practiced in Europe and Australia, building on the health promoting schools framework, tends to be based on 'bottom up' health promotion principles including; empowerment, autonomy, democracy, local adaptability and ownership, and apply a more flexible and non-prescriptive approach to implementation. This is in contrast with school-based programmes from the US, where the majority of the evidence originates. The US interventions adopt a more top-down, manualised approach, with pre-packaged programmes emphasising prescriptive training and a strict requirement for programme fidelity. Weare and Nind (2011) contend that whilst the health promoting schools approach is generally seen as providing essential supportive structures, positive climates, empowered communities and end-user involvement, this style of approach makes it more difficult to evaluate and demonstrate measurable change. They propose balancing this style with some more focused and prescriptive elements. As part of this, greater clarity around the operationalisation of what is to be implemented and how it should be implemented is recommended in order to achieve optimum results. Samdal and Rowling (2013) have also commented that while general guidelines articulating the principles of a health promoting schools approach have been provided, the lack of specific implementation guidelines makes it difficult for schools to identify how to achieve a health promoting schools approach and results in a wide array of practices across schools and countries.

The challenge of evaluating comprehensive health promoting schools interventions has highlighted the need for research approaches that take into account the contextual and dynamic nature of the school setting (Dooris and Barry 2013; Rowling 2002; Parson and Stears 2002). The evaluation of a health promoting schools approach requires careful documentation of actual implementation, assessing the role of contextual factors in the school's ecology that facilitate effective change processes, and measuring multiple outcomes using a variety of measures drawn from diverse sources. Adopting a health promoting schools approach brings a shift away from delivering single discrete interventions measuring their 'linear' impact on individual students. Dooris and Barry (2013) argue that implementation and evaluation strategies are needed that will capture the synergistic interaction and impact on multiple interdependent interventions and systems operating at different levels within the context of the whole school setting. This underscores the need for methodological approaches that acknowledge the importance of a non-linear process of change involving multi-interdependent systems.

Samdal and Rowling (2013) call for greater attention to creating a science base for the health promoting schools approach. They argue that the absence of effective comprehensive implementation guidance for health promoting schools comprises the efficacy outcomes of whole school change for health and learning. Dooris and Barry (2013) point to the limitations of the traditional evaluation approach to researching implementation in settings, due to the methodological challenges involved in capturing complexity and determining the extent of systems change and transformation within the settings approach. Weiner and colleagues (2009) have called for a stronger knowledge and theory base to guide implementation of complex innovations and interventions in organisational settings. Greenhalgh et al. (2005) have also argued for the development of theory driven research, a focus on process rather than 'package', a greater emphasis on ecological analyses, a common language, measures and tools, collaboration and coordination, multi-method research and participation between researchers and practitioners. To date, however, there has been a relative paucity of research on implementation within a health promoting schools approach and the quality of implementation necessary for positive outcomes to be produced and sustained.

15.2 Programme Implementation

Fixsen et al. (2005) describe implementation as a specified set of activities designed to put into practice an activity or programme of known dimensions. Durlak (1998) defines implementation as what the intervention consists of in practice and how much it is delivered according to how it is designed. From a research perspective, implementation research enhances our ability to map the critical connections between the local context, intervention activities and the intended intermediate and long-term outcomes. Understanding the implementation process is, therefore, critical to the effective adoption, replication and dissemination of interventions and facilitates the translation of research into effective practice and the development of practice-based evidence (Dooris and Barry 2013). Implementation information allows for greater understanding of the internal dynamics and operations of interventions, how the intervention components fit together, how the implementers and intervention recipients/users interact and the obstacles they face and resolve in the process. Implementation data is also critical to interpreting positive or negative outcomes as they strengthen any conclusions that can be made about the intervention's role in producing change and inform the replication and maintenance of interventions across settings (Barry 2007; Barry and Jenkins 2007; Barry et al. 2005; Mihalic 2002; Greenberg et al. 2001; Durlak 1998). To assess implementation adequately, information is needed about specific interventions activities, how they are delivered, and the characteristics of the context or setting in which the intervention is conducted. Within a health promoting schools approach, understanding the school context and the wider implementation system is critical as the school setting itself constitutes the focus of the intervention.

From the research conducted to date, it is clear that implementation represents a complex interaction between characteristics of the implementation system, characteristics of the implementer, and various aspects of the setting and organisational context in which the intervention is implemented (Dariotis et al. 2008; Greenberg et al. 2005; Chen 1998). A conceptual model of the implementation system has been proposed by Chen (1998) and expanded by Greenhalgh et al. (2005). It outlines five dimensions which play an integral role in influencing implementation; the characteristics of the implementer (e.g. knowledge, skills, motivation), implementing organisation (e.g. structure, ethos, history, resources), interventions activities (e.g. delivery of specific actions, quality and availability of training, resources etc.), participants (identifying, recruiting, engaging and retaining the participation of the target population), and the specific context (environment, local policies, agencies, collaboration etc.) that may affect the quality of the intervention's implementation.

The necessity for a multilevel ecological framework for understanding implementation is underscored in the literature, including the importance of assessing the context specific factors that influence the quality of implementation in the local setting (Durlak and DuPre 2008). This includes organisational structures and policies, readiness to implement the interventions, both in terms of general organisational capacity and intervention specific capacity, mobilisation of support, and generally determining the ecological fit of the intervention in the local context.

A health promoting schools approach, like all setting approaches, involves a complex dynamic of group behaviours and system changes within the school and in collaboration with external stakeholders (Deschesnes et al. 2003; Whitelaw et al. 2001). This dynamic interplay between organisations and individuals requires implementation theory and research to capture the synergistic interaction and impact of multiple interdependent interventions and systems operating at different levels and spheres within the context of the specific school setting.

15.3 Implementation Components Critical to Health Promoting School Practice

Building on scientific knowledge of implementation theory, Samdal and Rowling (2011) conducted a meta-analysis of the literature of implementation studies of health promoting schools. The purpose of this analysis was to identify implementation components critical to health promoting schools practice. A total of eight sources fulfilled the inclusion criteria and the narrative synthesis identified eight implementation components. These components were grouped subsequently into three categories; School Leadership, Establishing Readiness for Change and Organisational Context (Samdal and Rowling 2013). An overview of each of these implementation components will now be provided.

15.3.1 School Leadership

The first category of school leadership covers the two components of *Leadership and Management* and *Policy and Institutional Anchoring*. Leadership and policy anchoring play important roles in terms of providing direction and support for the health promoting change process in school.

15.3.1.1 Leadership and Management Practices

A balance of leadership and management has been found to be core to achieving organisational development and change. The primary function of leadership is to stimulate readiness and motivation for change as well as providing role modelling and support for staff in their change agent roles. The development and nurturing of a professional learning community is found to be a functional vehicle for the leadership role. In its management role leadership needs to put in place practices and structures that can facilitate organisational development and change. Such practices and structures are related to resource allocation for professional development and time frames for teacher collaboration and exchange.

15.3.1.2 Policy and Institutional Anchoring

A key strategy for school change for improved health outcomes is the development and/or review of school/district level policies. Inclusion of concrete actions in the school policy ensures that priority will be given from the leadership in terms of facilitation and resource allocation. In addition, the statements in the policy document commit all stakeholders to work towards achieving the aims stated and thereby ensuring that effort is given to the agreed aims.

15.3.2 Establishing Readiness for Change

The second category of establishing readiness for change encompasses the components of *Preparing and Planning for School Development, Professional Development and Learning* and *Student Participation*. Creating readiness is concerned with providing direction and vision by identifying how the innovation fits with the overall values and aims of the school. This approach is considered a prerequisite for implementation success.

15.3.2.1 Preparing and Planning for School Development

Preparing a health promoting school approach requires a focus on programme theory and/or programme planning models. The health promoting school approach builds on the belief that by achieving change in the school environment, change can also be achieved at individual level. Thus, focusing on the school as a setting and making structural changes anchored in policies is recommended as an efficient approach. In this initial phase it is important to identify concrete policies, structures and practices for the complete implementation approach. This is intended to help anchor the health promoting school approach in the school organisation. Time and effort should be given to develop readiness including a commitment for change in staff, students, parents and other relevant stakeholders in the local community. The establishment of a coordinating committee with representation from all relevant stakeholders is recommended. This team will identify, in collaboration with stakeholders and based on local needs, topics / areas to address and develop actions to initiate the organisational change process.

15.3.2.2 Professional Development and Learning

In any school-based change process the teachers are core change agents. The staff's competence and understanding of what to achieve and how to achieve it, have been found to be critical to the success and impact of the organisational change process. Partners such as health promotion specialists or advisors in the local community may provide crucial technical support. Professional development and learning may be seen to constitute a core base for building organisational capacity for change. Such capacity has been found to be core for developing necessary understanding, motivation and skills for the implementation of the health promoting school approach as well as generating a general attitude and competence for undertaking organisational change processes. Not only teaching staff, but all relevant partners in the health promoting school process need to be integrated in a professional development and learning process.

15.3.2.3 Student Participation

Student participation is a core element of the health promoting school. Student participation may be seen as a means and a goal to maximise motivation for health and learning. Students who experience that their contributions are sought and valued may increase their intrinsic motivation and thereby academic achievement and wellbeing. Teacher provision of support to facilitate students' participation in decisions where students are being heard and also have the skills to listen to others' arguments is found to empower students to achieve learning goals and develop self-reliance in their thinking.

15.3.3 Organisational Context

The third category addresses the organisational context and includes the components of *Relational and Organisational Support Context, Partnerships and Networking* and *Sustainability*. These components are about structures and practices that enable the school to conduct the specific and needed actions.

15.3.3.1 Relational and Organisational Support Context

Relational and organisational support includes the development of support structures, strategies and practices which facilitate smooth and efficient implementation of actions and activities. The climate and culture provides exchange of experiences, role modelling and relational support. Organisational structures including timetabling, physical environment and fiscal resources constitute organisational support. It is suggested that the development of relational and organisational support brings together all other seven components for implementation of health promoting schools. The close collaboration with school leadership is vital, as the leadership is both in control of the resources and structures and represents an important capacity to align and stimulate the staff.

15.3.3.2 Partnerships and Networking

Establishing effective partnerships is critical to the implementation of the health promoting school. For effective partnerships, collaborations between school and community are needed that complement and enhance each other and evolve into comprehensive, integrated approaches. Moving in this direction through policies and systemic change processes help address the fragmentation of programmes and services that frequently evolves with health sector agencies and provides opportunities for greater participation with parents.

15.3.3.3 Sustainability

Sustainability involves building the capacity of an organisation and individuals to move beyond thinking about a time limited project, to reflecting on how new approaches and practices can be built into school priorities. Actions to facilitate sustainability include long-term anchoring of the initiative in policy plans and ongoing resources for professional development, monitoring performance of agreed actions and evaluating progress. A critical point in achieving sustainability is to address it when working on each of the other seven implementation components.

The results from Samdal and Rowling's meta-analysis represent a unique opportunity to identify a comprehensive theoretical base for implementation of health promoting schools. Samdal and Rowling (2011) contend that understanding core mechanisms of each component is vital to the effectiveness of implementing health promoting schools and that global testing of these components is now required. The remainder of this chapter examines research on the implementation of an emotional wellbeing programme for children in Irish primary schools. This study is presented as an example of a multi-method evaluation approach, combining the use of a clustered RCT design with qualitative approaches documenting the process of implementation, and case studies exploring the contextual factors that influence process and outcomes. Key findings from the evaluation of the programme are presented and the implications for researching the implementation of interventions within a health promoting schools approach are considered.

15.4 Zippy's Friends

In February 2007, an international emotional wellbeing programme was introduced into Irish primary schools on a pilot basis as part of the Social Personal and Health Education curriculum[1] (NCCA 1999). The programme, *Zippy's Friends*, is a universal intervention that aims to promote the emotional wellbeing of children age 6–9 years of age. It is designed to help all children, with different abilities and backgrounds, in diverse countries and cultures to expand their range of effective coping skills and to promote varied and flexible ways of coping with problems of day to day life (Bale and Mishara 2004). The programme has been translated into 12 different languages and to date over 712,000 children across 27 countries worldwide have participated in the programme. The Zippy's Friends programme is based upon Lazarus and Folkman's (1984) theoretical framework of coping. Lazarus and Folkman define coping as "*constantly changing cognitive and behavioural efforts to manage specific external and/or internal demands that are*

[1] The Social Personal and Health Education Curriculum is a compulsory subject in primary schools in Ireland which focuses on the development of a broad range of skills relevant to children's health and wellbeing.

appraised as taxing or exceeding the resources of the person" (p. 1410). The underlying hypothesis in developing Zippy's Friends is that if children learn at a young age to expand their repertoire of coping abilities, they will be less likely to develop serious problems in childhood, adolescence and even adult life when they are confronted with the inevitable occurrence of stressful situations (Mishara and Ystgaard 2006). Numerous studies have indicated that having a repertoire of coping skills at a young age can 'buffer' or moderate the effects of negative life stress on the development of psychological maladjustment (Dubow and Tisak 1989; Spivack and Shure 1974).

The Zippy's Friends programme is centered around a set of six illustrated stories about a group of children, their families, friends and an imaginary stick insect called *Zippy*. The 24 sessions are divided into six modules, each focusing on a particular theme: Feelings, Communication, Making and Breaking Relationships, Conflict Resolution, Dealing with Change and Loss and General Coping Skills. Throughout the programme the emphasis is on helping children to find their own solutions and to expand their range of coping strategies so that they have more options to choose from and can learn to master effective means of coping with difficult situations. Throughout the programme, great importance is attached to the interpersonal mutual aspects of coping, or seeking and giving social support. Furthermore, Zippy's Friends aims to give the children better skills in communication, conflict resolution, self-assertion, co-operation, self-control and empathy. This is in line with previous research that has found that programmes focusing on a single phenomenon have proved less effective than those that integrate training of various competencies (Weissberg and Elias 1993). The emphasis of Zippy's Friends on the promotion of resourcefulness and generic coping and competence skills is a key component of the health promoting schools framework, that is, the development of personal health skills as part of the whole school approach.

To date, the programme has been evaluated in Norway, Denmark, Lithuania, the UK, and Canada. Key findings from these studies highlight the significant positive effects of the programme on children's coping strategies (Holen et al. 2012; Mishara and Ystgaard 2006), emotional literacy skills (Holmes and Faupel 2004, 2005), social skills (Mishara and Ystgaard 2006), externalising behaviour (Mishara and Ystgaard 2006; Holmes and Faupel 2004, 2005), improved autonomy (Dufour et al. 2011) and the impact of mental health difficulties in daily life (Holen et al. 2012). One of the limitations of the evaluation studies conducted to date is that they provide little information about the process of implementation across the diverse school settings and factors that affected the quality of implementation. These limitations weaken the conclusions that can be drawn regarding the outcomes produced and the degree to which the 'implementation system' affected programme outcomes. The Irish evaluation of Zippy's Friends was designed to build on previous evaluations and develop an understanding of factors affecting the adoption and implementation of the programme and the supports necessary to strengthen future implementation and sustainability of the programme within the Irish education system (Clarke and Barry 2010; Clarke 2011).

15.5 Implementation and Evaluation of Zippy's Friends in Ireland

In consultation with the Department of Education and Skills, the Zippy's Friends programme was piloted in DEIS (designated disadvantaged) schools in the West of Ireland *(DEIS schools: Delivering Equality of Opportunity in Schools: an action plan for social inclusion)* (Department of Education and Science 2005). Economically disadvantaged children are considered especially at risk for the development of mental health problems because of the greater number of negative or undesirable life events and adverse conditions (risks factors) to which they are exposed (Fryers et al. 2005; Keenan et al. 1997; McLeod and Shanahan 1996; McLoyd 1998). This over-abundance of negative life events can place demands on them that exceed their coping resources (Sterling et al. 1985).

The specific aims of this evaluation were to:

- Determine if an international evidence-based programme could be adapted and successfully implemented in the local context of disadvantaged primary schools in Ireland
- Assess the immediate and long-term impact of the programme on the children's emotional and behavioural wellbeing and coping skills
- Examine the process of implementation and the relationship between this process and the outcomes of the programme.

This study employed a cluster randomised controlled trial design with assessments carried out before (T1), during (T2 & T3), immediately after (T4) and at 12 months post-implementation (T5). The 24 week programme was piloted with children in first class (age 6–7 years). In January 2008, prior to the random assignment of schools to intervention and control status, the Department of Education and Science required that teachers in the intervention group would be randomly assigned to one of two groups: full implementation and partial implementation. Teachers in the full implementation group (Intervention Type I) were asked to implement the programme as faithfully as possible. Teachers in the partial implementation group (Intervention Type II) were told that they could use the programme as a resource. They were not required to implement all aspects of the Zippy's Friends lessons and they had the freedom to combine / supplement Zippy's Friends lessons with other SPHE resources such as *Walk Tall, RSE* and the *Stay Safe Programme*. The purpose of assigning teachers to full and partial implementation was to determine if there was a difference in programme outcomes between the two intervention groups. Specifically, the Department of Education and Science wished to determine if the Zippy's Friends could be successfully used as a resource in Irish primary schools rather than as a discrete programme that required faithful replication.

Implementation of the programme was coordinated by the Health Promotion Service of the Health Service Executive (HSE) in Ireland. Health Promotion Specialists provided a two day training workshop on programme implementation for the intervention teachers. During programme implementation, the Health

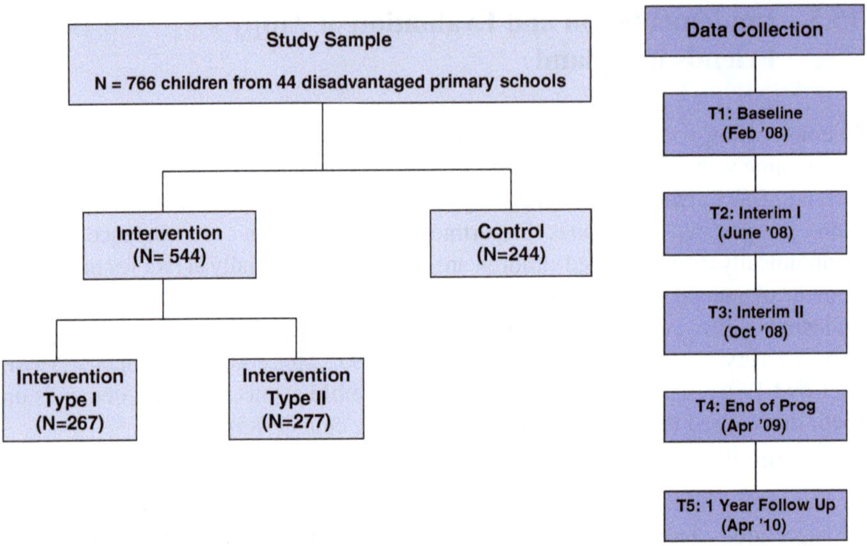

Fig. 15.1 Study sample and data collection points

Promotion Specialists were also engaged in the provision of ongoing support to teachers through school visits and an interim group meeting. An overview of the study sample and data collection points is presented in Fig. 15.1.

A total of 766 pupils and 42 teachers from 44 disadvantaged schools were randomly assigned to control and intervention groups. Impact and outcome measures were used to assess the effect of the programme on the children's emotional literacy skills (Faupel 2003), emotional and behavioural wellbeing (Goodman 1997) and coping skills (Ryan-Wenger 1990; Williams et al. 1989). The process of programme implementation was monitored and documented using a range of qualitative and quantitative measures including structured observations, weekly questionnaires completed by teachers, focus group review sessions with teachers, child participatory workshops, an Ethos Questionnaire and Social, Personal and Health Education Questionnaire completed by teachers. The purpose of these measures were to: (i) understand the context within which the programme was implemented (ii) determine the level of programme fidelity and quality of implementation across the intervention group (iii) ascertain the strengths and weaknesses of the programme (iv) understand the interaction between characteristics of the implementation system, characteristics of the implementer and various aspects of the setting and organisational context in which the programme was implemented. In addition, case studies of two specific schools participating in the programme were carried out. The purpose of these case studies was to examine the implementation of the programme in a rural and urban disadvantaged school within the context of a whole school approach. Specifically, this study sought to determine the ethos and environment of the schools, their links with the local community, the degree of parental involvement and factors that influenced programme implementation in the local context (Clarke et al. 2010). Key findings related to programme efficacy and implementation will now be presented. Full details of the results are reported elsewhere (Clarke and Barry 2010; Clarke 2011).

15.6 Zippy's Friends Evaluation Findings

15.6.1 Programme Fidelity

On completion of each session, teachers completed a programme fidelity checklist, indicating what parts of each session they fully implemented, partially implemented or omitted. Prior to examining programme efficacy, a comparison between the level of programme fidelity across teachers in Intervention Type I (full implementation) and Intervention Type II (partial implementation) was carried out. Results from the teachers' weekly questionnaires indicated that programme fidelity was high among both intervention groups. The teachers in Intervention Type I (full implementation group) fully implemented 86.4 % of the programme and the teachers in Intervention Type II (partial implementation) fully implemented 86.6 % of the programme. These findings correspond with the fidelity results from the structured observations which found similar high levels of programme fidelity across both intervention groups. Given that there was no difference in programme fidelity across the two intervention groups, data from the Emotional Literacy Checklist (Faupel 2003) and the Strengths and Difficulties Questionnaire (Goodman 1997) were analysed by comparing the intervention group (full and partial implementation groups combined) with the control group's results.

15.6.2 Programme Efficacy

15.6.2.1 Emotional Literacy Skills

Repeated measures analysis of covariance (ANCOVA) conducted at the individual level[2] indicated that the programme had an overall significant positive effect on the children's emotional literacy skills. Post-intervention scores from the Emotional Literacy Checklist showed a significant increase in the intervention group's Total Emotional Literacy score ($p<0.01$) when compared with the control group. Specifically, there was a significant increase in the intervention group's Self-Awareness, Self-Regulation and Motivation scores ($p<0.01$) between pre- and post-intervention (Clarke 2011). These positive findings are supported by qualitative findings from child participation workshops and teacher focus group review sessions carried out at the interim and post-intervention (Clarke 2011). Twelve month follow up results revealed that the programme did not have a lasting effect on

[2] Multi-level modelling has since been applied to the data and results are forthcoming.

children's Total Emotional Literacy score as measured by the Emotional Literacy Checklist. Long term findings were evident for one Emotional Literacy subscale, Motivation ($p<0.001$).

15.6.2.2 Emotional and Behavioural Wellbeing

Data from the Strengths and Difficulties Questionnaire were also analysed at the individual level using repeated measures analysis of covariance (ANCOVA). Results showed that the programme did not have a significant impact on the children's Total Difficulties score between pre- and post intervention or between pre-intervention and 12 months follow-up. Examination of the subscales, however, revealed that at post-intervention there was a significant decrease in the intervention group's Hyperactivity score ($p<0.05$) when compared with the control group. Twelve month follow up results revealed that the programme did not have a lasting effect on children's Hyperactivity score.

15.6.2.3 Impact of Programme Fidelity on Programme Outcomes

In order to determine the impact of fidelity on programme outcomes, fidelity results were analysed according to teachers who implemented the programme with high fidelity (≥90 % of the programme) and low fidelity (≤75 % of the programme). Using the children's Emotional Literacy Checklist (Faupel 2003) scores, as measured by the teachers, paired samples t-tests revealed no significant change in the Emotional Literacy scores of children in the low fidelity group between pre- and post-interventions ($p=.860$). In contrast to this, there was a significant improvement in the Total Emotional Literacy score among children the high fidelity group ($p=.000$). These results indicate that the programme had a significant positive effect on children's emotional literacy skills when implemented with high fidelity and had no effect when implemented with low fidelity. These findings are consistent with a number of other studies, which found that greater fidelity is associated with better outcomes across a diverse set of prevention and promotion models such as social skills training interventions (Kam et al. 2003; Wilson and Lipsey 2000; Gresham et al. 1993; Botvin et al. 1990), coordinated community based prevention programmes (Pentz et al. 1990) and classroom ecology interventions (Harachi et al. 1999).

15.6.3 Programme Implementation

As mentioned previously, an important and unique aspect of this study was the use of a process evaluation with multiple methods as a means to understanding the context within which the programme was implemented. Findings in relation to factors that affected programme implementation, teachers' recommendations, and the school context within which the programme was implemented will now be presented:

15.6.3.1 Factors Affecting Programme Implementation

Through the use of the teachers' weekly questionnaires, class observations, interim and end of programme focus group review sessions with intervention teachers and child participatory workshops, this study identified factors related to the programme, the implementer and the recipients which facilitated and hindered programme implementation. Firstly, in relation to the programme, the teachers repeatedly made reference to the active engagement of the children through child-centered learning techniques and the relevance of the content to the children's lives. Looking at the most successful modules (Module 5: Dealing with Change and Loss and Module 3: Making and Breaking Friendship), the suitability of the content to the children's daily lives, the child-centered activities used in the sessions (role play, making puppets, drawing pictures) and the practical nature of the lessons, were the most frequently reported positive aspects of these modules by the teachers. Of the modules that received the lowest ratings (Module 1: Feelings and Module 2: Communication), lack of activities and over-use of teacher talk were consistently reported. These results demonstrate that the children responded better when they were away from their desks and actively involved in a variety of activities such as role play / drawing / engaging in group discussions. This finding is in line with previous research carried out on educational interventions regarding the need for varied teaching methods (Warwick et al. 2005; Nation et al. 2003; Tobler and Stratton 1997; Dusenbury and Falco 1995).

Another element of the programme that facilitated programme implementation was the provision of teacher training. All of the teachers that implemented the Zippy's Friends programme commented on the value of teacher training. Several teachers also spoke about the importance of the ongoing support which they received from the Health Promotion Specialists throughout the year. The teachers' views are supported by research relating to programme implementation and sustainability which have shown that high quality teacher training combined with the provision of ongoing technical support increases providers' (i) knowledge of how the programme works and what is necessary to implement the programme effectively (ii) ability to deal with implementation challenges in a timely manner and (iii) understanding and acceptance of the intervention (Hallam et al. 2006; Kam et al. 2003; Bishop and Roberts 2005; Lewis et al. 1990). Fixsen and colleagues (2005) state that the conventional approach of 'train them and send them on their way' is ineffective in promoting high quality implementation. Instead, effective programme adoption and implementation requires initial training that is interactive and engaging, provides opportunities for behavioural rehearsal and is followed up with ongoing coaching, technical assistance and support (Bumbarger et al. 2010).

Regarding the teacher themselves, their involvement with the lesson in terms of taking part in the role plays and sharing their personal experiences with the children was regarded as factors that greatly facilitated child participation. Whilst the authors are not aware of previous research which specifically points to the importance of teacher participation, the development and maintenance of a safe, supportive learning environment where children feel cared for and respected is regarded as one of the key components necessary for the promotion of social and emotional learning (Zins et al. 2004). It is clear from this study that the teachers' participation had a significant impact on the creation of a supportive environment which enhanced child participation.

Finally, in terms of the recipients, the teachers regarded the children's enjoyment of the programme as a salient facilitating factor. Teachers spoke about the children's enthusiasm for the lessons and the knock-on effect this had on their motivation to teach the programme. Although not widely discussed in the literature, the children's enjoyment and enthusiasm for the programme appears to influence teachers' perception of and attitude towards the intervention, factors which are known to facilitate/hinder quality of implementation (Domitrovich et al. 2008).

It is important to note that the findings from the teacher weekly questionnaires, focus group review sessions and class observations in relation to programme implementation are in line with findings from child participatory workshops which were carried out in a sample of intervention schools at post-interventions. When the children were asked about what they had learned at post-intervention, key themes from each module were recalled – 'Feelings', 'Friendship', 'Caring for others', 'Not fighting/bullying others'. 'Listening, 'Coping with your problems'. The accurate recollection of the programme content by the children suggests that they were engaged and actively learning throughout the programme. Findings in relation to what the children liked the most about the Zippy's Friends programme included the stories, pictures to accompany them, the activities and sharing their problems in the circle. These findings from the child participatory workshops further accentuate the importance of activity-based child centered learning in the implementation of the Zippy's Friends programme.

Factors that hindered programme implementation were mostly concerned with structural and school level factors. Time was the most frequently reported problem associated with implementing the programme. Teachers found it difficult to find time during the day when all of the children were present in the classroom and not attending learning support classes. Time of implementation during the school year (commencing the programme half way through the academic year) was also problematic for several teachers, particularly for teachers implementing the programme in the multi-grade setting. Lack of space in the room was noted as a barrier by some teachers. Small classrooms meant that some teachers were not able to put the children into a circle for Zippy time. Teachers found this to have a negative effect on child participation throughout the programme.

15.6.3.2 Teachers' Recommendations

The teachers made a number of recommendations to enhance the future implementation and sustainability of Zippy's Friends in Ireland. Firstly, teachers recommended the need for more hands on activities and also the use of multi-media activities to further engage the children and enhance their learning. Secondly, teachers commented that a once-off programme in primary school was not sufficient and advocated the need to implement the programme throughout the school from junior infants to sixth class. Several teachers spoke about the importance of a whole school approach and whole school training so that all teachers are aware of and able to implement the strategies during class time and out in yard. One teacher wrote: "*Will work best if taken on as a whole school approach with everyone being aware of the strategies*". The third recommendation was the need for parental involvement in the programme. Teachers suggested the use of a workbook at home that would reinforce strategies taught in school in the home environment. One teacher remarked: "*I felt it needed something to let the parents know what we had discussed so that they could even reinforce it at home*".

15.6.3.3 School Context

In order to understand the environmental context within which the programme was implemented, teachers in both intervention (N=30) and control groups (N=10) completed the Ethos Questionnaire. This questionnaire was specifically concerned with: school policies, the promotion of positive mental health throughout the school, the implementation of the Social, Personal and Health Education (SPHE) curriculum, school ethos, support from community services, parental involvement and barriers that exist in the promotion of positive mental health throughout the school. In addition to the Ethos Questionnaire, case studies of two contrasting schools (a large urban school with a multicultural profile in an area of multiple disadvantage and a smaller, almost monocultural rural school) were employed. The case study method explored the views of teachers, pupils, parents and key informants of the wider community on the delivery of the programme within the context of a whole school approach.

The results from the Ethos Questionnaire indicated that both intervention and control schools provided a positive and supportive school environment for the children and that teachers within the schools worked towards providing for the children's needs. Three areas that both intervention and control schools scored poorly were highlighted through the Ethos results. Firstly, the needs of staff were not reported as being a high priority of schools in this study. The majority of schools did not have a policy on staff health and welfare. In addition, staff were unlikely to seek help when feeling stressed and most teachers said support was not available for staff involved in stressful incidents. Secondly, links with the wider school community were not reported as being well established for control and intervention schools. Despite the fact that most teachers said that schools were receptive to approaches from community agencies in relation to health matters, both intervention and control schools reported low levels of collaboration with these agencies. Thirdly, whilst most teachers reported that parents were interested and supportive of the school and its governance, fewer teachers reported the active involvement of a broad range of parents in school life. Furthermore, teachers also noted the lack of opportunities given to parents to participate and learn about the content of the school's SPHE curriculum.

Additional findings from the case studies revealed the reality of programme implementation in two disadvantaged schools and exemplified how the local contexts for programme implementation can differ within a relatively small regional area. With regard to the community context, the data indicated contrasting levels of community engagement and parental involvement in the two schools. The perceived lack of a cohesive community context and low levels of parental involvement in the large urban school contributed to a more challenging environment for programme implementation. In relation to organisational practices and processes, differences emerged between the teachers in the two schools regarding the perceived school ethos and environment, particularly with regard to school practices in supporting children during stressful periods and developing positive open relationships between staff, children and parents. Concerning characteristics of the implementers, despite similar levels of reported implementation fidelity, the teachers in the small rural school had a much more positive view of the programme and reported positive programme influences on the classroom atmosphere and on their own capacity to help the children. These results were not evident from the urban school where teachers' attitudes were much less positive. Further details of the case study findings are reported in Clarke et al. (2010).

15.7 Lessons Learned in the Context of Health Promoting Schools

The key findings from the evaluation of the Zippy's Friends programme in Ireland are significant in terms of their application to health promoting schools. The results from this study highlight the significant positive effect of the Zippy's Friends programme on children's emotional literacy skills and hyperactivity. These findings are particularly relevant given the fact that evidence suggests the foundations of good mental health throughout life are laid in the early years and that without intervention, emotional and behavioural problems in young children may be less amenable to intervention after 8 years (Tennant et al. 2007; Leckman and Leventhal 2008; Eron 1990). This points to the need for evidence-based strategies that promote children's emotional wellbeing, such as those employed in the Zippy's Friends programme, to be implemented as part a health promoting schools approach in primary schools. Furthermore, given the increasing evidence of the strong positive associations between interventions that teach social and emotional competence and their impact on a wide range of educational gains including improved school attendance, higher motivation, higher morale and improved academic achievement (Durlak et al. 2011; Payton et al. 2008; Zins et al. 2004; Catalano et al. 2002; Durlak and Wells 1997; Durlak 1995), the integration of evidence-based mental health promotion interventions as part of health promoting schools is ever more salient.

Twelve month follow up results revealed that the programme did not have a lasting effect on children's emotional literacy skills or behavioural wellbeing as measures by the Emotional Literacy Checklist and Strengths and Difficulties Questionnaire. These findings point to the need for a more long term sustained approach. This is in line with recommendations from systematic reviews regarding the need for interventions to operate for a lengthy period of time and embed work within a health promoting school approach which includes features such as changes to the curriculum, improving school ethos, teacher education, liaison with parents, parenting education, community involvement and coordinated work with outside agencies (Weare and Nind 2011; Adi et al. 2007; Stewart-Brown 2006; Wells et al. 2003). Furthermore, the request from teachers for (i) another 'Zippy's Friends' programme in the senior end of primary school (ii) whole school teacher training and (iii) parental involvement highlights the recognition among teachers that one off interventions are insufficient in sustaining children's competencies and that a long term, developmental approach involving all key stakeholders in necessary.

Analysis of programme implementation in the context of its impact on children's outcomes reveals two important findings. Firstly, the fact that teachers in intervention Type II, who were given the option to implement the programme as a resource, implemented the programme with high fidelity (86.5 %) indicates the level of teacher commitment and support for the programme. Samdal and Rowling (2011) report that the advancement of understanding, motivation and skills through professional development and learning are essential to the implementation of the health promoting schools approach. The results from this study point to the level of buy-in and commitment from teachers as a result of the training and support provided by the Health Promotion Specialists. Teachers themselves commented on the importance of teacher training and ongoing support in facilitating programme implementation. The fidelity results thus support the important role professional development and learning play in implementation as highlighted by Samdal and Rowling (2011). In addition, strong programme adherence amongst the teachers who were given the option to use the programme as a resource indicates the likelihood of future faithful replication and sustainability of the programme in the Irish school setting.

Secondly, the programme had a significant positive impact on children's emotional literacy skills and hyperactivity when implemented with high fidelity (>90 %), however, children who received less than 75 % of the programme showed no improvements in their emotional literacy skills between pre- and post-intervention. In terms of health promoting schools, these findings point to the importance of quality of implementation in achieving programme outcomes. In addition, these findings support Dooris and Barry's (2013) call for the need to assess implementation when evaluating health promoting schools in order to understand how outcomes are produced and to protect against the dangers of a Type III error in reporting evaluation findings, that is, the intervention as delivered is of such poor quality as to invalidate the outcomes.

The results from process evaluation provide an insight into factors which facilitate and hinder implementation. These results have implications not just for the future implementation of Zippy's Friends but also additional interventions implemented as part of the health promoting schools approach. The structured nature of the programme, the suitability of the content for the children, teacher involvement during the lessons, the variety of engaging activities, teacher training and the provision of ongoing support were cited as factors that facilitated programme implementation. These factors are in line with components Samdal and Rowling (2011) identified as necessary for health promoting schools, specifically, student participation and professional development and learning. In addition, the issue of time, which was recognised by teachers as a barrier to implementation, points to the need for organisational support at school level to facilitate implementation. Relational and organisational support was another core component identified by Samdal and Rowling (2011) in the implementation of health promoting schools. Furthermore, it is suggested that strategies to promote positive mental health and emotional wellbeing using a health promoting schools approach could assist in ameliorating the barrier of time.

Finally, the results from the Ethos Questionnaire and the case studies of two schools provide an insight into factors operating at school level that can affect the implementation of a mental health promotion intervention. In general, the results from the Ethos Questionnaire confirm the lack of supports structures in place to assist teachers in addressing their own needs. Teachers as drivers of change in schools must be supported in their role and it is, therefore, suggested that this could be incorporated as a key component of teacher training and the provision of ongoing support in schools. Given that technical support has been identified as one of the most important dimensions of the implementation support system (Samdal and Rowling 2011; Mihalic et al. 2004; Kam et al. 2003), the delivery of pre-service and in-service teacher training and ongoing high quality support needs to be addressed at a national level. Weist and Paternite (2006) contend that the provision of strong training, technical assistance and pragmatic ongoing support increases the likelihood of positive outcomes being achieved. These outcome findings in turn fuel advocacy and policy agendas, which subsequently leads to increased resources which are strategically applied to expand and improve the quality of services. The provision of training and support, according to Weist and Paternite, is therefore, a critical element of the 'snowballing' process of growth and improvement of services in schools, across the community and beyond.

The results also illustrate the need for greater collaboration between schools and key stakeholders within the local community. As recommended in the health promoting school approach, schools must view their role as being a vital part of the wider community, in reaching out to and receiving support from parents and local agencies. Samdal and Rowling (2011) contend that collaborative working is at the core of health promoting schools approach. It enables communities to draw upon its diverse and unique strengths and ensures that the identified health needs and strategies to address health are relevant to and owned by the community. The results from this study underscore the value of providing schools with the resources and training they require to help them work collaboratively with pupils, parents and the broader community.

The case study findings highlight the uniqueness of both schools with their differing community histories, cultures, local politics and organisational capacity and structures. The capacity development needs of both schools were quite different, particularly with regard to community and parental involvement. In relation to the teachers, it was evident that those in the large urban school were less positive about the programme. Although both schools received the same level of training and support, stronger links with local support agencies and staff were reported by the rural school. The less positive attitudes of the teachers in the urban school would suggest the need for ongoing technical assistance and support in addressing perceived barriers to effective implementation. Overall, the case study findings indicated that the two schools were at very different stages of 'readiness' in terms of implementing the intervention within a whole school context. The findings point to the important influence of contextual factors on programme implementation in disadvantaged school settings, including characteristics of the local community context, level of parental involvement, school ethos and practices and teachers' attitudes to the intervention. The results point to the need to adapt strategies for school organisational change to support implementation according to the requirements of each school.

15.8 Conclusion

The presentation of evaluation findings on the implementation of an emotional well-being intervention in primary schools provides valuable insights into the process of programme delivery and the range of factors in the school setting that influence programme implementation. The results from the Zippy's Friends study highlight the importance of embedding evidence-based interventions within a whole school context. As part of this, it is necessary to give due attention to both the core elements of the programme itself and the broader whole school implementation components as outlined by Samdal and Rowling (2011). In line with this, and as recommended by Dooris and Barry (2013), it is necessary to design evaluation studies that move beyond the conventional outcome-focused approach to a more systems-based approach that account for non-linear causality and seek to embrace and elucidate the inter-relationships, interactions and synergies within the whole settings approach. Dooris and Barry (2013) state that in so doing, it will be important to integrate health measures with measures relating to the core business of the setting, to use multi-method approaches and to recognise the synergistic effects of combining methods to answer a range of evaluation questions.

Acknowledgements We are grateful to the staff and pupils of all the schools who participated in the study and to the Health Promotion Specialists and members of the National Advisory Board for their support and contribution to the research. The evaluation of the Zippy's Friends programme was funded by the Health Service Executive Population Health Directorate - Health Promotion, the National Office for Suicide Prevention Ireland, and the Irish Research Council for Humanities and Social Services. The views expressed in this paper are solely those of the authors.

References

Adi, Y., Killoran, A., Janmohamed, K., & Stewart-Brown, S. (2007). *Systematic review of the effectiveness of interventions to promote mental wellbeing in primary schools: Universal approaches which do not focus on violence or bullying*. London: National Institute for Clinical Excellence.

Bale, C., & Mishara, B. (2004). Developing an international mental health promotion programme for young children. *International Journal of Mental Health Promotion, 6*(2), 12–16.

Barry, M. M. (2007). Generic principles of effective mental health promotion. *International Journal of Mental Health Promotion, 9*(2), 4–16.

Barry, M., & Jenkins, R. (2007). *Implementing mental health promotion*. Oxford: Churchill Livingstone/Elsevier.

Barry, M. M., Domitrovich, C., & Lara, M. A. (2005). The implementation of mental health promotion programmes. *Promotion & Education,* (Suppl. 2), 30–35.

Barry, M. M., Clarke, A. M., Jenkins, R., & Patel, V. (2013). A systematic review of the effectiveness of mental health promotion interventions for young people in low and middle income countries. *BMC Public Health, 13*, 835.

Bishop, B., & Roberts, C. (2005). The process of embedding and sustaining mental health promotion programs in school contexts. *Community Psychologist, 38*(1), 14–16.

Botvin, G. J., Baker, E., Dusenbury, L., Tortu, S., & Botvin, E. M. (1990). Preventing adolescent drug abuse through a multimodal cognitive-behavioral approach: Results of a 3-year study. *Journal of Consulting and Clinical Psychology, 58*(4), 437–446.

Browne, G., Gafni, A., Roberts, J., Byrne, C., & Majumdar, B. (2004). Effective/efficient mental health programs for school-age children: A synthesis of reviews. *Social Science & Medicine, 58*(7), 1367–1384.

Bumbarger, B., Perkins, D., & Greenberg, M. (2010). Taking effective prevention to scale. In B. Doll, W. Pfohl, & J. Yoon (Eds.), *Handbook of youth prevention science*. New York: Routledge.

Catalano, R. F., Berglund, L., Ryan, A. M., Lonczak, H. S., & Hawkins, J. (2002). Positive youth development in the United States: Research findings on evaluations of Positive Youth Development Programmes. *Prevention and Treatment, 5*, Article 15.

Chen, H. (1998). Theory-drive evaluations. *Advances in Educational Productivity, 7*, 15–34.

Clarke, A. M. (2011). *An evaluation of the Zippy's Friends emotional wellbeing programme for primary schools in Ireland*. PhD Thesis, Discipline of Health Promotion, National University of Ireland Galway.

Clarke, A. M., & Barry, M. M. (2010). *An evaluation of the Zippy's Friends emotional wellbeing programme for primary schools in Ireland: End of programme report*. Galway: Health Promotion Research Centre, National University of Ireland Galway.

Clarke, A. M., O'Sullivan, M., & Barry, M. M. (2010). Context matters in programme implementation. *Health Education, 110*(4), 273–293.

Dariotis, J., Bumbarger, B., Duncan, L., & Greenberg, M. (2008). How do implementation efforts relate to program adherence? Examining the role of organizational, implementer, and program factors. *Journal of Community Psychology, 36*(6), 744–760.

Department of Education and Science. (2005). *Delivering equality of opportunity in schools. An action plan for education inclusion*. Athlone: Department of Education and Science.

Deschesnes, M., Martin, C., & Hill, A. J. (2003). Comprehensive approaches to school health promotion: How to achieve broader implementation? *Health Promotion International, 18*(4), 387–396.

Domitrovich, C. E., Bradshaw, C. P., Poduska, J. M., Hoagwood, K., Buckley, J. A., Olin, S., et al. (2008). Maximizing the implementation quality of evidence-based preventive interventions in schools: A conceptual framework. *Advances in School Mental Health Promotion, 1*(3), 6–28.

Dooris, M., & Barry, M. M. (2013). Overview of implementation in health promoting settings. In O. Samdal & L. Rowling (Eds.), *The implementation of health promoting schools: Exploring the theories of what, why and how*. Oxfordshire: Routledge.

Dubow, E. F., & Tisak, J. (1989). The relation between stressful life events and adjustment in elementary school children: The role of social support and social problem-solving skills. *Child Development, 60*(6), 1412–1423.

Dufour, S., Denoncourt, J., & Mishara, B. L. (2011). Improving children's adaptation: New evidence regarding the effectiveness of Zippy's Friends, a school mental health promotion program. *Advances in School Mental Health Promotion, 4*(3), 18–28.

Durlak, J. (1998). Why program implementation is important. *Journal of Prevention and Intervention in the Community, 17*(2), 5–18.

Durlak, J. A. (1995). *School-based prevention programs for children and adolescents*. London: Sage.

Durlak, J. A., & DuPre, E. P. (2008). Implementation matters: A review of research on the influence of implementation on program outcomes and the factors affecting implementation. *American Journal of Community Psychology, 41*(3–4), 327–350.

Durlak, J. A., & Wells, A. M. (1997). Primary prevention mental health programs for children and adolescents: A meta-analytic review. *American Journal of Community Psychology, 25*(2), 115–152.

Durlak, J. A., Weissberg, R. P., Dymnicki, A. B., Taylor, R. D., & Schellinger, K. B. (2011). The impact of enhancing students' social and emotional learning: A meta-analysis of school-based universal interventions. *Child Development, 82*(1), 405–432.

Dusenbury, L., & Falco, M. (1995). Eleven components of effective drug abuse prevention curricula. *The Journal of School Health, 65*(10), 420–425.

Eron, L. (1990). Understanding aggression. *Bulletin of the International Society for Research on Aggression, 12*, 5–9.

Faupel, A. (2003). *Emotional literacy: Assessment and intervention – Ages 7 to 11*. London: nferNelson.

Fixsen, D. L., Naoom, S. F., Blase, K. A., Friedman, R. M., & Wallace, F. (2005). *Implementation research: A synthesis of the literature*. Tampa: University of South Florida, Louis de la Parte Florida Mental Health Institute, National Implementation Research Network.

Fryers, T., Melzer, D., Jenkins, R., & Brugha, T. (2005). The distribution of the common mental disorders: Social inequalities in Europe. *Clinical Practice and Epidemiology in Mental Health, 1*(1), 14.

Goleman, D. (1996). *Emotional intelligence*. London: Bloomsbury.

Goodman, R. (1997). The strengths and difficulties questionnaire: A research note. *Journal of Child Psychology and Psychiatry, and Allied Disciplines, 38*(5), 581–586.

Greenberg, M., Domitrovich, C., & Bumbarger, B. (2001). The prevention of mental disorders in school-aged children: Current state of the field. *Prevention and Treatment, 4*(1), 1–52.

Greenberg, M. T., Domitrovich, C. E., Graczyk, P. A., & Zins, J. E. (2005). *The study of implementation in school-based preventive interventions: Theory, research, and practice*. Washington, DC: Centre for Mental Health Services, Substance Abuse and Mental Health Administration, US Department of Health and Human Services.

Greenhalgh, T., Robert, G., Bate, P., Kyrakidou, O., Macfarlane, F., & Peacock, R. (2005). *Diffusion of innovations in Health Service Organizations: A systematic literature review*. Oxford: Blackwell Publishing.

Gresham, F. M., Gansle, K. A., & Noell, G. H. (1993). Treatment integrity in applied behavior analysis with children. *Journal of Applied Behavior Analysis, 26*(2), 257–263.

Hallam, S., Rhamie, J., & Shaw, J. (2006). *Evaluation of the primary behaviour and attendance pilot* (Research Report RR717). Nottingham: DfES Publications.

Harachi, T. W., Abbott, R. D., Catalano, R. F., Haggerty, K. P., & Fleming, C. B. (1999). Opening the black box: Using process evaluation measures to assess implementation and theory building. *American Journal of Community Psychology, 27*(5), 711–731.

Harden, A., Rees, R., Shepherd, J., Brunton, G., Oliver, S., & Oakley, A. (2001). *Young people and mental health: A systematic review of research on barriers and facilitators*. London: EPPI Centre.

Holen, S., Waaktaar, T., Lervag, A., & Ystgaard, M. (2012). The effectiveness of a universal school-based programme on coping and mental health: A randomised, controlled study of Zippy's Friends. *Educational Psychology, 32*(5), 657–677.

Holmes, D., & Faupel, A. (2004). *Zippy's Friends: Interim report year 1 (2003–2004)*. Southampton: Southampton Psychology Service.

Holmes, D., & Faupel, A. (2005). *Zippy's Friends: Southampton evaluation report, year 2 (2004–2005)*. Southampton: Southampton Psychology Service.

Jane-Llopis, E., Barry, M., Hosman, C., & Patel, V. (2005). Mental health promotion works: a review. *Promotion and Education,* (Suppl. 2).

Kam, C. M., Greenberg, M. T., & Walls, C. T. (2003). Examining the role of implementation quality in school-based prevention using the PATHS curriculum. *Prevention Science, 4*(1), 55–63.

Keenan, K., Shaw, D. S., Walsh, B., Delliquadri, E., & Giovannelli, J. (1997). DSM-III-R disorders in preschool children from low-income families. *Journal of the American Academy of Child and Adolescent Psychiatry, 36*(5), 620–627.

Keyes, C. L. M. (2007). Promoting and protecting mental health as flourishing: A complementary strategy for improving national mental health. *American Psychologist, 62*(2), 95–108.

Kovess-Masfety, M., Murray, M., & Gureje, O. (2005). Positive mental health. In H. Herrman, S. Saxena & R. Moodie (Eds), *Promoting mental health: Concepts, emerging evidence practice*. A report of the World Health Organization, Department of Mental Health and Substance Abuse in collaboration with the Victorian Health Promotion Foundation and University of Melbourne. Geneva: WHO.

Lazarus, R. S., & Folkman, S. (1984). *Stress, appraisal, and coping*. New York: Springer.

Leckman, J. F., & Leventhal, B. L. (2008). Editorial: A global perspective on child and adolescent mental health. *Journal of Child Psychology and Psychiatry, and Allied Disciplines, 49*(3), 221–225.

Lewis, C., Battistich, V., & Schaps, E. (1990). School-based primary prevention: What is an effective program? *New Directions for Child Development,* (50), 35–59.

Lister-Sharp, D., Chapman, S., Stewart-Brown, S., & Sowden, A. (1999). Health promoting schools and health promotion in schools: Two systematic reviews. *Health Technology Assessment, 3*(22), 1–207.

McLeod, J. D., & Shanahan, M. J. (1996). Trajectories of poverty and children's mental health. *Journal of Health and Social Behavior, 37*(3), 207–220.

McLoyd, V. C. (1998). Socioeconomic disadvantage and child development. *American Psychologist, 53*(2), 185–204.

Mihalic, S. (2002). *Blueprints for violence prevention violence initiative: Summary of training and implementation (Final Process Evaluation Report)*. Boulder: University of Colorado at Boulder, Centre for the Study and Prevention of Violence.

Mihalic, S., Irwin, K., Fagan, A., Ballard, D., & Elliott, D. (2004). *Successful program implementation: Lessons from Blueprints*. Washington, DC: US Department of Justice, Office of Justice Programs, Office of Juvenile Justice and Delinquency Prevention.

Mishara, B., & Ystgaard, M. (2006). Effectiveness of a mental health promotion program to improve coping skills in young children: Zippy's Friends. *Early Childhood Research Quarterly, 21*(1), 110–123.

Nation, M., Crusto, C., Wandersman, A., Kumpfer, K. L., Seybolt, D., Morrissey-Kane, E., et al. (2003). What works in prevention: Principles of effective prevention programs. *American Psychologist, 58*(6–7), 449–456.

National Council for Curriculum Assessment. (1999). *Social, personal and health education curriculum*. Dublin: Government Publications.

Parson, C., & Stears, D. (2002). Evaluating health promoting schools: Steps to success. *Health Education, 102*(1), 7–15.

Payton, J., Weissberg, R. P., Durlak, J. A., Dymnicki, A. B., Taylor, R. D., Schellinger, K. B., et al. (2008). *The positive impact of social and emotional learning for kindergarten to eight-grade students: Findings from three scientific reviews*. Chicago: Collaborative for Academic, Social, and Emotional Learning.

Pentz, M. A., Trebow, E. A., Hansen, W. B., MacKinnon, D. P., Dwyer, J. H., & Johnson, C. A. (1990). Effects of program implementation on adolescent drug use behavior. *Evaluation Review, 14*, 264–289.

Rowling, L. (2002). Mental health promotion. In L. Rowling, G. Martin, & L. Walker (Eds.), *Mental health promotion and young people: Concepts and practice*. Sydney: McGraw-Hill.

Rowling, L., & Samdal, O. (2011). Filling the black box of implementation for health-promoting schools. *Health Education, 111*(5), 347–362.

Ryan-Wenger, N. M. (1990). Development and psychometric properties of the Schoolagers Coping Strategies Inventory. *Nursing Research, 39*(6), 334–349.

Salovey, P., & Mayer, J. D. (1990). Emotional intelligence. *Imagination, Cognition, and Personality, 9*, 185–211.

Samdal, O., & Rowling, L. (2011). Theoretical and empirical base for implementation components of health-promoting schools. *Health Education, 111*(5), 367–390.

Samdal, O., & Rowling, L. (2013). *The implementation of health promoting schools: Exploring the theories of what, why and how*. Oxfordshire: Routledge.

Spivack, G., & Shure, M. B. (1974). *Social adjustment of young children: A cognitive approach to solving real life problems*. London: Jossey Bass.

Spivack, G., Platt, J. J., & Shure, M. B. (1976). *The problem solving approach to adjustment*. San Francisco: Jossey Bass.

Steiner, C., & Perry, P. (1997). *Achieving emotional literacy*. London: Bloomsbury Publishing.

Sterling, S., Cowen, E. L., Weissberg, R. P., Lotyczewski, B. S., & Boike, M. (1985). Recent stressful life events and young children's school adjustment. *American Journal of Community Psychology, 13*(1), 87–99.

Stewart-Brown, S. (2006). *What is the evidence on school health promotion in improving health or preventing disease and, specifically, what is the effectiveness of the health promoting schools approach?* (Health Evidence Network report). Copenhagen: WHO Regional Office for Europe. http://www.euro.who.int/__data/assets/pdf_file/0007/74653/E88185.pdf. Accessed 30 June 2013.

Tennant, R., Goens, C., Barlow, J., Day, C., & Stewart-Brown, S. (2007). A systematic review of reviews of interventions to promote mental health and prevent mental health problems in children and young people. *Journal of Public Mental Health, 6*(1), 25–32.

Tobler, N. S., & Stratton, H. H. (1997). Effectiveness of school-based drug prevention programs: A meta-analysis of the research. *Journal of Primary Prevention, 18*(1), 71–128.

Warwick, I., Aggleton, P., Chase, E., Schagen, S., Blenkinsop, S., Schagen, I., et al. (2005). Evaluating healthy schools: Perceptions of impact among school-based respondents. *Health Education Research, 20*(6), 697–708.

Weare, K., & Markham, W. (2005). What do we know about promoting mental health through schools? *Promotion & Education, 12*(3–4), 118–122.

Weare, K., & Nind, M. (2011). Mental health promotion and problem prevention in schools: What does the evidence say? *Health Promotion International, 26*(Suppl. 1), i29–i69.

Weiner, B. J., Lewis, M. A., & Linnan, L. A. (2009). Using organization theory to understand the determinants of effective implementation of worksite health promotion programs. *Health Education Research, 24*(2), 292–305.

Weissberg, R. P., & Elias, M. J. (1993). Enhancing young people's social competence and health behavior: An important challenge for educators, scientists, policymakers, and funders. *Applied and Preventive Psychology, 2*(4), 179–190.

Weissberg, R. P., Caplan, M., & Harwood, R. L. (1991). Promoting competent young people in competence-enhancing environments: A systems-based perspective on primary prevention. *Journal of Consulting and Clinical Psychology, 59*(6), 830–841.

Weist, M. D., & Paternite, C. E. (2006). Building an interconnected policy-training-practice-research agenda to advance school mental health. *Education and Treatment of Children, 29*(2), 173–196.

Wells, J., Barlow, J., & Stewart-Brown, S. (2003). A systematic review of universal approaches to mental health promotion in schools. *Health Education, 103*(4), 197–220.

Whitelaw, S., Baxendale, A., Bryce, C., MacHardy, L., Young, I., & Witney, E. (2001). 'Settings' based health promotion: A review. *Health Promotion International, 16*(4), 339–353.
Williams, D. T., Wetton, N., & Moon, A. (1989). *A way in: Five key areas of health education.* London: Health Education Authority.
Wilson, S. J., & Lipsey, M. W. (2000). *Effects of school violence prevention programs on aggressive and disruptive behaviour: A meta-analysis of outcomes evaluations.* Nashville: Centre for Evaluation Research and Methodology, Vanderbilt Institute for Public Policy.
Zins, J., Weissberg, R., Wang, M., & Walberg, H. (2004). *Building academic success on social and emotional learning: What does the research say?* New York: Teachers College Press.

Part IV
The Challenges of Evaluation and Evidence

Chapter 16
Health-Promoting Teaching Strategies in Schools – A Review of the Literature and Recommendations for Teacher Education

Edith Flaschberger and Lisa Gugglberger

Abstract While comprehensive approaches towards school health promotion appear promising, they often lack adaptability to the schools. Research on school health promotion rarely includes attention to the core processes of schools, namely teaching and learning. The objective of this chapter is to present recommendations for health-promoting teaching strategies, which also incorporate the goals of student achievement.

To develop an understanding of what health-promoting teaching and learning processes could look like, an understanding of effective teaching strategies in terms of student achievement is needed. In addition, the interrelatedness of health and well-being on the one hand and teaching and learning on the other requires research. A narrative literature review was conducted primarily using the database ASSIA (including ERIC).

Health and achievement complement each other. Teaching strategies should thus include a health perspective to further learning outcomes and well-being of students. Several areas of action are outlined and a model is presented. To ensure sustainability of health promotion efforts and to make them more compatible to school life, it is advisable to integrate a health perspective in the core processes of schools. Professional learning communities could be a form of professional development to promote both student achievement and health.

Keywords Teaching strategies • Health and well-being • Academic achievement • Literature review

---

E. Flaschberger, Ph.D. (✉) • L. Gugglberger, Ph.D.
Ludwig Boltzmann Institute Health Promotion Research, Ludwig Boltzmann Gesellschaft GmbH, Untere Donaustraße 47/3.OG/B09, 1020 Vienna, Austria
e-mail: edith.flaschberger@lbihpr.lbg.ac.at; lisa.gugglberger@lbihpr.lbg.ac.at

V. Simovska and Patricia Mannix-McNamara (eds.), *Schools for Health and Sustainability: Theory, Research and Practice*, DOI 10.1007/978-94-017-9171-7_16,

16.1 Introduction and Objectives

Health has been an issue in schools for a long time. In the late 18th century, teaching about cleanliness to prevent the spread of infectious diseases was already part of the school curricula. For many decades, health education as a behaviour-oriented approach to school health has been the only way of addressing the issue. With the Ottawa Charter (WHO 1986), the settings approach to health promotion was introduced (cf. Denman et al. 2002).

St Leger and colleagues (2010) define health promotion in the school setting as any activity undertaken to improve and/or protect the health of all school users including activities relating to healthy school policies, the school's social and physical environment, the curriculum, community links and health services (St Leger et al. 2010, p. 2). They point out that the terms "health education" and "health promotion" are used interchangeably, which can constitute difficulties in understanding. In our definition, health promotion and health education in schools are related, but distinct concepts. Health education is a more classroom- and topic-based approach, focusing on changing behaviour through providing health information and building up health competences. Health promotion is defined as a more comprehensive approach, with an emphasis on change towards healthy settings, including a setting- and a behaviour orientation. Consequently, health education is seen as part of health promotion. This is in accordance with the understanding of Nutbeam (1998) and St Leger and colleagues (2010). Tones (2005) describes health promotion as a function of health education and healthy public policy, also bringing together health education with an approach related to changes in the environment.

Whole-school approaches that use a coordinated and integrated multi-level approach implemented over a longer period of time, have shown to be most effective regarding outcomes related to health (cf. Stewart-Brown 2006). Studies have shed light on the numerous difficulties teachers and school leaders encounter when it comes to implementing a comprehensive whole-school approach of school health promotion and there are references to an "implementation gap" (Samdal and Rowling 2013; Gugglberger and Dür 2011; Roberts-Gray et al. 2007). Some of the problems in realising a programme in actual school life originate from the lack of a culture of collaboration in schools and a lack of mutual commitment to health promotion, both of which are essential for achieving changes in the whole school setting (e.g., Flaschberger et al. 2012). Moreover, it has to be acknowledged that resources in schools worldwide are often scarce and consequently spending time, money, and energy on school health is therefore seen as not being conducive to the main aim of academic preparation of students (cf. Rosas et al. 2009).

Increasingly, the core processes of schools – teaching and learning – became important elements of school health promotion (e.g., Leurs et al. 2005). One of the main arguments is that health and education influence each other. Some studies have focused on the connection between health and education (cf. Suhrcke and de Paz Nieves 2011). The effect of education on health is well researched for both developing and industrialised countries, i.e. higher educated people tend to have better health than those with lower educational attainment. The effect of health on education has not been studied as thoroughly. There are, however, some results indicating that health may determine educational results, in developing as well as industrialised countries (Suhrcke and de Paz Nieves 2011, p. 22).

Overall, the literature is rather sparse when it comes to studies that have systematically looked at the core processes of schools in relation to health. Our research question is therefore: what can teachers do to promote health during everyday teaching and learning processes in the classroom?

Single teachers cannot implement health promotion on their own. What one teacher can do, however, is to create a health-promoting learning environment in class and thus contribute to school health promotion in a pivotal area, namely the core processes of school. For school health promotion to become a sustainable endeavour, it is necessary to focus on the core processes of schools and thus health promotion to become more relevant to teachers. Knowledge on how teaching and learning processes can be made more effective by focusing on health and well-being could enforce overall commitment to the concept on a school level, which is crucial for sustainability of health-promoting changes in the school setting.

A health-promoting way of teaching should not only produce more health and well-being, but also better learning outcomes. Otherwise it would be counterproductive to the functioning of schools and therefore not beneficial to the (future) health of students in a wide sense. Consequently, it has two – possibly conflicting – aims: (1) fulfil (externally specified) achievement goals and (2) be health promoting, empowering, and participative or in other words help students to meet their self-specified (health-related) needs.

The objective of this study is to develop recommendations on health-promoting teaching strategies, which can be used for designing pre-service teacher training, as well as for continuous professional development of teachers and for further research.

16.2 Method

This study is based on a narrative literature review (cf. Cronin et al. 2008), primarily performed via the database ASSIA (Applied Social Sciences Index and Abstracts), including results from ERIC (Education Resources Information Center), in 2011. The database was chosen as it covered literature from social sciences as well as specifically educational research. Titles and abstracts were screened and relevant articles selected. Additional literature was obtained from hand searching and cross-referencing with reference lists in identified articles. The search in the databases was done for English publications; for hand searching German sources were also included.

In order to be able to develop recommendations for teacher behaviour supporting both health and well-being of their students and their academic achievement, it is crucial to first develop an understanding of what makes teaching strategies effective. Some key factors from educational science are identified here. In a second step, an understanding about the interrelatedness of health, teaching/learning-processes and achievement can be developed and consequently compared to state-of-the-art recommendations for effective teaching. The literature review was thus conducted in two separate steps:

1. Review on effective teaching and learning;
2. Review on the interrelatedness of health, academic achievement and teaching and learning processes.

The literature review on effective teaching and learning was conducted through a search for the keywords "effective teaching" or "effective learning" and "school" in ASSIA in autumn of 2011. The included publication years were 2007–2011 as 5 years were presumed to be an adequate time frame for covering current and timely recommendations. Three factors were identified and will be presented in the first part of the results section of this chapter (3.1).

Literature on the interrelatedness of health, academic achievement and teaching and learning processes was searched for in a separate step. For this search, the keywords "health" and "well-being" have been used in combination with "effective teaching", "effective learning", "academic achievement", "teacher behaviour", "teaching style", "learning style", "differentiated instruction", "individualised learning", "individualised instruction", and "personalised learning". Furthermore, the combination of "physical activity" and "academic achievement" was searched for. The database ASSIA was used for this search and the publication years 1995–2011 were included as we assumed that research in this area is not as advanced and thus a longer time frame needs to be covered. Six factors were identified and will be presented in the second part of the results section of this chapter (3.2).

First, the retrieved articles were screened and articles relevant to the research focus as identified by the authors were included in the analysis. Exclusion criteria were: sole focus on early childhood education and students at university-level, focus on specific minority groups or students with special needs, and focus on the non-Western world.

Second, the articles were analysed and important aspects were identified and grouped. Subsequently, the various aspects are not presented in order of importance, but similar recommendations found in the literature are grouped.

The literature review has obvious limitations and cannot be viewed as exhaustive, as the review is not a systematic review, and as the search was only performed for the above-mentioned keywords. Otherwise a rather explorative approach was applied. The study is, however, designed to lay the groundwork for further investigations into the matter of health-promoting teaching and learning.

16.3 Results

The two literature reviews will be presented separately, first, describing three factors that were derived from the search on the effectiveness of teaching and learning processes, second, describing six factors from the search on the interrelatedness of health and performance (for an overview see Table 16.1).

Table 16.1 Overview of the results according to the two reviews

1. Effective teaching and learning	2. Interrelatedness of health and academic achievement
Creating a positive learning environment	Respect, support, participation, and fairness
Focusing on motivating students to learn	Increasing motivation through the fulfilment of psychological needs
Differentiated approach to teaching	Strengthening students' sense of coherence
	Promoting physical activity in the classroom
	Psychosocial health in class
	Health promotion for teachers

16.3.1 Effective Teaching and Learning in School

16.3.1.1 Creating a Positive Learning Environment

The literature on effective teaching and learning very often includes references to the creation of a positive learning environment for students. Classroom and school learning climate are said to have direct effects on student outcomes and may be equally important as students' family background – or even more important than the socioeconomic background (Willms et al. 2009). An article on the quality of teaching reviews the literature on the topic and also stresses the importance of a safe and stimulating learning climate in class (Van de Grift 2007).

It is thus recommended for teachers to make an active effort in supporting students and treating them with respect. Teacher-student relationships of high quality are seen as promoting student achievement (O'Connor and McCartney 2007). Effective teachers were found to act more fairly towards their students and to demonstrate more respect for them (Stronge et al. 2007). Positive reinforcements in the form of praise should be applied more often than corrective statements (Conroy et al. 2008). Moreover, teacher support in the form of creating a caring atmosphere as well as showing personal interest in the students and involvement was connected with more student engagement and thus better learning outcomes (e.g., Klem and Connell 2004; Skinner and Belmont 1993).

Another important factor for student engagement and therefore increased chances of student success are teachers' high expectations for their students' academic success and their behaviour in general (Willms et al. 2009; Stronge et al. 2007). If teachers believe in their students and actively show their support, learning outcomes are likely to improve accordingly.

Effective classroom management minimising disturbances and optimising qualitative teaching/learning time is also an important feature of a positive classroom climate. Effective use of time and structured and adaptive teaching are seen as critical characteristics of successful schools (Scheerens 1992; Slavin 1994, cited in Willms et al. 2009). Students who describe their classroom disciplinary climate as positive are one-and-a-half times more likely to report high levels of interest, motivation, and enjoyment in learning (Willms et al. 2009, p. 35). Effective classroom management involves the development of classroom rules and consequences for breaking them (Freiberg 2002). However, classroom management should be dealt with in an informative rather than controlling manner, so as not to decrease students' interest and motivation (Kunter et al. 2007; Koestner et al. 1984).

A well-structured (Klem and Connell 2004) and well-organised (Stronge et al. 2007) learning environment with clear rules and stable routines (Kunter et al. 2007) is important for enhancing student engagement and academic achievement respectively. According to a review on the literature regarding the topic by Van de Grift (2007), teachers who explicitly model, scaffold, explain strategies, give corrective feedback and ensure that children master the material taught contribute highly to the academic success of their pupils (p. 135).

16.3.1.2 Focusing on Motivating Students to Learn

Students motivated to learn are obviously ideal for a positive climate in classrooms and learning outcomes. Student engagement can be divided into three dimensions: social engagement (participation and sense of belonging), academic engagement (attendance), and intellectual engagement (Willms et al. 2009, p. 10). Intellectual engagement means that students are interested and motivated in their classes, enjoy them and see them relevant to their everyday life. Relevance, meaningfulness and authenticity are portrayed as essential characteristics of effective classrooms (Willms et al. 2009). This definition of intellectual engagement evokes associations with applications of the Self-Determination Theory in the school setting, which is also important in terms of student engagement: Deci and colleagues (1991) suggest that in order to get the best results for individuals as well as for society, it is important to increase intrinsic motivation by promoting an interest in learning and a valuing of education in students. However, in many classrooms, controlling conditions undermine natural tendencies to enjoy learning. These controlling measures are often well-intended, as many teachers believe that students learn better under external control. In contrast, an intrinsically motivating learning environment increases positive student learning outcomes, particularly pertaining to tasks requiring conceptual or deeper learning (cf. Niemiec and Ryan 2009).

Using assessment methods that have a positive effect rather than discouraging low-achieving students, is also discussed in the literature on effective teaching. Formative assessment is more important for improving learning than summative assessment like external tests and grades. In cultures that focus on rewards and competition in the classroom, students focus on obtaining the best marks rather than on improving their learning, often avoiding difficult tasks and asking questions (Black and Wiliam 1998). If assessment is used to help students collect their thoughts, articulate what they have found, and speculate about where they are and where they might go, students become more self-directed in learning (cf. Willms et al. 2009, p. 35). Transparency in the evaluation of students, fairness in assessing them and proper and timely feedback leads to cognitive achievement gains (Lavy 2011; Stern 2010). Applying diverse methods for (formative) assessment is also seen as a relevant approach (Stern 2010).

Especially regarding assessment, participation of students is portrayed as important in terms of effective teaching and learning processes. Students should be able to co-create assessment criteria and therefore guide their own learning (Willms et al. 2009). Black and Wiliam (1998) also claim that self- and peer-assessment are an integral part of formative assessment. To be able to self-assess their learning, students need to have a clear picture of the targets their learning is meant to attain. When students acquire such an overview, they become more committed and more effective as learners (Black and Wiliam 1998, p. 144). Flutter and Rudduck (2004) also stress the importance of teaching and learning to be a "joint endeavour" between students and teachers (p. 13). For them, student participation in the classroom can have many positive effects, such as helping students to see learning as a serious matter, to understand learning processes and thus to develop metacognitive skills. Students' perceived autonomy can be supported by giving them a voice and choice regarding their academic activities (Niemiec and Ryan 2009). Additionally, teachers can also profit from promoting a more collaborative learning atmosphere in class, as they get feedback, the chance to improve teacher-student relationships and to develop new ideas for teaching (Flutter and Rudduck 2004).

16.3.1.3 Differentiated Approach to Teaching

According to research, students' chances of being intellectually engaged in school increase when their classroom climate reflects an appropriate level of learning challenge (Willms et al. 2009).

Good teaching practice is portrayed as being responsive to students' needs and heterogeneous student, classroom and school background factors (Schleicher 2011). Using a differentiated approach to teaching and supporting individualised learning for students respectively is a learner-centred approach and can be called state-of-the-art in educational science. Some of the rationales most often mentioned for this are the social and economic needs of the present century, requiring all young people to be supported in learning and promoting their skills and talents (cf. e.g., Willms et al. 2009; Lawrence-Brown 2004). Furthermore, differentiated teaching has proven to be effective for increasing learning outcomes in various studies (cf. Stronge et al. 2007).

Differentiated teaching means ensuring that what a student learns, how he/she learns it, and how the student demonstrates what he/she has learned is a match for that student's readiness level, interests, and preferred mode of learning (Tomlinson 2004, p. 188). Interest is also seen as important for student motivation. In accordance, learning is taken more seriously if students detect relevance of the content to everyday life (Flutter and Rudduck 2004). To foster relevance, it is necessary for teachers to understand students' goals, interests and needs, and then to link school tasks to those goals, interests and needs (Assor et al. 2002, p. 265). Rock and colleagues (2008) also mention the importance of relating teaching to real-life experiences. Moreover, they claim that teachers can heighten student learning by giving support that challenges students to work slightly above what they can do by themselves. However, not only teachers should be responsible for the success of differentiated learning environments; as Tomlinson (2004) suggests, students should contribute equally and engage in teamwork, which is relating to concepts of participation and empowerment.

A variation of methods or teaching styles is often recommended in relation to the concept of differentiation. Van de Grift (2007), for example, talks about a "balance of activities" (p. 130), as for some students, especially younger or low-achieving ones, a direct teaching style is more effective than teaching styles where students have to be more active. However, positive effects for many subgroups of students could be found for "modern forms of teaching" as well (Lavy 2011). Inquiry-based teaching practices that include a more student-centred, active learning concept, are recommended especially for teaching science (cf. Kazempour 2009). However, in contradiction to the results presented above, Lavy (2011) found no effects on student performance for the element of "instilment of the capacity for individual study" (p. 24).

Other authors mention peer learning and peer tutoring, i.e. collective learning, as beneficial for learning outcomes, especially for students who have learning difficulties (e.g., Schleicher 2011). The above-mentioned formative assessment can also be understood as part of a differentiated approach to teaching, as it is conducted so that all pupils have an opportunity to think and to express ideas (Black and Wiliam 1998, p. 147) and the teacher is not only interacting with a few students in class who, for example, are quick enough to give answers. Furthermore, specific feedback on strengths and weaknesses should be provided for each student rather than giving out overall marks. An important feature of formative assessment is giving students not only feedback, but also providing them with support and opportunities to improve their work (Black and Wiliam 1998).

16.3.2 Relations Between Health, Academic Achievement, and Processes of Teaching and Learning

Generally it can be said that what improves learning outcomes is also good for health and well-being. Some authors have already embraced this view and mention connections between health and academic achievement. However, even where no such connection is explicitly mentioned, results can be related at least theoretically. Practically all of the elements conducive to effective teaching can be found in the following section with a connection to health and well-being.

16.3.2.1 Respect, Support, Participation, and Fairness

The importance of a positive classroom environment is not only part of the recommendations for effective teaching and learning, but is also seen as essential with regard to health and well-being of students. Supportive classroom management does not rely on elements of external regulation and thus has positive, not only motivational, but also emotional consequences (Kunter et al. 2007; Bergin and Bergin 1999). Van de Grift directly addresses health and academic achievement: "A safe and orderly climate is not only good for pupils' health and well-being, but also for pupils' results" (Van de Grift 2007, p. 130).

Encouraging and respectful teachers have been found to contribute to students' well-being (e.g., Engels et al. 2004). Respect is described as a critical component of school climate, especially for students in secondary schools (LaRusso et al. 2008). Moreover, LaRusso and colleagues (2008) make a connection with health behaviours insofar as they claim that a respectful school climate promotes healthy behaviours and mental health in students. School connectedness and positive experiences with teachers and peers are related to the development of self-confidence and strong emotional bonds (Freeman et al. 2011).

Social support from teachers is thus not only conducive to students' academic achievement, but can also increase happiness in students (Natvig et al. 2003a). Teachers can demonstrate support by showing their students positive attention and interest, and providing warmth and care in the everyday life at school (Danielsen et al. 2009, p. 305). Chu and colleagues (2010) reported that social support from teachers and school personnel had the strongest association with children's and adolescents' well-being. Therefore, it can be assumed that a supportive environment – even if it is "only" perceived as supportive – can be a significant contributor to student well-being. Van Petegem and colleagues (2008) claim that course content is also relevant regarding student well-being, but conclude that – independent of course content – teachers who are understanding and supportive are most important. Slightly less encouraging for the association between social support and well-being is the study of Torsheim and Wold (2001), whose findings suggest that as long as the level of support is above a minimum level, additional increases in support may have little beneficial effect for buffering stress and reducing somatic complaints (p. 300). Teachers who value their students' perspectives and encourage participation are said to be more likely to create respectful learning environments in schools (LaRusso et al. 2008). Willms and colleagues also point out that participation and engagement in learning are key to both individual and collective well-being (Willms et al. 2009, p. 7). Equally portrayed as beneficial in terms of learning outcomes, literature relates student participation to fostering health and well-being. A study on Irish students aged 10–17 years, using a self-completion questionnaire, showed that there are strong associations between participating in making school rules and perceived academic achievement. Additionally, being encouraged to express views in class is strongly associated with a positive perception of academic achievement (de Róiste et al. 2012, p. 97). In the same study, the authors reported a positive association between being encouraged to express views in class and positive school perceptions and positive health and well-being across age groups and in both genders. Natvig and colleagues (2003b) name group work and class discussions as the participatory learning methods that have the strongest effect in student perceptions of social support. In addition, the authors found indications that those kinds of teaching methods can have an indirect effect on reducing stress in students. Cooperative learning environments can have positive effects on student-student-relationships, which is especially important since peer support can positively influence the psychosocial health of students (cf. Stewart and Suldo 2011; Natvig et al. 2003a, b).

16.3.2.2 Increasing Motivation Through the Fulfilment of Psychological Needs

In a study on students attending vocational or technical schools, desire to learn was associated with higher levels of well-being (Van Petegem et al. 2008). Accordingly, learning motivation is not only seen as important for student achievement, but also for their well-being. The above already briefly mentioned Self-Determination Theory (SDT), which was initially developed by psychologists Edward L. Deci and Richard M. Ryan during the 1970s and 1980s, seems to be of great value when talking about health-promoting teaching and learning processes. The basic assumption of this theory of motivation is that there are three psychological needs: the needs for autonomy, competence, and relatedness. These basic human needs are considered to be innate, and they are said to specify the necessary conditions for psychological health or well-being (cf. e.g., Deci and Ryan 2000; Ryan and Deci 2000). The SDT perspective implies that people will tend to pursue goals, domains, and relationships that allow or support their need satisfaction. Psychological well-being is experienced to the extent that people are successful in finding such opportunities (Deci and Ryan 2000, p. 230).

Autonomy in this context means volition or the will to self-organise and to experience integration and freedom. It literally means "self-rule" (La Guardia 2009, p. 92) and refers to being self-initiating and self-regulating one's actions (Deci et al. 1991, p. 327). Harré and Bullen (2010, p. 236) use an extended concept of autonomy, namely integrity, which is supposed to capture not only self-direction in behaviour, but also the notion of having one's values and actions in alignment. Competence concerns effectance or the feeling of being competent and able to accomplish demands; it means understanding how to attain various outcomes (Deci and Ryan 2000; Deci et al. 1991). The need for relatedness refers to the need to feel belongingness and connectedness and to develop satisfying relationships in one's social surroundings that are characterised by security (Ryan and Deci 2000; Deci et al. 1991).

Social environments that provide the opportunity to satisfy the three psychological needs promote motivation and performance (Deci et al. 1991) and healthy psychological development (La Guardia 2009). Intrinsic motivation and more autonomous forms of extrinsic motivation were found to be related to positive student learning outcomes in several studies (Grolnick et al. 1991; for an overview see: Deci et al. 1991). Furthermore, positive coping with failure and enjoyment of school were positively related to more autonomous forms of motivation (cf. Ryan and Connell 1989). Deci and colleagues (1991) summarise that students with more intrinsic motivation for schoolwork are more likely to stay in school, to perform well, and to be well-adjusted than students with more externally controlled regulatory styles.

In the school context, supports for competence can be, for example, an optimal balance of challenges or positive feedback; examples for supports for relatedness can be peer acceptance or a caring attitude of the teacher (cf. e.g., La Guardia 2009). Van Ryzin and colleagues (2009) also stress that feeling supported and accepted is important for students' success in school – as well as for well-being and general adjustment. Promoting the satisfaction of the needs for competence and relatedness is seen as facilitating motivation, whereas for the development of intrinsic motivation and more self-determined forms of motivation respectively, autonomy-supportive rather than controlling interpersonal contexts are crucial. Accordingly, positive feedback is only beneficial when given on self-initiated educational tasks; praising students for doing what they were told to do does not necessarily have the same effect, even though feelings of competence may be promoted. Nevertheless, furthering feelings of competence and relatedness have irrespectively been associated with increased motivation in students in various studies (cf. Deci et al. 1991).

SDT states that, for example, the offer of rewards, the setting of a deadline, the use of threats or other events signifying some kind of external control can have negative effects on motivation. However, there are research studies showing that the interpersonal context can have moderating effects regarding external events unfavourable to self-determined regulatory styles. Study results also indicate that giving students the opportunity to participate in the decision-making regarding their school work can benefit the self-determined regulation of those educational activities and thus learning outcomes (cf. e.g., Ryan and Deci 2000; Deci et al. 1991). While examples for autonomy-supporting teacher behaviour given in most of the literature include giving students choices about when to do what, others state that while providing choice is good, it is not always the most effective tool to increase feelings of autonomy in students (Reeve et al. 2003; Assor et al. 2002). Assor and colleagues (2002) name fostering relevance, allowing criticism and encouraging independent thinking as the most effective methods of autonomy-enhancing teacher behaviour. As previously stated, teachers should take an active-empathic role towards their students to link school tasks to students' goals, interests, and needs, possibly after supporting them to identify their goals (Assor et al. 2002, p. 265). It is recommended to let the students understand the connections between their interests and goals and schoolwork, promoting feelings of relevance and thus feelings of autonomy, which in turn increase self-determined forms of motivation, school engagement, and well-being (Assor et al. 2002). Related to this are results of a study on students in Flanders stating that students show most interest and commitment to subjects they perceive as useful, up-to-date and in accordance with their views on the world (Engels et al. 2004).

The afore-mentioned moderating effects of otherwise autonomy-suppressing situations can additionally be achieved by teachers creating a non-pressuring classroom climate that increases feelings of competence in students. Teachers' language and style are said to be crucial for feelings of either autonomy or control. Internalisation of the value of an activity, which is not considered interesting, can help students' willingness to engage in the activity. This internalisation of values regarding education can be supported by the afore-mentioned fostering of relevance, the provision of choices while applying minimal pressures, and the acknowledgement of students' feelings of not liking it (Reeve et al. 2003; Deci et al. 1991). In a SDT-related study, students' perceptions of autonomy in class and support from adults and peers in school were each found to have an independent, positive effect on student engagement, which in turn has a positive impact on adjustment (Van Ryzin et al. 2009, p. 7). Danielsen and colleagues (2010) showed in their study of Norwegian HBSC[1]-data that caring and autonomy-supporting teacher behaviour related to self-reported academic initiative in class. Moreover, the concept of school connectedness, which also combines positive associations with health and academic achievement, can be related to SDT (cf. Waters et al. 2010). Samdal and Rowling (2010) connect SDT to student participation as a core element of health-promoting schools.

16.3.2.3 Strengthening Students' Sense of Coherence

Another construct which seems to be valuable for health-promoting teaching and learning is Antonovsky's sense of coherence (SOC) in his salutogenic, health-oriented model – as opposed to a pathogenic, disease-oriented approach. SOC is defined as a generalised orientation toward the world which perceives it, on a continuum, as comprehensible, manageable and meaningful (Antonovsky 1996, p. 15). It is designed to explain successful coping with stressors (Modin et al. 2011). In a meta-review, it was stated that the SOC strongly relates to health, especially mental health, and that it can be seen as a health-promoting resource (cf. Eriksson and Lindström 2006). Certain life events lead to a stronger SOC in a person, allowing the person to apply the resources appropriate to stressors in any given situation (Antonovsky 1996). According to Antonovsky's theory, SOC is strengthened by three kinds of life experiences: consistency/predictability, underload/overload balance and participation in socially valued decision-making (Modin et al. 2011; Antonovsky 1996).

As students' SOC is still in its developmental phase, the school environment can have a negative or positive influence on it. The notion of comprehensibility is strengthened, when a school class is experienced as predictable, characterised by continuity and security, and confidence-boosting. Experiencing the situation in school as manageable is furthered by an optimal balance of demands, as well as by the actual availability of necessary resources. The notion of meaningfulness increases

[1] *Health Behaviour in School-Aged Children*-Study.

motivation and is, for instance, brought upon by participation in decision-making (Modin et al. 2011).

The theoretical and empirical research on the construct of the SOC suggests that in terms of health and well-being, it is important to create a learning environment characterised by comprehensibility, manageability and meaningfulness, which coincides with the recommendations above concerning environments conducive to student achievement: a positive atmosphere in the classroom with clear objectives and an informative classroom-management, increasing student motivation including a certain level of student participation, and a differentiated approach to teaching, preventing overworked as well as under-challenged students.

Differentiated teaching or similar strategies can additionally lead to the reduction of educational disparities. Reducing inequalities is seen as a principle or important element of health promotion (cf. WHO 2009). Many of the afore-mentioned teaching strategies can be seen as ways to reduce the achievement gap between low- and high-achieving students. Therefore, implementing effective learning in classrooms leads to better attainment also for those who are not among the class champions, concurrently fulfilling one of the principles of health promotion, especially considering that higher education leads to better health.

16.3.2.4 Promoting Physical Activity in the Classroom

While there is ample evidence that physical activity improves health, there are not as many studies on the effect of physical activity on academic achievement. Still, several studies focused on the interrelatedness of physical activity and learning outcomes, and most of these studies reported potentially positive, yet not fully substantiated effects of increased physical activity on student achievement (cf. e.g., Bailey et al. 2009; Taras 2005). More recent sources claim that school-based interventions designed to boost physical activity during everyday school life were found to improve academic performance outcomes like achievement, cognitive performance and concentration in many studies (cf. Barr-Anderson et al. 2011). Most research on increasing physical activity shows that academic achievement is improved or at least not deteriorating (cf. Bailey et al. 2009), but on the whole, evidence is considered to be limited still. What can be claimed, however, is that physical activity programmes can promote the development of social skills and mental health and the reduction of risky behaviours (Taras 2005), which could very well indirectly affect students' learning outcomes. Blom and colleagues (2011) describe their study results indicating that students who were more fit were less likely to miss school and do poorly on standardised tests (p. 17).

As there is evidence that short bouts – around 5 min – of exercise can have (short-term) effects on cognitive performance (Barr-Anderson et al. 2011) and increasing physical activity during the school day is considered health-promoting, the concept of including physical activity in class is furthered. An example is the programme called "The Class Moves!®", which was developed in the Netherlands in the late 1980s. The programme consists of a number of physical activity exercises for primary school children, aiming at various improvements, including increasing student relaxation, concentration, physical awareness and sensorimotor development. An evaluation of a pilot study on its implementation in Scotland and Wales confirmed adaptability and enjoyment of the programme; teachers and students alike attested to positive outcomes in the areas of increased student concentration and relaxation (Lowden et al. 2001).

Not only relaxation exercises and physical activity breaks are recommended, but also learning through physical activity. Kottmann and colleagues (2005) distinguish between physical activity related to themes and physical activity related to methods. The former relates to the concept of making cognitive content accessible through physical activity, e.g., by experiencing distance, dimensions or shapes through movement or by doing plays for deepening understanding of stories. The latter means using movement as part of a learning method through dissolving the seating arrangements in class, e.g., for doing group work, having several "work stations", and going to the school library (Kottmann et al. 2005). Although those kinds of concepts for integrating physical activity are not yet fully scientifically evaluated to date, they seem to be a promising endeavour pertaining to the improvement of academic performance and the provision of more physical activity during the school day.

16.3.2.5 Psychosocial Health in the Class

In a recent report on data of Canada's Health Behaviour in School-Aged Children (HBSC) Study, links between lower levels of academic achievement and an increased incidence of emotional problems were found (Freeman et al. 2011, p. 59). Mental health problems are brought in connection with lower levels of academic achievement in many studies (cf. e.g., Guzman et al. 2011). It is therefore recommended, also for the benefit of addressing issues of educational inequality, to focus on the promotion of psychosocial health in schools.

Approaches to promote psychosocial health in schools are relatively promising, especially when implemented through a whole-school approach over a longer period of time (Stewart-Brown 2006; Jané-Llopis and Barry 2005). Osher and colleagues (2004) claim that mental health promotion, as well as the above-mentioned individualised instruction, can increase "success and motivation" in schools (p. 16). One kind of approach in psychosocial health promotion that can also be applied in the classroom is the well-known life skills approach (cf. e.g., Mangrulkar et al. 2001). The goal of this approach is to increase students' resilience, i.e., to promote their ability to deal effectively with the challenges of everyday life and to cope with difficult life events. Life skills include social and interpersonal skills (e.g., communication and negotiation/refusal skills), cognitive skills (e.g., decision-making, self-evaluation), and emotional coping skills (e.g., managing stress) (Mangrulkar et al. 2001). Results of programme evaluations indicate that life skills development can delay the onset of drug use, prevent high-risk sexual behaviours, teach anger management, improve academic performance, and promote positive social adjustment (Mangrulkar et al. 2001, p. 5). Tones (2005) describes the promotion of life skills in the school context as a training for empowerment, which is one of the goals or principles of health promotion.

In addition, several studies of different school-based anti-violence measures provide evidence of improvements regarding academic achievement, discipline and behaviour in class (e.g., Luiselli et al. 2005; Twemlow et al. 2001). Aggression and violence are also named as part of the health disparities which contribute to the achievement gap (Basch 2011). Not only high-level violence is seen as a problem, but also low-level violence (behaviours such as bullying, peer sexual harassment and the psychological maltreatment of students by teachers) (Dupper and Meyer-Adams 2002). Apart from exhibiting lower health and well-being, students exposed to low-level violence are also affected by diminished academic achievement; thus the creation of a positive school climate is in order (Rothon et al. 2011; Dupper and Meyer-Adams 2002). Moreover, not only victims of bullying experience negative consequences, but also their bullies have, for example, lower school adjustment (Dake et al. 2003). While most articles focus on peer bullying, the issue of teacher bullying has recently gotten more attention as an equally serious problem (Zerillo and Osterman 2011).

16.3.2.6 Health Promotion for Teachers

Studies show that there are negative consequences for teaching effectiveness if teachers themselves are not healthy and feeling well (cf. e.g., Klusmann et al. 2006; Maslach and Leiter 1999). This can be seen as especially important in light of study results clearly placing the teacher at the centre of crucial factors contributing to student academic achievement. Research indicates that teachers, who are on the verge of burnout, also change their behaviour towards their students, which in turn has a negative impact on student behaviour and academic achievement. The social relationship between teacher and students is negatively affected by teachers not feeling well and exhibiting burnout-related symptoms (e.g., Klusmann et al. 2006).

Positive student-teacher relations are not only important for student achievement; they are also related to teachers' job satisfaction on the individual teacher level, thus contributing to teacher well-being (Schleicher 2011). The link between high teacher self-efficacy beliefs on the one hand, and school functioning, student motivation and achievement, and less teacher stress, burnout and attrition on the other hand, is reported in the literature (for an overview see: Skaalvik and Skaalvik 2007; Caprara et al. 2006). Furthermore, SDT also applies to teachers: if teachers feel strong external pressures and thus cannot satisfy their psychological needs, they also tend to be more controlling towards their students and exhibit less well-being themselves (cf. e.g., Niemiec and Ryan 2009). In contrast, Felner and colleagues (2007) claim that teachers who are working in well-established teams with adequate organisational support in smaller learning communities, experience higher levels of motivation, and student engagement and achievement is increased at the same time.

16.4 Discussion

This literature review confirms that health and education obviously complement one another, i.e., it is advisable to incorporate a health perspective in processes of teaching and learning for the sake of favourable results in the areas of health and student achievement alike. Both aims stated in the introduction, i.e. attaining student academic achievement and supporting students in meeting their own (health-related) needs, can be addressed via the explicated interrelated theories and research results leading to several areas of recommendations, so there seem to be no conflicting aims of the two approaches. Figure 16.1 is a model displaying the results of this chapter. It shows possible elements of health- promoting teaching that can also promote academic achievement. The boxes in a darker shade (middle and right columns) represent aspects found in the review on the interrelatedness of health and academic achievement; the lighter shade (left column) indicates elements of effective teaching strategies.

The results of this study point to at least three important issues that need to be further discussed here: (1) the interconnectedness of health and education, (2) the possible use of Self-Determination Theory (SDT), and (3) the changing teachers' role.

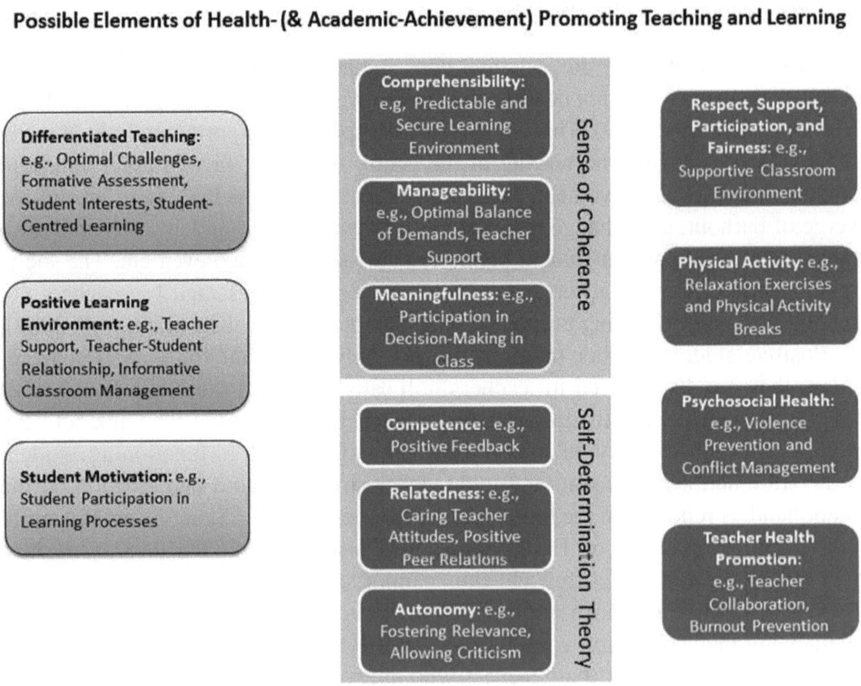

Fig. 16.1 Elements of health-promoting teaching that can also promote academic achievement

16.4.1 The Interconnectedness of Health and Education

First, arguments about the interconnectedness of health and education are usually brought forward by the health sector. One of these arguments is "Healthy students learn better" (St Leger et al. 2010, p. 7). It can be seen as a more deficit-oriented approach from the educational perspective, as it is usually followed by explications of health risks and health-compromising behaviours and their negative effects on students' academic achievement. Moreover, it is acknowledged that health can improve school effectiveness and that some elements of effective schools can also promote health by providing students with opportunities to build their health assets (St Leger et al. 2010, p. 7). However, there are usually no relationships addressed regarding specific structures or processes to do this, especially using scientific data.

This chapter's aim was to bring about the possibility to address health promotion from "the other side", meaning that educational strategies that have proven to be effective in terms of learning outcomes also have relevance to health issues. Effective teaching and learning strategies operate with aspects that are also essential for fostering health and well-being in students. An example would be the application of formative assessment that helps students to achieve better academic results, as they are encouraged and motivated in their attempts to improve, and they concurrently also develop more well-being.

Furthermore, teaching and learning processes can benefit from health-promoting activities; for example, measures to reduce bullying and violence contribute to better educational attainment via higher attendance rates and a more positive classroom climate. This is again more consistent with the above-mentioned deficit-oriented approach, but still focuses on the promotion of a positive learning environment.

If an understanding of these relationships exemplified above can be attained for all teachers, the chances are much higher that whole-school approaches to school health including setting components can be institutionalised on a larger scale, as health becomes an important "ingredient" of teaching and learning processes. It is not something separate or additional any longer, but one of the aspects that aid schools in achieving their goals. Valli and Buese (2007) refer to studies indicating that teachers are motivated to enact changes if they believe in them and have sufficient support, even if there is considerable work intensification involved.

Additionally, it becomes more and more important for schools to convey social skills to young people (e.g., Stern 2010). By emphasising social learning, psychosocial aspects of health are also covered. A focus on holistic forms or views of education is seen as a rather new perspective in educational science and practice, but it is based on scientific findings, such as neurobiological ones, that stress the interdependence of emotional, social and rational processes (cf. Lovat et al. 2011).

16.4.2 The Possible Use of Self-Determination Theory

Second, Self-Determination Theory (SDT) and the work of Deci and Ryan (cf. e.g. Deci and Ryan 2000) have been central in this literature review as in our view, they provide a good reference point for combining health promotion and school improvement. Through focusing on the three fundamental psychological needs proposed – autonomy, competence, and relatedness –, it is possible to increase (intrinsic) motivation and thus educational outcomes as well as well-being or mental health. SDT-related studies show that teachers that are pressured by demands in the school organisation and superiors, thus experience a lower level of self-determination themselves, and tend to be more controlling towards their students (cf. Deci et al. 1991). Therefore, it seems beneficial not only to integrate this knowledge into initial teacher education, but also to implement it in school development efforts. Ideally those two areas are brought together in the form of professional learning communities or similar groups of learning and exchange that aim towards a mutual approach to school improvement, rather than placing the pressure or blame on individual teachers.

16.4.3 The Changing Role of Teachers

Finally, it has to be kept in mind that teachers' roles have expanded over the last decades (e.g., Valli and Buese 2007), and that many teachers feel overwhelmed by additional tasks that are presented to them. Bartlett (2004) stresses that schools that have integrated the expanded teaching roles into the regular structures of the working day are more likely to sustain the engagement and commitment of their teachers (p. 568). Including a health perspective in schools could also be seen as an expansion in teachers' duties. Therefore, it seems important to (a) integrate this perspective into core processes of schools and (b) support teachers in their attempts to do so, for example, by supporting the creation of professional learning communities that embrace the concept of health promotion. Furthermore, the aspect of teacher health and well-being should be considered to be important as well, as many teachers work overtime and experience overstrain (cf. Bartlett 2004), which is neither favourable for effective teaching practice nor sustainable health promotion efforts nor teacher satisfaction and thus teacher retention.

On a similar note, recommendation for a differentiated teaching style can also cause high levels of stress, as the teachers' roles change and expand through this rather new teaching concept. Depending on the context – for example, if there is organisational support or if the pressures from state or district level involve high stakes-testing and standardisation –, teachers can profit from differentiation in their classrooms or they suffer from it. Collaboration between teachers becomes important for this kind of change as well, adding to new professional lives, including chances and difficulties (cf. Valli and Buese 2007).

The present chapter tries to convey that according to our literature review, teaching practices that are conducive to academic achievement are also beneficial for student health and well-being. Furthermore, there seems to be a benefit from many health-promoting activities in terms of improving academic achievement.

It has to be stressed though that this literature analysis does not explicitly discriminate between determinant and outcome, i.e., between the teaching processes and their outcomes. Therefore, it has to be clear that health in this study can be either seen as determinant for student academic achievement or a welcome byproduct of teaching styles promoting academic achievement. The present study does not allow for definite statements about these directions of causality, which can be seen as a limitation. Accordingly, there are no connections between the different concepts shown in the model (see Fig. 16.1). However, the literature study can be used as a basis for further research in this area.

16.5 Conclusion

The present review has produced an overview of a variety of elements that teachers can incorporate in their teaching strategies, both conducive to students' health and their academic achievement (see Fig.16.1).

The value of health promotion in teaching and learning processes should thus be made clear to teacher candidates and in-service teachers of all school types and subjects. It is what connects health and school issues. Therefore, it might in the end lead the way to truly health-promoting schools that combine not only a behaviour and a setting component in terms of health promotion, but are also characterised by a way of organising teaching and learning processes that is both beneficial for academic achievement and health.

While many authors promote a shift in priorities in school systems from a more academic towards a more holistic orientation including health (e.g., Langille and Rodgers 2010), it seems that this is a long-term goal. To attain this goal, it might be helpful to demonstrate the importance of health and well-being for academic success first. In order to concurrently avoid a duplication of teacher efforts and to increase their sustainability, it is recommended to integrate school health as a component of school improvement (cf. Rosas et al. 2009).

Therefore, professional development recommended for school improvement might also be important for health promotion. As already mentioned above, professional learning communities (cf. e.g., Stoll et al. 2006) or professional learning teams (Griffin et al. 2010) could help to put health promotion in the spotlight, for reasons of promoting health and well-being, but – as this chapter shows – at the same time for supporting the schools' main mission of improving student outcomes. Research indicates that teacher collaboration and professional learning communities can lead to better outcomes in the fields of teacher health and teacher effectiveness, provided that teachers are supported in these endeavours (cf. e.g., Felner et al. 2007). Schleicher (2011) states the importance of teacher collaboration for teacher effectiveness as well. It seems that this aim coincides with an identified weakness in school health promotion and the recommendation to improve collaboration structures in schools for aiding institutionalisation of health promotion. Of course, it is not enough to just expose pre- and in-service teachers to traditional training including a teamwork-component. The establishment of conditions in schools conducive to the establishment of collaboration structures is also necessary, as well as promoting ways of professional learning that are collaborative in nature.

So, for the sake of health and academic achievement, state-of-the-art teaching methods with a health perspective should be applied as well as supported by favourable conditions in teacher education, school administration and planning, and in everyday practice at schools.

Acknowledgements The authors would like to thank Wolfgang Dür and Rosemarie Felder-Puig for their feedback on the manuscript and their former colleague Waldemar Kremser, whose preceding work on the topic has contributed to this chapter.

References

Antonovsky, A. (1996). The salutogenic model as a theory to guide health promotion. *Health Promotion International, 11*(1), 11–18.

Assor, A., Kaplan, H., & Roth, G. (2002). Choice is good, but relevance is excellent: Autonomy-enhancing and suppressing teacher behaviours predicting students' engagement in schoolwork. *British Journal of Educational Psychology, 72*(2), 261–278.

Bailey, R., Armour, K., Kirk, D., Jess, M., Pickup, I., Sandford, R., & BERA Physical Education and Sport Pedagogy Special Interest Group. (2009). The educational benefits claimed for physical education and school sport: An academic review. *Research Papers in Education, 24*(1), 1–27.

Barr-Anderson, D. J., AuYoung, M., Whitt-Glover, M. C., Glenn, B. A., & Yancey, A. K. (2011). Integration of short bouts of physical activity into organizational routine. A systematic review of the literature. *American Journal of Preventive Medicine, 40*(1), 76–93.

Bartlett, L. (2004). Expanding teacher work roles: A resource for retention or a recipe for overwork? *Journal of Education Policy, 19*(5), 565–582.

Basch, C. E. (2011). Healthier students are better learners: A missing link in school reforms to close the achievement gap. *Journal of School Health, 81*(10), 593–598.

Bergin, C., & Bergin, D. A. (1999). Classroom discipline that promotes self-control. *Journal of Applied Developmental Psychology, 20*(2), 189–206.

Black, P., & Wiliam, D. (1998). Inside the blackbox – Raising standards through classroom assessment. *Phi Delta Kappan, 80*(2), 139–144, 146–148.

Blom, L. C., Alvarez, J., Zhang, L., & Kolbo, J. (2011). Associations between health-related physical fitness, academic achievement and selected academic behaviors of elementary and middle school students in the State of Mississippi. *ICHPER-SD Journal of Research, 6*(1), 13–19.

Caprara, G. V., Barbaranelli, C., Steca, P., & Malone, P. S. (2006). Teachers' self-efficacy beliefs as determinants of job satisfaction and students' academic achievement: A study at the school level. *Journal of School Psychology, 44*(6), 473–490.

Chu, P. S., Saucier, D. A., & Hafner, E. (2010). Meta-analysis of the relationships between social support and well-being in children and adolescents. *Journal of Social and Clinical Psychology, 29*(6), 624–645.

Conroy, M. A., Sutherland, K. S., Snyder, A. L., & Marsh, S. (2008). Classwide interventions. Effective instruction makes a difference. *Teaching Exceptional Children, 40*(6), 24–30.

Cronin, P., Ryan, F., & Coughlan, M. (2008). Undertaking a literature review: A step-by-step approach. *British Journal of Nursing, 17*(1), 38–43.

Dake, J. A., Price, J. H., & Telljohann, S. K. (2003). The nature and extent of bullying at school. *Journal of School Health, 73*(5), 173–180.

Danielsen, A. G., Samdal, O., Hetland, J., & Wold, B. (2009). School-related social support and students' perceived life satisfaction. *Journal of Educational Research, 102*(4), 303–320.

Danielsen, A. G., Wiium, N., Wilhelmsen, B. U., & Wold, B. (2010). Perceived support provided by teachers and classmates and students' self-reported academic initiative. *Journal of School Psychology, 48*(3), 247–267.

De Róiste, A., Kelly, C., Molcho, M., Gavin, A., & Nic Gabhainn, S. (2012). Is school participation good for children? Associations with health and wellbeing. *Health Education, 112*(2), 88–104.

Deci, E. L., & Ryan, R. M. (2000). The "what" and "why" of goal pursuits: human needs and the self-determination of behaviour. *Psychological Inquiry, 11*(4), 227–268.

Deci, E. L., Vallerand, R. J., Pelletier, L. G., & Ryan, R. M. (1991). Motivation and education: The self-determination perspective. *Educational Psychologist, 26*(3&4), 325–346.

Denman, S., Moon, A., Parsons, C., & Stears, D. (2002). *The health promoting school. Policy, research and practice*. London/New York: Routledge Falmer.

Dupper, D. R., & Meyer-Adams, N. (2002). Low-level violence. A neglected aspect of school culture. *Urban Education, 37*(3), 350–364.

Engels, N., Aelterman, A., Van Petegem, K., & Schepens, A. (2004). Factors which influence the well-being of pupils in Flemish secondary schools. *Educational Studies, 30*(2), 127–143.

Eriksson, M., & Lindström, B. (2006). Antonovsky's sense of coherence scale and the relation with health: A systematic review. *Journal of Epidemiology and Community Health, 60*(5), 376–381.

Felner, R. D., Seitsinger, A. M., Brand, S., Burns, A., & Bolton, N. (2007). Creating small learning communities: Lessons from the project on high-performing learning communities about "what works" in creating productive, developmentally enhancing, learning contexts. *Educational Psychologist, 42*(4), 209–221.

Flaschberger, E., Nitsch, M., & Waldherr, K. (2012). Implementing school health promotion in Austria. Experiences from a pilot training course. *Health Promotion Practice, 13*(3), 364–369.

Flutter, J., & Rudduck, J. (2004). *Consulting pupils. What's in it for schools?* Abingdon/New York: Routledge.

Freeman, J. G., King, M., Pickett, W., Craig, W., Elgar, F., Janssen, I., & Klinger, D. (2011). *The health of Canada's young people: A mental health focus*. Ottawa: Public Health Agency of Canada.

Freiberg, H. J. (2002). Essential skills for new teachers. *Educational Leadership, 59*(6), 56–60.

Griffin, P., Murray, L., Care, E., Thomas, A., & Perri, P. (2010). Developmental assessment: lifting literacy through professional learning teams. *Assessment in Education: Principles, Policy and Practice, 17*(4), 383–397.

Grolnick, W. S., Ryan, R. M., & Deci, E. L. (1991). Inner resources for school achievement: Motivational mediators of children's perceptions of their parents. *Journal of Educational Psychology, 83*(4), 508–517.

Gugglberger, L., & Dür, W. (2011). Capacity building in and for health promoting schools: Results from a qualitative study. *Health Policy, 101*(1), 37–43.

Guzman, M. P., Jellinek, M., George, M., Hartley, M., Squicciarini, A. M., Canenguez, K. M., Kuhlthau, K. A., Yucel, R., White, G. W., Guzman, J., & Murphy, J. M. (2011). Mental health matters in elementary school: First-grade screening predicts fourth grade achievement test scores. *European Child and Adolescent Psychiatry, 20*(8), 401–411.

Harré, N., & Bullen, P. (2010). Personal project commitment in adolescence: The role of relatedness, competence and integrity. *European Journal of Social Psychology, 40*(2), 235–251.

Jané-Llopis, E., & Barry, M. M. (2005). What makes mental health promotion effective? *Promotion and Education, 12*(Suppl. 2), 47–54.

Kazempour, M. (2009). Impact of inquiry-based professional development on core conceptions and teaching practices: A case study. *Science Educator, 18*(2), 56–68.

Klem, A. M., & Connell, J. P. (2004). Relationships matter: Linking teacher support to student engagement and achievement. *Journal of School Health, 74*(7), 262–273.

Klusmann, U., Kunter, M., Trautwein, U., & Baumert, J. (2006). Lehrerbelastung und Unterrichtsqualität aus der Perspektive von Lehrenden und Lernenden. *Zeitschrift für Pädagogische Psychologie, 20*(3), 161–173.

Koestner, R., Ryan, R. M., Bernieri, F., & Holt, K. (1984). Setting limits on children's behavior: The differential effects of controlling vs. informational styles on intrinsic motivation and creativity. *Journal of Personality, 52*(3), 233–248.

Kottmann, L., Küpper, D., & Pack, R. P. (2005). *Bewegungsfreudige Schule. Schulentwicklung bewegt gestalten – Grundlagen, Anregungen, Hilfen*. Gütersloh: Bertelsmann Stiftung.

Kunter, M., Baumert, J., & Köller, O. (2007). Effective classroom management and the development of subject-related interest. *Learning and Instruction, 17*(5), 494–509.

La Guardia, J. G. (2009). Developing who I am: A self-determination theory approach to the establishment of healthy identities. *Educational Psychologist, 44*(2), 90–104.

Langille, J. L. D., & Rodgers, W. M. (2010). Exploring the influence of a social ecological model on school-based activity. *Health Education & Behavior, 37*(6), 879–894.

LaRusso, M. D., Romer, D., & Selman, R. L. (2008). Teachers as builders of respectful school climates: Implications for adolescent drug use norms and depressive symptoms in high school. *Journal of Youth and Adolescence, 37*(4), 386–398.

Lavy, V. (2011). *What makes an effective teacher? Quasi-experimental evidence* (NBER Working Paper No. 16885). Cambridge, MA: National Bureau of Economic Research.

Lawrence-Brown, D. (2004). Differentiated instruction: Inclusive strategies for standards-based learning that benefit the whole class. *American Secondary Education, 32*(3), 34–62.

Leurs, M. T. W., Schaalma, H. P., Jansen, M. W. J., Mur-Veeman, I. M., St Leger, L. H., & de Vries, N. (2005). Development of a collaborative model to improve school health promotion in the Netherlands. *Health Promotion International, 20*(3), 296–305.

Lovat, T., Dally, K., Clement, N., & Toomey, R. (2011). Values pedagogy and teacher education: Reconceiving the foundations. *Australian Journal of Teacher Education, 36*(7), 31–44.

Lowden, K, Powney, J., Davidson, J., & James, C. (2001). *The Class Moves!® Pilot in Scotland and Wales. An evaluation* (SCRE Research Report No. 100). Edinburgh: The Scottish Council for Research in Education.

Luiselli, J. K., Putnam, R. F., Handler, M. W., & Feinberg, A. B. (2005). Whole-school positive behaviour support: Effects on student discipline problems and academic performance. *Educational Psychology, 25*(2–3), 183–198.

Mangrulkar, L., Whitman, C. V., & Posner, M. (2001). *Life skills approach to child and adolescent healthy human development*. Washington, DC: Adolescent Health and Development Unit, Division of Health Promotion and Protection, Pan American Health Organization.

Maslach, C., & Leiter, M. P. (1999). Teacher burnout: A research agenda. In R. Vandenberghe & A. M. Huberman (Eds.), *Understanding and preventing teacher burnout: A sourcebook of international research and practice* (pp. 295–303). Cambridge: Cambridge University Press.

Modin, B., Östberg, V., Toivanen, S., & Sundell, K. (2011). Psychosocial working conditions, school sense of coherence and subjective health complaints. A multilevel analysis of ninth grade pupils in the Stockholm area. *Journal of Adolescence, 34*(1), 129–139.

Natvig, G. K., Albrektsen, G., & Qvarnstrom, U. (2003a). Associations between psychosocial factors and happiness among school adolescents. *International Journal of Nursing Practice, 9*(3), 166–175.

Natvig, G. K., Albrektsen, G., & Qvarnstrom, U. (2003b). Methods of teaching and class participation in relation to perceived social support and stress: modifiable factors for improving health and wellbeing among students. *Educational Psychology, 23*(3), 261–274.

Niemiec, C. P., & Ryan, R. M. (2009). Autonomy, competence, and relatedness in the classroom: Applying self-determination theory to educational practice. *Theory and Research in Education, 7*(2), 133–144.

Nutbeam, D. (1998). Health promotion glossary. *Health Promotion International, 13*(4), 349–364.
O'Connor, E., & McCartney, K. (2007). Examining teacher-child relationships and achievement as part of an ecological model of development. *American Educational Research Journal, 44*(2), 340–369.
Osher, D., VanAcker, R., Morrison, G. M., Gable, R., Dwyer, K., & Quinn, M. (2004). Warning signs of problems in schools: Ecological perspectives and effective practices for combating school aggression and violence. *Journal of School Violence, 3*(2), 13–37.
Reeve, J., Nix, G., & Hamm, D. (2003). Testing models of the experience of self-determination in intrinsic motivation and the conundrum of choice. *Journal of Educational Psychology, 95*(2), 375–392.
Roberts-Gray, C., Gingiss, P. M., & Boerm, M. (2007). Evaluating school capacity to implement new programs. *Evaluation and Program Planning, 30*(3), 247–257.
Rock, M. L., Gregg, M., Ellis, E., & Gable, R. A. (2008). REACH: A framework for differentiating classroom instruction. *Preventing School Failure, 52*(2), 31–47.
Rosas, S., Case, J., & Tholstrup, L. (2009). A retrospective examination of the relationship between implementation quality of the coordinated school health program model and school-level academic indicators over time. *Journal of School Health, 79*(3), 108–115.
Rothon, C., Head, J., Klineberg, E., & Stansfeld, S. (2011). Can social support protect bullied adolescents from adverse outcomes? A prospective study on the effects of bullying on the educational achievement and mental health of adolescents at secondary schools in East London. *Journal of Adolescence, 34*(3), 579–588.
Ryan, R. M., & Connell, J. P. (1989). Perceived locus of causality and internalization: Examining reasons for acting in two domains. *Journal of Personality and Social Psychology, 57*(5), 749–761.
Ryan, R. M., & Deci, E. L. (2000). Self-determination theory and the facilitation of intrinsic motivation, social development, and well-being. *American Psychologist, 55*(1), 68–78.
Samdal, O., & Rowling, L. (2010). Theoretical and empirical base for implementation components of health-promoting schools. *Health Education, 111*(5), 367–390.
Samdal, O., & Rowling, L. (Eds.). (2013). *The implementation of health promoting schools. Exploring the theories of what, why and how*. Abingdon/New York: Routledge.
Scheerens, J. (1992). *Effective schooling: Research, theory, and practice*. London: Cassell.
Schleicher, A. (2011). Lessons from the world on effective teaching and learning environments. *Journal of Teacher Education, 62*(2), 202–221.
Skaalvik, E. M., & Skaalvik, S. (2007). Dimensions of teacher self-efficacy and relations with strain factors, perceived collective teacher efficacy, and teacher burnout. *Journal of Educational Psychology, 99*(3), 611–625.
Skinner, E. A., & Belmont, M. J. (1993). Motivation in the classroom: reciprocal effects of teacher behavior and student engagement across the school year. *Journal of Educational Psychology, 85*(4), 571–581.
Slavin, R. E. (1994). Quality, appropriateness, incentive, and time: A model of instructional effectiveness. *International Journal of Educational Research, 21*(2), 141–157.
St Leger, L., Young, I., Blanchard, C., & Perry, M. (2010). *Promoting health in schools. From evidence to action*. Saint Denis Cedex: IUHPE.
Stern, T. (2010). *Förderliche Leistungsbewertung*. Vienna: Österreichisches Zentrum für Persönlichkeitsbildung und soziales Lernen.
Stewart, T., & Suldo, S. (2011). Relationships between social support sources and early adolescents' mental health: the moderating effect of student achievement level. *Psychology in the Schools, 48*(10), 1016–1033.
Stewart-Brown, S. (2006). *What is the evidence on school health promotion in improving health or preventing disease and, specifically, what is the effectiveness of the health-promoting school approach?* Copenhagen: WHO Regional Office for Europe.
Stoll, L., Bolam, R., McMahon, A., Wallace, M., & Thomas, S. (2006). Professional learning communities: A review of the literature. *Journal of Educational Change, 7*(4), 221–258.

Stronge, J. H., Ward, T. J., Tucker, P. D., & Hindman, J. L. (2007). What is the relationship between teacher quality and student achievement? An exploratory study. *Journal of Personnel Evaluation in Education, 20*(3–4), 165–184.

Suhrcke, M., & de Paz Nieves, C. (2011). *The impact of health and health behaviours on educational outcomes in high-income countries: A review of the evidence*. Copenhagen: WHO Regional Office for Europe.

Taras, H. (2005). Physical activity and student performance at school. *Journal of School Health, 75*(6), 214–218.

Tomlinson, C. A. (2004). Sharing responsibility for differentiating instruction. *Roeper Review, 26*(4), 188–189.

Tones, K. (2005). Health promotion in schools: The radical imperative. In S. Clift & B. B. Jensen (Eds.), *The health promoting school: International advances in theory, evaluation and practice* (pp. 23–40). Copenhagen: Danish University of Education Press.

Torsheim, T., & Wold, B. (2001). School-related stress, school support, and somatic complaints: A general population study. *Journal of Adolescent Research, 16*(3), 293–303.

Twemlow, S. W., Fonagy, P., Sacco, F. C., Gies, M. L., Evans, R., & Ewbank, R. (2001). Creating a peaceful school learning environment: A controlled study of an elementary school intervention to reduce violence. *American Journal of Psychiatry, 158*(5), 808–810.

Valli, L., & Buese, D. (2007). The changing roles of teachers in an era of high-stakes accountability. *American Educational Research Journal, 44*(3), 519–558.

Van de Grift, W. (2007). Quality of teaching in four European countries: A review of the literature and application of an assessment instrument. *Educational Research, 49*(2), 127–152.

Van Petegem, K., Aelterman, A., Van Keer, H., & Rosseel, Y. (2008). The influence of student characteristics and interpersonal teacher behaviour in the classroom on student's wellbeing. *Social Indicators Research, 85*(2), 279–291.

Van Ryzin, M. J., Gravely, A. A., & Roseth, C. J. (2009). Autonomy, belongingness, and engagement in school as contributors to adolescent psychological well-being. *Journal of Youth and Adolescence, 38*(1), 1–12.

Waters, S., Cross, D., & Shaw, T. (2010). Does the nature of schools matter? An exploration of selected school ecology factors on adolescent perceptions of school connectedness. *British Journal of Educational Psychology, 80*(3), 381–402.

Willms, J. D., Friesen, S., & Milton, P. (2009). *What did you do in school today? Transforming classrooms through social, academic, and intellectual engagement. First national report.* Toronto: Canadian Education Association.

World Health Organization (WHO) (1986). *Ottawa charta for health promotion*. Ottawa: WHO Regional Office.

World Health Organization (WHO) (2009). *Milestones in health promotion. Statements from global conferences*. Geneva: WHO.

Zerillo, C., & Osterman, K. F. (2011). Teacher perceptions of teacher bullying. *Improving Schools, 14*(3), 239–257.

Chapter 17
Evaluation Research in Health Promoting Schools and Related Challenges

Hege E. Tjomsland, Bente Wold, Rune Johan Krumsvik, and Oddrun Samdal

Abstract Research indicates that the main reason why teachers change their practice is the recognition of improvements in students' educational outcomes (Guskey TR, Teach Teach 8(3):381–391, 2002). To ensure widespread dissemination of health promoting schools, evidence that the health promoting school approach can be a crucial vehicle for enhancing both students' health- and educational outcomes are therefore warranted. This advocates study designs and methods that take into account the multifaceted, whole-school and context specific characteristics of health promoting schools. In this chapter, we therefore first discuss specific challenges in health promoting school research, and secondly, we propose an evaluation design combining the advantages of different research methodologies to examine the health promoting school's effectiveness in creating "better schools through health".

Keywords Health promoting school evaluation • Challenges • Design

17.1 Introduction

While the aim of public health and health promotion is the health gain of individuals, the educational system is concerned with development of knowledge and attitudes important to prepare students for their future role as contributors to society (Samdal 2008). Health aspects are therefore mainly of concern to teachers and school leaders out of a functional perspective, i.e. the function health may have for students'

H.E. Tjomsland, Ph.D. (✉) • R.J. Krumsvik, Ph.D.
Department of Education, University of Bergen, Bergen, Norway
e-mail: Hege.Tjomsland@iuh.uib.no; Rune.Krumsvik@psych.uib.no

B. Wold, Ph.D. • O. Samdal, Ph.D.
Department of Health Promotion and Development, University of Bergen, Bergen, Norway
e-mail: Bente.Wold@iuh.uib.no; Oddrun.Samdal@uib.no

V. Simovska and Patricia Mannix-McNamara (eds.), *Schools for Health and Sustainability: Theory, Research and Practice*, DOI 10.1007/978-94-017-9171-7_17,

academic performance and educational attainment. Relevant health aspects in such a functional perspective may include adequate nutrition and physical activity, development of social skills and promotion of students' well-being and enjoyment of school. Although there is a difference between the aims of the public health sector and the educational sector, there seems to be an overlap between some of the objectives of the two sectors regarding the intermediate outcomes addressed to achieve both their aims. Examples of such objectives are the development of students' self-esteem, social competence, action competence and healthy lifestyles. Given the shared objectives, the health promoting school may seem as a viable road to take, not only for representatives from the public health sector but also for teachers and school leaders, if they perceive that the health promoting school helps them achieve their educational mission.

"School improvement" refers to sustained change in learning conditions and other internal conditions with the ultimate aim of achieving educational goals more effectively (Reynolds et al. 2001). Dadaczynski and Paulus (2013) found in their study from Germany that schools which adopted a variant of the health promoting school (the good healthy school) were efficient in systematically improving the educational quality in school through the implementation of structured and evidence based health actions. In view of this, they argue that health can be seen as a key driver for school quality, and that health related measures may support schools in realizing its educational mission (Dadaczynski and Paulus 2013). Similarly, Rowling (2005) notes, that the health promoting school may represent a school improvement initiative because it aims to maximize health and educational outcomes through an integration of policy and practice from both sectors.

However, the educational system in most European countries has been under immense pressure to improve test scores in literacy and numeracy during the past decade. In this current school political climate, many teachers and school leaders are not convinced that scarce resources should be allocated to improving students' health, health behaviors and well-being (Tjomsland et al. 2009b). Teachers and school leaders are however inclined to change their practice if they believe that it may benefit their students' academic performance and educational attainment (Fullan 2007). Thus, while the race towards better academic performances may thwart teachers' motivation to promote students' health and well-being, it may also motivate them for the implementation of health promotion practices if researchers produce evidence that validate the Vilnius resolution that schools in fact become "better through health" (Ragaišienė 2009).

Attempting to produce the aforementioned evidence, it is important to develop evaluation designs that take into account the multifaceted, whole-school and context specific characteristics of the health promoting school. In this chapter, we therefore first discuss why it is important to study the effectiveness of the health promoting school in improving students' health and educational outcomes. We further point to challenges in health promoting school research; and finally, we propose an evaluation design which combines the advantages of different research methodologies to study the health promoting school's effectiveness in creating "better schools through health".

17.2 The Health Promoting School: Effect Outcomes and Process Outcomes

"The Vilnius resolution" points as mentioned above to improved health and educational outcomes as valuable health promoting school outcomes. Similarly, evaluation research from Norwegian health promoting schools indicates that the proposed relationship between health and educational outcomes motivated educational staff to adopt, implement and sustain the health promoting school (Tjomsland et al. 2010; Viig et al. 2010). More specifically, Tjomsland and colleagues (2010) found that educational staff perceived that the implementation of daily physical activity improved students' concentration and academic performance. Such outcome perceptions among the staff motivated them, in turn, to sustain health promotion practices as part of the school's daily fabric over a 10 years period. Green and Kreuter (2005) argue that students, teachers and parents ought to be involved in determining what constitutes successful outcomes of school health promotion. Following their reasoning, the Norwegian findings indicate that students' improved health and educational outcomes constitute valued indicators of success in health promoting school research.

A review of health and health behaviors' impact on educational outcomes indicates that there is evidence, in particular from the North American context, supporting a link between childhood and adolescent health and educational outcomes (Suhrcke and de Paz Nieves 2011). The studies in this review found a negative correlation between (1) risky health behaviors and (ill) health conditions and (2) education in terms of both educational attainment and academic performance (Suhrcke and de Paz Nieves 2011). Two other meta reviews (Stewart-Brown 2006; Lister-Sharp et al. 1999) further reveal that comprehensive school health programs like the health promoting school may be effective in improving students' health, for example in terms of increased physical activity levels and decreased substance misuse. Rather than focusing only on the health outcomes of health promoting schools, evaluations of health promoting schools ought to also explore if there is a causal link between the implementation of health promotion in schools and educational outcomes. If future studies support the aforesaid study by Dadaczynski and Paulus (2013), which indicates that the health promoting school may help schools realizing its educational mission, evidence of how the health promoting school is effective in producing this outcome is warranted. What are for example the concrete health promotion strategies and actions implemented at the school level which materialize into health and educational outcomes at the student level?

Producing such evidence is possible if health promoting school evaluation research attends not only to the impact and effect of the health promoting school, but also to the processes involved in producing the anticipated outcomes (Carlsson and Simovska 2012; Inchley et al. 2007). Because health promotion actions must first be adopted and implemented by the staff before it can be effective on the student level, and because the research body on implementation indicates that higher levels of implementation are associated with more positive outcome effects than lower levels of implementation (Durlak and DuPre 2008; Deschesnes et al. 2003), an exploration only of outcome effects in health promoting school research may be misleading. Moreover, since implementation research from the educational setting suggests that teachers are prone to adjusting and modifying programs and packages to fit the individual needs of their students (Fullan 2007; Dusenbury et al. 2005), implementation research in this setting seems particularly important. For example, if teachers do not implement the agreed upon actions because they perceive a mismatch between their professional aims and the health promoting school, this may decrease the likelihood of identifying outcome effects at the student level since part of the student body may not have been exposed to health promotion actions. Whereas implementation research providing concise descriptions of what was done, and why, is essential to make a valid assessment of the effectiveness of the health promoting school in producing health and educational outcomes, it is equally important to generate knowledge of how the approach is effective in creating such outcomes (Mukoma and Flisher 2004). Without an exploration of the implementation quality, Dooris and Barry (2013) note that the health promoting school may be judged as ineffective even if the lack of outcome effects is a result of short comings in the delivery process.

17.3 Evaluation Research

Varying definitions of evaluation exists. Nutbeam (1998), for example, suggests that at the core of evaluation in health promotion is an assessment of the extent to which an action achieves a valued outcome and the process by which the outcomes are achieved, while Nevo (2006) refers to evaluation in education as the collection of systematic information regarding the nature and quality of educational objects. According to Green and Kreuter (2005), there are three reasons for evaluation. First, evaluation data may be used by researchers to determine whether improvements in health and well-being are linked to a specific program or intervention. In the school setting, producing evidence concerning whether or not the health promoting school actually makes schools better in promoting academic performances and educational attainments, is as suggested above, essential for widespread dissemination of the approach. Second, Green and Kreuter (2005) note that evaluation may guide practitioners' use of a program. This is also relevant for teachers and school leaders in health promoting schools since systematic use of self-evaluation seems to produce an effective mix of stability and change in teachers' health promotion practices conducive to health and educational outcomes at the student level (Simovska et al. 2010; Tjomsland et al. 2010). Third, evaluation results can be used by elected officials to demonstrate that a given program served its purposes and citizens. At the national level, evidence of the health promoting school's effectiveness in improving students' health and educational outcomes seems crucial in order to improve the probability that school authorities write the health promoting school approach into national curricula and school policy documents charging all schools to work with health promotion.

17.3.1 Evaluation Research on "Health Promotion in Schools" and "The Health Promoting School"

The existent research body on school health promotion is large. However, it is important to distinguish between research on "health promotion in schools" and research on "the health promoting school as a setting based policy" (Lister-Sharp et al. 1999). The two concepts are often used interchangeably even if they have different ideological and epistemological bases that affect the way of working with or in schools, as well as the nature of evidence and the criteria for success (Rowling and Jeffreys 2006; Stewart-Brown 2006; Lister-Sharp et al. 1999). Whereas research on "health promotion in school" tends to focus on specific health promoting practices and its effect on health behaviors for example in terms of substance use, exercise and mental health promotion, health promoting school evaluation research also needs to attend to change in organizational structures and students' educational outcomes. There is a growing body of evidence of the effectiveness of "health promotion in schools" (Kirby et al. 2012; Suhrcke and de Paz Nieves 2011; Tang et al. 2009). Fewer studies, however, assess the impact of the health promoting school (Deschesnes et al. 2010; Inchley et al. 2007; Mukoma and Flisher 2004). The paucity of health promoting school evaluation studies may be owing to the infancy of the field, as well as to the whole school approach generating questions regarding what should be evaluated and what constitutes indicators of success in health promoting school evaluation research (St Leger et al. 2007; Dooris 2006; Deschesnes et al. 2003).

17.3.1.1 Challenges in Health Promoting School Evaluation Research

Within the culture of evidence-based medicine, interventions which have not been subject to randomized controlled trials are generally regarded as unproven (Moore et al. 2003). Given however, that the settings approach to health promotion moves health out of the professional action frame into organizations such as the school, stringent research designs may be economically and scientifically challenging since health promoting schools develop context specific aims and actions (Dooris 2006; Rowling 2005; St Leger 2004). Hence, experimental designs, such as the randomized controlled trials, may be misleading in health promoting school evaluation research since it is not a standardized pre-packaged program. Stringent designs may not be able to pick up the process of active participation by the participants, and the complex interaction of factors that may contribute to change in the way staff practices health promotion, and through this, produce health and educational outcomes (Stewart-Brown 2006).

Additionally, the use of control groups may be problematic in schools due to the possibility of cross-contamination from educational staff, students and parents in one school to control schools in other regions (Laurence et al. 2007). It may further be problematic in school experimental designs to determine if outcome effects are a result of the implemented health promotion actions or of other significant variables influencing students, such as social class, family socialization and friends (Turunen et al. 2006).

17.3.1.2 Recommendations for Health Promoting School Evaluation Research

While the research community still debates the form and focus of evaluation research of health promoting schools, a growing number of studies have identified fields that need attention in health promoting school evaluations (St Leger et al. 2007). For example, Inchley et al. (2007) note that it is essential to shift the current focus on individual level outcomes to measures of success also at the school level. Likewise, Rowling and Jeffreys (2006) object to evaluations of school health promotion that view schools as sites where measurements occur across schools for specific individual health behavior outcomes. Instead, health promoting school research should attend to how health promotion strategies have been adjusted to fit the 'growth' state of each school, and thus work differently in each context (Rowling and Jeffreys 2006). Accordingly, health promoting school research needs to, in addition to measures of students' health and educational outcomes (Inchley et al. 2007), also explore changes in visions and policies, changes in the environment, in event rates and in teachers' practices. Previous studies from health promoting schools have for example identified reduced vandalism at school and more engaged student councils as valuable outcomes at the school level, whereas staff's perception of enhanced concentration and reduced discipline problems among students have been identified as valuable outcomes at the teacher level (Samdal and Rowling 2013; Viig 2010; Tjomsland 2009). Adding to this, measures of school connectedness, school satisfaction and truancy at the student level also constitute significant indicators of success provided that a growing number of studies indicate that a personalized and caring learning climate in school is one potential contributor to students' academic achievements (Hattie 2009; Murray et al. 2007).

Finnish health promoting school research has further shown that participatory action research may be particularly useful for health promoting school research (Turunen et al. 2004). Provided that educational staff are used to exploring new methods and distinguishing good practice from bad, the school setting may in fact be particularly susceptible for such research methods. There is also an increased interest in research methods involving educational staff in other areas of educational research (Cain 2011). This research body indicates that action research involving educational staff may generate detailed and inspiring narratives of change in educational practice that can lead to a realization of educational ideals (Cain 2011). Participatory action research then, seems to have the potential to increase the probability that scientific knowledge is translated into information that help educational staff adapt the new way of being and doing in health promoting schools (Mukoma and Flisher 2004).

Finally, evaluations of setting based health promotion interventions may benefit from theory based evaluation (Dooris and Barry 2013). In particular, "theories of

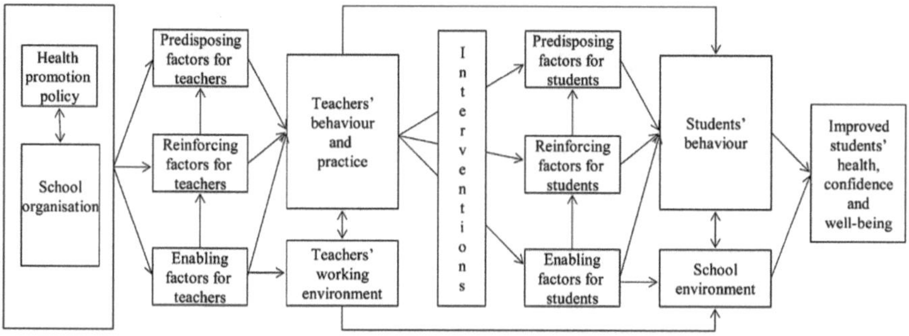

Fig. 17.1 The Norwegian evaluation model building on Green and Kreuter's (2005) Model for Planning, Implementing and Evaluating health promotion programs. (Reproduced from Viig and Wold 2005)

change" may be useful frameworks for health promoting school research by directing attention to the causal chain between individual and organizational factors that lead to behavioral and organizational change. Dooris and Barry's (2013) recommendations corroborate with research findings from Norway indicating that the application of a model for planning, implementing and evaluating health promotion was useful for the evaluation of the Norwegian schools enrolled in the European Network of health promoting schools (Viig 2010; Tjomsland 2009).

17.4 An Evaluation Example from Norway

The Norwegian evaluation research, carried out by the University of Bergen, included both implementation research as well as effect research. The evaluation employed multiple data collection methods in terms of surveys, in-depth interviews, school documents, and focus group interviews. The evaluation, which extended through a decade, aimed to determine whether improvements in health and educational outcomes among the students could be linked to the implementation of the health promoting school approach. The Norwegian evaluation model (Fig. 17.1) developed by Viig and Wold (2005) is an expansion of Green and Kreuter's (2005) model for planning, implementing, and evaluating health promotion programs. It helps to identify factors enabling people to gain control over conditions that influence their health and quality of life, and it takes into account the principles of participation and empowerment which are central in health promotion (WHO 1986). The Norwegian version of the model (Fig. 17.1) includes a double loop implementation process at school level assuming that there are factors predisposing, enabling and reinforcing teachers' behaviors and practices which in turn influence students' health and educational outcomes. Teachers first need to be motivated, trained and supported by the school leaders before they can implement health promotion actions involving the students. Likewise, organizational factors in terms of professional development activities, curriculum time and involvement of collaborating network and partners need to be established, and finally leadership reinforcement is needed to ensure that the agreed upon actions are implemented and sustained (Samdal and Rowling 2013).

Viig (2010) and Tjomsland (2009) found that the application of such a model helped to identify factors facilitating or hindering schools journey towards health promoting school implementation, because it helped to organize variables into sequences of cause and effect that seemed to influence behavioral and environmental change in school. Based on this grouping of cause and effect, it is possible to identify activities that apply mechanisms to achieve change, first in teacher attitudes and practice and thereafter in student behavior and perceptions. Samdal and Rowling (2013) argue that additional loops of regional, national and international policies and practices can be added to the double-loop model to better understand how contexts external to the school also have an impact on the school level practices. For example, if the national curriculum and national polices emphasize the importance of health promoting schools activities to students' learning, schools are more likely to give priority to implementation of the health promoting school initiative.

Moreover, the evaluation results from Norway indicated that the transformation of schools into health promoting settings could have been more effective if it had been run systematically in accordance with organizational theories (Wold and Samdal 1999). However, in some schools, the process taking place fulfilled the criteria of successful implementation of organizational change. Here, the school leaders set aside time to motivate and establish commitment in the staff to health promotion, and they also ensured agreement among the staff concerning what health promotion actions and strategies should be implemented. In these schools, the implementation process was characterized by high agreement among the staff and a low degree of change, which according to implementation theories, is more likely to produce high and sustained outcome effects in student behavior (Wold and Samdal 1999).

17.5 The Sustainability of the Health Promoting School and Evaluation Research

Building health promoting schools are not done overnight. Inchley and colleagues (2007) therefore call for a greater recognition of the time it takes to achieve change in health and educational outcomes. This raises the question as to *when* it is appropriate to conduct health promoting school evaluation research. Murray et al. (2007) found in their review of coordinated school health programs and academic achievements that the strongest evidence arose from projects that had a history of many years of planning, implementing and follow-up. Given also that students and teachers come and go, and that differences between the cohorts affect research results, Hargreaves and Goodson's (2006) claim that studies of change processes in school should be performed from a longitudinal perspective, seems prudent also for health promoting school research. While cross-sectional investigations reflect the participants' points of view and experiences at an exact moment of time, longitudinal research can examine how educational staff works over time to improve health and educational outcomes among students (Deschesnes et al. 2010). Longitudinal research designs therefore comprise an essential component of health promoting school research. This perspective is supported by the Norwegian health promoting school research indicating that 10 and 14 year follow up studies were more efficient in providing evidence of the effectiveness of the health promoting school than 3 year follow up studies (Viig et al. 2011; Tjomsland et al. 2009a, b).

17.5.1 A Mixed Methods (MM) Design

The above reasoning calls for a variety of methodological approaches in health promoting school research to capture both the process of rethinking school's practice in a health promoting direction as well as to examine the effects of such work over time (Stewart-Brown 2006). Whereas Greene et al. (2007) argue that evaluation studies in general benefit from pragmatic standpoints and the adoption of mixed methods, the holistic nature of the health promoting school urges the use of methodological pluralism and the combination of both qualitative and quantitative approaches (Pommier et al. 2010). The collection and analyses of data from multiple sources and levels may yield the most comprehensive and in-depth understanding of if and how health promoting schools produce change in students' health, health behaviors and well-being as well as students' educational attainment and academic performance.

17.5.1.1 Data Collection Methods in Health Promoting School Evaluation Research

Taken together then, we have argued that health promoting school research ought to attend to implementation outcomes as well as effect outcomes. We have also made the argument that a mixed method longitudinal research design may be particularly suitable to examine if and how the implementation and integration of health promotion into school life produce health and educational outcomes at the student level. Against this background, we propose a number of data collection methods to identify implementation outcomes as well as effect outcomes in health promoting schools.

Firstly, *qualitative interviews* either in terms of focus group interviews or in-depth interviews are considered highly suitable for an exploration of practice (Elias et al. 2003). A good description of what is actually taking place in health promoting schools is essential in order to understand low or high outcome effects at the student level. Further, interview data may also provide a good basis for the development of questionnaire surveys for all students and teachers by ensuring a good fit between the survey items and context specific developments within each school. The focus group interview is particularly suitable when researching students' experiences because there is reduced pressure on the informants to respond to each questions, and because peer support help to level out the power imbalance between the adult researcher and the child in one-to-one interviews (Hennessy and Heary 2005). Although, one-to-one interviews may be more suitable to explore sensitive issues, since children may be afraid to speak out in a group setting.

Secondly, *school documents* may serve as useful indicators of official activity within institutions since it provides the researcher with a sense of history related to the context being studied (Hatch 2002). For evaluative purposes in health promoting schools, various types of school documents are relevant. An analysis of developmental plans and curricula outlines may for example provide insights into whether or not health promoting policies has been formalized into school policy documents. This is important because a formalization of health promotion into schools' curricula seems to be an essential strategy for making the health promoting school an integral part of school life (Samdal et al. 2010; Hoyle et al. 2008). Likewise, activity plans and annual evaluation outlines offer insights into the actions and strategies implemented, and to what extent these follow the school's curriculum outlines. In addition, school's web site may be a powerful indicator of whether educational staff believe that schools in fact become better through health. For example, the Norwegian evaluation showed that the most successful and dedicated health promoting schools signaled staff's commitment to students' learning, health and well-being on the school's web site (Tjomsland et al. 2009a, b). As such, school documents can give a good indication of whether health promoting actions and strategies actually have been implemented and sustained into the everyday lives of the schools. The collection of school documents may also be valuable at the national level to examine whether national school policy documents emphasize the importance of health promoting school activities to students' educational outcomes. If the analysis of national policy documents indicates that health promotion is a prioritized area at the national level, higher levels of implementation may be expected at school level, and more positive outcome effects among students.

Thirdly, *survey data* are suitable to examine students' health and health behaviors as well as students' educational outcomes since questionnaires represent a cost-effective way of collecting data from a large number of students about their subjective experiences of health, well-being, self-esteem, lifestyle, self-regulated learning and academic achievements. When measuring students' educational outcomes through surveys, it is important to distinguish between students' *subjective* and *objective* learning outcomes. Subjective learning outcomes reflect students' own perception of their learning outcomes, while objective learning outcomes reflect students' actual learning outcomes. Subjective learning outcomes can be measured by using methods that assess the individual's subjective experience (e.g., surveys and interviews), while the measurement of objective learning outcomes requires some form of objective behavioral measurement (e.g., standardized tests, experimental methods, observations). For example in Norway, annual national tests among 5th, 8th and 9th graders provide information on students' basic skills in English, reading and mathematics. The national tests are not designed to reveal whether individual students improve from one year to another. However, the tests reflect improvements and development in school as well as information about how the school performs compared to the national sample (www.udir.no). In health promoting school research, national statistics like this, may for example be used to explore whether students in health promoting schools produce better test scores than students in schools which have not transformed into health promoting settings.

Preferably, however, following the same sample over time is recommended to monitor possible changes in health and educational outcomes. Samdal and colleagues' (2010) study from the Norwegian context demonstrates this design with one baseline measurement and one follow-up. At the same point in time, identical surveys about subjective health, health behaviors, school well-being and self-reported academic achievements were distributed to students in health promoting schools as well as to a referent group consisting of a national representative sample of students in the same age group. The student responses were then compared to examine if the health promoting school students' responses differed from the responses of the students in the referent group.

Finally, the quantitative component in a mixed method research design may benefit from an extended selection cohorts quasi-experimental design since it is not possible or desirable to use a random selection of control schools (Olweus 2005). In this design, the groups to be compared belong to the same schools (for example, the grade 5 cohort at Time 1, with no intervention, compared with the grade 4 cohort at Time 2, with 8 months of intervention, recruited from the same schools). This design implies that several grade levels are examined at baseline and that the lower grade levels are followed up after the intervention and compared with baseline data from higher grade levels. As a result, the school constitutes its own control since baseline data from a higher grade level may be regarded as control group data for the younger students (who will be at the same grade level after the intervention has been carried out as the "control" group students were before the intervention). This design may be useful both for a school's own evaluation of effectiveness of the health promoting school intervention as well as for a national or regional evaluation of a health promoting school implementation.

17.6 Conclusion

To ensure further dissemination of health promoting schools, we have argued that it is crucial to examine if the health promoting school enhances students' educational outcomes through students' improved health, health behaviors and well-being. We have therefore suggested that health promoting school evaluation designs should study the proposed causal relationship between an implementation and integration of health promotion into school life and educational and health outcomes among the students. We have further suggested that a mixed method design may be particularly suitable in health promoting school evaluations since it allows for an exploration of both implementation outcomes as well as effect outcomes. Whereas the qualitative component may explore the processes taking place within a specific school context during the implementation period, the quantitative component may examine the proposed relationship between health and educational outcomes among the students. Finally, we have underscored that due to the long term effort of transforming a school into a health promoting setting, a longitudinal design in health promoting school evaluation research is warranted.

References

Cain, T. (2011). Teachers' classroom-based action research. *International Journal of Research and Method in Education, 34*(1), 3–16.

Carlsson, M., & Simovska, V. (2012). Exploring learning outcomes of school-based health promotion—A multiple case study. *Health Education Research, 27*(3), 437–447.

Dadaczynski, K., & Paulus, P. (2013). Germany: Anschub de – Alliance for sustainable school health and education. In O. Samdal & L. Rowling (Eds.), *The implementation of Health Promoting Schools. Exploring the theories of what, why and how* (pp. 139–141). London: Routledge.

Deschesnes, M., Martin, C., & Hill, A. J. (2003). Comprehensive approaches to school health promotion: How to achieve broader implementation? *Health Promotion International, 18*(4), 387–396.

Deschesnes, M., Trudeau, F., & Kébé, M. (2010). Factors influencing the adoption of a health promoting school approach in the province of Quebec, Canada. *Health Education Research, 25*(3), 438–450.

Dooris, M. (2006). Healthy settings: Challenges to generating evidence of effectiveness. *Health Promotion International, 21*(1), 55–65.

Dooris, M., & Barry, M. M. (2013). Overview of implementation in health promoting settings. In O. Samdal & L. Rowling (Eds.), *The implementation of health promoting schools. Exploring the theories of what, why and how* (pp. 14–33). London: Routledge.

Durlak, J. A., & DuPre, E. P. (2008). Implementation matters: A review of research on the influence of implementation on program outcomes and the factors affecting implementation. *American Journal of Community Psychology, 413*(4), 327–350.

Dusenbury, L., Brannigan, R., Hansen, W. B., Walsh, J., & Falco, M. (2005). Quality of implementation: Developing measures crucial to understanding the diffusion of preventive interventions. *Health Education Research, 20*(3), 308–313.

Elias, M. J., Zins, J. E., Graczyk, P. A., & Weissberg, R. P. (2003). Implementation sustainability and scaling up of social emotional and academic interventions in public schools. *School Psychology Review, 3*, 303–319.

Fullan, M. (2007). *The new meaning of educational change* (4th ed.). New York: Teachers College Press.

Green, L., & Kreuter, M. (2005). *Health program planning, an educational and ecological approach* (4th ed.). New York: Mc Graw Hill.

Greene, J. C., Benjamin, L., & Goodyear, L. (2007). The merits of mixed methods in evaluation. *Evaluation, 7*(1), 25–44.

Guskey, T. R. (2002). Professional development and teacher change. *Teachers and Teaching, 8*(3), 381–391.

Hargreaves, A., & Goodson, I. (2006). Educational change over time? The sustainability and nonsustainability of three decades of secondary school change and continuity. *Educational Administration Quarterly, 42*(1), 3–41.

Hatch, J. A. (2002). *Doing qualitative research in education settings*. Albany: State University of New York Press.

Hattie, J. A. C. (2009). *Visible learning. A synthesis of over 800 meta-analyses relating to achievement*. London/New York: Routledge.
Hennessy, E., & Heary, C. (2005). Exploring children's views through focus groups. In S. Greene & D. Hogan (Eds.), *Researching children's experience approaches and methods* (pp. 236–251). London: Sage.
Hoyle, T. B., Samek, B. B., & Valois, R. B. (2008). Building capacity for the continuous improvement of health-promoting schools. *Journal of School Health, 78*(1), 1–8.
Inchley, J., Muldoon, J., & Currie, C. (2007). Becoming a health promoting school: Evaluating the process of effective implementation in Scotland. *Health Promotion International, 22*(1), 65–71.
Kirby, J., Levin, K. A., & Inchley, J. (2012). Associations between the school environment and adolescent girls' physical activity. *Health Education Research, 27*(1), 101–114.
Laurence, S., Peterken, R., & Burns, C. (2007). Fresh Kids: The efficacy of a health promoting schools approach to increasing consumption of fruit and water in Australia. *Health Promotion International, 22*(3), 218–226.
Lister-Sharp, D., Chapman, S., Stewart-Brown, S., & Sowden, A. (1999). Health promoting schools and health promotion in schools: Two systematic reviews. *Health Technology Assessment, 3*, 1–207.
Moore, L., Graham, A., & Diamond, I. (2003). On the feasibility of conducting randomised trials in education: Case study of a sex education intervention. *British Educational Research Journal, 29*(5), 673–689.
Mukoma, W., & Flisher, A. J. (2004). Evaluations of health promoting schools: A review of nine studies. *Health Promotion International, 19*(3), 357–368.
Murray, N. G., Low, B., Hollis, C., Cross, A., & Davis, S. (2007). Coordinated school health programs and academic achievement: A systematic review of the literature. *The Journal of School Health, 77*(9), 589.
Nevo, D. (2006). Evaluation in education. In I. F. Shaw, J. C. Greene, & M. M. Mark (Eds.), *Handbook of evaluation, policies, programs and practices* (pp. 441–460). London: Sage.
Nutbeam, D. (1998). Evaluating health promotion – Progress, problems and solutions. *Health Promotion International, 13*(1), 27–42.
Olweus, D. (2005). A useful evaluation design, and effects of the Olweus Bullying Prevention Program. *Psychology, Crime & Law, 11*(4), 389–402.
Pommier, J., Guével, M. R., & Jourdan, D. (2010). Evaluation of health promotion in schools: A realistic evaluation approach using mixed methods. *BMC Public Health, 10*(1), 43. 1–12.
Ragaišienė, I. (2009). *Better schools through health: The 3rd European conference on health promoting schools*. Vilnius: State Environmental Health Centre (VASC).
Reynolds, D., Teddlie, C., Hopkins, D., & Stringfield, S. (2001). *Linking school effectiveness and school improvement*. London: Routledge Falmer.
Rowling, L. (2005). Dissonance and debates encircling 'health promoting schools'. *Health Promotion Journal Australia, 16*(1), 55–57.
Rowling, L., & Jeffreys, V. (2006). Capturing complexity: Integrating health and education research to inform health-promoting schools policy and practice. *Health Education Research, 21*(5), 705–718.
Samdal, O. (2008). *International encyclopedia of public health* (Vol. 5). San Diego: Academic.
Samdal, O., & Rowling, L. (2013). Cross fertilization of national experiences and need for future developments. In O. Samdal & L. Rowling (Eds.), *The implementation of health promoting schools. Exploring the theories of what, why and how* (pp. 149–161). London: Routledge.
Samdal, O., Viig, N. G., & Wold, B. (2010). Health promotion integrated into school policy and practice: Experiences of implementation in the Norwegian network of health promoting schools. *Journal of Child and Adolescent Psychology, 1*(2), 43–72.
Simovska, V., Dadaczynski, K., Viig, N. G., Tjomsland, H. E., Bowker, S., Woynarowska, B., et al. (2010). *HEPS tool for schools. A guide for school policy development on healthy eating and physical activity*. Woerden: NIGZ.
St Leger, L. (2004). What's the place of schools in promoting health? Are we too optimistic? *Health Promotion International, 19*(4), 405–408.

St Leger, L., Kolbe, L. J., Lee, A., McCall, D., & Young, I. (2007). School health promotion, achievements, challenges and priorities. In D. V. McQueen & C. M. Jones (Eds.), *Global perspectives on health promotion effectiveness*. New York: Springer.

Stewart-Brown, S. (2006). *What is the evidence on school health promotion in improving health or preventing disease and, specifically, what is the effectiveness of the health promoting schools approach?* Copenhagen: WHO Regional Office for Europe.

Suhrcke, M., & de Paz Nieves, C. (2011). *The impact of health and health behaviors on educational outcomes in high income countries: a review of the evidence*. Copenhagen: WHO Regional Office for Europe.

Tang, K. C., Nutbeam, D., Aldinger, C., St Leger, L., Bundy, D., Hoffmann, A. M., … & Heckert, K. (2009). Schools for health, education and development: A call for action. *Health Promotion International, 24*(1), 68–77.

Tjomsland, H. E. (2009). *Health promotion with teachers. An evaluation of the Norwegian network of health promoting schools*. Bergen: University of Bergen.

Tjomsland, H. E., Wold, B., & Iversen, A. C. (2009a). The Norwegian network of health promoting schools: A three-year follow-up study of teacher motivation, participation and perceived outcomes. *Scandinavian Journal of Educational Research, 53*(1), 89–102.

Tjomsland, H. E., Larsen, T. B., Viig, N. G., & Wold, B. (2009b). A fourteen year follow-up study of health promoting schools in Norway: Principals' perceptions of conditions influencing sustainability. *The Open Education Journal, 2*, 54–64.

Tjomsland, H. E., Larsen, T., Samdal, O., & Wold, B. (2010). Sustaining comprehensive physical activity practice in elementary school: A case study applying mixed methods. *Teachers and Teaching: Theory and Practice, 16*(1), 73–95.

Turunen, H., Tossavainen, K., & Vertio, H. (2004). How can critical incidents be used to describe health promotion in the Finnish European Network of Health Promoting Schools? *Health Promotion International, 19*(4), 419–427.

Turunen, H., Tossavainen, K., Jakonen, S., & Vertio, H. (2006). Did something change in health promotion practices? A three-year study of Finnish European Network of Health Promoting Schools. *Teachers and Teaching: Theory and Practice, 12*(6), 675–692.

Viig, N. G. (2010). *Tilrettelegging for laereres deltakelse i helsefremmende arbeid*. Bergen: Universitetet i Bergen.

Viig, N. G., & Wold, B. (2005). Facilitating teachers' participation in school-based health promotion – A qualitative study. *Scandinavian Journal of Educational Research, 49*(1), 83–109.

Viig, N. G., Tjomsland, H. E., & Wold, B. (2010). Program and school characteristics related to teacher participation in school health promotion. *The Open Education Journal, 3*, 10–20.

Viig, N. G., Fosse, E., Samdal, O., & Wold, B. (2011). Leading and supporting the implementation of the Health promoting schools program in Norway. *Scandinavian Journal of Educational Research, 56*(6), 1–14.

Wold, B., & Samdal, O. (Eds.). (1999) *Health promotion among young people: The development of healthy school environments. Dissemination of experiences from the Norwegian schools in the Norwegian schools in the European Network of Health Promoting Schools (ENHPS)* (HEMIL Report Series, 7). Bergen: Research Centre for Health Promotion, University of Bergen.

World Health Organization (WHO). (1986). Ottowa charter for health promotion. *Health Promotion International, 1*(4), iii–v.

Chapter 18
Mixed Methods' Contribution to the Evaluation of Health Promotion Initiatives in the School Setting

Marie-Renée Guével, Jeanine Pommier, and Didier Jourdan

Abstract The development of suitable approaches for evaluating health promotion in schools to produce useful evidence to enhance sustainability and transferability is still a major topic of discussion. There is currently a trend among health promotion researchers to develop evaluation approaches that are able to measure the impact of an initiative as well as to understand how this impact is obtained in order to inform the implementation of sustainable health promotion initiatives by practitioners and decision-makers from both health and education sectors. This chapter sets out to illustrate the contribution of mixed methods to take into account the complexity of school health promotion initiatives to help address the challenges faced by the field of school health promotion especially those related to evaluation, sustainability and transferability. Empirical data gathered from an intervention research implemented in the French context are used to highlight the interest of such a research strategy.

Keywords Evaluation • School health promotion • Mixed methods • Teacher training

M.-R. Guével, Ph.D. (✉) • J. Pommier, Ph.D.
Department of Social and Human Sciences, EHESP (French School of Public Health), Rennes, France
e-mail: Marie-Renee.Guevel@ehesp.fr; Jeanine.Pommier@ehesp.fr

D. Jourdan, Ph.D.
Faculty of Education, University Blaise Pascal Clermont-Ferrand, Clermont-Ferrand, France
e-mail: Didier.JOURDAN@univ-bpclermont.fr

V. Simovska and Patricia Mannix-McNamara (eds.), *Schools for Health and Sustainability: Theory, Research and Practice*, DOI 10.1007/978-94-017-9171-7_18,

18.1 Introduction

The contribution of health promotion to the health and well-being of pupils has been widely recognized (Hamel et al. 2001; Li et al. 2009; St Leger et al. 2007) as well as the key role played by schools (St Léger 2004; OECD 2010). Recent publications from the World Health Organization (WHO) and from the Organisation for Economic Co-operation and Development (OECD) showed a very close link between health and education. The work from the WHO, carried out by Suhrcke and de Paz Nieves (2011), tends to show the negative influence of a "bad" health status (obesity, sleeping issues, anxiety and depression) and of health behaviours (smoking, alcohol and cannabis consumption) on the academic achievements of children and youth (Suhrcke and de Paz Nieves 2011). From a complementary perspective the OECD (2010) highlights education's contribution both to health improvements and to the strengthening of civic and social involvement. It also points out that education, within national school systems, cannot alone achieve these goals of health improvement and social cohesion; the key role of families and the community also need to be recognised (OECD 2010).

Therefore, in the past decades, schools have been a major setting for implementing health promotion initiatives and the scientific literature tends to show some positive achievements of such initiatives on pupils' health and schools' organization (Stewart-Brown 2006; St Leger and Young 2009). The synthesis carried out by Stewart-Brown (2006) shows that health promoting school approaches were found to have a beneficial effect on: the social and physical environment of the school, staff development, provision of school meals, provision of exercise programmes, and the school's social climate. The author also highlights that positive impacts on health-related behaviours were identified in some studies but not all. She also mentions there was some evidence that mental and social well-being may benefit from the development of school health promotion initiatives. However, most of the studies were small-scale and their quality was variable. In their literature review, Mũkoma and Flisher (2004) specify that impact can also be observed at a school level in terms of political and organizational development to support health promotion implementation, its integration within the curriculum and the involvement of parents and community (Mukoma and Flisher 2004). Moreover, Lee and his colleagues (2005) highlight that links can be found between school health promotion and issues related to school improvement and school effectiveness, showing that health promotion can both benefit health and education sectors (Lee et al. 2005).

The literature also presents a body of evidence regarding the factors that influence the quality of health promotion initiatives implemented in the school setting (St Leger et al. 2007). Literature reviews carried out in the past few years agree on several factors as essential for the success and the quality of these initiatives (Lister-Sharp et al. 1999; Stewart-Brown 2006; St Leger et al. 2007; St Leger and Young 2009; St Leger 2005; Peters et al. 2011). They identified three main aspects:

- Involvement of the school community as a whole and addressing all the aspects of school life;
- Addressing the school social environment (relationships between pupils and staff, among pupils, among staff and between parents and schools);
- Development of children's life skills.

Some authors also add factors such as the length of the initiative, the staff's collective work, the institutional support and actors' support and training (St Leger 1999; Han and Weiss 2005; Peters et al. 2011).

In spite of these results, the development of suitable approaches for evaluating health promotion in schools in order to produce useful evidence to enhance sustainability and transferability is still a major topic of discussion (St Leger et al. 2007). There is currently a trend among health promotion researchers to develop evaluation approaches that are able to measure the impact of an initiative as well as to understand how this impact is obtained in order to inform the implementation of sustainable health promotion initiatives for both health and education sectors (Rowling 2009). Furthermore, school health promotion initiatives can be considered as complex systems, i.e. systems that consist of multiple components; that can only be understood by observing the interactions of these components; and are open-systems interacting with and influenced by their environment (Burton 2002). Implementing such initiatives may mean school communities and their partners changing some of their working habits and adapting them to the specificities of health promotion within the school setting (Mérini et al. 2010).

Against this backdrop, this chapter sets out to illustrate the contribution of mixed methods to the evaluation of school health promotion initiatives. The purpose is to take into account their complexity to inform the development of sustainable initiatives.

After presenting a brief overview of the challenges and issues related to health promotion evaluation and more precisely to the evaluation of school health promotion, we will discuss the opportunity of using mixed methods in evaluation approaches to embrace some of these challenges. We will then illustrate this discussion using the example of the evaluation of a French initiative for school health promotion. The evaluation design will be described and some major results will be presented. The chapter will conclude with some comments regarding the use of mixed methods in designing methods to evaluate school health promotion.

18.2 Issues in Evaluating School Health Promotion – How can Mixed Methods help?

As mentioned in the introduction, evaluation is still a major topic of discussion within the field of health promotion. Various evaluation approaches have been used in health promotion (Tones and Tilford 2001). They are influenced by the multidisciplinary nature of health promotion and refer to various traditions. In this section, the main issues raised by evaluation in the field of health promotion and school health promotion are first reviewed. The opportunity of using mixed methods is then discussed through the presentation of what some authors considered as a third methodological tradition (Johnson et al. 2007).

18.2.1 Issues Raised by Evaluation in the field of Health Promotion and School Health Promotion

According to the definition given by the WHO, evaluation aims to produce information that can be used by those who have an interest in the improvement and effectiveness of interventions (WHO 1998). However, evaluation in the field of health promotion has raised particular issues (Rootman et al. 2001).

These issues are illustrated by Merzel and D'Afflitti (2003) who conducted a systematic literature review of 32 community-based health promotion programmes. They identified five main issues: (1) methodological issues including the choice of the unit of analysis (individuals, communities, etc.) and design and sampling issues; (2) the influence of secular trends and the difficulty of separating the impact of health promotion programmes from these trends; (3) smaller-than-expected effects, i.e. relatively small effects are to be expected from community-level programmes; (4) limitations of the health promotion programmes including their duration, insufficient tailoring to reflect local conditions and the difficulty for community-level programmes to ensure sufficient community penetration; and (5) limitations of theory because of the complexity of conceptualizing the relationship between multiple interventions and multiple levels of influence which makes it difficult to develop integrated explanatory theories as well as testable models (Merzel and D'Afflitti 2003). Other authors also pointed out further issues such as the complexity of the causality between a health promotion programme and its effects, and the unsuitability of the experimental evaluation process of the health promotion values enshrined in the Ottawa Charter, i.e. the holistic nature of health promotion interventions and the values of participation, collaboration and empowerment (Nutbeam 1998; Tones and Tilford 2001). Potvin and colleagues identified three main challenges for those evaluating health promotion programmes: (1) defining the activity to be evaluated in order to raise relevant evaluation questions, (2) implementing an appropriate, rigorous research methodology, and (3) producing relevant knowledge for actions (Potvin et al. 2008).

More specifically, regarding the evaluation of health promotion initiatives in the school setting and the type of evidence produced by these studies to inform practitioners and policy makers, Rowling and Jeffreys (2006) argued for considering research from both education and health sectors and for trying to articulate evidence from both sectors in order to inform planning and strengthen partnerships. These authors also pointed out the importance of considering contextual issues such as school practice and local policy factors, when evaluating school health promotion interventions (Rowling and Jeffreys 2006).

The specificities of health promotion initiatives and especially within the school setting lead us to consider methodological approaches that could potentially capture the complexity of such initiatives.

Chen (1997) suggested that there are three types of configuration depending on programme evaluation contexts. In the first configuration, the programme evaluation context requires intensive information, has low availability of credible information and has a highly open programme system. In this type of configuration, it is more appropriate to use qualitative methods. In the second configuration, the evaluation context requires extensive, precise information, has high availability of credible information and has a closed programme system. This would require a quantitative approach. The third configuration concerns programme evaluation contexts requiring information that is both intensive and extensive, that provide high access to some information but low access to other information and have the characteristics of both open and closed systems. In this case, the use of mixed methods is the most appropriate (Chen 1997). Due to their complexity, most health promotion interventions in school settings can be considered as an example of this last case. Moreover, mixed methods and methodological pluralism are increasingly used within the field of health promotion (Nutbeam 1998; Tones and Tilford 2001). In the next section of this chapter, we propose a brief overview of this emerging methodological field.

18.2.2 Mixed Methods: A Combination of Qualitative and Quantitative Methodological Approaches

In the past, quantitative and qualitative methods have been blended by researchers in various research fields, but it is only recently that this association was conceptualized as mixed methods research. Some authors have qualified this emergent field as a third methodological tradition (the two others being the qualitative and quantitative traditions) (Johnson et al. 2007).

Generally, the first reason for using mixed method is to offset the weaknesses of both quantitative and qualitative approaches. Therefore, this methodological approach makes it possible to integrate several perspectives and presents an added value for the study of complex interventions. This is particularly true in school health promotion where interdisciplinarity is a key concept and where there is a degree of uncertainty of the outcomes, that may only be long term and with a non-linear relationship between programme and outcomes. Moreover, there is a strong interaction with the school context as well as the involvement of multiple stakeholders (teachers, parents, children, etc.) (Jourdan 2011). Therefore, using more than one method within a research project may produce a more complete picture of the phenomena being studied (Morse 2003) and may help answer questions that cannot be answered by one approach alone.

Creswell and Plano Clark (2007) defined mixed methods research as the combination of quantitative and qualitative approaches that provide a better understanding of research problems than either approach alone. The literature shows that mixed methods research provides more comprehensive evidence for studying a research problem than either quantitative or qualitative research alone; encourages researchers to collaborate; encourages the use of multiple worldviews or paradigms; and is 'practical' in the sense that the researcher is free to use all possible methods to address a research problem (Creswell and Plano Clark 2007). Although, there is a consensus on what should be mixed, i.e. qualitative and quantitative approaches, debates still exist on when approaches should be mixed (at the data collection stage, the data analysis stage or throughout the research project), why they should be mixed and the reasons for choosing a mixed methods approach (driven by the research questions or the evaluator's philosophical stand) (Johnson et al. 2007).

The mixed methods approach can vary in design depending on how the qualitative and quantitative approaches are combined. Creswell and Plano Clark (2007) classified the mixed methods designs into four major types:

- Triangulation: its purpose is to obtain a more complete understanding of a phenomenon from two databases, to corroborate results from different methods or to compare multiple levels within a system;
- Embedded: one data set provides a supportive, secondary role in a study based primarily on the other data type, its purpose is to address different questions that call for different methods or to enhance an experiment by improving recruitment procedures, examining the intervention process or explaining reactions to participation;
- Explanatory: a two-phase mixed methods design where qualitative data helps to explain or build upon initial quantitative results, for example to help explain quantitative results that need further exploration or to purposefully select best participants for qualitative study;
- Exploratory: the results of the first method (qualitative) help to develop or form the basis of the second method (quantitative), for example to first explore new variables, theories and hypotheses, to develop an instrument or a typology that is not available or to assess whether qualitative themes can be generalized to a population.

Creswell and Plano Clark (2007) identified three questions linked to the choice of a mixed methods research design: the level of interaction between the quantitative and qualitative strands, the priority of the strands and their timing.

Regarding the integration of results from both methods, two main approaches are presented by Creswell and Plano Clark (2011). The first one is merging the data that are collected concurrently using either side-by-side comparison, joint display (i.e. a cross table) or data transformation (*quantitising* – data collected through qualitative methods, converted into numerical codes to be statistically analysed – or *qualitising* – data collected with a quantitative method, and converted into narrative data analysed qualitatively). The second approach mentioned by Creswell and Plano Clark (2011) is connecting the data that are collected sequentially. One of the main issues when integrating results from both qualitative and quantitative methods is the possibility of discrepant results. Four strategies are then proposed in the literature to find a solution (Pluye et al. 2009):

- Reconciliation, when the data can be re-analysed and interpreted plausibly;
- "Bracketing", which provides, when the results are irreconcilable, extreme cases to determine the image of a statistical confidence interval;
- Initiation of new perspectives from data analysis, a new research project or a new data collection based on a new research question;
- And exclusion, that is to say, the omission of part of the data, for example, if the results are incomplete or if some data are not sufficiently robust in terms of validity.

As a developing field, challenges are faced both by researchers and practitioners regarding epistemological foundations, technical issues and regarding skills, time or resources. Nevertheless, in this quite early stage of development, mixed methods, as a methodological tool, may help to resolve some of the issues related to the evaluation of school health promotion initiatives, especially with regard to appraising a complex phenomenon and providing different points of view about it. In the next section of this chapter, we will illustrate this potentiality by presenting how mixed methods were used to evaluate a health promotion initiative implemented in French primary schools.

18.3 Use of Mixed Methods in the Evaluation of a School Health Promotion Initiative – An Example from France

In this section, the French health promotion initiative studied, the evaluation framework and the evaluation design implemented are described and illustrated by some of the first results.

18.3.1 The Health Promotion Initiative Studied

The French system is national and centralised. Schools set a low priority on health promotion (Pommier et al. 2009). Professionals in the workplace are not always aware of their health promotion role (Jourdan et al. 2002). A health promotion initiative tailored to the French context was therefore developed to address health promotion issues within school settings and to equip school staff to implement health promotion policy. This initiative aimed to promote children's social, emotional and physical health by contributing to their well-being at school (Samdal et al. 1998) and enhancing their life skills (WHO 1993, 1999). Its objective was to encourage the development of sustainable health promotion projects in school settings by empowering local actors and by "mobilizing" existing resources. The main strategy was the development of teachers' health promotion practices and a health promotion environment within schools. The initiative takes into account the most recent international publications and data concerning the development of school health promotion approaches (Tones and Tilford 2001; St Leger et al. 2007; Stewart-Brown 2006; St Leger and Young 2009). This implies the development of a progressive sustainable intervention:

- Taking into account the development of the children;
- Linking health to educational issues as well as integrating them into on-going school activities and existing policies;
- Communication with parents and communities;
- Training and support of school staff and accessibility of resources and other methodological tools.

It also takes into account the special features of the French system. The intervention is a combination of top-down and bottom-up approaches and therefore the characteristics of the actions implemented in each school may vary (Grieg Viig and Wold 2005).

Figure 18.1 presents the theory-of-change model underlying this health promotion intervention (Knowlton and Phillips 2009). It suggests that the strategies developed through the intervention (teacher training, school team support, resources and tools, and institutional lobbying) can positively influence teachers' health promotion practices (Goigoux 2007) and the schools' health promotion environment and enhance the well-being of both children and teachers, improve the relationship between schools and families (Hamel et al. 2001; Schoonbroodt and Gélinas 1996), develop children's knowledge, attitudes and skills about health (WHO 1993) and eventually improve children's social, emotional and physical health (WHO 1993). This model is based on the assumption that the outcomes and strategies interact with the general and local contextual factors and the way in which the intervention is implemented (i.e. rules, organizational structure and personnel who are responsible for managing the programme) (Chen and Rossi 1983).

A 4-year pilot study (2003–2007) was carried out in 21 schools (Mérini et al. 2009; Mérini et al. 2010; Simar and Jourdan 2010a; Simar and Jourdan 2010b; Simar and Jourdan 2011). During this pilot stage, there were in-depth interviews with the intervention designers and those involved locally, observations were made, documents were collected and questionnaires were filled in by children, teachers and parents. The initiative was then proposed in 2007 to all 31 French teacher training institutes. These institutions have the authority and legitimacy to sustain such research initiatives. Ten institutes in 10 different French regions agreed to participate. Six regions were able to gain institutional support and set up a regional team to implement the intervention and to collect data. Within those six regions, a total of 115 schools were given institutional support for participating and reaching out to

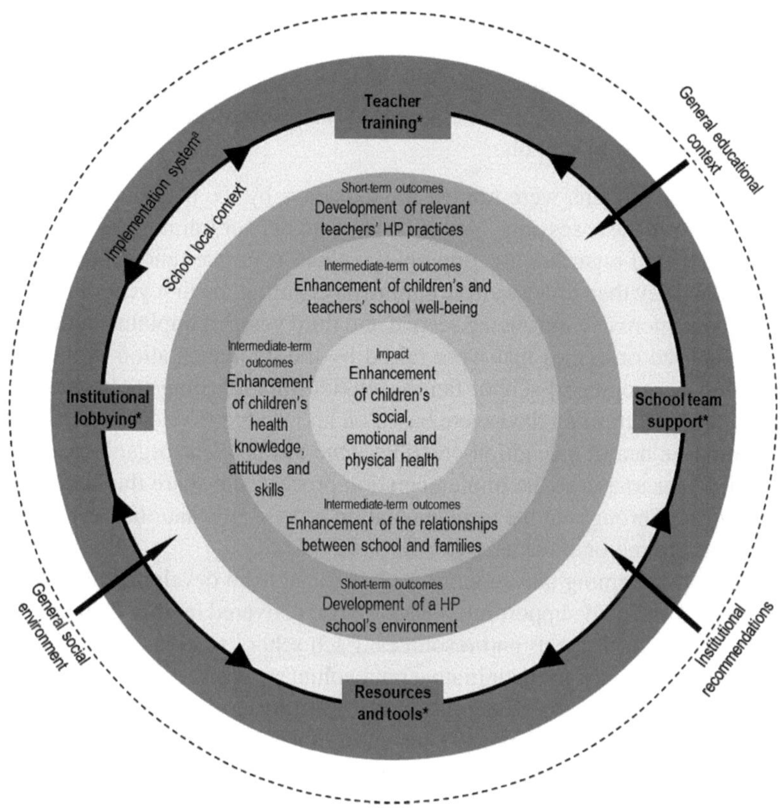

Fig. 18.1 Theory-of-change model of a health promotion intervention in a school setting – Extract from (Pommier et al. 2010). [a]Implementation system – an intervention once enacted must be carried out through an implementation system that includes rules, organizational structures and personnel who have been given the responsibility to administer the intervention (Chen and Rossi 1983). *Strategies; HP: health promoting

approximately 650 teachers and 11,000 children and their families. The implementation phase started in 2008 and lasted 3 years.

A national team, with two senior researchers and a coordinator, was in charge of coordinating both the implementation and the evaluation phases. In each region, a team was set up including teacher trainers, educational advisors, school nurses and doctors as well as members of local non-governmental health promotion organisations. The regional teams were assigned with five objectives:

1. Implement regional support services with representatives from the key institutions involved;
2. Develop training sessions for school staff;

3. Support schools in the implementation of health promotion projects within their community;
4. Provide teaching tools to schools; and
5. Develop local partnerships in communities to sustain the implementation of health promotion in schools.

These regional teams were trained for four days by the national team, prior to implementation. These sessions focused on how to deliver training, advice and support to school staff regarding the principles, values, resources and evaluation of the intervention. They then underwent two more days during the first year of implementation and one more day during the second and third years of implementation. These sessions focused on issues that were raised by the implementation in the schools (partnership development, school-family relationship, teaching tools, and conflict management). All training days were based on knowledge development and experience sharing on health promotion practices. One last day was organized at the end of the third year to assess the implementation process and share the first results of the evaluation. Throughout the implementation process, the national team also provided support to regional teams.

Based on this training and support, each regional team developed its own health promotion training and support intervention to be delivered in their own region taking into account local needs and resources. Each school received a set of teaching tools. Schools took part to the initiative on a voluntary basis.

The overall initiative was supervised by a scientific committee formed by health and educational experts and practitioners. An ethics committee was also created. The research project was registered at a national ethics committee (registered number 1332359). In each region, a regional steering committee was also set up including school board representatives, parent representatives, other regional relevant actors, and members of the regional team.

Figure 18.2 presents the implementation system of this initiative. It shows the different stakeholders involved at the national, regional and local levels, their role and the way they interact as well as the data collection for the follow up and evaluation.

18.3.2 The Evaluation Design – Using a Mixed Methods Approach

The evaluation framework chosen is based on the "theory-driven" approach to evaluation defined by Chen and Rossi (1983). This approach "is not the global conceptual scheme of the grand theorists, but more prosaic theories that are concerned with how human organizations work and how social problems are generated [...]. What we are strongly advocating is the necessity for theorizing, for constructing plausible and defensible models of how programmes can be expected to work before evaluating them" (Chen and Rossi 1983, p. 285).

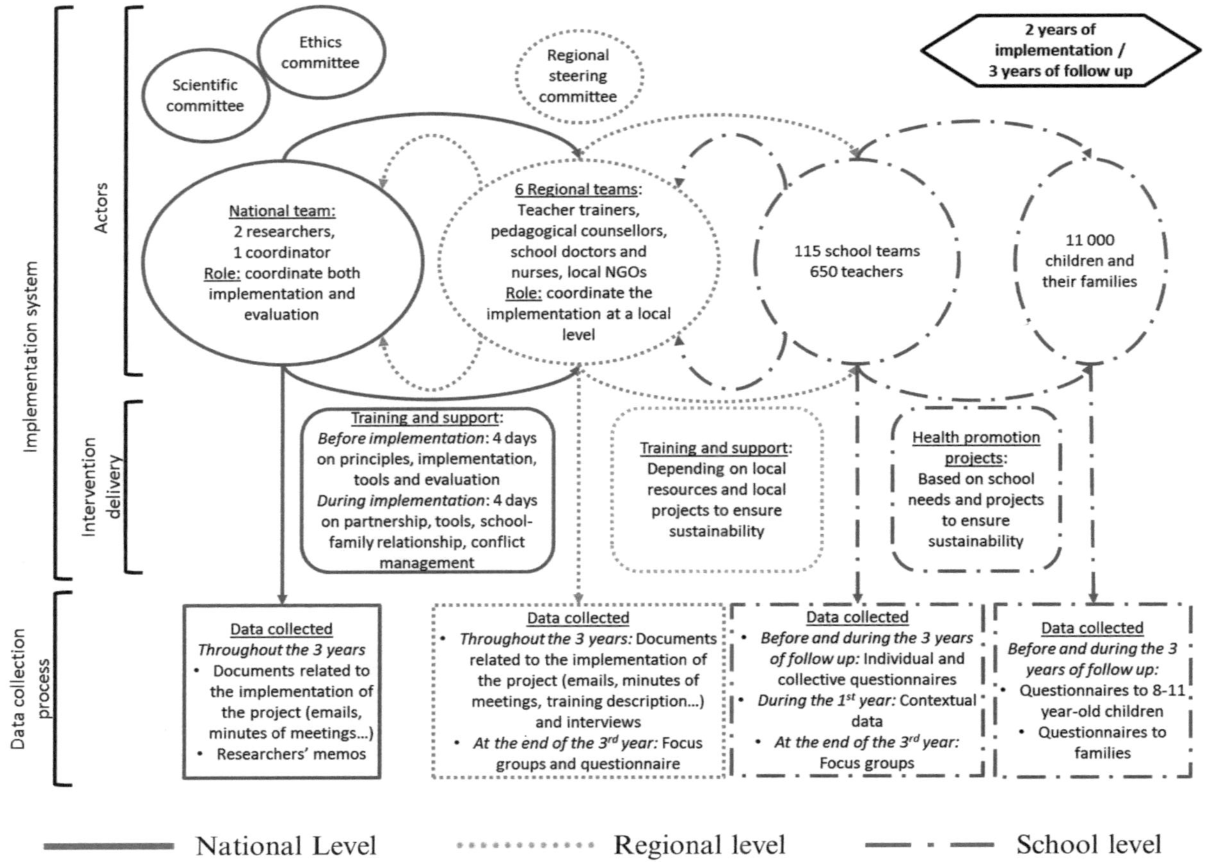

Fig. 18.2 Implementation system and data collection process

The evaluation of the French health promotion initiative described previously required both intensive and extensive information: intensive information is related to a need for stakeholders to have an in-depth understanding both of the context and of the activities implemented in order to better know what the ingredients are that support the development of school health promotion in the French context. As for the extensive information, the same stakeholders may need to know what the outcomes on children and families are. Thus, evaluation will be required to provide both types of information and in this case, the use of mixed methods is recommended (Chen 1997). In this section, the overall evaluation design implemented will first be described. We will then focus on three specific aspects to illustrate how both qualitative and quantitative approaches can be used.

18.3.2.1 An Overview of the Design Implemented

Two main sets of evaluation questions were raised:

1. What are the factors that allow the school community to develop a health promotion approach?
2. How do the strategies developed through the intervention influence the development of teachers' health promotion practices and the schools' health promotion environment? How do these practices affect well-being in the schools? What is the influence of the intervention on the children's perceived life skills?

According to the factors that influence the choice of a mixed methods design as defined by Creswell and Plano Clark (2007), the evaluation design implemented is based on an embedded design: QUAN(qual). The evaluation questions focus on quantitative data to measure changes and qualitative data plays a supportive role in exploring health promotion practices and contexts to better understand the QUAN data. Data are collected concurrently: quantitative numerical data are collected from questionnaires and forms and qualitative data (text data, transcripts and memos) from open-ended questions included in questionnaires, forms and from semi-directed interviews and focus groups. The data are analysed using quantitative (univariate, multivariate and multilevel analysis) and qualitative analysis (content analysis). In addition, qualitative data were *quantitised* – i.e. numerical conversion of qualitative codes (Tashakkori and Teddlie 2003) – in order to be included in the quantitative analysis. The interpretation is quantitative, qualitative and combined where the quantitative results are clarified by the qualitative results, in order to generalize the findings, predict and interpret theory. Figure 18.3 presents this mixed methods embedded design and summarizes the data collection and analysis procedures and products as well as the QUAN(qual) interpretation stage. Qualitative and quantitative methods are mixed throughout all phases of the project from the design stage through data collection to data interpretation.

Regarding data collection, Fig. 18.2 also shows the data collected at the national, regional, schools and children levels. At the national level, documents related to the implementation of the intervention and researchers' memos were collected over

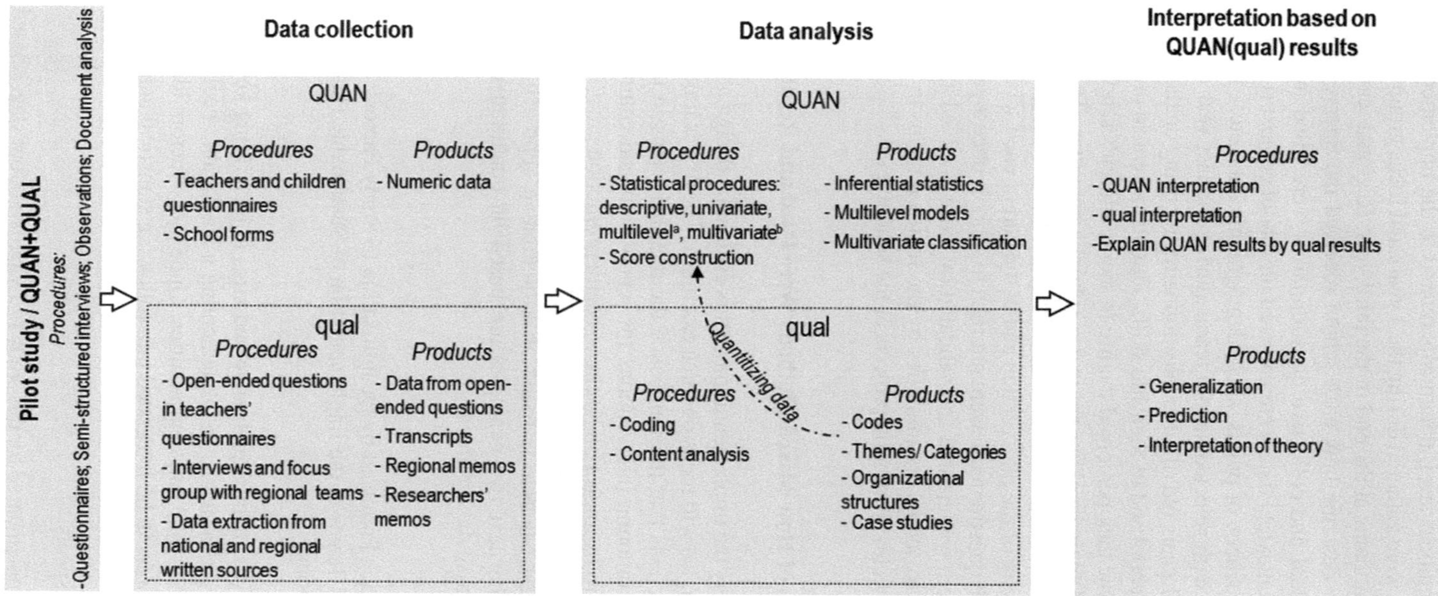

Fig. 18.3 Mixed methods embedded design of the research: data collection, analysis and interpretation procedures and products – Adapted from (Pommier et al. 2010)
QUAN: quantitative, qual: qualitative
[a]: regression (logistic, linear…)
[b]: principle component analysis, multiple correspondent analysis, classification

the 3 years. At the regional level, documents related to the implementation of the regional intervention were gathered during the 3 years and focus groups were organised with the regional teams at the end of the third year. Regional teams also filled in a questionnaire to describe how each school received the intervention. At the school level, individual and collective questionnaires were filled in by teachers and school teams. Contextual data were gathered on school context (i.e. size of the school, number of teachers, of pupils, socio-demographic data, etc.). Focus groups were also organised with selected school teams during the last year of follow up. Children from 8 to 11 years old filled in questionnaires focusing on their perception of their life in their school and of their life skills. Parents were also invited to fill in a questionnaire on how they perceived the life in their children's school, their relationships with it and their involvement in the school's activities. These questionnaires were collected from parents and children in a 3 year multiple time series design, at the beginning of the intervention and at the end of each school year.

The results of each questionnaire were returned to schools once a year. A specific user-friendly document was created and validated by those involved and the results were communicated to school communities by the regional teams to contribute to the development of health promotion projects at the school level.

18.3.2.2 Examples of How Mixed Methods Were Used and Results

To illustrate concretely the implementation of this evaluation design and the kind of results we can obtain, we will focus on three examples. The first is linked to the factors and their interactions that may influence the regional teams in implementing sustainable health promotion interventions for school staff and communities. The second is in relation to its impact on teachers' health promotion practices. The third is linked to the factors that may influence how children perceived their school social environment. The first example is based on a qualitative approach followed by *quantitisation* and the next two examples are based on quantitative approaches completed by qualitative data sources.

A Qualitative Approach Followed by Quantitisation to Study Factors Influencing Regional Teams in the Implementation of Sustainable Health Promotion Interventions

This part of the study is mainly qualitative and aims at identifying key factors that may influence the regional teams in implementing a sustainable health promotion intervention for school staff and communities. Once these factors identified, they were *quantitised* in order to be integrated into following statistical analysis.

a) Methodological Aspects

The specific design implemented for this study is divided into four stages based on the realist evaluation cycle suggested by Pawson and Tilley (see Box 18.1). The first stage

Box 18.1: The Realistic Evaluation Framework

The realistic evaluation framework was developed by Pawson and Tilley (1997). This is a framework that takes into account the complexity of social interventions in specific contexts focusing on how they work, for whom and under what circumstances. It aims: (1) to understand the mechanisms through which health promotion interventions produce change; (2) to understand the contextual conditions necessary to trigger these mechanisms; and (3) to develop outcome pattern predictions according to the context and mechanisms triggered. These are the three guiding themes of the research strategy defined by Pawson and Tilley. According to these authors, in a realistic evaluation approach, the outcomes of a health promotion programme are explained by the action of specific mechanisms in specific contexts. It is thus essential in this type of evaluation approach to identify the mechanisms involved, i.e. what, within the programme, produces change. The idea is to determine "which individuals, subgroups and locations might benefit most readily from the programme, and which social and cultural resources are necessary to sustain the changes" (Pawson and Tilley 1997, p. 85). They name these configurations "context-mechanism-outcome pattern configurations" (CMO configurations). Realistic evaluators can then identify, modify, test and refine the CMO configurations. For these authors, a mechanism is "not a variable but an *account* of the make-up, behaviour and interrelationships" of the processes which are responsible for the change, "a mechanism is thus a theory" (Pawson and Tilley 1997, p. 68). CMO configurations are developed both on the basis of the literature and using the point of view of the stakeholders/participants of the intervention who play a key role in confirming, refuting or refining the theory. The realistic evaluation framework does not require the use of a specific method. Indeed, Pawson and Tilley (1997) acknowledge that, when it comes to the choice of method, realistic evaluation can be based on methodological pluralism and thus on both qualitative and quantitative approaches.

aims at identifying contexts and potential mechanisms in literature. The second stage aims at organising them for use as a starting point for the analysis of the empirical data collected at the regional level – interviews, focus groups with regional teams and data from national and regional written sources, i.e. emails, minutes of meetings, training description, etc. (see Fig. 18.2) – within the third stage. Each specific context, mechanism and outcome was then transformed into variables and *quantitised* for each region in order to integrate them into following statistical analysis. For example, a numerical value was assigned to each mechanism according to its degree of activation in each region. Finally, based on a transversal qualitative analysis (i.e. comparing the six regions considering them as six different contexts) of the contexts, mechanisms and outcomes identified, stage four aims at proposing a "context-mechanism-outcome pattern configurations" that outlines the main elements that may influence the development of health promotion interventions based on training and support for schools (Guével et al. 2013a).

b) Main Results

The "context-mechanism-outcome pattern configurations" proposed outlines the main elements that may influence the development of health promotion interventions based on training and support for school staff.

To achieve this outcome, according to the results of this study, the following contexts and their interactions need to be taken into account:

- A high local commitment, especially from the head of the department of education as well as from the municipalities;
- A stable regional team able to train and support schools;
- A stable district and school organization;
- Regional teams that are trained and supported.

The results also show the key mechanisms triggered:

- At an individual level of the regional team members: knowledge development on school health promotion approaches, improvement of self-efficacy regarding the implementation of such approaches, development of a reflective practitioner perspective in this area, and development of their motivation and of their conviction regarding the values promoted;
- At an inter-individual level: development of opportunities to exchange with peers, shared values and knowledge within and between the regional teams;
- At a collective level: development of partnership inside and outside the educational department.

A Quantitative Approach Completed by Qualitative Data Sources

Two examples are given, one at the teacher level, focusing on teachers' self-reported health promotion practices and another one at children level, focusing on how children perceived their school social environment. Both examples are based on a quantitative approach completed by qualitative data sources.

a) Teachers' Self-Reported Health Promotion Practices

This part of the study mainly uses a quantitative approach and aims at assessing the impact of the intervention on teachers' health promotion practices. Open-ended questions were used to describe the different themes addressed by teachers and the way they have done it. The next stage of analysis will lead us to integrate in multilevel analysis quantitative data from the teacher's questionnaire as well as qualitative data that would have been *quantitised* both from the teachers' questionnaire (i.e. from open-ended questions) and from the qualitative analyse carried out at the regional level in order to better take into account the specificity of the teachers' regional context.

i) Methodological Aspects

A questionnaire was designed to collect data on teachers' attitudes to health promotion, on their own practices and factors that might influence them (facilitators, barriers, etc.), on their motivation, interest in health promotion and perceived self-efficacy in health promotion, as well as on their perception of the life in their school (school climate, perceived violence, etc.). This questionnaire was primarily developed in 1991 in a study on teachers' practices and attitudes to health promotion (Jourdan et al. 2002); it was amended and used in the pilot study. Both closed- and open-ended questions were used.

Initially, 115 schools were involved, however at the end of the first year; our population was reduced to 100 schools for various reasons. The participating schools were divided into two groups: the first group (group 1) received training and support from the first year, the second group (group 2) received training and support from the second year. Group 1 consists of 62 schools and group 2 of 38 schools. This difference between the two groups can be explained by the fact that most of the schools that left the project were part of this group 2. In addition, initially, the regional teams had encountered difficulties in recruiting schools agreeing to take part in the evaluation design from the first year and receive training and support only from the second year. However, the two groups are not significantly different in terms of size, geographical location, socioeconomic status and average number of teachers per school (Table 18.1).

A first set of analysis was carried out on the data collected before the implementation (Guével et al. 2010). A second set of analysis was carried out after 1 year of implementation, i.e. after 1 year of training and support for schools in group 1 and before group 2 starts to receive training and support. In this set of analysis, answers from the two groups were compared in order to identify on which aspects, training and support may have had an influence. In the first place, teachers' questionnaires were analysed about whether teachers reported having health promotion practices and about their perception of the climate in their school.

Table 18.1 Schools' characteristics and comparison between the two groups of schools

Schools' characteristics		Total % (n)	Group 1 % (n)	Group 2 % (n)	Group 1/ Group 2[a]
Size	Small (≤3 classes)	35 (35)	35 (22)	34 (13)	ns
	Medium (4–7 classes)	53 (53)	55 (34)	50 (19)	
	High (≥8 classes)	12 (12)	10 (6)	16 (6)	
Location	Rural	38 (38)	42 (26)	32 (12)	ns
	Urban	62 (62)	58 (36)	68 (26)	
Socio-economic status	Privileged	24 (24)	26 (16)	21 (8)	ns
	Medium	37 (37)	34 (21)	42 (16)	
	Underprivileged	39 (39)	40 (25)	37 (14)	
Mean number of teachers per school (mean ± sd)		5.72 ± 3.09	5.34 ± 2.80	6.34 ± 3.45	ns

[a]Mean test or chi-squared test, significance was set at $p < 0.05$. *ns* non-significant

Qualitative data coming from the regional focus groups as well as from the written sources collected have been content analysed and then transformed into quantitative data that will be integrated into the quantitative database. Furthermore, in the teachers' questionnaires, the open-ended questions concerning teachers' attitudes to health promotion and the activities they have implemented will be *quantitised* for the quantitative analysis.

ii) Main Results

Three hundred and nineteen teachers responded and were included in the first set of analysis (response rate of 54.5 %). Seventy-one per cent of teachers reported having implemented health promotion activities during the previous school year. According to the content analysis of open-ended questions, these activities were mainly related to health topics (nutrition, hygiene and dental health), to citizenship, community and environmental issues. Physical education as well as arts disciplines were most frequently cited as media. The four main factors cited by teachers as influencing their practices in health promotion were their own interest (63 %), the school syllabus and the collective reflection at the school level (60 %), the fact that health promotion is part of their mission (58 %) and the school climate (39 %). Eighty-five per cent of teachers declared working with partners: their fellow teachers (61 %), other members of the school team (23 %), associations (23 %) and families (21 %). The four main difficulties encountered by these teachers were the lack of time (43 %), the lack of training (30 %), material issues for organizing activities (29 %) and the lack of tools (23 %). Of the 29 % of teachers who reported not having implemented health promotion activities, 56 % of them reported having done it in the past. The main obstacles encountered were the lack of training (45 %), the lack of time (44 %), the lack of tools (36 %) and the lack of experience (32 %).

For the second set of analysis, at the end of the first year of implementation, 168 teachers returned their questionnaire (response rate: 30.1 %). Teachers from group 1 have had implemented more health promotion activities than teachers from group 2 (88 % vs. 74 %, $p=0.0281$). Regarding how much though was put into the issue of implementing health promotion activities, teachers from group 1 appeared to be more advanced in their reflection ($p=0.0025$). As Fig. 18.4 shows, 66 % of teachers from group 1 reported to be at stage 1 (considered to be the highest).

Teachers from group 1 seemed to have a more positive perception of the life in their school than those from group 2, especially regarding the atmosphere among pupils ($p=0.0341$), the relationship between pupils and teachers ($p=0.0191$), the relationships with parents ($p=0.0031$), the relationship between adults ($p=0.0107$) and the violence perceived ($p < 0.0001$). Figure 18.5 shows the percentage of teachers in each group who selected the modality *"Very good"* for the first four proposals mentioned above and the modality *"Not at all"* for the proposal on perceived violence.

b) School Evolution Based on Children's Perception of Their Life at School

This part of the study aims at studying how schools benefit from the intervention from the children's point of view. It is based on the evolution of children's perception of their life at school. The approach used is mainly quantitative. In the next stage of

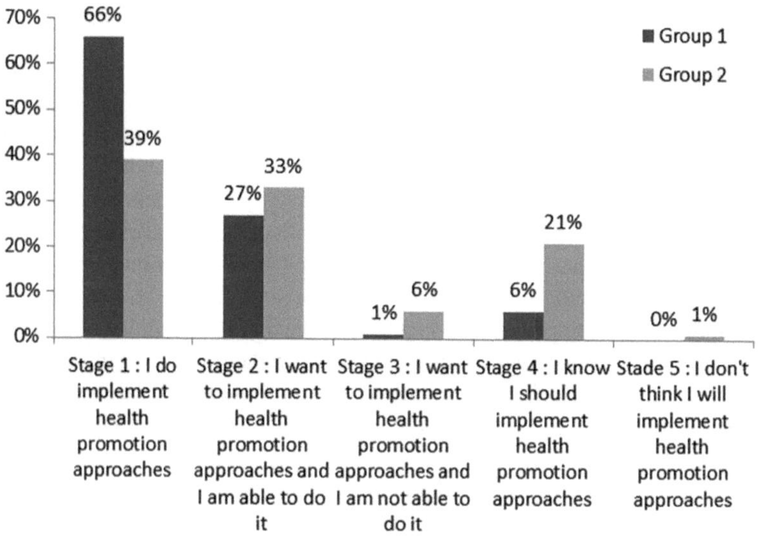

Fig. 18.4 Teachers' distribution according to their stage of reflection regarding health promotion approaches

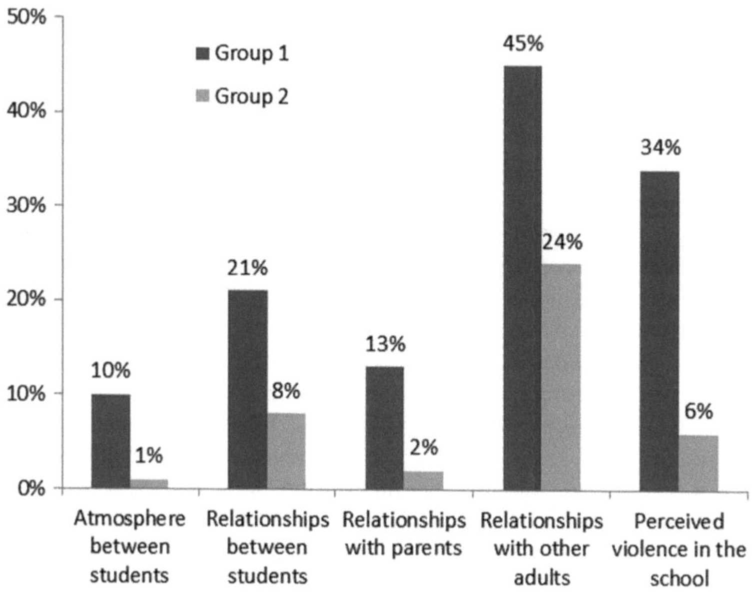

Fig. 18.5 Percentage of teachers that chose the highest modality regarding their perception of the life in their school

analysis, *quantitised* data from the qualitative analyse carried out at the regional level as well as data from teachers' and school teams' questionnaires will be integrated in order to better take into account the specificity of the school context.

i) Methodological Aspects

Data were collected from 8 to 11 year-old children before the implementation of the intervention, 1 and 2 years after the beginning of the implementation. To be included elementary schools (i.e. schools with children between 8 and 11 years of age) needed to have returned questionnaires at least twice. In the end, 45 schools were included in this sub-study, representing approximately 3,500 children.

A children's questionnaire was designed to collect data on children's perception of their life in school and life skills (WHO 1999). Children's perception of their life at school was studied through questions on the school climate and on their perception of their relationship with other children, teachers and adults working in the school. This part of the questionnaire was based on the questionnaire developed by Debarbieux at the European observatory of school violence (Debarbieux 1996; Debarbieux and Blaya 2001) which was adapted and used in the pilot study (Younès et al. 2011).

Three indicators were developed and their evolution through the implementation of the intervention was studied: *Perceived violence* (score of four variables, Cronbach alpha=0.64), *Perceived relationships with others* (score of eight variables, Cronbach alpha=0.74) and the variable *Feeling at school.*

In the first place, mean and Kruskall-Wallis tests were carried out as well as multiple correspondence analysis and hierarchical clustering on principal components (Escofier and Pagès 2008).

ii) Main Results

Table 18.2 presents the results for these three indicators before the implementation of the intervention, 1 and 2 years after the beginning of it. *Perceived violence* is relatively low; through the three-time data collection, the average is a little more than 4 out of 12, i.e. children perceived little violence in their schools. Throughout the implementation, this score slightly increased (from 4.31 to 4.39, i.e. perceived violence is higher) but it is not statistically significant. *Perceived relationships with others* is high with an average over 15 (out of 21) through the three times of data collection i.e. children had a quite good perception of their relationships with others in their school. Throughout the implementation, this score slightly decreased (from 16.10 to 15.82) but it is not statistically significant. Finally, children tend to felt well or very well at school. Even if, throughout the implementation, children feel a little less well at school as this variable goes up from 1.67 to 1.78 where 1 means *children feel very well at school* and 2 means *children feel well at school* ($p=0.0184$).

Following multiple correspondence analysis and hierarchical clustering on principal components analysis, four school profiles emerged reflecting how children's perceptions of their social environment at school may evolve during the implementation of the intervention in their school:

- First profile – "*No change*": *Perceived relationships with others* and *Perceived violence* have not changed through the implementation. *Perceived violence* was below average before the implementation (20 schools out 45).

Table 18.2 Descriptive results regarding the indicators of children's perception of school before the implementation of the project in schools, 1 and 2 years after the beginning of the implementation

		Mean	SD	Min.	Max.	Comp.[a]
Perceived violence	Before the implementation	4.31	0.86	2.41	6.18	ns
	A year after the beginning of the implementation	4.64	0.89	3.00	6.56	
	Two years after the beginning of the implementation	4.39	1.04	1.64	7.00	
Perceived relationships with others	Before the implementation	16.10	1.04	12.63	18.00	ns
	A year after the beginning of the implementation	15.89	1.11	12.82	17.59	
	Two years after the beginning of the implementation	15.82	1.01	13.75	17.70	
Feeling at school	Before the implementation	1.67	0.21	1.41	2.40	p=0.0184
	A year after the beginning of the implementation	1.76	0.30	1.13	2.71	
	Two years after the beginning of the implementation	1.78	0.21	1.27	2.20	

SD standard-deviation, *Min.* minimum, *Max.* maximum, *Comp.* comparison within the three data collection time points

[a]Mean or Kruskall-Wallis tests, significance was set at $p<0.05$. *ns* non-significant

- Second profile – *"No change with a good situation throughout the implementation"*: *Perceived relationships with others* was above average before the implementation. *Perceived relationships with others* and *Perceived violence* were above average after the implementation (12 schools out 45).
- Third profile – *"Decreased"*: *Perceived relationships with others*, *Perceived violence* and *Feeling at school* have decreased through the implementation. *Feeling at school* was below average after the implementation (9 schools out 45).
- Fourth profile – *"Good situation at the end of the implementation"*: *Perceived relationships with others*, *Perceived violence* and *Feeling at school* were above average after the implementation. Values are missing regarding the evolution of the indicators (4 schools out 45).

This classification illustrates that from the children's point of view, schools may evolve in different manners: some positively (i.e. they have higher score than average) and some negatively (i.e. their scores decreased during the implementation phase). These results will be further explored to identify the school and regional contextual factors – such as at the regional level, the support from the community or from regional institutions; at the school level, its size and location as well as its degree of involvement in health promotion and at the classroom level, teachers' collective and individual health promotion practices – that may explain why we observed some schools with a positive evolution and others with a negative one. These analyses are currently underway (Guével et al. 2013b).

18.4 Concluding Comments Regarding Mixed Methods and School Health Promotion Evaluation

This chapter was focused on the mixed methods contribution to the development of school health promotion. Empirical data gathered from a study implemented in the French context were used to highlight the interest of such a research strategy (Creswell 2009). In this concluding paragraph, we would like to summarize the key contributions of mixed methods regarding the sustainability and transferability of school health promotion and the necessity of addressing the different types of evidence that can be produced to ensure successful implementation and sustainability.

In her literature review, Stewart-Brown (2006) stated "It is becoming increasingly clear that research on promoting health requires a variety of methodological approaches, including process- and outcome-based evaluation, and quantitative and qualitative methods." (Stewart-Brown 2006, p. 16) In combining both approaches, mixed methods offer a framework to think and to carry out evaluation designs. Such designs could potentially propose an answer to these challenges in order to improve the sustainability of health promotion initiatives by identifying the key factors linked with the efficacy of an intervention (both its success and its sustainability). They may also help in identifying the factors related to the transferability of an intervention from one context to another by taking into account qualitative data that might explain why a same intervention may work in one setting and not in another one. As illustrated in this chapter, work is still ongoing to improve the integration of the results of both approaches. As mentioned mixed methods is a research strategy currently under-development; however regarding school health promotion evaluation, mixed methods may help to take into account the complexity of the initiatives carried out (by providing different points of view) as well as to face challenges both from educational and health sectors (by providing evidence that might be acceptable for both sectors). For example, the possibility of thoroughly exploring the regional context in France with a qualitative approach and then, of integrating the key factors identified into a quantitative analysis in order to explain the factors influencing teachers' health promotion practices is an important contribution to a better understanding of how school health promotion can be developed in France. Both results from the qualitative component and from the quantitative analysis will be valuable for French practitioners and decision-makers that wish to support school health promotion in their own context.

Moreover, mixed methods encourage the use of multiple sources of information that provide a more comprehensive understanding of the processes underlying the development of health promotion approaches in the school setting. In the example presented above, this was facilitated through a close partnership between researchers and practitioners: practitioners were involved since the beginning and participated in the construction of the initiative; moreover, data collected at both levels were included to have a more complete picture of the implementation. The aim was also to better secure the sustainability of the project once the research is over.

To go further, this type of design leads us to a broader discussion on the nature of the evidence in the field of school health promotion. Indeed in the first chapter of his book, Donaldson sets out his framework stating that there are many ways to establish what is required for evidence to be judged credible in an evidence based global society (Donaldson 2009). Julnes' put the emphasize on "actionable" rather than "best" evidence within a context (Julnes and Rog 2009). Based on Donaldson's and Julnes' work, three sources of evidence can be identified and applied to the field of health promotion: scientific evidence (outcome based evidence), "professional" contextual evidence (practice based evidence) and "critical" review, i.e. an ethical approach of the intervention. Mixed methods by its ability to provide different perspectives on a same phenomenon and to take into account its complexity can contribute to the production of credible evidence related to philosophy, theory and practice in the evaluation context (Mertens and Hesse-Bider 2013). Therefore, their use has the potential to lead to a better understanding of health promotion initiatives and eventually to their strengthening.

This last point is essential when thinking about the transferability of school health promotion initiatives from one context to another, even within a same country; as our example has shown, outcomes may differ from one region to another and even from one school to another. The methodology for producing transferable knowledge is then of importance and some authors argue to focus on different modes of complementary or integrative studies combining qualitative and quantitative methods (Cambon et al. 2012). To be concrete, for example, if other French schools inspectors or educational advisors or school nurses or doctors or any other decision-makers wish to support the development of school health promotion, they could gain a better understanding of the initiative and how it might produce the desired outcomes in a specific context. The possibility of capturing both intensive and extensive information by using mixed methods in the evaluation design is therefore an important added-value to the evidence produce and to the possibility of using this evidence to invent new initiatives in new context.

Mixed methods have already a great potential to help address the challenges faced by the field of school health promotion especially those related to evaluation, sustainability and transferability, what we have tried to point out in this chapter.

Acknowledgements The authors are very grateful to the National Institute for Health Education and Prevention (Inpes) for its financial support. They also wish to thank all those who took part in the project especially the members of the regional teams and the school communities involved. They thank Julie Pironom (IUFM d'Auvergne – Université de Blaise Pascal, France) for her help with the statistics analysis.

References

Burton, C. (2002). Introduction to complexity. In K. Sweeney & F. Griffiths (Eds.), *Complexity and health care. An introduction* (pp. 1–18). Oxon: Redcliffe Medical Press Ltd.

Cambon, L., Minary, L., Ridde, V., & Alla, F. (2012). Transferability of interventions in health education: A review. *BMC Public Health, 12*(497). doi: 10.1186/1471-2458-12-497.

Chen, H. T. (1997). Applying mixed methods under the framework of theory-driven evaluation. *New Directions for Evaluation, 74*, 61–72.

Chen, H. T., & Rossi, P. H. (1983). Evaluating with sense – The theory driven approach. *Evaluation Review, 7*(3), 283–302.

Creswell, J. W. (2009). *Research design: Qualitative, quantitative and mixed methods approaches* (3rd ed.). Thousand Oaks: SAGE Publications.
Creswell, J. W., & Plano Clark, V. L. (2007). *Designing and conducting mixed methods research* (1st ed.). Thousand Oaks: SAGE Publications.
Creswell, J. W., & Plano Clark, V. L. (2011). *Designing and conducting mixed methods research* (2nd ed.). Thousand Oaks: SAGE Publications.
Debarbieux, E. (1996). *La violence en milieu scolaire, Tome 1*. Nanterre: ESF.
Debarbieux, E., & Blaya, C. (2001). *Violence à l'école et politiques publiques*. Nanterre: ESF.
Donaldson, S. I. (2009). In search of the blueprint for an evidence-based global society. In S. I. Donaldson, C. A. Christie, & M. M. Mark (Eds.), *What counts as credible evidence in applied research and evaluation practice?* (pp. 2–18). Thousand Oaks: SAGE Publications.
Escofier, B., & Pagès, J. (2008). *Analyses factorielles simples et multiples: objectifs, méthodes et interprétation* (4th ed.). Paris: Dunod.
Goigoux, R. (2007). Un modèle d'analyse des activités des enseignants. *Education et didactique, 3*, 3–11.
Grieg Viig, N., & Wold, B. (2005). Facilitating teachers' participation in school-based health promotion – A qualitative study. *Scandinavian Journal of Educational Research, 49*(1), 83–109.
Guével, M. R., Pommier, J., & Jourdan, D. (2010, September). *L'apport des méthodes mixtes à l'évaluation de l'effet d'un dispositif de formation d'enseignants en éducation à la santé*. AREF 2010, Genève (Suisse). URL: https://plone.unige.ch/aref2010
Guével, M. R., Pommier, J., Jourdan, D. (2013a, August). *School health promotion – Contexts and mechanisms influencing the implementation of health promotion approaches*. 21st IUHPE world conference on health promotion. Pattaya (Thailand). URL: http://www.iuhpeconference.net/en/index.php
Guével, M. R., Pironom, J., Pommier, J., Jourdan, D. (2013b, August). Etude de la diversité des effets observés suite à la participation d'écoles à un projet de promotion de la santé. 21st IUHPE world conference on health promotion. Pattaya (Thailand). Pattaya (Thaïlande). URL: http://www.iuhpeconference.net/en/index.php
Hamel, M., Blanchet, L., & Martin, C. (2001). *6, 12, 17 Nous serons bien mieux! Les déterminants de la santé et du bien-être des enfants d'âge scolaire*. Québec: Publications du Québec.
Han, S., & Weiss, B. (2005). Sustainability of teacher implementation of school-based mental health programs. *Journal of Abnormal Child Psychology, 33*(6), 665–679.
Johnson, R. B., Onwuegbuzie, A. J., & Turner, L. A. (2007). Toward a definition of mixed methods research. *Journal of Mixed Methods Research, 1*(2), 112–133.
Jourdan, D. (2011). *Health education in schools – The challenge of teacher training*. Saint-Denis: Inpes.
Jourdan, D., Piec, I., Aublet-Cuvelier, B., Berger, D., Lejeune, M. L., Laquet-Riffaud, A., Geneix, C., & Glanddier, P. Y. (2002). Education à la santé à l'école: pratiques et représentations des enseignants du primaire. *Santé Publique, 14*(4), 403–423.
Julnes, G., & Rog, D. (2009). Evaluation methods for producing actionable evidence. In S. I. Donaldson, C. A. Christie, & M. M. Mark (Eds.), *What counts as credible evidence in applied research and evaluation practice?* (pp. 96–131). Thousand Oaks: SAGE Publications.
Knowlton, L. W., & Phillips, C. C. (2009). *The logic model guidebook – Better strategies for great results*. Thousand Oaks: SAGE Publications.
Lee, A., St Léger, L., & Moon, A. (2005). Evaluating health promotion in schools: A case study of design, implementation and results from the Hong Kong School Award Scheme. *Promotion & Education, 12*(3–4), 123–130.
Li, J., Mattes, E., Stanley, F., McMurray, A., & Hertzman, C. (2009). Social determinants of child health and well-being. *Health Sociology Review, 18*(1), 3–11.
Lister-Sharp, D., Chapman, S., Stewart-Brown, S., & Sowden, A. (1999). Health promoting schools and health promotion in schools: Two systematic reviews. *Health Technology Assessment, 3*(22), 1–218.
Mérini, C., Victor, P., & Bizzoni-Prévieux, C. (2009). Through a functioning typology in the school teams. In P. Masson & M. Pilo (Eds.), *Partnership in education. Theoretical approach*

and case studiesLe partenariat en éducation. Approache théorique et études de cas (pp. 19–31). Lille: Book Edition.
Mérini, C., Victor, P., & Jourdan, D. (2010). Le travail des enseignants en éducation à la santé: analyse des dynamiques collectives du dispositif. *Travail et Formation en Education, 6*. URL: http://tfe.revues.org/1334
Mertens, D. & Hesse-Bider, S.(2013). Mixed methods and credibility of evidence in evaluation. *New Directions for Evaluation, 2013*(138), 5–13. doi: 10.1002/ev.20053.
Merzel, C., & D'Afflitti, J. (2003). Reconsidering community-based health promotion: Promise, performance, and potential. *American Journal of Public Health, 93*(4), 557–574.
Morse, J. M. (2003). Principles of mixed methods and multimethod. In A. Tashakkori & C. Teddlie (Eds.), *Handbook of mixed methods in social and behavioral research* (1st ed., pp. 189–208). Thousand Oaks: SAGE Publications.
Mukoma, W., & Flisher, A. J. (2004). Evaluations of health promoting schools: A review of nine studies. *Health Promotion International, 19*(3), 357–368.
Nutbeam, D. (1998). Evaluating health promotion – Progress, problems and solutions. *Health Promotion International, 13*(1), 27–44.
OECD. (2010). Improving health and social cohesion through education. *Educational Research and Innovation*. doi:10.1787/9789264086319-en. OECD Publishing.
Pawson, R., & Tilley, N. (1997). *Realistic evaluation*. London: SAGE Publications.
Peters, L. W. H., Kok, G., TenDam, G. T. M., Buijs, G. J., & Paulussen, T. G. W. M. (2011). Effective elements of school health promotion across behavioral domains: A systematic review of reviews. *BMC Public Health, 9*(182). doi: 10.1186/1471-2458-9-182.
Pluye, P., Grad, R. M., Levine, A., & Nicoleau, B. (2009). Understanding divergence of quantitative and qualitative data (or results) in mixed methods studies. *International Journal of Multiple Research Approaches, 3*(1), 58–72.
Pommier, J., Jourdan, D., Vandoorne, C., Pioreka, B., & Simoes de Carvalho, G. (2009). School health promotion: Organisation of services and roles of health professionals in seven European countries. *European Journal of Public Health, 20*(2), 182–188.
Pommier, J., Guével, M. R., & Jourdan, D. (2010). Evaluation of health promotion in schools: A realistic evaluation approach using mixed methods. *BMC Public Health, 10*(43). doi: 10.1186/1471-2458-10-43.
Potvin, L., Bilodeau, A., & Gendron, S. (2008). Trois défis pour l'évaluation en promotion de la santé. *Promotion & Education, 15*(1), 17–21.
Rootman, I., Goodstadt, M., Hyndman, B., McQueen, D. V., Potvin, L., Springett, J., & Ziglio, E. (2001). *Evaluation in health promotion: Principles and perspectives*. Copenhagen: WHO Regional Publications.
Rowling, L. (2009). Strengthening "school" in school mental health promotion. *Health Education, 109*(4), 357–368.
Rowling, L., & Jeffreys, V. (2006). Capturing complexity: Integrating health and education research to inform health-promoting schools policy and practice. *Health Education Research, 21*(5), 705–718.
Samdal, O., Nutbeam, D., Wold, B., & Kannas, L. (1998). Achieving health and educational goals through schools – A study of the importance of the school climate and the students satisfaction with school. *Health Education Research, 13*(3), 383–397.
Schoonbroodt, C., & Gélinas, A. (1996). Envisager la prévention par le changement émergent: apprendre à gérer ce qui fait problème. *Education santé, 108*, 3–10.
Simar, C., & Jourdan, D. (2010a). Education à la santé à l'école: étude des déterminants des pratiques des enseignants du premier degré. *Revue des Sciences de l'Education, 36*(3), 739–760.
Simar, C., & Jourdan, D. (2010b). Education et santé: étude de l'impact d'un dispositif de formation et d'accompagnement sur l'implication des enseignants dans une démarche de promotion de la santé. *Revue Recherches et Educations, 3*. URL: http://rechercheseducations.revues.org/561
Simar, C., & Jourdan, D. (2011). Analyse de l'activité d'enseignants du primaire en éducation à la santé. *Nouveaux Cahiers de la Recherche en Education, 14*(2), 7–26.

St Leger, L. H. (1999). The opportunities and effectiveness of the health promoting primary school in improving child health – A review of the claims and evidence. *Health Education Research, 14*(1), 51–69.
St Léger, L. (2004). What's the place of schools in promoting health? Are we too optimistic? *Health Promotion International, 19*(4), 405–408.
St Leger, L. (2005). Protocoles et lignes directrices pour les Ecoles promotrices de santé. *Promotion & Education, 12*(3–4), 145–147.
St Leger, L., & Young, I. M. (2009). Creating the document 'Promoting health in schools: From evidence to action'. *Global Health Promotion, 16*(4), 69–71.
St Leger, L., Kolbe, L., Lee, A., McCall, D. S., & Young, I. M. (2007). School health promotion: Achievements, challenges and priorities. In D. V. McQueen & C. M. Jones (Eds.), *Global perspectives on health promotion effectiveness* (pp. 107–124). Saint-Denis: Springer.
Stewart-Brown, S. (2006). *What is the evidence on school health promotion in improving health or preventing disease and, specifically, what is the effectiveness of the health promoting schools approach?* Copenhagen: WHO.
Suhrcke, M., & de Paz Nieves, C. (2011). *The impact of health and health behaviours on educational outcomes in high-income countries: A review of the evidence*. Copenhagen: WHO Regional Office for Europe.
Tashakkori, A., & Teddlie, C. (2003). *Handbook of mixed methods in social and behavioral research* (1st ed.). Thousand Oaks: SAGE Publications.
Tones, K., & Tilford, S. (2001). *Health promotion: Effectiveness, efficiency and equity* (3rd ed.). Cheltenham: Nelson Thornes.
WHO. (1993). *Life skills education in school*. Geneva: WHO.
WHO. (1998). *Health promotion evaluation: Recommendations to policy-makers*. Geneva: WHO.
WHO. (1999). *Partners in life skills education – Conclusions from a United Nations Inter-Agency Meeting*. Geneva: WHO.
Younès, N., Debarbieux, E., & Jourdan, D. (2011). Le climat scolaire à l'école primaire – Etude de l'influence des variables de milieu sur sa perception par les élèves de 6 à 8 ans. *International Journal of Violence and School, 12,* 107–127. URL: http://www.ijvs.org/3-6224-Article.php?id=94&tarticle=0

Chapter 19
Developing and Testing a Health Promotion Theory—An Example of Creating a Model of School Staff's Occupational Well-Being

Terhi Saaranen, Kerttu Tossavainen, Marjorita Sormunen, Sari Laine, and Hannele Turunen

Abstract Developing a theory is a long-term, multi-phased process that can be used to produce models applicable to workplace health promotion in school communities. The study described in this chapter employed the action research method to empirically test a theoretical model for the occupational well-being of school staff. Inductive and deductive research approaches were applied to the development in that both quantitative and qualitative research materials (national and international data) and analysis methods were methodically utilised during 2002–2012 (until 2014). By developing and testing the theory, we produced a middle-level theory and the *Content model for the promotion of school community staff's occupational well-being,* which continues to be tested and developed. The content model is theoretically clear and flexible. It contains the premises for planning comprehensively through four aspects, and it serves as a suitable model for implementing and evaluating the development of school staff's occupational well-being, which also affects the learning, health and well-being of children and adolescents. The content model can therefore be used in various situations. It has been applied to, and tested in, comprehensive and upper secondary school communities, but it can also be applied to other schools and work communities where its functionality will also be tested in the future.

Keywords Health promotion • Theory • School • Staff • Well-being

---

T. Saaranen, Ph.D. (✉) • K. Tossavainen, Ph.D. • M. Sormunen, Ph.D.
S. Laine • H. Turunen, Ph.D.
Department of Nursing Science, Faculty of Health Sciences,
University of Eastern Finland, Kuopio, Finland
e-mail: terhi.saaranen@uef.fi; kerttu.tossavainen@uef.fi;
marjorita.sormunen@uef.fi; sari.laine@uef.fi; hannele.turunen@uef.fi

V. Simovska and Patricia Mannix-McNamara (eds.), *Schools for Health and Sustainability: Theory, Research and Practice*, DOI 10.1007/978-94-017-9171-7_19,

19.1 Introduction

Health promotion theories and models are needed to explain and predict health behaviour, as well as for the design and evaluation of interventions. Polit and Beck (2011) determined the concept of theory as an abstract generalisation that presents a systematic explanation of the relationship between phenomena and the model as a symbolic representation of concepts or variables and interrelationships among them. However, the theories and models that are used in health promotion textbooks and in related education remain largely unused in practice. Many published articles related to health behaviour interventions (Lippke and Ziegelmann 2008) and, generally, health promotion literature lacks a theoretical background or rationale (Van den Broucke 2012).

Examples of theories or models that are applied to health promotion are available from several viewpoints and disciplines. To name a few, a theory on the sense of coherence by Antonovsky (1996) has been largely applied to health promotion. The theory uses a central focus on salutary, rather than risk factors, and focuses on seeing people as an entirety rather than identifying them merely by their disease. Bandura (1998), in turn, examined the area of health promotion from the lens of social cognitive theory and called for further progress in building new structures for health promotion. More recently, Rütten et al. (2009) transferred the social psychological theory of Von Wright (1976) into assessing and developing the engagement of organisations in health promotion policy, and at the same time, testing the new theoretical model. The model was again used and tested in an intervention study related to the promotion of physical activity of elderly people (Rütten et al. 2012). Health promotion can, therefore, be seen as a continuously developing field and an ideal testing ground for scrutinising and applying different theoretical structures.

In addition to theories and models applied from other disciplines, there are health promotion models, such as Tannahill's model, created in mid-1980s (Tannahill 2008), and Tones and Tilford's empowerment model (1994), which offer a relatively large framework for health promotion. However, more specific models and theories are needed. For example, schools as settings for health promotion have been, and currently are, highly valued arenas for reaching children and adults, and, therefore, function as specific settings for health promotion (WHO 1986) where developed theories and models are justified and needed. One example of a school-related health promotion framework that has resulted from a settings perspective is an eco-holistic approach to the health-promoting school. The schools promote the health of all those who work and learn in them and collaborate with parents and the surrounding community (Weare 1998, 2000). Similarly, the ecological model by Lohrmann (2010) derived from coordinated school health ideology extends the health-promoting activities from classrooms to the greater community, but it also specifies the infrastructure within which the people in the schools operate.

In addition to concepts developing over time, methodological development is natural and follows the policies and needs set by national and international directions, health problems, and economical situations. To develop school communities' workplace health promotion, intervention studies are needed to create multidisciplinary and sufficiently extensive approaches from which practical theory and related models can be developed. Tested theories and produced models can help school health promotion, including workplace health promotion, in two ways: first, the model, developed specifically for workplace health promotion, helps to identify or describe the factors that have been proven to be important in enhancing or inhibiting complex interventions in workplaces. Second, the model offers a starting point for evaluating interventions and thus increases the dissemination of good practices, making them rooted in everyday actions (May et al. 2007), as well as examines the effectiveness and outcomes of interventions. This way, theory development and modelling help school community staff, including health promotion professionals, in planning, executing and evaluating the workplace development projects in school communities (Horn et al. 2004; see also Lippke and Ziegelmann 2008).

As mentioned, theoretical concepts evolve over time. In the area of school community staff health promotion, the ambiguity of concepts inhibits positive development. As Juniper (2011) argues, employee well-being has been shown to be subjective and multidimensional. The main reason for failure in employee well-being programmes seems to be the lack of agreement about what is meant by the concept 'employee well-being' at the start of the program (Juniper 2011). Furthermore, the concepts of' well-being' and 'workplace well-being' are used in varying ways in different situations depending on, for example, the aims, context, discipline and the focus of the study on well-being at workplace.

The variety of studies on workplace well-being and its concepts is a challenge, one that demands a strict definition of central concepts from an academic point of view (Ilmarinen et al. 2008). Juniper (2011) brings out the importance of putting primary emphasis on employees' own views of their well-being. By emphasising the position of employees, planning and approach are made directly relevant to them, and this involvement makes them also productive to employers. Academic research generally supports this idea that the healthier and happier people are, the more productive they are likely to be in their workplaces (Juniper 2011).

There is limited research available on holistic occupational well-being that takes into account the physical, mental, and social aspects. Furthermore, there is even less research on developing the occupational well-being of school staff where the topic would have been modelled or there would have been theoretical information to support the development (e.g., Konu et al. 2002). However, many studies have noted that there is need for developing the occupational well-being of school staff (e.g., Eaton et al. 2007).

This chapter describes and explains the development of a middle range theory to increase school staffs' occupational well-being, which refers to workplace health promotion in schools from the perspective of school staff. The approach of participatory action research, promoting a culture of information and knowledge sharing, has been used to develop and test the theory and to produce the model. Action research has become increasingly important in large or medium-sized projects (Thiollent 2011) and has turned out to be a focus of interest in health promotion research due to its appropriateness in capturing phenomena, in bringing out processes, and in including the engagement of both researchers and participants (Whitehead 2010). Based on previous research, the approach of action research facilitates the combination of practical knowledge and research-based knowledge in school health promotion intervention studies (e.g., Ozer et al. 2010; Sormunen et al. 2012) and, in a wider context, in the health sector and clinical nursing practices (Glasson et al. 2008; Soh et al. 2011). The so-called hypothesis of this study is that school staffs' occupational well-being can be increased by *actions for occupational well-being*, which is a fresh concept for maintaining one's ability to work (see Ilmarinen et al. 2008). Actions to promote school staffs' occupational well-being stem from promoting the health and resources of employees (e.g., personal health and physical and mental fitness), increase the functionality of a working community (e.g., work management and organisation, leadership, social support, information), development of professional competence (e.g., in-service education), and enhance work and working conditions (e.g., the physical working environment; physical, biological, and chemical factors, and safety at work; and working postures and equipment) (Saaranen et al. 2012a, b).

19.2 Development of the Middle Range Theory and the Model of Health Promotion

When developing a theory, the steps of generalising and finding out theoretical connections and testing the theory in practice follow those of creating conceptual meanings, contextualising and structuring. *Middle range theories* are used as aids in creating and developing everyday practices based on research information in health promotion and nursing (Smith 2008; Polit and Beck 2011). The purpose of a middle range theory is to describe, explain and predict, e.g., a clearly defined phenomenon of health promotion (Polit and Beck 2011). Middle range theories are also meant to be concrete. According to Cross (2010), health promotion theories are used to understand or examine health promotion practices and to provide a framework for conducting them systematically. However, Crosby and Noar (2010) indicate that the development of a theory on health promotion has not proceeded in a way that would correspond with the evolution of health promotion practices. Indeed, they maintain that when developing a theory, there should be more emphasis on rationalising the development in practical environments, the development should transcend the individual level and theory should be accessible to practitioners. In order to develop occupational well-being, work communities need the development of such a theory that produces models for those involved in the work community and others who act to promote occupational well-being. The resulting model must be such that it can be sufficiently generalised in different situations and contexts and also takes into account the special needs and premises of the employee and working place in question.

Developing a theory can be *inductive*, *deductive* or *inductive-deductive*. Developing an inductive theory is based on concrete data that relies on empirical knowledge and is linked to the phenomenon in question, in which case the methods of study are qualitative (e.g., ethnography, phenomenology, or grounded theory) (Polit and Beck 2011). When developing a deductive study, quantitative research methods are usually utilised. In this case, the developing process uses arguments as its starting point, testing these in several, carefully defined situations. The development of an inductive-deductive theory combines the previously mentioned research methods of developing a theory. The use of this method is typical, for instance, in intervention studies of health sciences, where employing mixed-methods research has become increasingly frequent (Polit and Beck 2011).

The *models* that result from the development of a theory function as theoretical frames of reference for, e.g., health promotion and health evaluation in different contexts. A theoretical model must be sufficiently carefully defined and it must suit its purpose (May et al. 2007). In action research, a model functions as a theoretical frame of reference for understanding multifaceted interventions. A model brings transparency and explanations to phenomena and their development that empirical action research reveals with its various methods.

Here, we use the example of developing a theory on promoting occupational well-being of school staff and modelling the theory based on many action research projects. In this context, what is meant by the word 'modelling' is producing *content model for the promotion of a school community staff's occupational well-being*. The purpose of such a model is to describe and develop occupational well-being of staff in school communities and to function as an instrument of evaluation.

19.2.1 Promoting Occupational Well-Being of School Staff—The Stages of Developing a Theory and a Model

Developing a theory for promoting occupational well-being of a school staff and crafting a content model linked to this have been a long-term process consisting of several stages. Developmental work began in the participatory action research project, *Promotion of school community staff's occupational well-being in co-operation with occupational health nurses*, in Eastern Finland in 2001–2004. Since then, development work on the theory has continued in international projects (*Teachers well-being project, 2004–2006*, and *Promotion of school community staff's occupational well-being—action research project in Finland and Estonia, 2009–2014*).

The method of action research using a triangulation of methodology and data has made it possible to not only develop practical aspects but to also develop and test a theory in which descriptive, explanatory and predictive middle range theory on occupational well-being of school staff has been developed. With the help of the theory, it is possible to describe, explain and direct practical matters, here the promotion of occupational well-being of school staff. By the concept school staff, it is meant here all the professional groups that are involved in a school community, such as teachers, principals/school management, special needs assistants, cooking staff and cleaning workers. The theory has been developed in cooperation with those involved with the practical aspects of school work (school staff and occupational health nurses) and researchers.

The premises of the theory that has been developed, *Promotion of school community staff's occupational well-being*, is promoting health and practices according to the formation of the inductive-deductive theory. Developing the theory is presented below in five stages (Fig. 19.1). The results are a situation-based middle range theory and the *Content model for the promotion of school community staff's occupational well-being* (Fig. 19.2) that can be utilised in describing, explaining, evaluating and directing the practices.

At the first stage in 2002, the aim of the study was to investigate school staffs' and occupational health nurses' evaluations of school community staff's occupational well-being, work ability maintenance and related factors. School community staff here refers to all occupational groups working in a school community (principals, teachers, school assistants, school nurses and other staff, e.g., cleaners and cooks) (Saaranen et al. 2005, 2006). The researcher gathered the data via themed, customised telephone interviews conducted by occupational health nurses (12 telephone interviews by an occupational health nurse) and group interviews of school staff (11 group interviews that included in total 66 interviewees), and these were analysed by using an inductive content analysis. As the unorganised reality functioned as the starting point, gathering data on the phenomenon to be studied inductively was required.

The customised phone interviews by occupational health nurses and the group interviews by school staff members resulted in bringing light to the factors of the occupational well-being in school communities: a *working community*'s positive atmosphere (e.g., open communication and working atmosphere), motivation for *work* and the quality of *working conditions* (e.g., rewarding work and conditions of physical workspaces), *professional* abilities and adequate *education* (e.g., adequate professional skills and opportunities for maintaining and developing one's professional skills) and *private life conditions* (e.g., domestic matters and good physical and psychological conditions) (Saaranen et al. 2006). Similarly, based on the results, the participants described the actions for maintaining occupational well-being as containing four aspects: work ability maintenance targeted to the *school community* (e.g., developing staff meetings and discussion in a working community), work ability maintenance targeted to school *work* and *working conditions* (e.g., surveying and developing working positions and conditions), work ability maintenance of school staff's *professional competence* (e.g., school surveys and organising training events) and work ability maintenance targeted to the *employee* (e.g., activities supporting mental and physical fitness) (Saaranen et al. 2005). The four aspects of actions maintaining occupational well-being were very similar as the four categories describing the contents of the concepts of occupational well-being.

Therefore, the qualitative data supported the paradigm of Finnish work ability maintenance, where keeping one's capacity to work is considered to be formed of the aspects of the *worker and work, working community, professional competence*

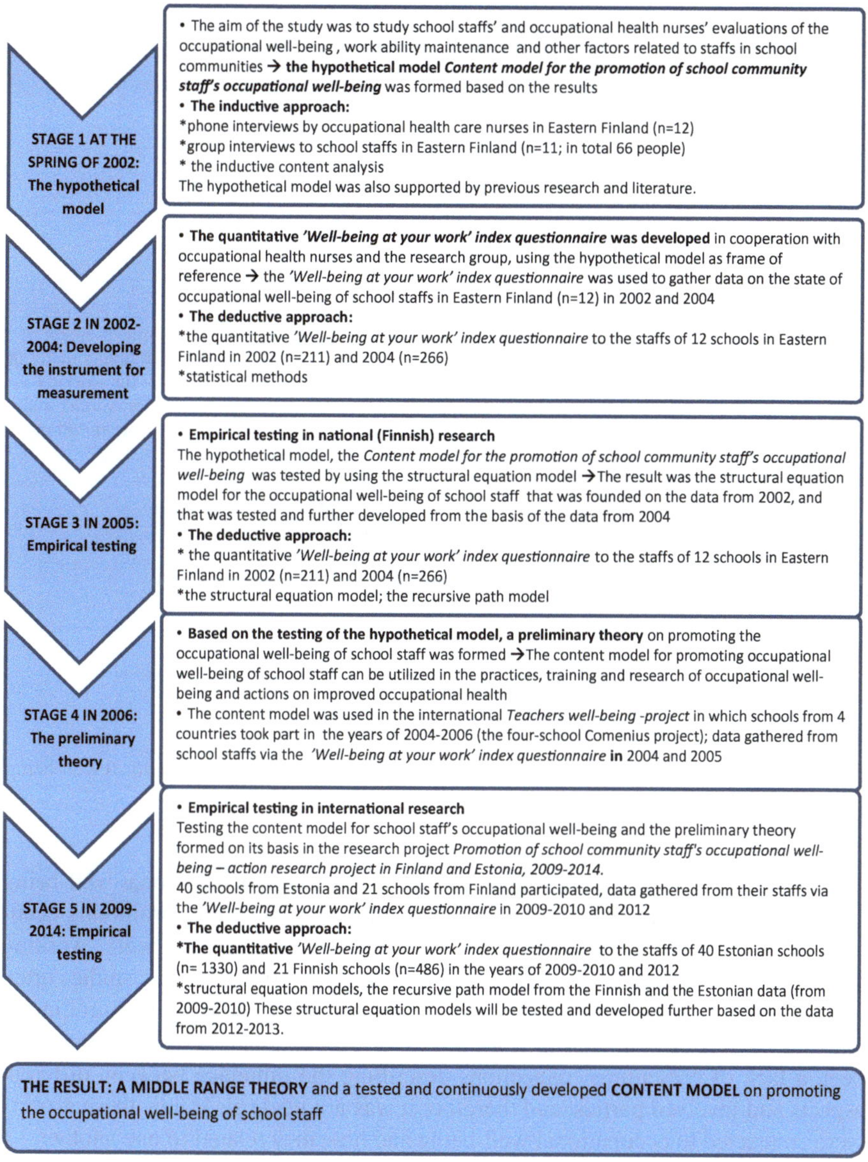

Fig. 19.1 Developing and testing the theory and the content model on promoting occupational well-being of school staff

and *working conditions* (Ilmarinen et al. 2008). The results from the interview data also reinforce the view that actions promoting work ability maintenance can be used to develop the occupational well-being of school staff in a wide-ranging way (see factors of the occupational well-being in school communities; Saaranen et al. 2005). Previous studies and literature from different areas also support this view. However,

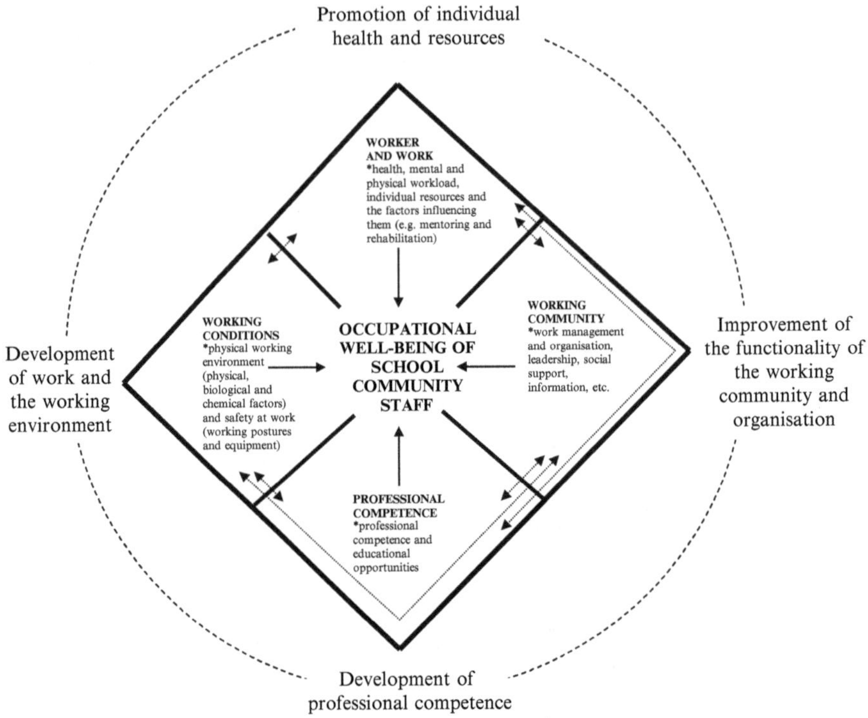

Fig. 19.2 Content model for the promotion of school community staff's occupational well-being

there was a shortage on such research information where occupational well-being has been developed in a wide-ranging way with the use of actions promoting occupational health that would take care of all its aspects (worker and work, working community, professional competence and working condition). Indeed, studies often touched upon only some of the previously mentioned areas. What is more, in the existing models of maintaining work ability or occupational well-being, there has been a lack of presenting connections describing the influence between different aspects and involved parties, and therefore it was not made clear how these aspects were connected to occupational well-being and how they related to one another.

By utilising previous research information and literature and the quantitative methods of the planning stage, meaning the results of the phone interviews by occupational health care nurses and those from the group interviews of the school staff members, a hypothetical model, the *Content model for the promotion of school community staff's occupational well-being* was formed at the initial stage of developing the theory. The goal of the hypothetical model was to thoroughly indicate out of which factors occupational well-being of school staff is formed and which components must be taken into account when promoting occupational well-being of school staff.

At the second stage, the hypothetical model functioned as the starting point and the goal was to define arguments and operationalising concepts in the model utilising the deductive approach. This is when the quantitative *'Well-being at your work'* index questionnaire was developed in cooperation with occupational health-care nurses and the research group. The research group of the project and the health-care nurses evaluated the *'Well-being at your work'* index questionnaire several times during its planning stage (e.g., by providing written feedback on the contents of the questionnaire in the form of an e-mail questionnaire from the health-care nurses to the researchers). The *'Well-being at your work'* index questionnaire was used to pre-test the staff of one secondary school that did not participate in the study (n = 14). At first, the questionnaire was used to find out background information (ten questions), satisfaction with occupational well-being and the available actions for occupational well-being (=actions maintaining one's ability to work) were recorded using four variables on a Likert scale (1–5; 1 = very poor... 5 = very good). Moreover, on the topic of one's satisfaction on the available actions for occupational well-being, one could make their responses more thorough by using their own words when filling in further details on their responses, and these were utilised when planning local development projects or when personalising projects to individual schools. After this, the different aspects of occupational well-being (working conditions, working community, worker and work and professional competence) were also examined with the use of the Likert scale (1–5) variables. The staff members were asked on their opinion on each separate variable (a total of 51 from the four aspects of occupational well-being) (1 = totally disagree and 5 = totally agree), and the need for development of each particular matter was evaluated (1 = requires a lot of development and 5 = does not require any development). Additionally, the respondents could provide further details on each aspects of occupational well-being (working conditions, worker and work, working community and professional competence) in two open-ended questions that also provided further information when planning local, school-based areas for development.

The questionnaire forms were coded and these were delivered personally to the schools' occupational health-care nurses at both periods of observation (2002 and 2004) and were then again brought by the nurses to the school staff members. The occupational health-care nurses also collected the anonymous question forms after they had been put in their envelopes and delivered them back to the researchers. The *'Well-being at your work'* index questionnaires were used to map the state of 12 school staff in the 2002 and the follow-up study was conducted at the same schools in 2004.

At the third stage, the hypothetical model was tested on a structural equation model (path model) that is typically used to test a model of theory (Polit and Beck 2011). Creating a path model proceeded according to the normal stages of a structural equation model, consisting of the phases of model specification, examination of identification, estimation of parameters, testing of hypotheses related to the model and other adequacy tests of the model (see Bollen 1989; Kline 2005).

Based on the data from the *'Well-being at your work'* index questionnaire from 2002 (n = 211; percentage of responses 78 %), a structural equation model describing the occupational well-being of school staff was produced. The model used the hypothetical *Content model for the promotion of school community staff's occupational well-being* as its theoretical framework. Testing was then continued so that the structural equation model from 2002 was tested further and the data from 2004 (n=266; percentage on responses 83 %) was used to develop it. The data were collected with the use of the same *'Well-being at your work'* index questionnaire at the same schools as in 2002. A close look at the structural equation models (see Saaranen et al. 2007: Occupational well-being of school staff Model 1 at the initial stage of 2002, p. 20, and occupational well-being of school staff Model 3 at the follow-up stage in 2004, p. 22) reveals that the explanatory variables have remained the same and that there have been some changes only in the relationship of interaction. The explanatory variables were from the different aspects of occupational well-being (working community, working conditions, professional competence, the worker and work), and this describes what a wide scope the topic of occupational well-being of school staff covers. The structural equation model on occupational well-being of school staff in 2004 confirmed the view that the hypothetical model (*Content model for the promotion of school community staff's occupational well-being*; see Saaranen et al. 2006) is sufficient in scope (including the aspects of working conditions, working community, worker and work, professional competence) when planning, carrying out and evaluating the occupational well-being of school staff.

At the fourth stage, a tentative theory based on the testing of the model was formed for the purpose of promoting occupational well-being of school staff. Based on the results, connections describing the influence between different aspects (working conditions, working community, worker and work, professional competence) or those involved were added in the *Content model for the promotion of school community staff's occupational well-being* (Fig. 19.2). The content model was then utilised in the *Teachers well-being project of 2004–2006*, in which four countries took part. The *Teachers well-being project* was a Comenius project participated in by four different school communities (in Finland, Ireland, Italy and Germany), and its goal was particularly to develop occupational well-being of school staff. In the school communities, information on the occupational well-being of staff was gathered in 2004 with the *'Well-being at your work'* index questionnaire. Based on the results, development projects on the occupational well-being of school staff were carried out in the school communities in question. A follow-up study was conducted in the school communities using the same indicators in 2005 (Saaranen et al. 2013). The model had to be tested in similar projects taking place in school environments both internationally and in Finland in order to test the reliability of the model and to develop the middle range theory to explain it.

At the fifth stage, developing the middle range theory and testing the model have been continued in the *Promotion of school community staff's occupational well-being action research project* in Finland and Estonia in 2009–2014. The goal of the project has been, and still is, to promote occupational well-being of school staff by using extensive actions that promote occupational well-being in earlier actions, maintaining one's ability to work in elementary schools in Finland and

Estonia, and to test and further develop the functionality and structure of the existing *Content model for the promotion of school community staff's occupational well-being* when using it to explain occupational well- being of school staff. So far, the results of the mapping of the preliminary stage have been reported (e.g., Saaranen et al. 2012a, b), and structural equation models have also been composed of the data from the preliminary stage. Structural equation models have been constructed on the occupational well-being of school staff and the factors that explain the phenomenon based on data from Estonia and Finland. Constructing the models at stage 5 has been similarly carried out as has been described at the stage 3 of empirical testing. The structural equation models support the previous ideas that the content model for occupational well-being of school staff consists of four different aspects. The explanatory factors came from the aspects of occupational well-being (working community, working conditions, professional competence, worker and work) based on both Finnish and Estonian data. A more detailed academic publication on these structural equation models is currently being constructed. Furthermore, quantitative follow-up data on Finnish and Estonian school communities have been collected with the *'Well-being at your work'* index questionnaire at the turn of 2012–2013. This will allow for the structural equation models based on the data from the preliminary mapping stage of the project to be further tested and developed. *Promotion of school community staff's occupational well-being—action research project in Finland and Estonia, 2009–2014* is part of the wider international Schools for Health in Europe research programme (2008-) that used to be known as the European Network of Health Promoting Schools (1991–2007) (see e.g., Hansen et al. 2009; Lepp et al. 2007).

19.3 Discussion

A middle range theory and a content model have been produced on the promotion of occupational well-being of school staff with the help of different stages of theory (Fig. 19.1; stages 1–5). The middle range theory and the model are useful in planning, carrying out and evaluating occupational well-being of staff in school communities. We will now move on to discuss the developed *Theory for promotion of school community staff's occupational well-being*, how it was tested and how the content model was formed stage by stage, and finally pointing out some viewpoints on the reliability and ethicalness of the development process.

19.3.1 Discussion of the Developed Model and Theory

Using an inductive approach and a qualitative research method during **the first stage** of researching occupational well-being is also supported by the fact that information available on the topic was scarce and fairly unorganised. As Juniper (2011) indicates, promoting occupational well-being is complicated by the variety of concepts on occupational well-being based on the subjective and multidimensional nature of the concepts. In fact, a school is a shared working and learning community for a number of different professional groups (e.g., teachers, cleaning staff, kitchen workers and property maintenance staff) and pupils, and it should support the health and ability to function of anyone working in it (Bonell et al. 2011). Crosby and Noar's (2010) views on developing a theory that is especially based on practices and made for practical situations also support the idea that the target group for research must be made actively involved in the developmental process. The starting point was that school communities need the development of such a theory that produces models for those working in school communities and others involved in promoting their occupational well-being, such as occupational health-care nurses and other professionals in occupational health care. Furthermore, the resulting model must be sufficiently generalisable to different school contexts and take into account the special needs and premises of workers and individual workplaces. The hypothetical model *Content model for the promotion of school community staff's occupational well-being* was created on the basis of the results that were gathered via the questionnaire, as well as with the help of previous research information and literature.

At the second stage of developing the theory, the goal was to find out the means and methods that could be used to systematically explain the hypothetical model. At this stage, the deductive approach was used. This method generally utilises the approaches of quantitative research. The *'Well-being at your work'* index questionnaire was developed to function as a tool that could be used for the thorough evaluation of the state of occupational well-being of staff and the needs for development, but one that could also be used for analysing specific aspects. The quantitative set of measures made it possible to use statistical methods when researching the different aspects of the factors influencing occupational well-being. Using statistical methods allowed for making the work community-based results applicable to practice, and enabled one to systematically observe the concepts and the factors between different aspects. Therefore, the questionnaire can also be utilised as an instrument for the evaluation of research and for development projects related to occupational well-being in the future.

When evaluating the applicability of a theory, one must be able to test it (May et al. 2009). **At the third stage** of developing the theory, the functionality and structure of the content model for the promotion of occupational well-being of school staff was tested by using the structural equation model in 2002 and in 2004. The results from the structural equation model confirmed the presumption (the hypothesis) that in order to promote occupational well-being of school staff members, four aspects must be taken into account, according to the content model: working condition, working community, worker and work, and professional competence. The different aspects of the content model did not contradict one another, but had explanatory effectiveness, and the hypothetical content model was shown to be sufficiently durable for formal testing (May et al. 2009). Therefore, at **the fourth stage**, the content model for occupational well-being of school staff was now utilized in the teachers' well-being project in 2004–2006, where the results on occupational well-being of school staff gained through the follow-up study were positive

(Saaranen et al. 2013). However, at this stage it was recognised that the content model required further testing and possible development.

At the fifth stage, testing and developing the theory and the content model empirically have been continued in the Estonian-Finnish international project. The premise for testing the content model was a view that the occupational well-being of school staff is affected by working conditions, working community, worker and work and professional competence, all of whose essence and interconnectivity are tested in the studies. The testing was conducted by using the structural equation model on the Lisrel programme. Here, the Finnish and Estonian data were processed separately. Constructing the structural equation model proceeded similarly as during the previous stage of empirical testing (see Fig. 19.1; stage 3). Both stages of testing the model (stages 3 and 5, see Fig. 19.1) confirmed the view that occupational well-being is a broad concept that is affected by all the four different aspects.

The empirical testing of and reporting on, the theory and the content model will be continued until 2014. This will result in a tested, middle range *Theory for the promotion of school community staff's occupational well-being* and the further developed *Content model for the promotion of school community staff's occupational well-being*. We can, however, already note that the produced middle range theory and content model have helped to promote the health of school communities and have developed occupational well-being in intervention studies in many ways. Firstly, the content model for promoting occupational well-being has made it possible to recognise or describe factors that have been indicated to be important in advancing or stopping the realisation of diverse interventions in school communities. Secondly, the model has offered a clear starting point for evaluating different projects' interventions that have promoted the distribution of successful practices and making them become everyday activities in school communities (May et al. 2007).

Promoting occupational well-being and health in other working communities in addition to school communities has emerged as a political area of concern both internationally and nationally, as the working population is aging and this has resulted in a pressure to continue careers longer than before (Commission of the European Communities 2007; Ministry of Employment and the Economy 2012; Ministry of Social Affairs and Health 2011). Occupational well-being and health promotion are founded on collaboration taking place in workplaces that is supported by competent occupational health care and up-to-date legislation (Ministry of Employment and the Economy 2012; Ministry of Social Affairs and Health 2011). Thus the results of the developed *Content model and theory for the promotion of school community staff's occupational well-being* produced evidence-based information for school staff, occupational health nurses, occupational health service administration, school administration, researcher, and educators on how occupational well-being can be promoted in practice. The developed model and theory can also benefit other workplaces and their staff by developing measures for the promotion of occupational well-being or workplace health promotion. However, in order to increase the validity of the content model and theory and to develop them further, they have to be tested in other, comparable developments projects in school communities and other workplaces.

19.3.2 *Ethicalness and Reliability in This Process*

This process (developing and testing the theory and the content model on promoting occupational well-being of school staff) has been carried out according to the **ethical** principles of The National Advisory Board on Research Ethics (2002), which are presented in the guidelines "Good Scientific Practice and Procedures for Handling Misconduct and Fraud in Science" and "Guidelines on Research Ethics of the Academy of Finland" (2003). Research permits have been obtained from different organisations of the projects as well as from individual respondents according to ethical guidelines in both Finnish and international projects.

The **reliability** of developing the theory and the content model is confirmed by the longevity of development work (2002–2014). There has been development work in national and international action research projects, and this has allowed for a natural integration of theory and practice and made it possible to develop both. However, employees' experiences of their occupational well-being vary widely on personal level and in different contexts (Juniper 2011), and therefore creating an unequivocal theory and content model on occupational well-being of school staff is challenging in the constantly changing work and school environment. Extensive health promotion models (see e.g., Tannahill 2008; Tones and Tilford 1994) offer successful frames of reference for health promotion, but these extensive models have also been criticised for being overtly theoretical for practical workers and for developing practices (e.g., Crosby and Noar 2010). The goal of the studies presented in this article has been to create and develop a middle range theory and a content model to serve entire school staff and the development work for occupational well-being as successfully and extensively as possible without forgetting, e.g., school-based special requirements. Indeed, the experiences and the gained research results from the projects have been positive, but further testing of the theory and the model in different school contexts will still be needed.

However, international projects have particularly proven to be challenging in the area of considering the points of view of ethics and reliability. There have been attempts in trying to take differences in culture and language into consideration at every stage of the research. For instance, in the Estonian-Finnish project, the *'Well-being at your work'* index questionnaire was translated into Estonian, and in the Comenius project, the questionnaire was available in English so that each school staff member was able to take part in the study. Indeed, research projects at the different stages of developing the theory (see stages 1–5) have been based on the participatory action research approach (Thiollent 2011), according to which every participant has an important role in the projects. The research projects and data collection have been voluntary and participants have had the option to cancel their participation at any given stage. However, the amounts of participants from school staff have been excellent in the different projects (see e.g., Saaranen et al. 2012a), and this also indicates that there is need for, and interest in, developing the occupational well-being of school staff.

There is also the example of the Estonian-Finnish project, where members of the research group have come from both countries, and this has made it easier to consider cultural differences and to carry out research in the school contexts of different countries. Problems with language among the school staff in Estonia and the Finnish members of the research group have been tackled, e.g., by employing an interpreter, which has made it possible to have active and profound discussions in one's own native tongue. Indeed, the cooperation between school staff and the research group has been realised in the project in a natural and planned manner.

Moreover, the successful interactive and social skills of the project coordinator, who is a worker in, and directs the practical aspects of, the project, have a great importance to the reliability and ethicality of the study (Casey 2011). For instance, a researcher from an action research project or an employee in a participating work community, such as a teacher or a principal, can function as the coordinator. For example, the coordinator of practical activities in the Comenius project (see stage 4) was a classroom teacher from a Finnish school. Similarly, in the Estonian-Finnish project (see stage 5), coordinators for the School for Health in Europe (SHE) network functioned in the role on school level in Estonia, while a member of the research group, a post doc researcher and a SHE coordinator served in the position in Finland. Their interactive skills in their own school staff's community and/or the school communities in the partner countries and the research group have had a positive, important meaning.

Members of the research group from the Department of Nursing Science in the University of Eastern Finland have been in charge of the coordination of all the aforementioned projects as a whole and have been responsible for the research. As the research group in the University of Eastern Finland has taken care of the questionnaires for the research and the analysis of the results, the school staffs have been able to free their resources into developing occupational well-being at the level of their school community. Successful cooperation and participatory working methods between the members of the research group, the practical coordinator/s involved in the projects and the school staff members have been a unifying force from the point of view of the reliability as well as the ethicality of research.

19.4 Conclusions

The content model for promoting occupational well-being of school staff, the method of the *'Well-being at your work'* index questionnaire that was developed in the process and the middle range theory closely related to these provide a concrete frame of reference to map the occupational well-being of school staff and to develop planning, carrying out and evaluating it for staff in health care and schools, researchers in the field and educators. The theory allows school staff to be actively involved in actions that develop occupational well-being and the decision-making processes that take into account the individual and communal special needs of staff and, e.g., the role of the occupational health care as a supporting, professional entity. To increase the reliability and usability of the theory, it will still be tested and developed further in an international research project (see Fig. 19.1; stage 5). In the future, the theory and the *'Well-being at your work'* index questionnaire must also be tested and developed when promoting occupational well-being in other work communities.

References

Academy of Finland. (2003). *Guidelines on research ethics*. http://www.aka.fi/Tiedostot/Tiedostot/Julkaisut/Suomen%20Akatemian%20eettiset%20ohjeet%202003.pdf. Accessed 7 June 2013.

Antonovsky, A. (1996). The salutogenic model as a theory to guide health promotion. *Health Promotion International, 11*(1), 11–18.

Bandura, A. (1998). Health promotion from the perspective of social cognitive theory. *Psychology & Health, 13*(4), 623–649.

Bollen, K. (1989). *Structural equations with latent variables*. New York: Wiley.

Bonell, C., Harden, A., Wells, H., et al. (2011). Protocol for a systematic review of the effects of school and school-environment interventions on health: Evidence mapping and syntheses. *BMC Public Health, 11*, 453. doi:10.1186/1471-2458-11-453.

Casey, M. (2011). Interorganisational partnership arrangements: A new model for nursing and midwifery education. *Nurse Education Today, 31*(3), 304–308.

Commission of the European Communities. (2007). *Communication from the commission to the council and the European parliament. Improving the quality of teacher education* (3.8.2009. COM (2007) 392 final). Brussels. Available at: http://www.atee1.org/uploads/EUpolicies/improving_the_quality_of_teacher_education_aug2007.pdf. Accessed 9 June 2014.

Crosby, R., & Noar, S. M. (2010). Theory development in health promotion: Are we there yet? *Journal of Behavioral Medicine, 33*(4), 259–263.

Cross, R. (2010). Health promotion theory, models and approaches. In D. Whitehead & F. Irvine (Eds.), *Health promotion & health education in nursing. A framework for practice* (pp. 22–44). Basingstoke, Hampshire: Palgrave Macmillan.

Eaton, D. K., Marx, E., & Bowie, S. E. (2007). Faculty and staff health promotion: Results from the school health policies and programs study 2006. *Journal of School Health, 77*(8), 557–566.

Glasson, J., Chang, E. M. L., & Bidewell, J. W. (2008). The value of participatory action research in clinical nursing practice. *International Journal of Nursing Practice, 14*(1), 34–39.

Hansen, S., Varava, L., & Streimann, K. (2009). Establishing the network of health promoting kindergartens and schools in Estonia in 2005–2009. In *Better schools through health: The third European conference on Health Promoting Schools, 15–17 June, Vilnius, Lithuania* (pp. 50–51). The conference book. Vilnius, Lithuania "Kriventa": State Environment Health Centre.

Horn, J. E., Taris, T. W., Schaufeli, W. B., & Schreurs, P. J. G. (2004). The structure of occupational well-being: A study among Dutch teachers. *Journal of Occupational and Organizational Psychology, 77*(3), 365–375.

Ilmarinen, J., Gould, R., Järvikoski, A., & Järvisalo, J. (2008). Diversity of work ability. In R. Gould, J. Ilmarinen, J. Järvisalo, & S. Koskinen (Eds.), *Dimensions of work ability. Results of the health 2000 survey* (pp. 13–24). Vaasa: Waasa Graphics Oy.

Juniper, B. (2011). Defining employee wellbeing. *Occupational Health, 63*(10), 25.

Kline, R. B. (2005). *Principles and practice of structural equation modeling* (2nd ed.). New York: Guidford Press.

Konu, A., Alanen, E., Lintonen, T., & Rimpelä, M. (2002). Factor structure of the school well-being model. *Health Education Research, 17*(6), 732–742.

Lepp, K., Villeruca, A., & Jociute, A. (2007). Self-assessment tool for dissemination of health-promoting schools on the school level: Collaboration between Estonia, Latvia and Lithuania. In V. Barnekow, G. Buijs, S. Clift, B. B. Jensen, P. Paulus, D. Rivett, & I. Young (Eds.), *Health-promoting schools: a resource for developing indicators* (pp. 95–101). European Network of Health Promoting Schools. Kailow Graphic. http://www.nepes.eu/files/Report%20 Health%20Promoting%20Schools.pdf. Accessed 9 June 2014.

Lippke, S., & Ziegelmann, J. P. (2008). Theory based health behavior change: Developing, testing, and applying theories for evidence-based interventions. *Applied Psychology: An International Review, 57*(4), 698–716.

Lohrmann, D. K. (2010). A complementary ecological model of the coordinated school health program. *Journal of School Health, 80*(1), 1–9.

May, C., Finch, T., Mair, F., et al. (2007). Understanding the implementation of complex interventions in health care: The normalization process model. *BMC Health Services Research, 19*(7), 148. doi:10.1186/1472-6963-7-148.

May, C., Mair, F., Finch, T., et al. (2009). Development of a theory of implementation and integration: normalization process theory. *Implementation Science, 4*, 29. http://www.implementationscience.com/content/4/1/29. Accessed 7 June 2013.

Ministry of Employment and the Economy. (2012). *National working life development strategy to 2020.* http://www.tem.fi/files/35434/Tyoelaman_kehittamisstrategia2020_A4_eng.pdf. Accessed 31 May 2013.

Ministry of Social Affairs and Health. (2011). *Työterveyshuolto ja työkyvyn tukeminen työterveysyhteistyönä. Työryhmän loppuraportti* [Occupational health services and support for work ability through occupational health cooperation. Final working group report.] Sosiaali-ja terveysministeriön selvityksiä 2011:6. Helsinki. http://www.stm.fi/c/document_library/get_file?folderId=2872962&name=DLFE-14934.pdf. Accessed 31 May 2013.

Ozer, E. J., Ritterman, M. L., & Wanis, M. G. (2010). Participatory action research (PAR) in middle school: Opportunities, constraints, and key process. *The American Journal of Community Psychology, 46*(1/2), 152–166.

Polit, D. F., & Beck, C. T. (2011). *Nursing research. Generating and assessing evidence for nursing practice* (9th ed.). Philadelphia: Lippincott Williams and Wilkins.

Rütten, A., Röger, U., Abu-Omar, K., & Frasha, A. (2009). Assessment of organizational readiness for health promotion policy implementation: Test of a theoretical model. *Health Promotion International, 24*(3), 243–251.

Rütten, A., Abu-Omar, K., Gelius, P., et al. (2012). Policy assessment and policy development for physical activity promotion: results of an exploratory intervention study in 15 European Nations. *Health Research Policy and Systems, 10*, 14. http://www.health-policy-systems.com/content/10/1/14. Accessed 7 June 2013.

Saaranen, T., Tossavainen, K., & Turunen, H. (2005). School staff members' and occupational health nurses' evaluation of the promotion of occupational wellbeing—With good planning to better practice. *Journal of Interprofessional Care, 19*(5), 465–479.

Saaranen, T., Tossavainen, K., Turunen, H., & Vertio, H. (2006). Occupational wellbeing in a school community—Staff's and occupational health nurses' evaluations. *Teaching and Teacher Education, 22*(6), 740–752.

Saaranen, T., Tossavainen, K., Turunen, H., Kiviniemi, V., & Vertio, H. (2007). Occupational well-being of school staff members: A structural equation model. *Health Education Research, 22*(2), 248–260.

Saaranen, T., Sormunen, M., Streimann, K., Pertel, T., Hansen, S., Varava, L., Lepp, K., Turunen, H., & Tossavainen, K. (2012a). The occupational well-being of school staff and maintenance of their ability to work in Finland and Estonia—Focus on the school community and professional competence. *Health Education (Special Issue), 112*(3), 236–255.

Saaranen, T., Pertel, T., Kalle, T., Hansen, S., Varava, L., Lepp, K., Turunen, H., & Tossavainen, K. (2012b). School staffs' experiences of work and working conditions in Finnish and Estonian schools. *Open Public Health Journal, 5*, 40–51.

Saaranen, T., Tossavainen, K., Ryhänen, E., & Turunen, H. (2013). Promoting the occupational well-being of teachers for the Comenius program. *International Journal of Higher Education, 2*(2), 159–174.

Smith, M. J. (2008). Disciplinary perspectives linked to middle range theory. In M. J. Smith & P. R. Liehr (Eds.), *Middle range theory for nursing* (pp. 1–11). New York: Springer.

Soh, K. L., Davidson, P. M., Leslie, G., & Rahman, A. B. A. (2011). Action research studies in the intensive care setting: A systematic review. *International Journal of Nursing Studies, 48*(2), 258–268.

Sormunen, M., Saaranen, T., Tossavainen, K., & Turunen, H. (2012). Process evaluation of an elementary school health learning intervention in Finland. *Health Education (Special Issue), 112*(3), 272–291.

Tannahill, A. (2008). Health promotion: The Tannahill model revisited. *Public Health, 112*(12), 1387–1391.

Thiollent, M. (2011). Action research and participatory research: An overview. *International Journal of Action Research, 7*(2), 160–174.

Tones, B. K., & Tilford, S. (1994). *Health education: Effectiveness, efficiency and equity* (2nd ed.). London: Chapman and Hall.

Van den Broucke, S. (2012). Theory-informed health promotion: Seeing the bigger picture by looking at the details. *Health Promotion International, 27*(2), 143–147.

Von Wright, G. H. (1976). Determinism and the study of man. In J. Manninen & R. Tuomela (Eds.), *Essays on explanation and understanding* (pp. 415–435). Dordrecht: D Reide.

Weare, K. (1998). *The health promoting school—An overview of concept, principles and strategies and the evidence for their effectiveness*. First workshop on practice of evaluation of the Health Promoting School—Models, experiences and perspectives, Bern/Thun, Switzerland (Executive summary, pp. 9–18).

Weare, K. (2000). *Promoting mental, emotional and social health. A whole school approach.* London: Routledge.

Whitehead, D. (2010). Setting based health promotion. In D. Whitehead & F. Irvine (Eds.), *Health promotion & health education in nursing. A framework for practice* (pp. 95–120). Basingstoke, Hampshire: Palgrave Macmillan.

WHO. (1986). *Ottawa charter for health promotion*. http://www.who.int/hpr/NPH/docs/ottawa_charter_hp.pdf. Accessed 7 June 2013.

Index

V. Simovska and Patricia Mannix-McNamara (eds.), *Schools for Health and Sustainability: Theory, Research and Practice*, DOI 10.1007/978-94-017-9171-7,

Q

R

MIX
Papier aus verantwortungsvollen Quellen
Paper from responsible sources
FSC® C105338

If you have any concerns about our products,
you can contact us on
ProductSafety@springernature.com

In case Publisher is established outside the EU,
the EU authorized representative is:
Springer Nature Customer Service Center GmbH
Europaplatz 3, 69115 Heidelberg, Germany

Printed by Libri Plureos GmbH
in Hamburg, Germany